Lecture Notes in Computer Science　　　9095

Commenced Publication in 1973
Founding and Former Series Editors:
Gerhard Goos, Juris Hartmanis, and Jan van Leeuwen

More information about this series at http://www.springer.com/series/7407

Ignacio Rojas · Gonzalo Joya
Andreu Catala (Eds.)

Advances in Computational Intelligence

13th International Work-Conference
on Artificial Neural Networks, IWANN 2015
Palma de Mallorca, Spain, June 10–12, 2015
Proceedings, Part II

 Springer

Editors
Ignacio Rojas
University of Granada
Granada
Spain

Gonzalo Joya
University of Malaga
Malaga
Spain

Andreu Catala
Polytechnic University of Catalonia
Vilanova i la Geltrú
Spain

ISSN 0302-9743 ISSN 1611-3349 (electronic)
Lecture Notes in Computer Science
ISBN 978-3-319-19221-5 ISBN 978-3-319-19222-2 (eBook)
DOI 10.1007/978-3-319-19222-2

Library of Congress Control Number: 2015939427

LNCS Sublibrary: SL1 – Theoretical Computer Science and General Issues

Springer Cham Heidelberg New York Dordrecht London

Printed on acid-free paper

Springer International Publishing AG Switzerland is part of Springer Science+Business Media
(www.springer.com)

Preface

We are proud to present the set of final accepted papers for the 13th edition of the IWANN conference "International Work-Conference on Artificial Neural Networks" held in Palma de Mallorca (Spain) during June 10–12, 2015.

IWANN is a biennial conference that seeks to provide a discussion forum for scientists, engineers, educators, and students about the latest ideas and realizations in the foundations, theory, models, and applications of hybrid systems inspired on nature (neural networks, fuzzy logic, and evolutionary systems) as well as in emerging areas related to the above items. As in previous editions of IWANN, it also aims to create a friendly environment that could lead to the establishment of scientific collaborations and exchanges among attendees. The proceedings will include all the presented communications to the conference. It has also foreseen the publication of an extended version of selected papers in a special issue of several specialized journals (such as Neurocomputing, Soft Computing, and Neural Processing Letters).

Since the first edition in Granada (LNCS 540, 1991), the conference has evolved and matured. The list of topics in the successive Call for Papers has also evolved, resulting in the following list for the present edition:

1. **Mathematical and theoretical methods in computational intelligence**. Mathematics for neural networks. RBF structures. Self-organizing networks and methods. Support vector machines and kernel methods. Fuzzy logic. Evolutionary and genetic algorithms.
2. **Neurocomputational formulations**. Single-neuron modelling. Perceptual modelling. System-level neural modelling. Spiking neurons. Models of biological learning.
3. **Learning and adaptation**. Adaptive systems. Imitation learning. Reconfigurable systems. Supervised, non-supervised, reinforcement, and statistical algorithms.
4. **Emulation of cognitive functions**. Decision Making. Multi-agent systems. Sensor mesh. Natural language. Pattern recognition. Perceptual and motor functions (visual, auditory, tactile, virtual reality, etc.). Robotics. Planning motor control.
5. **Bio-inspired systems and neuro-engineering**. Embedded intelligent systems. Evolvable computing. Evolving hardware. Microelectronics for neural, fuzzy, and bioinspired systems. Neural prostheses. Retinomorphic systems. Brain–computer interfaces (BCI). Nanosystems. Nanocognitive systems.
6. **Advanced topics in computational intelligence**. Intelligent networks. Knowledge-intensive problem-solving techniques. Multi-sensor data fusion using computational intelligence. Search and meta-heuristics. Soft Computing. Neuro-fuzzy systems. Neuro-evolutionary systems. Neuro-swarm. Hybridization with novel computing paradigms.
7. **Applications**. Expert Systems. Image and Signal Processing. Ambient intelligence. Biomimetic applications. System identification, process control, and manufacturing. Computational Biology and Bioinformatics. Parallel and Distributed Computing.

Human–Computer Interaction, Internet Modeling, Communication and Networking. Intelligent Systems in Education. Human–Robot Interaction. Multi-Agent Systems. Time series analysis and prediction. Data mining and knowledge discovery.

At the end of the submission process, and after a careful peer-review and evaluation process (each submission was reviewed by at least 2, and on the average 2.7, Program Committee members or additional reviewers), 100 papers were accepted for oral or poster presentation, according to the recommendations of reviewers and the authors' preferences.

It is important to note, that for the sake of consistency and readability of the book, the presented papers are not organized as they were presented in the IWANN 2015 sessions, but classified under 14 chapters. The organization of the papers is in two volumes arranged basically following the topics list included in the Call for Papers. The first volume (LNCS 9094), entitled "IWANN 2015. Advances on Computational Intelligence. Part I" is divided into eight main parts and includes the contributions on:

1. Computing Languages with Bio-Inspired Devices and Multi-Agent Systems (Special Session, organized by: M. Dolores Jiménez-López and Alfonso Ortega de la Puente)
2. Brain-Computer Interfaces: Applications and Tele-services (Special Session, organized by: Ricardo Ron Angevin and Miguel Angel Lopez)
3. Multi-Robot Systems: Applications and Theory (MRSAT) (Special Session, organized by: José Guerrero and Óscar Valero)
4. Video and Image Processing (Special Session, organized by: Enrique Domínguez and Jose Garcia)
5. Transfer Learning (Special Session, organized by: Luis M. Silva and Jorge M. Santos)
6. Structures, algorithms, and methods in artificial intelligence (Special Session, organized by: Daniela Danciu and Vladimir Răsvan)
7. Interactive and Cognitive Environments (Special Session, organized by: Wei Chen and Albert Samá)
8. Mathematical and theoretical methods in Fuzzy Systems

In the second volume (LNCS 9095), entitled "IWANN 2015. Advances on Computational Intelligence. Part II" is divided into six main parts and includes the contributions on:

1. Pattern Recognition
2. Embedded intelligent systems
3. Expert Systems
4. Advances in Computational Intelligence
5. Applications of Computational Intelligence
6. Invited Talks to IWANN 2015

In this edition of IWANN 2015, the plenary talks were given by Prof. Cristina Urdiales (The shared control paradigm for assistive and rehabilitation robots), Prof. Dan Ciresan (Deep Neural Networks for Visual Pattern Recognition), and finally by Prof. Andrea Cavallaro.

The 13th edition of the IWANN conference was organized by the University of Granada, University of Málaga, Polytechnic University of Catalonia, and University of the Balearic Islands, together with the Spanish Chapter of the IEEE Computational Intelligence Society. We wish to thank the University of the Balearic Islands for their support and grants.

We would also like to express our gratitude to the members of the different committees for their support, collaboration, and good work. We specially thank the Local Committe, Program Committee, the Reviewers, Invited Speaker, and Special Session Organizers. Finally, we want to thank Springer, and especially Alfred Hoffman and Anna Kramer for their continuous support and cooperation.

June 2015

Ignacio Rojas
Gonzalo Joya
Andreu Catala

Organization

Program Committee

Leopoldo Acosta	University of La Laguna, Spain
Vanessa Aguiar-Pulido	RNASA-IMEDIR, University of A Coruña, Spain
Arnulfo Alanis Garza	Instituto Tecnológico de Tijuana, Mexico
Ali Fuat Alkaya	Marmara University, Turkey
Amparo Alonso-Betanzos	University of A Coruña, Spain
Juan Antonio Alvarez-García	University of Seville, Spain
Jhon Edgar Amaya	University of Tachira (UNET), Venezuela
Gabriela Andrejkova	Pavol Jozef Šafarik University, Slovak Republic
Cesar Andres	Universidad Complutense de Madrid, Spain
Miguel Ángel López	University of Cádiz, Spain
Anastassia Angelopoulou	University of Westminster, UK
Plamen Angelov	Lancaster University, UK
Davide Anguita	University of Genova, Italy
Cecilio Angulo	Universitat Politècnica de Catalunya, Spain
Javier Antich	Universitat de les Illes Balears, Spain
Angelo Arleo	CNRS - University Pierre and Marie Curie Paris VI, France
Corneliu Arsene	SC IPA SA, Romania
Miguel Atencia	University of Málaga, Spain
Jorge Azorín-López	University of Alicante, Spain
Davide Bacciu	University of Pisa, Italy
Javier Bajo	Universidad Politécnica de Madrid, Spain
Juan Pedro Bandera Rubio	ISIS Group, University of Málaga, Spain
Cristian Barrué	Universitat Politècnica de Catalunya, Spain
Andrzej Bartoszewicz	Technical University of Lodz, Poland
Bruno Baruque	University of Burgos, Spain
David Becerra Alonso	University of the West of Scotland, UK
Lluís Belanche	Universitat Politècnica de Catalunya, Spain
Sergio Bermejo	Universitat Politècnica de Catalunya, Spain
Francesc Bonin	Universitat de les Illes Balears, Spain
Francisco Bonnín Pascual	Universitat de les Illes Balears, Spain
Julio Brito	University of La Laguna, Spain
Antoni Burguera	Universitat de les Illes Balears, Spain
Joan Cabestany	Universitat Politècnica de Catalunya, Spain
Inma P. Cabrera	University of Málaga, Spain
Tomasa Calvo	University of Alcalá, Spain

Nuno Lau   Universidade de Aveiro, Portugal
Amaury Lendasse   The University of Iowa, USA
Miguel Lopez   University of Granada, Spain
Otoniel López Granado   Miguel Hernandez University, Spain
Rafael Marcos Luque Baena   University of Málaga, Spain
Ezequiel López-Rubio   University of Málaga, Spain
Kurosh Madani   LISSI/Université PARIS-EST Creteil, France
Mario Martin   Universitat Politècnica de Catalunya, Spain
Bonifacio Martin Del Brio   University of Zaragoza, Spain
Jose D. Martin-Guerrero   University of Valencia, Spain
Luis Martí   Universidad Carlos III de Madrid, Spain
Francisco Martínez Estudillo   ETEA, Spain
José Luis Martínez Martínez   University of Castilla-La Mancha, Spain
José Fco. Martínez-Trinidad   INAOE, Instituto Nacional de Astrofísica,
                                     Óptica y Electrónica, Mexico

Miquel Massot   University of the Balearic Islands, Spain
Francesco Masulli   University of Genoa, Italy
Montserrat Mateos   Universidad Pontificia de Salamanca, Spain
Jesús Medina-Moreno   University of Cádiz, Morocco
Maria Belen Melian Batista   University of La Laguna, Spain
Mercedes Merayo   Universidad Complutense de Madrid, Spain
Gustavo Meschino   Universidad Nacional de Mar del Plata, Spain
Margaret Miro   University of the Balearic Islands, Spain
Jose M. Molina   Universidad Carlos III de Madrid, Spain
Augusto Montisci   University of Cagliari, Italy
Antonio Mora   University of Granada, Spain
Angel Mora Bonilla   University of Málaga, Spain
Claudio Moraga   European Centre for Soft Computing, Spain
Ginés Moreno   University of Castilla-La Mancha, Spain
Jose Andres Moreno   University of La Laguna, Spain
Juan Moreno Garcia   Universidad de Castilla-La Mancha, Spain
J. Marcos Moreno Vega   University of La Laguna, Spain
Susana Muñoz Hernández   Technical University of Madrid, Spain
Pep Lluís Negre Carrasco   University of the Balearic Islands, Spain
Alberto Núñez   Universidad de Castilla La Mancha, Spain
Manuel Ojeda-Aciego   University of Málaga, Spain
Sorin Olaru   "SUPELEC" École Supérieur d'Électricité, France
Iván Olier   The University of Manchester, UK
Madalina Olteanu   SAMM, Université Paris 1, France
Julio Ortega   Universidad de Granada, Spain
Alfonso Ortega de La Puente   Universidad Autónoma de Madrid, Spain
Alberto Ortiz   University of the Balearic Islands, Spain
Emilio Ortiz-García   Universidad de Alcalá, Spain

Claude Touzet   University of Provence, France
Olga Valenzuela   University of Granada, Spain
Óscar Valero   University of the Balearic Islands, Spain
Miguel Ángel Veganzones   Universidad del País Vasco (UPV/EHU), Spain
Francisco Velasco-Álvarez   Universidad de Málaga, Spain
Sergio Velastin   Kingston University, UK
Marley Vellasco   PUC-Rio, Brazil
Alfredo Vellido   Universitat Politècnica de Catalunya, Spain
Francisco J. Veredas   Universidad de Málaga, Spain
Michel Verleysen   Université catholique de Louvain, Belgium
Changjiu Zhou   Singapore Polytechnic, Singapore
Ahmed Zobaa   University of Exeter, UK

Additional Reviewers

Azorín-López, Jorge   Navarro-Ortiz, Jorge
Cortes, Ulises   Ortiz, Alberto
De La Cruz, Marina   Palomo, Esteban José
Fernandez-Blanco, Enrique   Peters, Peter
Garcia-Fidalgo, Emilio   Rodriguez, Juan A.
Georgieva, Petia   Rodriguez-Benitez, Luis
Luque-Baena, Rafael M.   Sánchez-Morillo, Daniel
López-Rubio, Ezequiel   Volosyak, Ivan
Martinez-Gomez, Jesus   Wang, Qi
Moreno Garcia, Juan   Wetzels, Mart

Contents – Part II

Embedded Intelligent Systems

Expert Systems

Advances in Computational Intelligence

Invited Talks to IWANN 2015

Contents – Part I

Multi-Robot Systems: Applications and Theory (MRSAT)

Video and Image Processing

Transfer Learning

Structures, Algorithms and Methods in Artificial Intelligence

Interactive and Cognitive Environments

Mathematical and Theoretical Methods in Fuzzy Systems

Pattern Recognition

Developing Gene Classifier System
for Autism Recognition

Tomasz Latkowski[1] and Stanislaw Osowski[1,2 (✉)]

[1] Warsaw University of Technology, Warsaw, Poland
tlatkowski@wat.edu.pl, sto@iem.pw.edu.pl
[2] Military University of Technology, Warsaw, Poland

Abstract. The paper presents comparison of few chosen approaches to recognition of autism on the basis of gene expression microarray. The important point in this task is selection of genes of the highest class discriminative ability. To solve the problem we have applied many selection methods, which are based on different principles. The limited set of genes in each method are selected for further analysis. In this paper we will compare the genetic algorithm and random forest in the role of final gene selection. The most important genes selected by each method are used as the input attributes to the support vector machine and random forest classifiers, cooperating in an ensemble. The final result of classification is generated by the random forest, performing the role of fusion system for an ensemble.

Keywords: Autism · Ensemble of classifiers · Gene selection · SVM · Random forest

1   Introduction

Autism spectrum disorder belongs to severe neurodevelopmental disorders with social and communication development [1]. The idea of using gene expression in autism recognition is relatively new and underexplored now [2]. The research is directed to identifying limited group of genes, which are strongly related to disease and allow recognizing the autistic data from the reference samples. The main problem is an ill-conditioning of the gene expression matrix. The number of genes is in the range of few dozens of thousands and number of observations (patients) very scarce (usually in the range of one hundred). Application of classical feature selection methods leads to poor results, not adequate to the real association of the genes with the class membership of data.

Actual approaches to the gene selection are concerned mainly with different types of cancer and apply many different methods, including neural networks and Support Vector Machines [3,4], linear regression methods [5], statistical tests [6,7], rough set theory [8], genetic algorithms [9], frequent itemset mining [10], as well as a combination of many selection methods [11,12].

In this paper we propose the application of multistage selection procedure to autism recognition. In the first stage we reduce the population of genes by eliminating

© Springer International Publishing Switzerland 2015
I. Rojas et al. (Eds.): IWANN 2015, Part II, LNCS 9095, pp. 3–14, 2015.
DOI: 10.1007/978-3-319-19222-2_1

the genes, which have very similar expression values for the autistic and reference (healthy) classes. In this way it is possible to halve the size of genes. In the next stage the genes are subject to selection according to the assumed measure of quality. Eight different methods of selection are applied in this stage. They include both wrapper and filter methods, such as Fisher discriminant, correlation of feature with a class, three statistical hypothesis tests, reliefF algorithm, application of recursive linear SVM feature elimination, and stepwise regression method [3]. The genes are ordered according to their discrimination ability related to the particular method. In further steps of analysis we limit their population to 100 the best. The important problem in this stage is to find the small number of the globally best genes, chosen as the most important in the repeated runs of the selecting procedure.

We will compare the application of few different approaches. In the first one we use the genetic algorithm [13] responsible for the selection of the most significant genes in the process of classification. The genetic algorithm determines automatically the optimal size of genes. In the next approach we use the random forest (RF) [14] applied sequentially for selection of the most important genes. The ordering of genes in this method depends on how the particular genes influence the process of classification. Their optimal number is determined by checking the accuracy of classification, corresponding to the particular population of genes.

As a result of such procedure we select eight different sets of genes, treated as the most significant in classification. Each of these sets represents the input attributes for the support vector machine (SVM) [15] and RF classifiers, forming an ensemble. The final decision of recognizing the autistic case from the reference one is done by fusing the outputs of the ensemble members. This is done by the random forest (RF), serving this time the role of integrator. For comparison we have checked also the direct application of best sets of genes selected by all methods as the input attributes to the random forest serving simultaneously the role of classifiers and an integrator.

The numerical experiments were performed on the NCBI base of autism [16]. They have shown very encouraging results. The classification results of application of individual methods have been significantly improved by such approach.

2   Gene Selection Methods

2.1   Data Base of Autism

In the numerical experiments we have used a publicly available database in GEO (NCBI) repository [16]. It contains 146 observations (patients) and 54613 genes. Among the patients there were 82 cases related to children with autism (treated as class 1) and 64 of healthy children forming the reference group (class 2). All subjects were males. Total RNA was extracted with Affymetrix Human U133 Plus 2.0 39 Expression Arrays. Children with autism were diagnosed by a medical professional according to the DSM-IV criteria and the diagnosis was confirmed by ADOS and ADI-R criteria [17]. The study population was primarily Caucasian and there were no group level differences in ethnicity.

2.2   Initial Step

The genes, regarded as important in class recognition, should show different expression value in class 1 and class 2. In the introductory step we use the median of their class value within all observations. Taking into account very high variance of these expressions within the same class we must be very careful in selecting the threshold. For example in the data base [16] the mean values of gene expression of the members of the autistic class extended from the smallest 4.96 to the largest 19000. At the same time standard deviations have varied up 1138. The curve presented in Fig. 1 shows that most genes depicts very similar ratio of the medians in both classes.

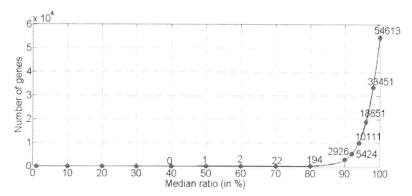

Fig. 1. The curve representing the population of eliminated genes at their differing median ratio in two classes

It is seen that there are no genes of the ratio below the value of 70%. The active range of reduction is from 80% to 100%. In our experiments we set the threshold on the level of 0.96. This value was chosen heuristically to reduce the number of less significant genes to a reasonable level. It resulted in elimination of 35762 genes. Therefore in further analysis we used only 18851 genes, for which the ratio of medians in both classes was below 0.96.

2.3   Selection Methods in the First Stage

The reduced set of genes was subject to further reduction by applying eight feature selection methods. They were chosen in a way to provide their highest possible statistical independence. The following methods have been selected: Fisher discriminant, correlation of feature with a class, three statistical hypothesis tests (Kolmogorov-Smirnov test, two-sample Student t-test and Kruskal-Wallis test), reliefF algorithm, application of recursive linear SVM feature elimination, and stepwise regression method [3].

In Fisher test the genes were selected according to the 2-class discrimination measure

$$S_{12}(g) = \frac{|c_1 - c_2|}{\sigma_1 + \sigma_2} \tag{1}$$

in which where c_1 and c_2 are the mean values for classes 1 and 2, respectively, while σ_1 and σ_2 the appropriate standard deviations. A large value of this measure indicates good class discriminative ability of the gene.

In correlation method we examine the correlation of gene g with a class. For two classes of data the correlation measure is given in the form

$$S(g) = \frac{\sum_{k=1}^{2} P_k (c_k - c)^2}{\sigma^2 \sum_{k=1}^{2} P_k (1 - P_k)} \tag{2}$$

where c and σ represent the mean value and standard deviation of gene g for all data, c_k is a mean value of the gene expression for the kth class data and P_k is a probability of kth class occurrence in the data set.

The statistical hypothesis tests applied in experiments compare the medians of the groups of data to determine if the samples come from the same population. Kolmogorov-Smirnov test is based on the distance between the cumulative distribution of samples $F_1(x)$ and $F_2(x)$ of the gene belonging to class 1 and 2. In Kruskal-Wallis test the ranks of the data rather than the numeric values of gene expressions are used. High distances in both tests mean good recognition ability of the particular gene. In a two-sample Student t-test the null hypothesis is checked if the data in the class 1 and 2 are independent random samples of normal distributions with equal means and equal, but unknown variances, against the alternative hypothesis that the means are not equal. Fulfilment of the alternative hypothesis means good class discrimination ability of the gene. The details of these statistics can be found in [18].

The reliefF algorithm checks the discrimination ability of the gene according to its highest correlation with the observed class while taking into account the distances between opposite classes. The quality of the gene is assessed according to how well its expression values distinguish between observations that are near to both opposite classes.

Stepwise linear regression is a systematic method based on adding and removing features to the set of input attributes according to their statistical significance in a regression. It compares the explanatory power of incrementally larger and smaller models based on the p value of F-statistics [18,19]. Based on the statistical result a decision is taken whether the gene should be included in a model or not.

In the SVM-recursive feature elimination method the linear SVM network is trained at application of all genes. The decision of accepting or removing the particular gene is taken on the basis of value of weight joining the gene with a SVM classifier. High weight means significant influence of the gene. The features are eliminated step by step, since the SVM is retrained at each step with the population of features becoming gradually smaller and smaller.

2.4 Final Stage of Selection

The genes assessed by the mentioned methods are arranged according to their discrimination abilities, from the highest to the lowest rank. Since the selection procedures were performed by using various methods the contents and order of genes may be different in each method. In further selection step we use only 100 the best genes selected in each method. We assume that the size of 100 genes is large enough to continue successfully our search of much smaller subset of genes of the highest class discrimination ability.

To find the reduced size of the most important genes in each method we have applied the next steps of selection. Two different approaches have been proposed for each set of 100 genes selected in the previous step. The first one makes use of the selection ability of the random forest [14]. Random forest constructs many decision trees at a training time and outputs the class that is the mode of the classes pointed by the individual trees. The learning data (usually 2/3 of data set) for each tree are selected randomly to learn the trees. At each node some number m of predictor variables (usually m is a square root of total number of predictors) are selected at random from all the predictor variables. The predictor variables that provide the best split, according to the objective function, is used to do a binary split on that node. At the next nodes, another m variables at random from all predictor variables are chosen and do the same.

The method enables to assess the importance of the predictors (genes). The importance of the particular gene is measured by taking into account its influence on the classification results, especially, how inclusion of this gene is important for getting the higher accuracy of class recognition. Generally, the gene importance is measured by the increase of the rate of prediction error for validation data if the values of this gene are permuted among the testing data. The out-of-bag prediction error is computed on this perturbed data set and compared to the error before perturbation. This measure is computed for every tree, then averaged over the entire ensemble and divided by the standard deviation over the entire ensemble. In this way the genes are ordered according to their statistical impact on the classification accuracy. The optimal population size of genes has been estimated trying different numbers of the best genes in the classification procedure and choosing one providing the best result for the validation data.

The second approach is based on application of the genetic algorithm (GA) for gene selection [13]. The limited number of the highest rank genes selected in all runs creates the input attributes (chromosome vectors) to the SVM classifiers. The genes are coded in a binary way. The value one means inclusion of the particular gene as the input signal to the classifier, zero – lack of such gene in the chromosome.

GA consists of selecting parents for reproduction, performing crossover with the parents, and applying mutation to the bits representing children. The process starts from a population of randomly generated chromosomes. For each chromosome the fitness function is determined (we assume fitness as the inverse of the classification error function) and then maximized (the error function minimized). In each generation, the fitness of every chromosome in the population is evaluated, multiple chromosomes are stochastically selected from the current population (based on their fitness function values and application of the fortune wheel), and modified (recombined and possibly mutated) to form a new population. The new population is then used in the next iteration of the algorithm.

Each binary chromosome is associated with the input vector **x** applied to the SVM classifier of Gaussian kernel (the value 1 - inclusion of the gene and zero – exclusion of gene). Two data sets are involved in GA based training: the learning set and the validation set. The classifier is trained on the learning data set and then tested on the validation data. The testing error function on the validation data forms the basis for the definition of the fitness function. On the basis of its value further genetic operations on the data are performed.

3   Recognition of Autism

After application of all selection stages we get eight sets of genes chosen as optimal by each of the applied methods. They form the input attributes to the SVM and random forest classifiers, responsible for recognition of data corresponding to the autistic individuals.

We have applied two types of classifiers to provide the maximum independence of their performance. The independence of classifiers is the basic condition of their proper cooperation in an ensemble. The genetic algorithm was cooperated with SVM in final selection. Therefore its results create the input to RF classifier, in order to provide higher independence. The results of selection at application of random forest form the input signals to SVM classifier of the Gaussian kernel. Thanks to this we enhance the independence of classifiers, which is an important point in fusing the results of an ensemble.

Our ensemble is composed of 16 units (2 approaches to fusing the results of 8 selection methods). The important point is to integrate their results into one common recognition verdict. We use the random forest in the role of integrator. The inputs to the RF are the output signals of the members of the ensemble. The decision trees of RF are responsible for creating the final decision regarding recognition of autism and the reference class data. The succeeding classification steps (classification and integration) are performed in the 10-fold cross validation mode followed by integration of their results into the final score. Fig. 2 presents the general scheme of applied procedure of autism recognition. The final recognition of the autistic or non-autistic class is the results of integration made by random forest network.

The dash lines represent the alternative approach to autism recognition. The genes selected by all 16 methods are put simultaneously to the input of RF, performing the role of classifier. It will be called the direct approach.

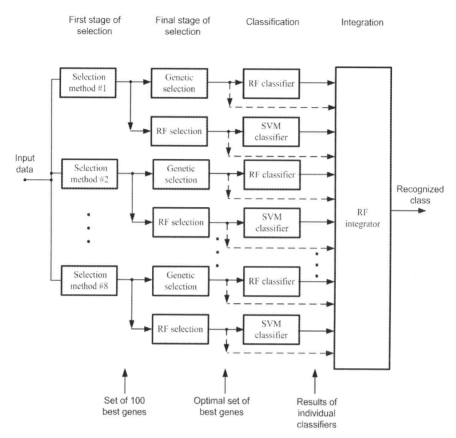

First stage of selection   Final stage of selection   Classification   Integration

Fig. 2. The general scheme of data processing in autism recognition

4   Results

The numerical experiments of autism recognition have been performed on the autism data [16] containing 146 observations and 54613 genes. All experiments have been performed using Matlab [19]. After initial step of selection the number of genes has been reduced to 18851. Further experiments have been done on this reduced set of genes. Application of 8 methods of feature selection has arranged the genes sequentially from the most to the least important. Since good generalization ability of classifier system needs selecting small set of the most significant genes in classification procedure, we reduced the number of genes to 100 of the highest importance in each selection approach. These reduced sets of genes were subject to the next stage of selection at application of two methods: the genetic algorithm combined with SVM and random forest.

The genetic algorithm cooperating with SVM of Gaussian kernel applied the elitist strategy of passing two fittest population members to the next generation. Thanks to this the fitness (the inverse of classification error) is never declined from one generation to the next. The algorithm created crossover children by combining pairs of parents in the current population using the roulette rule. The crossover probability applied in the solution was 0.8. The mutation of children was created by randomly changing the genes of individual parents. The assumed mutation rate was 0.02. After applying eight different selection procedures we got the optimal sets of genes containing possibly different contents. Table 1 presents the optimal number of genes selected in different methods. The selection methods have been denoted as follows: FDA - the Fisher discriminant analysis, RFA - the ReliefF algorithm, TT - the two-sample t-test, KST - the Kolmogorov-Smirnov test, KWT - the Kruskal-Wallis test, SWR - the stepwise regression method, COR - the feature correlation with a class, SVM - the SVM-RFE method.

Table 1. The size of the best genes in different selection methods by using genetic algorithm

FDA	RFA	TT	KST	KWT	COR	SWR	SVM
51	61	67	70	49	65	25	65

We observed significant diversity of the selected genes. The total set (the unique genes combined together) is composed of 345 the best (different) genes. The highest repetitions in 8 sets has shown the gene RFC2 (replication factor C (activator 1) 2), which was chosen commonly by 6 methods as a member of an optimal group composed of 100 best genes.

The analogous operation of choosing the optimal small set of the best genes has been repeated using random forest as the final selector. It was done also on the set of 100 best genes chosen by eight selection methods in first step of selection. Sequentially, we reduced the starting population by 10%, rejecting the genes of the smallest class discrimination ability. The typical sequence (from the left to right) presenting the importance of genes in the succeeding iterations are depicted in Fig. 3.

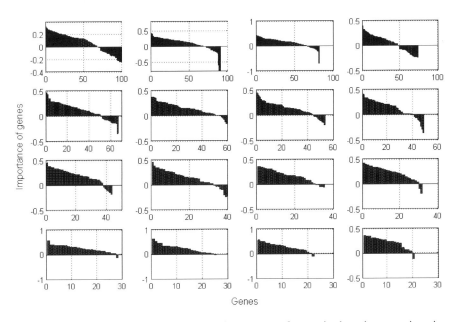

Fig. 3. The diagrams presenting the changing importance of genes in the subsequent iterations of RF selection performed on the set defined by RFA

We finish the selection at the number of genes providing the highest classification accuracy. As a result of such approach we have got 8 different sets of genes. Their populations defined in different selection approaches are presented in Table 2.

Table 2. The size of the best genes chosen in different selection methods by using RF

FDA	RFA	TT	KST	KWT	COR	SWR	SVM
17	17	41	11	26	33	26	51

This time the total set is composed of 169 the best (different) genes. According to this method of final selection the most often chosen gene was again RFC2 which appeared 6 times as a commonly chosen member of an optimal groups. Among the genes selected by RF there were 125 genes, which were also chosen by the genetic algorithm. Fig. 4a and b present the number of times the succeeding genes have been selected commonly in these 2 methods.

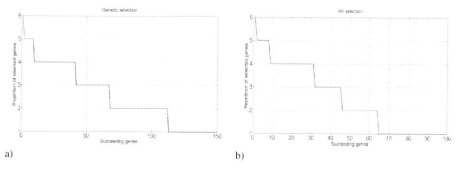

a)   b)

Fig. 4. Diagrams showing the number of times the best genes have been commonly selected by genetic algorithm (a) and random forest (b)

The selected genes formed the input attributes to the classifiers in the classification stage. Two types of experiments have been performed. In the first one, as presented in Fig. 2, the sets were treated independently and served as the input signals to the classifiers (SVM and RF). Then their results formed the input to RF, serving as an integrator.

In the second one (called direct approach) all genes selected by both methods created the common input signals to the RF classifier. It means, that the classification and integration were performed directly on the basis of the original values of gene expressions. In both approaches we have applied the 10-fold cross-validation procedure. It is a model validation technique for assessing how the results of a learned classification system will generalize to an independent data set. In the applied cross-validation approach the whole data set is split into k=10 approximately equal parts. Nine parts are used in learning and the last one in testing of the learned system. The procedure is repeated 10 times exchanging each time the testing part of data. Then the average error over the testing parts across all 10 trials is computed. The advantage of the method is that it matters less how the data sets are divided. Every data point gets to be in a test set exactly once, and gets to be in a training set k-1 times.

Table 3. The results of 10-fold cross-validation (mean error) in autism recognition at application of intermediate classifiers and direct approach (only testing data not taking part in learning)

		FDA	RFA	TT	KST	KWT	COR	SWR	SVM	Fusion of all methods	Fusion of best methods	Total fusion
Intermediate classifiers	RF	84.25 ±1.87	74.66 ±2.21	69.18 ±2.53	77.40 ±2.82	80.82 ±2.1	76.71 ±2.23	65.75 ±2.91	85.30 ±2.57	85.82 ±1.62	86.92 ±1.18	86.79 ±1.76
	GA	75.41 ±2.86	76.99 ±2.87	73.36 ±2.29	77.26 ±1.96	76.58 ±3.14	77.05 ±2.78	66.23 ±2.87	77.95 ±3.01	79.32 ±3.27	81.99 ±1.91	
Direct approach												82.93 ±2.88

Additionally we have repeated the whole 10-fold cross validation procedures 10 times, each time changing randomly the data set into 10 parts. The statistical results of application of both approaches, in the form of the mean percentage error and standard deviation are presented in Table 3. To check, how the low quality classifiers influence the efficiency of classification, we have made additional experiments excluding 2 least effective selection methods based on TT and SWR approaches.

As we can see direct application of selected genes to the classifier is not efficient (the relative accuracy after fusion equal 82.93%). However, application of the intermediate classifications has increased this accuracy in a significant way to the best value of 86.92%. The best result corresponds to the fusion of six, the most efficient selection methods. The detailed results of classification corresponding to the best approach (fusion of the best selection methods) are presented here in the form of confusion matrix, given in Table 4.

Table 4. The confusion matrix in autism recognition corresponding to the best approach

	Autistic class	Reference class
Autistic class	73	10
Reference class	9	54

The sensitivity of the system in recognition of autism is equal 0.89 and specificity 0.84. Fig. 4 presents the receiver operating characteristic (ROC) curve [20]. The area under curve (AUC) for autistic group is equal 0.86 and is close to 1. As it is seen, our classification system is much better than the random classifier (the diagonal line in the figure).

5   Conclusions

The paper has presented and compared different approaches to the autism recognition on the basis of gene expression microarray. The most important point in this proposition is selection of the most relevant genes which provide the highest efficiency of autism recognition. Eight different feature validation methods were investigated in gene selection. They arrange the order of genes according to their importance in autism recognition. Only the first 100 best genes in each method have been subject to further study.

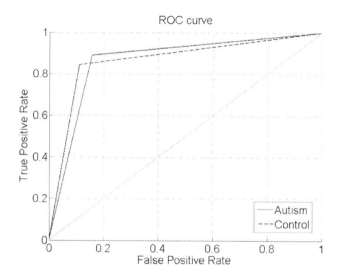

Fig. 5. The ROC curve corresponding to the best results of autism recognition

In the following stages of data processing we have chosen the locally optimal sets of genes and the optimality was found independently for each selection method. The final stage of data processing applied support vector machine and random forest as the classifiers. To get the highest possible accuracy of class recognition we have arranged all methods cooperating in an ensemble, integrated in one final result by the random forest.

The obtained results confirmed good performance of such a system. The average classification accuracy obtained on NCBI base [16] in the 10-fold cross validation mode was close to 87%. These results outperform the best result of 81.8% error reported in the most recent paper [21] for the autism data base [16].

References

1. Bailey, A., Philips, W.: Autism: toward an integration of clinical, genetic, neuropsychological and neurobiological perspectives. J. Child Psychol. Psychiatry **37**, 89–126 (1996)
2. Alter, M., Kharkar, R., Ramsey, K., Craig, D., Melmed, R., Grebe, T., Curtis-Bay, R., Ober-Reynolds, S., Kirwan, J., Jones, J., Blake-Turner, J., Hen, R., Stephan, D.: Autism and increased paternal age related changes in global levels of gene expression regulation. Plos One **6**, 1–10 (2011)
3. Latkowski, T., Osowski, S.: Data mining for feature selection in gene expression autism data. Expert Systems with Applications **42**(2), 864–872 (2015)
4. Guyon, I., Weston, A.J., Barnhill, S., Vapnik, V.: Gene selection for cancer classification using SVM. Machine Learning **46**, 389–422 (2002)
5. Huang, X., Pan, W.: Linear regression and two-class classification with gene expression data. Bioinformatics **19**, 2072–2078 (2003)
6. Mitsubayashi, H., Aso, S., Nagashima, T., Okada, Y.: Accurate and robust gene selection for disease classification using a simple statistics. Biomed. Informatics **391**, 68–71 (2008)

7. Garcia, V., Sanchez, J.S.: Mapping microarray gene expression data into dissimilarity spaces for tumor classification. Information Sciences **294**, 362–375 (2015)
8. Wang, X., Gotoh, O.: A Robust Gene Selection Method for Microarray-based Cancer Classification. Cancer Informatics **9**, 15–30 (2010)
9. Luque-Baena, R.M., Urda, D., Claros, M.G., Franco, L., Jerez, J.: Robust gene signatures from microarray data using genetic algorithms enriched with biological pathway keywords. Journal of Biomed. Informatics **49**, 32–44 (2014). doi:10.1016/j.jbi.2014.01.006
10. Seeja, K.R.: Feature selection based on closed frequent itemset mining: A case study on SAGE data classification. Neurocomputing **151**, 1027–1032 (2015)
11. Yang, F.: Robust feature selection for microarray data based on multicriterion fusion. IEEE Trans. Computational Biology and Bioinformatics **8**, 1080–1092 (2011)
12. Wiliński, A., Osowski, S.: Ensemble of data mining methods for gene ranking. Bulletin of the Polish Academy of Sciences **60**, 461–471 (2012)
13. Goldberg, D.: Genetic Algorithms in Search, Optimization, and Machine Learning. Addison-Wesley, Amsterdam (1989)
14. Breiman, L.: Random forests. Machine Learning **45**, 5–32 (2001)
15. Schölkopf, B., Smola, A.: Learning with kernels. MIT Press, Cambridge MA (2002)
16. NCBI base (2011). http://www.ncbi.nlm.nih.gov/sites/GDSbrowser?acc=GDS4431
17. Lord, C., Rutter, M.: Autism Diagnostic Interview-Revised: a revised version of a diagnostic interview for caregivers of individuals with possible pervasive developmental disorders. J. Autism Dev. Disord. **24**, 659–685 (1994)
18. Sprent, P., Smeeton, N.C.: Applied Nonparametric Statistical Method. Chapman & Hall/CRC, Boca Raton (2007)
19. Matlab user manual, Natick, USA: MathWorks (2013)
20. Tan, P.N., Steinbach, M., Kumar, V.: Introduction to data mining. Pearson Education Inc., Boston (2006)
21. Hu, V., Yinglei, L.: Developing a predictive gene classifier for autism spectrum disorders based upon differential gene expression profiles of phenotypic subgroups. North American Journal of Medicine and Science **6**, 107–116 (2013)

A Distributed Feature Selection Approach Based on a Complexity Measure

Verónica Bolón-Canedo[(✉)], Noelia Sánchez-Maroño,
and Amparo Alonso-Betanzos

Laboratory for Research and Development in Artificial Intelligence (LIDIA),
Computer Science Department, University of A Coruña, 15071 A Coruña, Spain
{vbolon,nsanchez,ciamparo}@udc.es

Abstract. Feature selection is often required as a preliminary step for many machine learning problems. However, most of the existing methods only work in a centralized fashion, i.e. using the whole dataset at once. In this paper we propose a new methodology for distributing the feature selection process by samples which maintains the class distribution. Subsequently, it performs a merging procedure which updates the final feature subset according to the theoretical complexity of these features, by using data complexity measures. In this way, we provide a framework for distributed feature selection independent of the classifier and that can be used with any feature selection algorithm. The effectiveness of our proposal is tested on six representative datasets. The experimental results show that the execution time is considerably shortened whereas the performance is maintained compared to a previous distributed approach and the standard algorithms applied to the non-partitioned datasets.

1 Introduction

The technological per-capita capacity to store information has almost doubled every 40 months since the 1980s, and for example in 2012, every day 2.5 exabytes of data were created. In this scenario, where data is not only big in volume, but also in complexity and variety, machine learning techniques have become indispensable when extracting useful information from huge amounts of meaningless data. If one analyzes the size (samples × features) of the datasets posted in the UCI Machine Learning Repository [1], it is easy to see that it has increased dramatically [2]. In the 1980s, the maximal size of the data was about 100; then in the 1990s, this number increased to more than 1500; and finally in the 2000s, it further increased to about 3 million. The proliferation of this type of datasets had brought unprecedented challenges to machine learning researchers. Learning algorithms can degenerate their performance due to overfitting, learned models decrease their interpretability as they are more complex, and finally speed and efficiency of the algorithms decline in accordance with size.

Machine learning can take advantage of feature selection to be able to reduce the dimensionality of a given problem. *Feature selection* (FS) is the process of detecting the relevant features and discarding the irrelevant and redundant ones,

© Springer International Publishing Switzerland 2015
I. Rojas et al. (Eds.): IWANN 2015, Part II, LNCS 9095, pp. 15–28, 2015.
DOI: 10.1007/978-3-319-19222-2_2

with the goal of obtaining a small subset of features that describes properly the given problem with a minimum degradation or even improvement in performance [3]. Feature selection, as it is an important activity in data preprocessing, has been an active research area in the last decade, finding success in many different real world applications, especially those related with classification problems.

Feature selection methods are divided into three categories: filters, wrappers and embedded methods. While wrappers involve optimizing a predictor as part of the selection process, filters rely on the general characteristics of the training data to select features with independence of any predictor. The embedded methods use machine learning models for classification, and then an optimal subset of features is built by the classifier algorithm. As stated in [4], even when the subset of features might be not so accurate as with embedded and wrapper methods, filters are preferable due to their computational and statistical scalability, so they will be the focus of this research. Traditionally, FS methods have been applied in a centralized manner, i.e. a single learning model to solve a given problem. However, there might be several reasons to use a FS method in a distributed way. First, nowadays the data are sometimes distributed in multiple locations. Second, and more common, when dealing with large amounts of data, most existing FS methods are not expected to scale well and their efficiency may significantly deteriorate or even become inapplicable. Therefore, a possible solution might be to distribute the data, run a FS method on each partition and then combine the results. There are two main techniques for partitioning and distributing data: vertically, i.e. by features, and horizontally, i.e. by samples. Distributed learning has been used to scale up datasets that are too large for batch learning in terms of samples [5–8]. While not common, there are some other developments that distribute the data by features [9–11]. Even less common, there are also approaches that address the distribution both vertically and horizontally [12].

This paper will be focused on the most common approach: distribution by samples. We will present a methodology in which several rounds of FS will be performed on different partitions of the data. Then, the partial outputs need to be combined into a single subset of relevant features. In our previous work [13], the distribution of the samples across the different nodes was performed randomly, which could lead to situations where some classes are not represented in all the nodes. Moreover, our previous approach used of a method to combine the partial outputs that involves classification algorithms, which introduced an important overhead. Different than our previous study, and trying to overcome these two disadvantages, in this work (i) we partition the data taking into account the class distribution, (ii) we propose a new method to combine the partial outputs which makes use of data complexity measures, leading to an impressive reduction in the time necessary for this task, and (iii) we present a case study trying to determine if, since we reduced the running time impressively, it is worth considering more computations trying to improve the accuracy. The experimental results from six different databases demonstrate that our new proposal can maintain the performance of both the original FS methods and our previous approach in [13], as well as showing important savings in running times.

2   Methods

2.1   Distributed Feature Selection Algorithm

The idea of distributing the data horizontally builds on the assumption that combining the output of multiple experts is better than the output of any single expert. The proposed methodology consists of performing several fast feature selectors on several partitions of the data, combining then the partial outputs into a single subset of features. The feature selection distributed algorithm is applied to the training dataset in several iterations or rounds. This repetition ensures capturing enough information for the combination stage. For this sake, at each round we start by dividing each dataset D into several small disjoint subsets D_i. Since the partition is being made by samples, it is necessary to bear in mind that random distributions of the data might imply that some of the classes are not represented exhaustively in all nodes. To solve this inconvenience, we divide the data maintaining the original class proportions in the training dataset, i.e. if a dataset has 70% of instances from the positive class and 30% from the negative class, in each partition of the dataset this distribution will be maintained. In this manner, we ensure that all the feature selectors in the different partitions are able to learn all classes.

After having the dataset partitioned into smaller subsets of data D_i, each feature selector is run on each of them, generating a corresponding selection in which the features selected to be removed receive a vote. At this point, a new round starts so a new partition of the dataset is performed and another round of voting is accomplished until reaching the predefined number of rounds. After all the small datasets D_i in each round have been used (which could be done in parallel, as all of them are independent of each other), the combination method builds the final selection S as the result of the filtering process that we will explain in detail in the next paragraph. This set S will be used to train a classifier C and to test its performance over a new set of samples (test dataset).

Combining the partial outputs on the different partitions of data is not an easy-to-solve question. Since at each round and partition, each feature to be removed receives a vote, at the end of the process we will have a number of votes for each feature which may range from 0 to v_{max}, being v_{max} the number of features or attributes in each dataset n times the number of rounds r. To decide which features to remove, it is necessary to estimate a threshold of votes. In our previous work [13], the method to calculate this threshold estimated the best value for the number of votes from its effect on the training set. Following the recommendations exposed in [14], we selected the number of votes taking into account two different criteria: the training error and the percentage of features retained. Both values must be minimized to the extent possible, by minimizing the fitness criterion $e[v]$:

$$e[v] \leftarrow \alpha \times error + (1 - \alpha) \times featPercentage \tag{1}$$

To calculate this criterion, a term α was introduced to measure the relative relevance of both values and was set to $\alpha = 0.75$ as suggested in [14], giving more influence to the classification error. Since the maximum number of votes v_{max} in some cases might be in the order of thousands, instead of evaluating all the possible values for the number of votes we opted for delimiting it into an interval $[minVote, maxVote]$ computed by using the mean and standard deviation such that $minVote = avg - 1/2std$ and $maxVote = avg + 1/2std$.

The drawback of our previous approach is that, by involving a classifier in the process of selecting the optimal threshold, in some cases the time necessary for this task was higher than the time required by the feature selection process, even without distributing the data, which introduced an important overhead in the running time. Furthermore, this fact made our methodology dependent on the classifier chosen, in a similar way than the wrapper approach does.

Trying to overcome the aforementioned problems, in this paper we propose to modify the function for calculating the threshold of votes by making use of data complexity measures, which have been proposed in the last few years in order to characterize the complexity of datasets beyond estimates of error rates [15]. Thus, instead of evaluating the merit of a candidate subset of features by its classification error rate, we propose to calculate the complexity of the dataset with the candidate features. The rationale behind this decision is that we assume that good candidate features would contribute to decrease the complexity and must be maintained, whilst bad candidate features would contribute to increase the complexity and must be discarded. Since our intention is to propose a framework that could be applicable to both binary and multiclass datasets, among the existing complexity measures, the Fisher discriminant ratio [15] was chosen, which is applicable to problems with any number of classes. Fisher's multiple discriminant ratio for C classes is defined as:

$$f = \frac{\sum_{i=1, j=1, i \neq j}^{C} p_i p_j (\mu_i - \mu_j)^2}{\sum_{i=1}^{C} p_i \sigma_i^2}, \tag{2}$$

where μ_i, σ_i^2, and p_i are the mean, variance, and proportion of the ith class, respectively. In this work we will use the inverse of the Fisher ratio, $1/f$, such that a small complexity value represents an easier problem. Therefore, we propose to replace the formula for calculating $e[v]$ defined in Eq. (1) with the one that we present in Eq. (3), expecting to achieve two important goals (1) a reduction in the time necessary to calculate the threshold and (2) a method independent on the classifier.

$$e[v] \leftarrow \alpha \times 1/f + (1 - \alpha) \times featPercentage \tag{3}$$

The algorithm for the whole methodology is detailed in Algorithm 1. At the end, the final selection S returned by the algorithm is applied to the training and test sets in order to obtain the ultimate classification accuracies. It must be noted that this algorithm can be used with any feature selection method, although the use of filters is recommended since they are faster than other techniques. Note that the method can be also applied on ranker methods, however it is required to establish another threshold to determine the number of features to be removed in each subset of data.

Algorithm 1. Pseudo-code for the proposed distributed methodology

---

Data: $\mathbf{d}_{(m \times n+1)}$ ← labeled training dataset with m samples and n input features

   X ←set of features, X = $\{x_1, \ldots, x_n\}$
   s ← number of submatrices of $\mathbf{d}$ with p samples
   V ← vector of votes
   r ← number of rounds
   α ← 0.75

Result: S ← subset of features $\backslash$S $\subset$ X

  /* Obtaining a vector of votes for discarding features   */

1   initialize the vector of votes V to 0, $|V|$=n
2   **for** *each round* **do**
3   Split $\mathbf{d}$ into s disjoint submatrices maintaining the class distribution
4   **for** *each submatrix* **do**
5   apply a feature selection algorithm
6   F ← features selected by the algorithm
7   E ← features eliminated by the algorithm $\backslash$E $\cup$ F = X
8   increment one vote in vector V for each feature in E
    end
   end

  /* Obtain threshold of votes, Th, to remove a feature   */

9   $minVote$ ← minimum threshold considered
10   $maxVote$ ← maximum threshold considered
11   **for** v ← *mixVote to maxVote with increment 5* **do**
12   F$_{th}$ ← subset of selected features (number of votes < v)
13    $1/f$ ← inverse of Fisher ratio computed on training dataset $\mathbf{d}$ using only features in F$_{th}$
14    $featPercentage$ ← percentage of features retained $\left(\frac{|F_{th}|}{|X|} \times 100 \right)$
15    $e[v]$ ← $\alpha \times 1/f + (1 - \alpha) \times featPercentage$
   end
16   Th ← $min(e)$, Th is the value which minimizes the function e
17   S ← subset of features after removing from X all features with a number of votes $\geq Th$

2.2 Experimental Setup

In order to test our distributed framework for feature selection, we have chosen the same six benchmark datasets as in our previous work [13], which are described in Table 1 depicting their properties (number of features, number of training and test samples and number of classes). These datasets can be considered representative of problems from medium to large size, since the horizontally distribution is not suitable for small-sample datasets. All of them can be free downloaded from the UCI Machine Learning Repository [1]. Those datasets originally divided into training and test sets were maintained, whereas, for the sake of comparison, datasets with only training set were randomly divided using the common rule 2/3 for training and 1/3 for testing. The number of packets (s) to divide the dataset in each round is also displayed in the last column of Table 1. This number was calculated trying to maintain a proportion between the number of samples and the number of features with the constraint of having, at least, three packets per dataset.

Table 1. Dataset description

Dataset	Features	Training	Test	Classes	Packets
Connect4	42	45038	22519	3	45
Isolet	617	6238	1236	26	5
Madelon	500	1600	800	2	3
Ozone	72	1691	845	2	11
Spambase	57	3067	1534	2	5
Mnist	717	40000	20000	2	5

The distributed approach proposed herein can be used with any feature selection method, although a subset of features is mandatory and so, a threshold is required for ranker methods. In this work, five well-known filters, based on different metrics, were chosen. While three of them return a feature subset (CFS, Consistency-based and INTERACT), the other two (ReliefF and Information Gain) are ranker methods so, as aforementioned, a threshold is necessary in order to obtain a subset of features. In this research we have opted for retaining the c top features, being c the number of features selected by CFS, since it is a widely-used method and, among the three subset methods chosen, it is the one which usually selects the larger number of features. It is also worth noting that although most of the filters work only over nominal features, the discretization step is done by default by Weka [16], working as a black box for the user.

- **Correlation-based Feature Selection** (CFS) is a simple filter algorithm that ranks feature subsets according to a correlation based heuristic evaluation function [17]. Theoretically, irrelevant features should be ignored and redundant features should be screened out.
- The **Consistency-based Filter** [18] evaluates the worth of a subset of features by the level of consistency in the class values when the training instances are projected onto the subset of attributes.
- The **INTERACT** algorithm [19] is based on symmetrical uncertainty (SU). The authors stated that this method can handle feature interaction, and efficiently selects relevant features. The first part of the algorithm requires a threshold, but since the second part searches for the best subset of features, it is considered a subset filter.
- **Information Gain** [20] is one of the most common attribute evaluation methods. This filter provides an ordered ranking of all the features and then a threshold is required.

– **ReliefF** [21] is an extension of the original Relief algorithm that adds the ability of dealing with multiclass problems and is also more robust and capable of dealing with incomplete and noisy data. This method may be applied in all situations, has low bias, includes interaction among features and may capture local dependencies which other methods miss.

3   Experimental Results

In this section we present and discuss the experimental results in terms of (a) the number of selected features; (b) the classification accuracy; and (c) the feature selection runtime. Three different approaches will be compared in the tables of this section: the centralized standard approach (C), the distributed approach presented in our previous work, which distributes the data randomly and merge the partial feature selection results by taking classification accuracy into account (D-Clas) and the distributed approach proposed herein, which distributes the data maintaining the class distribution and merges the partial outputs using data complexity measures (D-Comp). The name of the specific filter used will be added at the beginning. For example, the centralized approach employing CFS will be represented as CFS-C. Finally, the last subsection presents a case study to determine the most suitable interval for the threshold of votes.

3.1   Number of Selected Features

Table 2 shows the number of features selected by each approach. In the first block of the table we can see the features selected by the centralized and distributed "D-Comp" approaches, which are not dependent on the classifier. Then, the table visualizes the number of features selected by the distributed approach "D-Clas" for each classifier, since the stage devoted to finding the threshold of votes makes use of a given learning algorithm. As can be seen, the number of features selected by centralized and distributed approaches is similar, in some cases being even larger in the centralized approach (see Connect4 with INT or Cons). Therefore, we can affirm that applying a distributed approach does not imply a larger selection of features.

3.2   Classification Accuracy Results

This section presents the classification accuracy obtained by C4.5 [22], naive Bayes [23], k-NN [24] and SVM [25] classifiers both with the centralized and distributed approaches (Table 3). The best result for each dataset and classifier is highlighted in bold face, whilst the best result for each dataset (regardless of the classifier employed) is also shadowed. As expected, the results are very variable depending on the dataset and the classifier. For some datasets (Connect4 and Isolet) the highest accuracies are achieved by centralized approaches, although the results obtained by distributed approaches are only inferior in 1 or 2%. For other datasets (Madelon, Spambase and Mnist) the best results are reported

Table 2. Number of features selected by the different approaches tested

		Connect4	Isolet	Madelon	Ozone	Spambase	Mnist
	Full set	42	617	500	72	57	717
	CFS-C	7	186	18	20	19	61
	CFS-D-Comp	8	105	9	8	18	77
	INT-C	36	56	23	16	26	40
	INT-D-Comp	7	62	11	6	15	62
	Cons-C	39	11	22	16	20	18
	Cons-D-Comp	7	31	11	5	12	48
	IG-C	7	186	18	20	19	61
	IG-D-Comp	7	131	10	9	18	67
	ReliefF-C	7	186	18	20	19	61
	ReliefF-D-Comp	9	138	13	17	15	67
C4.5	CFS-D-Clas	9	132	14	14	19	90
	IG-D-Clas	9	142	15	13	19	79
	ReliefF-D-Clas	9	146	18	12	17	74
	INT-D-Clas	9	75	14	12	19	72
	Cons-D-Clas	9	32	15	6	18	50
NB	CFS-D-Clas	9	132	14	14	20	90
	IG-D-Clas	9	142	15	13	19	79
	ReliefF-D-Clas	9	146	18	13	19	74
	INT-D-Clas	9	75	14	12	19	72
	Cons-D-Clas	10	32	15	6	20	50
k-NN	CFS-D-Clas	9	131	14	14	19	90
	IG-D-Clas	10	142	15	13	19	79
	ReliefF-D-Clas	11	145	18	12	17	74
	INT-D-Clas	10	78	14	12	19	72
	Cons-D-Clas	10	34	15	6	18	50
SVM	CFS-D-Clas	9	137	14	14	19	90
	IG-D-Clas	9	142	15	13	20	79
	ReliefF-D-Clas	9	149	18	12	17	74
	INT-D-Clas	9	70	14	12	19	72
	Cons-D-Clas	9	28	15	6	18	50

by a distributed approach, improving in up to 6% the centralized approach (see Mnist). Finally, for Ozone dataset the three approaches tested obtain the highest accuracy. The important conclusion, however, is that by distributing the data there is not a significant degradation in classification accuracy. In fact, in some cases the accuracy is improved. It is worth mentioning, for example, the case of Isolet, in which Cons-D-Clas and Cons-D-Comp combined with SVM classifier report 68.12% and 60.49% accuracy, respectively, whilst the same filter method in the standard centralized approach degrades its performance until 31.17%, probably due to the small number of features selected by the centralized filter (see Table 2).

3.3   Runtime

Table 4 reports the runtime of the feature selection algorithms, both for centralized and distributed approaches. Notice that in both distributed approaches (D-Clas and D-Comp), the feature selection stage at each subset of data is the same, so the time required will be referred as "D" for both of them. Also, in the distributed approach, considering that all the subsets can be processed at the same time, the time displayed in the table is the maximum of the times required by the filter in each subset generated in the partitioning stage. In these

Table 3. Test classification accuracy. Best results are highlighted.

		Connect4	Isolet	Madelon	Ozone	Spambase	Mnist
C4.5	CFS-C	61.22	81.59	80.50	97.63	81.16	86.99
	CFS-D-Clas	61.25	**82.23**	76.88	95.86	79.27	88.65
	CFS-D-Comp	61.25	81.53	80.62	96.09	79.73	88.55
	INT-C	60.48	78.96	80.63	96.92	78.16	87.24
	INT-D-Clas	61.66	79.03	82.38	94.79	80.83	88.62
	INT-D-Comp	61.25	79.35	80.62	96.92	80.44	88.45
	Cons-C	60.49	56.00	80.63	**98.70**	84.62	87.00
	Cons-D-Clas	61.66	77.10	82.63	96.33	79.34	**90.46**
	Cons-D-Comp	61.25	72.87	80.62	97.75	85.27	89.46
	IG-C	**63.90**	81.40	72.75	98.22	83.83	87.83
	IG-D-Clas	62.34	81.08	79.63	97.87	**85.33**	87.88
	IG-D-Comp	62.27	79.41	80.62	97.87	81.68	87.77
	ReliefF-C	63.49	79.54	73.88	98.11	78.81	87.34
	ReliefF-D-Clas	63.00	80.56	**87.50**	98.46	84.75	87.95
	ReliefF-D-Comp	63.00	81.53	84.12	95.98	84.88	88.06
NB	CFS-C	60.28	75.05	71.75	78.22	57.69	71.88
	CFS-D-Clas	58.83	73.89	70.13	76.69	57.24	73.34
	CFS-D-Comp	58.83	**75.30**	70.50	77.63	58.87	73.49
	INT-C	53.85	71.26	70.00	78.22	57.95	70.94
	INT-D-Clas	59.16	70.75	70.13	75.03	74.77	71.06
	INT-D-Comp	58.83	69.60	70.00	76.21	78.42	71.30
	Cons-C	54.12	42.78	70.00	**98.70**	91.00	72.78
	Cons-D-Clas	59.16	69.92	70.38	73.25	**92.89**	**75.74**
	Cons-D-Comp	58.83	64.91	70.00	98.46	92.05	74.61
	IG-C	60.42	69.34	70.38	74.08	76.53	70.74
	IG-D-Clas	60.28	67.54	70.63	77.63	89.70	68.09
	IG-D-Comp	60.20	66.77	70.50	78.46	66.95	68.07
	ReliefF-C	60.42	62.67	68.63	71.36	41.85	69.82
	ReliefF-D-Clas	**60.50**	56.51	71.50	60.95	91.79	70.93
	ReliefF-D-Comp	**60.50**	53.69	**72.25**	66.86	92.05	70.89
kNN	CFS-C	53.90	56.00	85.63	96.45	79.14	87.93
	CFS-D-Clas	57.61	54.78	65.63	96.57	77.31	91.65
	CFS-D-Comp	53.68	54.65	88.50	94.56	79.92	91.57
	INT-C	58.27	52.92	88.75	94.44	79.73	86.87
	INT-D-Clas	57.61	49.84	71.75	95.27	76.86	91.79
	INT-D-Comp	53.68	50.42	88.75	94.79	76.92	91.72
	Cons-C	58.06	49.90	88.75	**98.70**	80.83	87.36
	Cons-D-Clas	57.61	58.31	71.63	95.27	77.38	**96.31**
	Cons-D-Comp	53.68	57.41	88.75	95.50	76.79	95.14
	IG-C	51.29	54.78	74.25	95.98	78.62	89.63
	IG-D-Clas	57.01	59.72	86.13	95.50	78.42	90.77
	IG-D-Comp	54.52	**61.83**	88.50	94.79	77.71	90.97
	ReliefF-C	**61.81**	59.14	75.25	95.98	76.99	89.97
	ReliefF-D-Clas	57.01	57.09	**90.88**	96.80	80.70	91.35
	ReliefF-D-Comp	57.01	55.93	88.38	96.33	80.18	91.77
SVM	CFS-C	**60.42**	83.45	66.50	**98.70**	**85.85**	79.58
	CFS-D-Clas	**60.42**	82.42	67.13	**98.70**	82.27	**81.52**
	CFS-D-Comp	**60.42**	82.30	66.75	**98.70**	85.46	81.49
	INT-C	**60.42**	73.83	66.38	**98.70**	80.31	78.54
	INT-D-Clas	**60.42**	78.00	**68.50**	**98.70**	81.49	80.84
	INT-D-Comp	**60.42**	75.18	66.38	**98.70**	81.10	80.87
	Cons-C	**60.42**	31.17	66.38	**98.70**	81.88	75.14
	Cons-D-Clas	**60.42**	68.12	66.50	**98.70**	81.94	80.85
	Cons-D-Comp	**60.42**	60.49	66.38	**98.70**	81.16	80.52
	IG-C	**60.42**	82.94	67.13	**98.70**	83.83	78.28
	IG-D-Clas	**60.42**	79.67	67.13	**98.70**	83.38	79.30
	IG-D-Comp	**60.42**	80.12	66.75	**98.70**	83.38	79.15
	ReliefF-C	**60.42**	**84.61**	67.50	**98.70**	81.94	75.43
	ReliefF-D-Clas	**60.42**	82.36	67.50	**98.70**	83.57	75.72
	ReliefF-D-Comp	**60.42**	81.98	67.25	**98.70**	83.77	75.86

Table 4. Maximum untime (hh:mm:ss) for the feature selection methods tested

	Connect4	Isolet	Madelon	Ozone	Spambase	Mnist
CFS-C	00:01:40	00:04:10	00:00:36	00:00:10	00:00:12	00:29:47
CFS-D	00:00:10	00:01:17	00:00:25	00:00:08	00:00:06	00:04:17
IG-C	00:01:37	00:02:51	00:00:41	00:00:09	00:00:11	00:24:11
IG-D	00:00:04	00:00:54	00:00:29	00:00:09	00:00:05	00:03:55
ReliefF-C	00:28:00	00:09:13	00:01:02	00:00:14	00:00:21	08:26:53
ReliefF-D	00:00:11	00:01:43	00:00:40	00:00:08	00:00:04	00:22:26
INT-C	00:01:52	00:03:16	00:00:40	00:00:09	00:00:13	00:52:25
INT-D	00:00:11	00:01:10	00:00:31	00:00:08	00:00:04	00:03:19
Cons-C	00:06:08	00:04:05	00:00:52	00:00:11	00:00:14	01:42:43
Cons-D	00:00:10	00:01:20	00:00:25	00:00:06	00:00:02	00:03:17

Table 5. Average runtime (hh:mm:ss) for obtaining the threshold of votes

Method	D-Clas-C4.5	D-Clas-NB	D-Clas-kNN	D-Clas-SVM	D-Comp
CFS	00:00:36	00:00:26	00:00:48	00:01:36	00:00:01
INT	00:00:31	00:00:24	00:00:50	00:01:23	00:00:01
Cons	00:00:29	00:00:23	00:00:46	00:01:41	00:00:01
IG	00:00:38	00:00:28	00:00:46	00:01:43	00:00:01
ReliefF	00:00:33	00:00:26	00:00:41	00:02:02	00:00:01

experiments, all the subsets were processed in the same machine, but the proposed algorithm could be executed in multiple processors. Please note that this filtering time is independent of the classifier chosen. The lowest time for each dataset is shadowed.

As expected, the advantage of the distributed approaches in terms of execution time over the standard method is significant. The time is reduced for all datasets and filters, except for Ozone with the IG filter, in which it is maintained. It is worth mentioning the important reductions when the dimensionality of the dataset grows. For Mnist dataset, which has 717 features and 40000 training samples, the reduction is more than notable. In fact, for ReliefF filter, the processing time is reduced from more than 8 hours to 22 minutes, proving the adequacy of the distributed approach when dealing with large datasets.

For the distributed approaches, it is necessary to take into account the time required to calculate the threshold to build the final subset of features. Since the distributed approach "D-Clas" makes uses of a classifier to establish the threshold, the time required by this approach depends highly on the classifier, whilst with the distributed approach "D-Comp" this time is independent of the classifier. Therefore, in Table 5, we can see the average runtime on all datasets for each filter and distributed approach. It is easy to note that the time required to find the threshold in the proposed distributed approach "D-Comp" is notable lower than the one in our previous approach "D-Clas", especially with the SVM classifier, in which in some cases the reduction goes from 2 minutes to 1 second. In light of these results, we can conclude that the distributed approaches performed successfully, since the running time was considerably reduced and the accuracy did not drop to inadmissible values. In fact, our approach is able to match and in some cases even improve the standard algorithms applied to the non-partitioned datasets. Moreover, it has been demonstrated that the distributed approach proposed in this paper, "D-Comp", outperforms our previous proposal "D-Clas", since the time for obtaining the threshold of votes was also reduced and the performance results are similar.

3.4   Case Study: Determining the Optimal Interval of Votes

Bearing in mind that with our proposed approach based on complexity measures the time required to find the threshold of votes has been significantly shortened,

Table 6. Test classification accuracy with different intervals of votes. Best results are highlighted.

		Connect4	Isolet	Madelon	Ozone	Spambase	Mnist
C4.5	CFS-D-Comp	61.25	**81.53**	80.62	96.09	79.73	88.55
	CFS-D-Comp+	61.20	79.35	79.62	97.63	78.68	88.15
	INT-D-Comp	61.25	79.35	80.62	96.92	80.44	88.45
	INT-D-Comp+	61.16	77.42	80.62	97.75	79.47	88.52
	Cons-D-Comp	61.25	72.87	80.62	97.75	85.27	**89.46**
	Cons-D-Comp+	61.16	73.06	80.62	97.75	83.90	89.06
	IG-D-Comp	62.27	79.41	80.62	**97.87**	81.68	87.77
	IG-D-Comp+	61.16	77.55	80.62	97.40	**85.40**	87.61
	ReliefF-D-Comp	63.00	**81.53**	84.12	95.98	84.88	88.06
	ReliefF-D-Comp+	**63.70**	80.37	77.50	97.28	81.75	87.39
NB	CFS-D-Comp	58.83	**75.30**	70.50	77.63	58.87	73.49
	CFS-D-Comp+	60.28	75.05	70.62	80.83	52.74	72.77
	INT-D-Comp	58.83	69.60	70.00	76.21	78.42	71.30
	INT-D-Comp+	60.28	68.63	70.00	84.26	88.98	71.19
	Cons-D-Comp	58.83	64.91	70.00	**98.46**	**92.05**	**74.61**
	Cons-D-Comp+	60.28	61.96	70.00	**98.46**	79.79	73.33
	IG-D-Comp	60.20	66.77	70.50	78.46	66.95	68.07
	IG-D-Comp+	60.28	65.55	70.50	79.17	90.35	69.20
	ReliefF-D-Comp	**60.50**	53.69	**72.25**	66.86	**92.05**	70.89
	ReliefF-D-Comp+	60.44	50.99	70.62	60.71	87.94	70.60
kNN	CFS-D-Comp	53.68	54.65	88.50	94.56	79.92	91.57
	CFS-D-Comp+	51.48	52.02	86.12	95.03	78.10	89.77
	INT-D-Comp	53.68	50.42	**88.75**	94.79	76.92	91.72
	INT-D-Comp+	50.38	54.14	**88.75**	95.62	77.25	91.26
	Cons-D-Comp	53.68	57.41	**88.75**	95.50	76.79	**95.14**
	Cons-D2	50.38	61.90	**88.75**	95.50	76.27	93.73
	IG-D-Comp	54.52	61.83	88.50	94.79	77.71	90.97
	IG-D2	50.38	**61.96**	88.50	95.27	78.03	90.23
	ReliefF-D-Comp	**57.01**	55.93	88.38	**96.33**	80.18	91.77
	ReliefF-D-Comp+	56.91	54.91	84.25	96.09	77.38	90.42
SVM	CFS-D-Comp	60.42	**82.30**	66.75	**98.70**	85.46	**81.49**
	CFS-D-Comp+	60.42	79.09	67.12	**98.70**	85.85	80.91
	INT-D-Comp	60.42	75.18	66.38	**98.70**	81.10	80.87
	INT-D-Comp+	60.42	74.15	66.38	**98.70**	82.46	80.30
	Cons-D-Comp	60.42	60.49	66.38	**98.70**	81.16	80.52
	Cons-D-Comp+	60.42	54.71	66.38	**98.70**	76.99	79.49
	IG-D-Comp	60.42	80.12	66.75	**98.70**	83.38	79.15
	IG-D-Comp+	60.42	78.64	66.75	**98.70**	83.51	78.78
	ReliefF-D-Comp	60.42	81.98	67.25	**98.70**	83.77	75.86
	ReliefF-D-Comp+	60.42	79.73	**67.88**	**98.70**	80.70	75.42

in this case study we would like to analyze if it is possible to increase the interval of possible number of votes. In Section 2.1 we have explained that this interval was set to $[avg \pm 1/2std]$, trying to avoid a high number of calculations which could lead to unaffordable computing times. However, since we have seen that this time is not prohibitive anymore, we performed some experiments setting this interval to $[avg \pm std]$. In Table 6 we can see the accuracy results comparing our distributed approach "D-Comp" and this new proposal that we have called "D-Comp+". If we compare both approaches for each combination of dataset, filter and classifier, it turns out that "D-Comp" outperforms or matches "D-Comp+" in 89 out of 120 cases. Regarding the number of selected features, "D-Comp" selects in almost all cases a slightly larger number of features than "D-Comp+", which apparently leads to better performances. In terms of running time, the computation of the optimal threshold does not take more than 1 second in any case, regardless of the interval chosen. Therefore, we can conclude that, although the low computational times required for finding the threshold would allow us to try a larger number of possible votes, it is better to maintain our original proposed approach in which the interval was set to $[avg \pm 1/2std]$. Moreover, if in the future we need to deal with datasets with millions of data, it is better to reduce the computation as much as possible.

4   Conclusions

Feature selection is usually applied in a centralized manner. However, if the data are distributed, feature selection may take advantage of processing multiple subsets in sequence or concurrently. The need to use distributed feature selection can be two-fold. On the one hand, with the advent of network technologies, the data are sometimes distributed in multiple locations and often with multiple parties. On the other hand, most existing feature selection algorithms do no scale well and their efficiency significantly deteriorates when dealing with large-scale data.

In this paper we propose a methodology for distributing the process of feature selection by tackling the most common distribution in the literature: the horizontal partition. A previous proposal was able to confront this problem, but presented certain drawbacks, mainly (1) the partitioning of the data did not take into account the class distribution, a fact which might lead to partitions in which a certain class is not represented and (2) the method has shown to be dependent on the classifier and time-consuming in the process of merging the partial results from the different partitions.

In this new proposal, we aimed at achieving a method able to overcome these drawbacks, especially in terms of running time, since in high dimensional datasets this will be a core issue. For this sake, we modify our previous methodology so that the partitions of the dataset maintain the original class distribution. Then, we propose a new methodology for combining the partial results from different partitions which makes use of data complexity measures. The rationale behind this was that features that contribute to decrease the complexity of a dataset must be maintained, whereas those that contribute to increase the complexity must be discarded. By using this new methodology, we were able to reduce significantly the running time while maintaining the classification performance. Moreover, our new approach is independent on the subsequent classifier. Finally, it is worth mentioning that the proposed method can be used with any feature selection algorithm without any modifications, so it could be seen as a general framework for distributed feature selection.

As future work, we plan to test the scalability properties of the proposed method with datasets larger than $100\,000$ samples. Moreover, it would be interesting to perform a sensitivity analysis on the value chosen for α. Finally, another line of future research would be trying other complexity measures.

Acknowledgments. This research has been economically supported in part by the Ministerio de Economía y Competitividad of the Spanish Government through the research project TIN 2012-37954, partially funded by FEDER funds of the European Union; and by the Consellería de Industria of the Xunta de Galicia through the research project GRC2014/035.

References

1. Bache, K., Lichman, M.: UCI machine learning repository (2013). http://archive.ics.uci.edu/ml (accessed January 2015)
2. Zhao, Z.A., Liu, H.: Spectral feature selection for data mining. Chapman & Hall/CRC (2011)
3. Guyon, I.: Feature extraction: foundations and applications, vol. 207. Springer, Heidelberg (2006)
4. Saeys, Y., Inza, I., Larrañaga, P.: A review of feature selection techniques in bioinformatics. Bioinformatics **23**(19), 2507–2517 (2007)
5. Chan, P.K., Stolfo, S.J.: Toward parallel and distributed learning by meta-learning. In: AAAI workshop in Knowledge Discovery in Databases, pp. 227–240 (1993)
6. Ananthanarayana, V.S., Subramanian, D.K., Murty, M.N.: Scalable, distributed and dynamic mining of association rules. In: Prasanna, V.K., Vajapeyam, S., Valero, M. (eds.) HiPC 2000. LNCS, vol. 1970, pp. 559–566. Springer, Heidelberg (2000)
7. Tsoumakas, G., Vlahavas, I.: Distributed data mining of large classifier ensembles. In: Proceedings Companion Volume of the Second Hellenic Conference on Artificial Intelligence, pp. 249–256 (2002)
8. Das, K., Bhaduri, K., Kargupta, H.: A local asynchronous distributed privacy preserving feature selection algorithm for large peer-to-peer networks. Knowledge and information systems **24**(3), 341–367 (2010)
9. McConnell, S., Skillicorn, D.B.: Building predictors from vertically distributed data. In: Proceedings of the 2004 Conference of the Centre for Advanced Studies on Collaborative Research, pp. 150–162. IBM Press (2004)
10. Skillicorn, D.B., McConnell, S.M.: Distributed prediction from vertically partitioned data. Journal of Parallel and Distributed computing **68**(1), 16–36 (2008)
11. Rokach, L.: Taxonomy for characterizing ensemble methods in classification tasks: A review and annotated bibliography. Computational Statistics & Data Analysis **53**(12), 4046–4072 (2009)

12. Banerjee, M., Chakravarty, S.: Privacy preserving feature selection for distributed data using virtual dimension. In: Proceedings of the 20th ACM international conference on Information and knowledge management, pp. 2281–2284. ACM (2011)
13. Bolón-Canedo, V., Sánchez-Maroño, N., Cerviño-Rabuñal, J.: Scaling up feature selection: a distributed filter approach. In: Bielza, C., Salmerón, A., Alonso-Betanzos, A., Hidalgo, J.I., Martínez, L., Troncoso, A., Corchado, E., Corchado, J.M. (eds.) CAEPIA 2013. LNCS, vol. 8109, pp. 121–130. Springer, Heidelberg (2013)
14. de Haro García, A.: Scaling data mining algorithms. Application to instance and feature selection. Ph.D. thesis, Universidad de Granada (2011)
15. Basu, M., Ho, T.K.: Data complexity in pattern recognition. Springer (2006)
16. Hall, M., Frank, E., Holmes, G., Pfahringer, B., Reutemann, P., Witten, I.H.: The Weka data mining software: an update. ACM SIGKDD Explorations Newsletter **11**(1), 10–18 (2009)
17. Hall, M.A.: Correlation-based feature selection for machine learning. Ph.D. thesis, The University of Waikato (1999)
18. Dash, M., Liu, H.: Consistency-based search in feature selection. Artificial intelligence **151**(1), 155–176 (2003)
19. Zhao, Z., Liu, H.: Searching for interacting features. In: IJCAI, vol. 7, pp. 1156–1161 (2007)
20. Hall, M.A., Smith, L.A.: Practical feature subset selection for machine learning. Computer Science **98**, 181–191 (1998)
21. Kononenko, I.: Estimating attributes: analysis and extensions of relief. In: Bergadano, F., De Raedt, L. (eds.) ECML 1994. LNCS, vol. 784, pp. 171–182. Springer, Heidelberg (1994)
22. Quinlan, J.R.: C4. 5: programs for machine learning. Morgan kaufmann (1993)
23. Rish, I.: An empirical study of the naive bayes classifier. In: IJCAI 2001 workshop on empirical methods in artificial intelligence, vol. 3, pp. 41–46 (2001)
24. Aha, D.W., Kibler, D., Albert, M.K.: Instance-based learning algorithms. Machine learning **6**(1), 37–66 (1991)
25. Vapnik, V.N.: Statistical learning theory. Wiley (1998)

Ensemble Feature Selection
for Rankings of Features

Borja Seijo-Pardo[✉], Verónica Bolón-Canedo, Iago Porto-Díaz,
and Amparo Alonso-Betanzos

Department of Computer Science, University of A Coruña,
Campus de Elviña s/n 15071, A Coruña, Spain
borja.seijo@udc.es

Abstract. In the last few years, ensemble learning has been the focus of much attention mainly in classification tasks, based on the assumption that combining the output of multiple experts is better than the output of any single expert. This idea of ensemble learning can be adapted for feature selection, in which different feature selection algorithms act as different experts. In this paper we propose an ensemble for feature selection based on combining rankings of features, trying to overcome the problem of selecting an appropriate ranker method for each problem at hand. The results of the individual rankings are combined with SVM Rank, and the adequacy of the ensemble was subsequently tested using SVM as classifier. Results on five UCI datasets showed that the use of the proposed ensemble gives better or comparable performance than the feature selection methods individually.

1 Introduction

Over the last few years, there has been a notable increase in the size of the datasets used in the area of machine learning. Fruit of this, *feature selection* has taken an important role in this field, as it allows to identify an ideal subset of relevant features of the data, eliminating irrelevant and redundant information. As a result, the size of the datasets is reduced, leading to a minor use of the storage size and improving the computational time of machine learning algorithms. Previous studies have shown that the performance of the classification models improves when irrelevant and redundant features are eliminated from the original dataset [7,28].

Machine learning methods have traditionally used a single learning model to solve a given problem. However, along the last few years, it has been observed that by using and combining different learning models on the same problem

---

This research has been economically supported in part by the Ministerio de Economía y Competitividad of the Spanish Government through the research project TIN 2012-37954, partially funded by FEDER funds of the European Union; and by the Consellería de Industria of the Xunta de Galicia through the research project GRC2014/035.

I. Rojas et al. (Eds.): IWANN 2015, Part II, LNCS 9095, pp. 29–42, 2015.
DOI: 10.1007/978-3-319-19222-2_3

better results could be obtained. This idea is based on the assumption that combining the output of multiple experts in a particular problem is better than the output of a single expert [18,19]. This combination of machine learning methods for solving problems is called *ensemble learning*. Typically, ensemble learning has been applied to classification, although it can be also thought as a means of improving other machine learning disciplines such as *feature selection*.

In feature selection, there are two different approaches when conducting the evaluation of the features of a dataset: (i) *individual evaluation* and (ii) *subset evaluation* [39]. In the first case, a ranking of features is returned by assigning a level of relevance to each of these features. In the second case, successive subsets of features are generated according to a predefined search strategy. The subsets are evaluated iteratively, according to an optimality criterion, until reaching the final subset of selected features. While the individual evaluation is incapable of removing redundant features, because redundant features are likely to have similar rankings, the subset evaluation approach can handle feature redundancy with feature relevance. However, methods in this framework can suffer from a problem of computational efficiency caused by searching through all feature subsets required in the subset generation step, and thus, this paper will be focused on the individual evaluation approach.

Aside from this classification, three major approaches can be distinguished based upon the relationship between a feature selection algorithm and the inductive learning method used to infer a model: (i) *filter methods*, (ii) *wrapper methods* and (iii) *embedded methods* [10]. Filter methods rely on the general characteristics of training data and carry out the feature selection process with independence of the induction algorithm. This model has the advantages of low computational cost and good generalization ability. On the contrary, wrapper methods involve a learning algorithm as a black box and then use its prediction performance to assess the relative usefulness of subsets of variables. This iteration with the classifier tends to give better performance results than other methods, but it is very time consuming and has the risk of overfitting. Finally, embedded methods search for an optimal subset of features into the classifier construction and can be seen as a search in the combined space of subsets and hypotheses. This approach is able to capture dependencies at a lower computational cost than wrappers, but may also suffer from overfitting. Thus, in this paper, and after a preliminary study involving the three approaches, filters and embedded methods were chosen because they allow for reducing the dimensionality of the data without compromising the time and memory requirements of machine learning algorithms.

So, the idea of this paper is to use an ensemble of filters and embedded methods to induce diversity, instead of a single method, which is the approach that has been mostly used in the past years. In this way, we aim at increasing the stability of the feature selection process, since the proposed method takes advantage of the strengths of the single selectors and overcomes their weak points.

The proposed ensemble combines the different rankings returned by each different feature selection method (a total of six methods that will be explained in detail later), and finally obtains a classification output for this unique ranking of features. Experimental validation of the methodology on a range of UCI datasets [2] using a *Support Vector Machine* [6] *(SVM)* shows the adequacy of the proposed ensemble, paving the way to its application to other real-world datasets. It is also worth mentioning that by using this ensemble the user is released from the task of choosing an adequate method for each scenario, since this approach provides acceptable results independently of the characteristics of the data.

The remainder of this paper is organized as follows: Section 2 provides an enumeration and description of previous related works. In Section 3 we introduce the proposed ensemble and its algorithm, as well as the individual ranker methods and the *SVM Rank* method used to join the individual rankers. Next, Section 4 describes the datasets, experimental design, and experimental results. Finally, in Section 5, we make our concluding remarks.

2   The Background

Feature selection has been applied in many machine learning and data mining problems. The aim of feature selection is to select a subset of features that minimizes the prediction error obtained by a given classifier. Previous works, as those presented by Guyon and Elisseeff [9] or Hall and Holmes [11] collect different approaches used for feature selection, including feature construction, feature ranking, multivariate feature selection, efficient search methods and feature validity assessment methods.

Along the last few years, it has been observed that, by using and combining different learning models on the same problem, better results could be obtained. This combination of machine learning methods for solving problems is called *ensemble learning*. Moreover, combining classifiers appears as a natural step forward when a critical mass of knowledge of single classifier models has been accumulated, and have been rapidly growing and enjoying a lot of attention from pattern recognition and machine learning communities [18].

Typically, ensemble learning has been applied to classification, where the most popular methods are *bagging* [5] and *boosting* [29]. Bagging creates an ensemble by training individual classifiers on bootstrap samples of the training set. Each bootstrap sample is generated by randomly selecting, with replacement, n instances from the training set where n is the size of the training set. As a result of the sampling with replacement procedure, each classifier is trained on the average of 63.2% of the training instances. The prediction of each classifier is combined using simple voting. On the other hand, in the boosting approach the sampling is proportional to an instance's weight. Bagging and boosting are two of the most well-known ensemble learning methods due to their theoretical performance guarantees and strong experimental results. Although these models are the most used to improve the classification results, new ensemble learning techniques on the feature subspace have been proposed. The *Random Subspace* [12] method is a simple random selection of feature subsets derived from the theory of stochastic discrimination. Optiz [24] describes an ensemble feature selection technique for neural networks called *Genetic Ensemble Feature Selection*. Another ensemble method for decision trees is called *Stochastic Attribute Selection Committees* [40], while *Multiple Feature Subsets* [3] is a combining algorithm for nearest neighbor classifiers. Finally, for steganalysis of digital media, an ensemble of classifiers implemented as random forests [16] has been proposed, since this ensemble is ideally suited for this kind of problems.

In recent works it is proposed to improve the robustness of a feature selection algorithm by using multiple feature selection evaluation criteria. Several studies have been performed in this general area, in order to achieve better classification accuracy. One of these studies [30] has been conducted on 21 UCI datasets [2], comparing five measures of diversity with regard to their possible use in ensemble feature selection. This study considers four search strategies for ensemble feature selection together with the simple random subspacing: genetic search, hill-climbing, and ensemble forward and backward sequential selection. Based on the idea of multiple feature selection evaluation criteria, many ensembles of feature selection methods have appeared. A *Multicriterion Fusion-based Recursive Feature Elimination* [38] *(MCF-RFE)* algorithm is developed with the goal of improving both the classification performance and the stability of the feature selection results. A feature ranking scheme for *Multi-layer Perceptron* [35] *MLP* ensembles is proposed, along with a stopping criterion based upon the *out-ofbootstrap (OOB)* estimate. Experimental results on benchmark data demonstrate the versatility of the MLP base classifier in removing irrelevant features.

Finally, there are some other works in which all the feature selection methods of the final ensemble are ranker methods. Diversity can be achieved by using various rankers, combined afterwards to yield more stable and robust results. Three commonly used filter-based feature ranking techniques for text classification problems were used by Olsson and Oard [23], where the combining methods employed are lowest, highest and average rank.

Wang et al. perform a few outstanding works in this area, providing two interesting studies. The first one examines the ensembles of 6 commonly used filter-based rankers [33] and the second one studies 17 different ensembles of feature ranking techniques [34], with 6 commonly-used rankers, the signal-to-noise filter technique *(S2N)* [37], and 11 threshold-based rankers. In this second work, the ensembles are composed of different numbers of rankers, ranging from 2 to 18 single feature selection methods. Also, other studies collect different methods to combine the single generated rankings, with the aim of obtaining a final ensemble. This combination of single rankings covers from simple –as mean, median, minimal, etc.– to more complex methods –as *Weighted mean aggregation* [1] *(WMA), Complete linear aggregation* [1] *(CLA)* and *Robust ensemble feature selection* [4] *Rob-EFS–*.

As stated before, the variability of results over different datasets is one of the problems of choosing a feature selection technique. The aim of this paper is to achieve a method that reduces the variability of the individual methods of feature selection, in order to take advantage of their individual strengths and overcome their weak points at the same time. There will be as many outputs as individual rankers employed and the result of the different methods will be combined using an innovative technique. In this research, the union method used to obtain a final ranking of features is *SVM Rank* [13]. From the generated ranking, a smaller subset of features will be obtained according to a threshold, and it will be used to be trained and classified by a *Support Vector Machine with Radial-Basis-Function* [25] *(SVM-RBF)*, in order to check the adequacy of the proposed ensemble in terms of classification error.

3   Ensemble Feature Selection

Real life datasets come in diverse flavors and sizes, and so their nature imposes several substantial restrictions for both learning models and feature selection algorithms [32]. Datasets may be very large in samples and number of features, and also there might be problems with redundant, noisy, multivariate and non-linear scenarios. Thus, most existing methods alone are not capable of con-fronting these problems, and something like "the best feature selection method" simply does not exist in general, making it difficult for users to select one method over another. In order to make a correct choice, a user not only needs to know the domain well and the characteristics of each dataset, but also is expected to understand technical details of available algorithms [21].

Besides, machine learning methods have come to be a necessity for many companies, in order to obtain useful information and knowledge from their increasingly massive databases. But a deep expertise in the field is needed to, for example, select the appropriate methods, tune the parameters, etc. As experts of this type are not universally available, more user-friendly methods are neces-sary. In this sense, a possible way to confront this situation is to use an ensemble of feature selection algorithms, which is our proposal in this paper. Specifically, we use methods that follow the ranking approach, i.e. they return an ordered ranking of all the features. Notice that methods that return a ranking of features are less computationally expensive than those which return a subset of selected features, and this is of vital importance when the current tendency is toward Big Data problems. Then, the outputs of all the components of the ensemble have to be combined in order to produce a common final output. The ensem-ble that we are proposing combines these rankings using *SVM Rank* [13], which is a SVM-based method for learning of ranking functions. Figure 1 shows the proposed approach.

The problem of ranking is formalized as follows: for a query q and a data collection $D = \{d_1, \ldots, d_n\}$, if a data d_i is ranked higher than d_j for an ordering r, i.e. $d_i <_r d_j$, then $(d_i, d_j) \in r$, otherwise $(d_i, d_j) \notin r$. The algorithm selects a ranking function f that maximizes:

$$\tau_S(f) = \frac{1}{n} \sum_{i=1}^{n} \tau(r_{f(q_i)}, r_i) \tag{1}$$

The function f must maximize (1) and must generalize well beyond the training data. Consider the class of linear ranking functions (2), where $\vec{w}$ is a weight

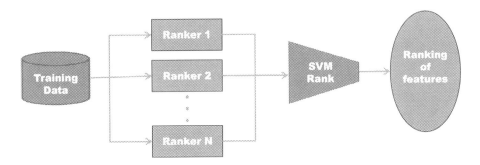

Fig. 1. The proposed ensemble method

vector that is adjusted by learning, and $\Phi(q, d)$ is a mapping onto features that describes the match between query q and data d.

$$(d_i, d_j) \in f_{\vec{w}}(q) \iff \vec{w}\Phi(q, d_i) > \vec{w}\Phi(q, d_j). \tag{2}$$

The task of the learner is to minimize the number of discordant ranking pairs. For the class of linear ranking functions (2), this is equivalent to finding the weight vector $\vec{w}$ so that the maximum number of the following inequalities (3 and 4) is satisfied.

$$\forall(d_i, d_j) \in r_1^* : \vec{w}\Phi(q_1, d_i) > \vec{w}\Phi(q_1, d_j) \tag{3}$$

$$\ldots$$

$$\forall(d_i, d_j) \in r_n^* : \vec{w}\Phi(q_n, d_i) > \vec{w}\Phi(q_n, d_j) \tag{4}$$

Unfortunately, this problem is known to be NP-hard, however it is possible to approximate the solution by introducing non-negative slack variables $\xi_{i,j,k}$ and minimizing the upper bound $\sum \xi_{i,j,k}$. Therefore, the above problem is optimized, obtaining the approximation shown in (5).

minimize: $\qquad V(\vec{w}, \vec{\xi}) = \dfrac{1}{2}\vec{w} \cdot \vec{w} + C \sum \xi_{i,j,k}$

subject to:

$$\forall(d_i, d_j) \in r_1^* : \quad \vec{w}\Phi(q_1, d_i) \geq \vec{w}\Phi(q_1, d_j) + 1 - \xi_{i,j,1} \tag{5}$$

$$\ldots$$

$$\forall(d_i, d_j) \in r_n^* : \quad \vec{w}\Phi(q_n, d_i) \geq \vec{w}\Phi(q_n, d_j) + 1 - \xi_{i,j,n}$$

$$\forall i \forall j \forall k : \qquad \xi_{i,j,k} \geq 0$$

C is a parameter that controls the trade-off between the margin size and the training error. By rearranging the constraints in (5) as

$$\vec{w}(\Phi(q_k, d_i) - \Phi(q_k, d_j)) \geq 1 - \xi_{i,j,k}, \tag{6}$$

it becomes equivalent to that of SVM classification on pairwise difference vectors $\Phi(q_k, d_i) - \Phi(q_k, d_j)$. For each test-model pair, features are calculated to measure the similarity between them. The ranking order of the model objects is also known. Thus, the input to the SVM learning algorithm, to learn the optimal ranking function, are the training data presented above. Given a new test q, the model objects can be sorted based on their value of

$$rsv(q, d_i) = \vec{w}\Phi(q, d_i) = \sum \alpha_{k,l}^* \Phi(q_k, d_l)\Phi(q, d_j). \qquad (7)$$

The $\alpha_{k,l}^*$ can be derived from the values of the dual variables at the solution.

Among the broad suite of feature selection ranking methods available in the literature, four filters and two embedded methods were chosen for this study:

- *Chi-Square* [20] (filter): this univariate filter is based on the χ^2 statistic and evaluates each feature independently with respect to the classes. The higher the value of chi-squared, the more relevant is the feature with respect to the class.
- *Information Gain* [27] (filter): this is one of the most common univariate methods of evaluation attributes. This filter evaluates the features according to their information gain and considers a single feature at a time.
- *mRMR* [26] (filter): the *minimum Redundancy Maximum Relevance* filter uses mutual information to select the features that have the highest relevance with the target class and are also minimally redundant, i.e., selects the features that are maximally dissimilar to each other.
- *ReliefF* [17] (filter): this method is an extension of the original Relief algorithm [15]. The original Relief works by randomly sampling an instance from the dataset and then locating its nearest neighbor from the same and opposite class. The values of the attributes of the nearest neighbors are compared to the sampled instance and used to update relevance scores for each attribute. The rationale is that a useful attribute should differentiate between instances from different classes and have the same value for instances from the same class. ReliefF adds the ability of dealing with multiclass problems and is also more robust and capable of dealing with incomplete and noisy data. This method may be applied in all situations, has low bias, includes interaction among features and may capture local dependencies which other methods miss.
- *SVM-RFE* [8] (embedded): *Recursive Feature Elimination for Support Vector Machines (SVM-RFE)* trains a SVM classifier iteratively with the current set of features and removes the least important features indicated by the weights in the SVM solution.
- *FS-P* [22] (embedded): *Feature Selection - Perceptron (FS-P)* is an embedded method based on a perceptron. A perceptron is a type of artificial neural network that can be seen as the simplest kind of feedforward neural network: a linear classifier. The basic idea of this method consists on training a perceptron in the context of supervised learning. The interconnection weights are used as indicators of which features could be the most relevant and provide a ranking.

This set of ranker methods was selected because (i) they are based on different metrics so they ensure a great diversity in the final ensemble; and (ii) they are widely used by researchers in feature selection.

In this study, R rankings are generated using the aforementioned feature selection methods, all of them with the same training data. The pseudo-code of this approach can be seen in Algorithm 1.

Algorithm 1. Pseudo-code of the proposed ensemble method

Data: R — number of ranker methods
Data: T — threshold of the number of features to be selected

Result: P — classification prediction

1 **for** *each r from 1 to R* **do**
2   Obtaining ranking A_r using method r
3 **end**
4 A = combining rankings A_r with SVM Rank
5 A_t = Select T first attributes from A
6 Build classifier SVM-RBF with the selected attributes A_t
7 Obtain prediction P

The A_r outputs obtained from the different methods are combined using the *SVM Rank* union method to obtain a single ranking list. Since the individual methods used for feature selection are rankers, it is necessary to establish a threshold T in order to obtain a practical subset of features. After obtaining this practical subset of features A_t, a *Support Vector Machine with Radial-Basis-Function* [25] *(SVM-RBF)* has been implemented for checking the adequacy of the proposed ensemble in terms of classification error.

4   Experimental Study

The performance of the proposed ensemble method is tested on five well-known datasets which are listed in Table 1. The number of samples ranges from 1 484 to 67 557 and the number of features ranges from 8 to 617. These datasets conform an interesting benchmark to check the adequacy of the proposed ensemble.

The experiment performed consisted of a comparison between the use of different feature selection methods individually and the use of an ensemble. As described in the previous section, six well-known methods have been chosen, including four filter methods (Chi-Square, InfoGain, mRMR and ReliefF) and two embedded methods (SVM-RFE and FS-P). Please note that all the feature selection methods used in this paper are rankers, i.e. they do not select a subset of features, but they sort all the features. Therefore it is necessary to establish a threshold in order to obtain a practical subset of features. In this study, we have opted for the threshold $\log_2(n)$ –where n is the number of features for a

Table 1. Datasets employed in the experimental study. All of them are freely available in [2].

Dataset	Samples	Features	Classes
Yeast	1 484	8	10
Spambase	4 601	57	2
Madelon	2 400	500	2
Connect4	67 557	42	3
Isolet	7 797	617	26

given dataset– since (i) selecting $\log_2(n)$ number of metrics for software quality prediction is recommended in [34] and (ii) a recent study [14] showed that it was appropriate to use $\log_2(n)$ as the number of features when using WEKA [36] to build Random Forests learners. In addition to this, we have opted for three more different thresholds of number of features selected to have a baseline with which to compare the threshold $\log_2(n)$. Thus, different thresholds have been analyzed and compared in this research, including the subsets containing $\log_2(n)$, 10%, 25% and 50% of features.

In the first place, an exhaustive analytical study was performed, where all possible combinations of 4, 5 and 6 rankers were tried, combining then the partial rankings by using *SVM Rank* to obtain a single final ranking. From this preliminary study, whose results cannot be displayed for reasons of space, it was observed that the ensemble of 6 rankers obtained, on average, the best errors. Therefore, in order to release the user from the task of deciding which feature selection methods are more appropriate, the combination of 6 rankers is utilized in the next part of the experiment.

A Support Vector Machine (SVM) has been chosen for checking the adequacy of the proposed ensemble in terms of classification error. A 10-fold cross validation was performed, which consists of dividing the dataset into 10 subsets and repeating the learning process 10 times. Each time, 1 subset is used as the test set and the other 9 subsets are put together to form the training set. Finally, the average error across all 10 trials is computed.

The next four tables (Tables 2, 3, 4 and 5) display the average test errors. Having 10 different errors as a result of the 10-fold cross validation, a Kruskal-Wallis test was applied to check if there were significant differences for a level of significance $\alpha = 0.05$. Then, a multiple comparison procedure (Tukey's) [31] was applied and those algorithms whose error average test results are not significantly worse than the best are labeled with a cross.

The experimental results demonstrated the adequacy of the proposed ensemble, since they matched or improved upon the results achieved by the feature selection methods alone. In fact, one can see that the ensemble is not significantly different to the lowest average error for every dataset with a threshold of $\log_2(n)$ and 10 % (respectively indicated in Tables 2 and 3). As the threshold

Table 2. $\log_2(n)$ threshold: average estimated percentage test errors. The cross shows results that are not significantly different than the best.

Ranker method	Yeast	Spambase	Madelon	Connect4	Isolet
Ensemble	$55.52^{\dagger} \pm 4.73$	$11.39^{\dagger} \pm 2.34$	$33.46^{\dagger} \pm 4.24$	$31.14^{\dagger} \pm 0.51$	$48.99^{\dagger} \pm 2.31$
Chi-Square	$58.76^{\dagger} \pm 3.92$	$13.17^{\dagger} \pm 1.02$	$33.92^{\dagger} \pm 4.64$	$30.81^{\dagger} \pm 0.58$	$50.99^{\dagger} \pm 3.95$
InfoGain	$58.76^{\dagger} \pm 3.92$	$13.39^{\dagger} \pm 1.24$	$33.83^{\dagger} \pm 4.70$	$30.76^{\dagger} \pm 0.54$	58.23 ± 6.83
mRMR	$53.90^{\dagger} \pm 3.52$	22.78 ± 2.24	$42.17^{\dagger} \pm 2.77$	32.29 ± 0.49	$43.50^{\dagger} \pm 1.92$
ReliefF	$58.76^{\dagger} \pm 3.92$	20.08 ± 3.14	$33.21^{\dagger} \pm 4.24$	$30.70^{\dagger} \pm 0.50$	59.47 ± 1.71
SVM-RFE	$55.86^{\dagger} \pm 4.44$	$12.50^{\dagger} \pm 1.41$	$33.50^{\dagger} \pm 4.35$	33.92 ± 0.59	$48.12^{\dagger} \pm 5.98$
FS-P	$54.38^{\dagger} \pm 3.11$	$12.17^{\dagger} \pm 1.52$	$33.50^{\dagger} \pm 4.42$	34.18 ± 0.60	61.86 ± 1.33

Table 3. 10% threshold: average estimated percentage test errors. The cross shows results that are not significantly different than the best.

Ranker method	Yeast	Spambase	Madelon	Connect4	Isolet
Ensemble	$53.24^{\dagger} \pm 4.62$	$11.39^{\dagger} \pm 2.34$	$35.46^{\dagger} \pm 4.05$	$31.14^{\dagger} \pm 0.51$	$50.13^{\dagger} \pm 2.03$
Chi-Square	$59.97^{\dagger} \pm 5.18$	$13.17^{\dagger} \pm 1.02$	$34.29^{\dagger} \pm 3.34$	$30.81^{\dagger} \pm 0.58$	75.06 ± 1.93
InfoGain	$55.13^{\dagger} \pm 4.99$	$13.39^{\dagger} \pm 1.24$	$33.62^{\dagger} \pm 3.50$	$30.76^{\dagger} \pm 0.54$	$48.62^{\dagger} \pm 2.30$
mRMR	$55.13^{\dagger} \pm 4.99$	22.78 ± 2.24	46.42 ± 3.46	32.29 ± 0.49	$47.15^{\dagger} \pm 1.71$
ReliefF	$55.13^{\dagger} \pm 4.99$	20.08 ± 3.14	$33.17^{\dagger} \pm 3.13$	$30.70^{\dagger} \pm 0.50$	58.38 ± 2.23
SVM-RFE	$54.31^{\dagger} \pm 6.50$	$12.50^{\dagger} \pm 1.41$	$31.71^{\dagger} \pm 2.56$	33.92 ± 0.59	$51.58^{\dagger} \pm 3.33$
FS-P	$54.66^{\dagger} \pm 4.33$	$12.17^{\dagger} \pm 1.52$	$33.96^{\dagger} \pm 2.94$	34.18 ± 0.60	64.95 ± 4.53

Table 4. 25% threshold: average estimated percentage test errors. The cross shows results that are not significantly different than the best.

Ranker method	Yeast	Spambase	Madelon	Connect4	Isolet
Ensemble	$53.24^{\dagger} \pm 4.62$	$19.19^{\dagger} \pm 2.00$	$36.21^{\dagger} \pm 4.42$	$26.96^{\dagger} \pm 0.67$	48.77 ± 2.11
Chi-Square	$59.97^{\dagger} \pm 5.18$	$18.38^{\dagger} \pm 1.12$	37.21 ± 2.85	$26.41^{\dagger} \pm 0.56$	60.34 ± 1.92
InfoGain	$53.17^{\dagger} \pm 6.89$	$17.67^{\dagger} \pm 1.90$	$36.08^{\dagger} \pm 2.37$	$26.34^{\dagger} \pm 0.53$	$43.68^{\dagger} \pm 1.94$
mRMR	$53.17^{\dagger} \pm 6.89$	$15.65^{\dagger} \pm 1.95$	49.83 ± 3.11	32.05 ± 0.47	$46.72^{\dagger} \pm 2.71$
ReliefF	$53.17^{\dagger} \pm 6.89$	$19.34^{\dagger} \pm 3.69$	$35.58^{\dagger} \pm 2.74$	$25.55^{\dagger} \pm 0.46$	54.08 ± 3.64
SVM-RFE	$53.56^{\dagger} \pm 6.45$	$18.04^{\dagger} \pm 3.99$	$31.12^{\dagger} \pm 2.89$	32.79 ± 1.32	$42.20^{\dagger} \pm 1.77$
FS-P	$53.03^{\dagger} \pm 4.51$	$14.45^{\dagger} \pm 5.53$	$33.08^{\dagger} \pm 1.76$	34.00 ± 0.64	61.45 ± 4.34

is increased, the results obtained are not as positive. Despite this, the proposed ensemble obtains favorable results in 4 out of the 5 datasets studied when the threshold is fixed to 25 % (indicated in Table 4). Finally, when the threshold is increased to 50 % (Table 5), only 2 out of the 5 datasets studied have results not significantly different from the lowest average error. Even so, in the 3 datasets in which significative differences between the ensemble method and the best single method, it can be seen that the estimated percentage error of the ensemble is lower than the one presented by several single rankers.

Table 5. 50% threshold: average estimated percentage test errors. The cross shows results that are not significantly different than the best.

Ranker method	Yeast	Spambase	Madelon	Connect4	Isolet
Ensemble	$53.24^{\dagger} \pm 4.62$	16.93 ± 1.91	39.29 ± 2.65	25.27 ± 0.59	$43.21^{\dagger} \pm 3.01$
Chi-Square	$59.97^{\dagger} \pm 5.18$	$16.71^{\dagger} \pm 1.88$	39.13 ± 3.37	$24.94^{\dagger} \pm 0.70$	50.03 ± 2.20
InfoGain	$54.05^{\dagger} \pm 4.31$	16.95 ± 1.28	39.00 ± 2.78	$24.82^{\dagger} \pm 0.50$	$37.98^{\dagger} \pm 2.36$
mRMR	$54.05^{\dagger} \pm 4.31$	$13.17^{\dagger} \pm 1.43$	39.17 ± 2.84	30.94 ± 0.55	46.75 ± 3.36
ReliefF	$54.05^{\dagger} \pm 4.31$	$16.17^{\dagger} \pm 1.15$	$38.21^{\dagger} \pm 2.44$	$23.51^{\dagger} \pm 0.50$	50.33 ± 3.45
SVM-RFE	$52.57^{\dagger} \pm 6.65$	18.11 ± 1.66	$32.71^{\dagger} \pm 2.94$	31.96 ± 2.06	$37.70^{\dagger} \pm 3.12$
FS-P	$54.80^{\dagger} \pm 5.87$	16.06 ± 4.18	$34.92^{\dagger} \pm 2.73$	32.61 ± 1.13	48.39 ± 4.95

However, if we focus on the behavior of the feature selection rankers individually (six last rows of each table), none of the six methods tested was able to significantly outperform the results obtained by the ensemble for all combinations. This fact proves that, although in some specific cases there is a single method that performs better than the ensemble, there is not a better feature selection ranker in general, and the ensemble seems to be the most reliable alternative when a feature selection process has to be carried out. Moreover, notice the adequacy of using *SVM Rank* as a method to combine different rankings.

A last experiment was performed, consisting of the analysis of the behavior of the different thresholds, with independence of the actual feature selection methods. Table 6 displays the average test errors obtained with the different thresholds. A Kruskal-Wallis test was applied to check if there were significant differences between them for a level of significance $\alpha = 0.05$. Then, a multiple comparison procedure (Tukey's) was applied and those algorithms whose error average test results are not significantly worse than the best are labeled with a cross.

Table 6. Ensemble methods: average estimated percentage test errors. The cross shows results that are not significantly different than the best.

Threshold	Yeast	Spambase	Madelon	Connect4	Isolet
$\log_2(n)$	$55.52^{\dagger} \pm 4.73$	$11.39^{\dagger} \pm 2.34$	$33.46^{\dagger} \pm 4.24$	31.14 ± 0.51	48.99 ± 2.31
10 %	$53.24^{\dagger} \pm 4.62$	$11.39^{\dagger} \pm 2.34$	$35.46^{\dagger} \pm 4.05$	31.14 ± 0.51	50.13 ± 2.03
25 %	$53.24^{\dagger} \pm 4.62$	19.19 ± 2.00	$36.21^{\dagger} \pm 4.42$	$26.96^{\dagger} \pm 0.67$	48.77 ± 2.11
50 %	$53.24^{\dagger} \pm 4.62$	16.93 ± 1.91	39.29 ± 2.65	$25.27^{\dagger} \pm 0.59$	$43.21^{\dagger} \pm 3.01$

This analysis demonstrated that an optimal threshold value does not exist such that its results stand out over the others. The four thresholds analyzed in this research show very similar results, since each one of the thresholds was significantly better than the others in 3 out of the 5 datasets. Thus, it can be concluded that the most appropriate threshold depends on the nature of the datasets and their features. In this regard, the users cannot be released from this decision, and must select an appropriate threshold according to the particularities of each specific dataset.

5   Conclusions and Discussion

Real life datasets may be very large in number of samples or features, and their classification can be hindered by phenomena such as redundancy, noise or non-linearity of the data. For these reasons, some feature selection methods are not able to confront these problems so it is the responsibility of the user to decide which one to use in a particular situation. In this paper, an ensemble for feature selection was presented. The idea is to use an ensemble of methods rather than a single method, in order to take advantage of their individual strengths and overcome their weak points at the same time. This will have the added benefits of releasing the user from the task of knowing technical details about the existing algorithms, since more user-friendly methods are in need nowadays.

The particularity of the proposed ensemble is that it works with ordered rankings of features, which is the most natural approach for feature selection methods, and is less computationally expensive than subset approaches, an important fact in the actual scenarios of high dimensionality. Six well-known algorithms which follow the individual evaluation philosophy were chosen to form part of the ensemble, and the individual rankings were combined with *SVM Rank*. The experiments on five datasets, considered representative of problems from medium to large size, showed that our proposal was able to obtain the best average results regardless of the dataset and threshold chosen. Notice the implications of this result, since it can release the user from the task of deciding which feature selection method is more appropriate for a given problem.

References

1. Abeel, T., Helleputte, T., Van de Peer, Y., Dupont, P., Saeys, Y.: Robust biomarker identification for cancer diagnosis with ensemble feature selection methods. Bioinformatics **26**(3), 392–398 (2010)
2. Asuncion, A., Newman, D.: UCI machine learning repository (2007). http://archive.ics.uci.edu/ml/datasets.html
3. Bay, S.D.: Combining nearest neighbor classifiers through multiple feature subsets. In: ICML, vol. 98, pp. 37–45. Citeseer (1998)
4. Ben Brahim, A., Limam, M.: Robust ensemble feature selection for high dimensional data sets. In: 2013 International Conference on High Performance Computing and Simulation (HPCS), pp. 151–157. IEEE (2013)
5. Breiman, L.: Bagging predictors. Machine Learning **24**(2), 123–140 (1996)
6. Cortes, C., Vapnik, V.: Support-vector networks. Machine Learning **20**(3), 273–297 (1995)
7. Gao, K., Khoshgoftaar, T.M., Wang, H.: An empirical investigation of filter attribute selection techniques for software quality classification. In: IEEE International Conference on Information Reuse & Integration, IRI 2009, pp. 272–277. IEEE (2009)

8. Guyon, I., Weston, J., Barnhill, S., Vapnik, V.: Gene selection for cancer classification using support vector machines. Machine Learning **46**(1–3), 389–422 (2002)
9. Guyon, I., Elisseeff, A.: An introduction to variable and feature selection. The Journal of Machine Learning Research **3**, 1157–1182 (2003)
10. Guyon, I., Gunn, S., Nikravesh, M., Zadeh, L.: Feature extraction. Foundations and applications (2006)
11. Hall, M.A., Holmes, G.: Benchmarking attribute selection techniques for discrete class data mining. IEEE Transactions on Knowledge and Data Engineering **15**(6), 1437–1447 (2003)
12. Ho, T.K.: The random subspace method for constructing decision forests. IEEE Transactions on Pattern Analysis and Machine Intelligence **20**(8), 832–844 (1998)
13. Joachims, T.: Optimizing search engines using clickthrough data. In: Proceedings of the Eighth ACM SIGKDD International Conference on Knowledge Discovery and Data Mining, pp. 133–142. ACM (2002)
14. Khoshgoftaar, T.M., Golawala, M., Van Hulse, J.: An empirical study of learning from imbalanced data using random forest. In: 19th IEEE International Conference on Tools with Artificial Intelligence, ICTAI 2007, vol. 2, pp. 310–317. IEEE (2007)
15. Kira, K., Rendell, L.A.: The feature selection problem: traditional methods and a new algorithm. In: AAAI, pp. 129–134 (1992)
16. Kodovsky, J., Fridrich, J., Holub, V.: Ensemble classifiers for steganalysis of digital media. IEEE Transactions on Information Forensics and Security **7**(2), 432–444 (2012)
17. Kononenko, I.: Estimating attributes: analysis and extensions of RELIEF. In: Bergadano, F., De Raedt, L. (eds.) ECML 1994. LNCS, vol. 784, pp. 171–182. Springer, Heidelberg (1994)
18. Kuncheva, L.: Combining Pattern Classifiers: Methods and Algorithms. Wiley-Interscience (2004)
19. Kuncheva, L., Whitaker, C.: Measures of diversity in classifier ensembles and their relationship with the ensemble accuracy. Machine Learning **51**(2), 181–207 (2003)
20. Liu, H., Setiono, R.: Chi2: feature selection and discretization of numeric attributes. In: 2012 IEEE 24th International Conference on Tools with Artificial Intelligence, pp. 388–388. IEEE Computer Society (1995)
21. Liu, H., Yu, L.: Toward integrating feature selection algorithms for classification and clustering. IEEE Transactions on Knowledge and Data Engineering **17**(4), 491–502 (2005)
22. Mejía-Lavalle, M., Sucar, E., Arroyo, G.: Feature selection with a perceptron neural net. In: Proceedings of the International Workshop on Feature Selection for Data Mining, pp. 131–135 (2006)
23. Olsson, J., Oard, D.W.: Combining feature selectors for text classification. In: Proceedings of the 15th ACM International Conference on Information and Knowledge Management, pp. 798–799. ACM (2006)
24. Opitz, D.W.: Feature selection for ensembles. In: AAAI/IAAI, pp. 379–384 (1999)
25. Park, J., Sandberg, I.W.: Universal approximation using radial-basis-function networks. Neural Computation **3**(2), 246–257 (1991)
26. Peng, H., Long, F., Ding, C.: Feature selection based on mutual information criteria of max-dependency, max-relevance, and min-redundancy. IEEE Transactions on Pattern Analysis and Machine Intelligence **27**(8), 1226–1238 (2005)
27. Quinlan, J.: Induction of decision trees. Machine Learning **1**(1), 81–106 (1986)
28. Rodríguez, D., Ruiz, R., Cuadrado-Gallego, J., Aguilar-Ruiz, J.: Detecting fault modules applying feature selection to classifiers. In: IEEE International Conference on Information Reuse and Integration, IRI 2007, pp. 667–672. IEEE (2007)

29. Schapire, R.: The strength of weak learnability. Machine Learning **5**(2), 197–227 (1990)
30. Tsymbal, A., Pechenizkiy, M., Cunningham, P.: Diversity in search strategies for ensemble feature selection. Information Fusion **6**(1), 83–98 (2005)
31. Tukey, J.W.: Comparing individual means in the analysis of variance. Biometrics, 99–114 (1949)
32. Tuv, E., Borisov, A., Runger, G., Torkkola, K.: Feature selection with ensembles, artificial variables, and redundancy elimination. The Journal of Machine Learning Research **10**, 1341–1366 (2009)
33. Wang, H., Khoshgoftaar, T.M., Gao, K.: Ensemble feature selection technique for software quality classification. In: SEKE, pp. 215–220 (2010)
34. Wang, H., Khoshgoftaar, T.M., Napolitano, A.: A comparative study of ensemble feature selection techniques for software defect prediction. In: 2010 Ninth International Conference on Machine Learning and Applications (ICMLA), pp. 135–140. IEEE (2010)
35. Windeatt, T., Duangsoithong, R., Smith, R.: Embedded feature ranking for ensemble mlp classifiers. IEEE Transactions on Neural Networks **22**(6), 988–994 (2011)
36. Witten, I.H., Frank, E.: Data Mining: Practical machine learning tools and techniques. Morgan Kaufmann (2005)
37. Yang, C.-H., Huang, C.-C., Wu, K.-C., Chang, H.-Y.: A novel GA-taguchi-based feature selection method. In: Fyfe, C., Kim, D., Lee, S.-Y., Yin, H. (eds.) IDEAL 2008. LNCS, vol. 5326, pp. 112–119. Springer, Heidelberg (2008)
38. Yang, F., Mao, K.: Robust feature selection for microarray data based on multi-criterion fusion. IEEE/ACM Transactions on Computational Biology and Bioinformatics (TCBB) **8**(4), 1080–1092 (2011)
39. Yu, L., Liu, H.: Efficient feature selection via analysis of relevance and redundancy. The Journal of Machine Learning Research **5**, 1205–1224 (2004)
40. Zheng, Z., Webb, G.I.: Stochastic attribute selection committees. Springer (1998)

A Medical Case-Based Reasoning Approach Using Image Classification and Text Information for Recommendation

Sara Nasiri[✉], Johannes Zenkert, and Madjid Fathi

Institute of Knowledge Based Systems and Knowledge Management,
University of Siegen, Siegen, Germany
{sara.nasiri,johannes.zenkert}@uni-siegen.de,
fathi@informatik.uni-siegen.de

Abstract. The combination of visual and textual information in a CBR system is a promising concept to overcome the limitations of existing medical CBR systems, which are mainly focused on evaluation of new and existing cases with data mining, clustering techniques or statistical analysis of patient's health condition parameters. The advantage of our proposed medical CBR system, called DePicT, is the knowledge based recommendation, which utilizes case-based reasoning through analyzing image and text from patient health records. DePicT can find a solution regarding patient's problem description even with partly missing information. It uses image interpretation parameters and profiles of word associations in the feature selection and case matching process to find similar cases for recommendation.

Keywords: Case based reasoning · Feature selection · Case matching · Similarity · Image interpretation · Word association · Recommendation

1 Introduction

In the last decades many medical assistance systems have been developed and the interest in computer-aided problem solving in medical and health care is constantly growing. Most of the software systems in this domain focus on decision support and recommendation of effective medication for patients. Diagnose and treatments have been improved by the retrieval and usage of various medical data with medical assistance systems. Latest designs and developments in this area focus on integration of data mining in case-based reasoning (CBR) systems, e.g. a proposed model in prognosis and diagnosis of chronic diseases [1]. CBR is also used for diagnostic screening of children with development delay [2]. Furthermore, a hybrid case-based architecture can improve learning of new adapted knowledge and can be applied in multiple disease diagnosis [3]. The combination of statistical analysis and case-based reasoning can facilitate a better medical diagnosis.

Within the case comparison mechanism of CBR, feature selection and similarity measurement play an important role to retrieve cases. In this paper, **DePicT**

© Springer International Publishing Switzerland 2015
I. Rojas et al. (Eds.): IWANN 2015, Part II, LNCS 9095, pp. 43–55, 2015.
DOI: 10.1007/978-3-319-19222-2_4

(**De**tect and **P**redict diseases using **i**mage **c**lassification and **T**ext Information from patient health records) uses image interpretation and word associations for feature selection and recommendation of medical solutions. We see the combination of text and image as a competitive advantage and bringing this information together in our proposed system can solve problems of different areas of applications.

This paper is structured as follows. Section 2 is a literature review of related work about existing CBR recommendation approaches in medical assistance systems. Section 3 introduces the concept of our proposed knowledge-based system DePicT. In Section 4 we want to give a detailed description of the feature selection mechanisms including image and text processing within the recommendation process. Section 5 provides conclusions and future work in this domain.

2   Related Works

A typical approach for the recommendation process in medical applications, and especially in medical assistance systems, is a CBR methodology. CBR is a concept, and algorithm for problem solving using existing cases from the knowledge database in similarity and comparison analysis. CBR is applied in various problem-solving domains and "it is appropriate in medicine for some important reasons; cognitive adequateness, explicit experience, duality of objective and subjective knowledge, automatic acquisition of subjective knowledge, and system integration" [4] and also "provide a service or services in an Ambient Assisted Living that it will evolve as a consequence of its interaction with the user and the environment, being effective even when the user changes its habits or routines" [5]. The evolution of medical CBR applications and systems during the last decades is summarized in Table 1. These applications are mainly developed as mechanisms for making recommendations based on previous cases. With a data set of illnesses, the corresponding symptoms and the suggested drugs, systems try to imitate an examination of the patient's illness by a medical expert like a doctor or specialist, including their advice as the output of expert knowledge from the system's knowledge base. Systems compare features of an unsolved case to the features from a compendium of existing cases in the database, and present the recommendation of a found data record if the case fits to the current problem. For example; CARE-PARTNER "assists clinicians with the long-term follow-up (LTFU) of stem cell transplant patients once they have returned to their home communities" [7]. This system proposed recommendations diagnosis, lab result interpretation, and treatment planning and it includes 1109 diseases, 452 functions (also known as signs and symptoms), 1152 labs, 547 procedures, 2684 medications, and 460 sites [7]. The 4DSS is "a hybrid case-based reasoning (CBR) system that detects problems in blood glucose control and suggests personalized therapeutic adjustments to correct them" [7]. It has a case base which is including 80 cases which contains a specific blood glucose control problem experienced by a T1D patient. And for that problem, therapeutic adjustment is recommended from physician and after making the therapeutic adjustment, the patients clinical outcome is added in the case [7]. The screening system for screening developmentally delayed children is used by professional medical personnel to assess the appropriateness with ten sets of similar cases. It was evaluated with 210 cases in a selected screening center in 2003 [2]. Moreover, "next-generation CBR (GCBR) proposes a revised case-based recommender mechanism and it is applicable to various real world applications, particularly case-based recommender mechanisms, and can serve as a new problem-solving paradigm" [4].

In general, CBR has different steps for creating the knowledge base which is first indexing of cases and then are case retrieval, adaptation, testing and storage [4]. In the following section we want to introduce the concept of our medical CBR system DePicT.

Table 1. Medical CBR Systems [2][3][6][7]

System	Author, Year	Description
CASY - Heart failure diagnosis	Koton, 1988	Ambitious attempt to solve the adaptation task by general adaptation operators.
MEDIC - memory organization	Turner, 1988	It is a schema-based diagnostic reasoner on the domain of pulmonology.
NIMON - A renal function monitoring system	Wenkebach, Pollwein & Finsterer, 1992	It was developed that daily prints a renal report that consists of 13 measured and 33 calculated parameter values. However, the interpretation of all reported parameters is quite complex and needs special knowledge of the renal physiology.
FLORENCE - Nursing healthcare planning	Bradburn & Zeleznikow, 1993	Health care planning in a broader sense, for nursing, which is a less specialized field. A combination of a rule-based and a cased-based approach.
GS.52 - A prototype-based expert system and serves for dysmorphic syndromes	Gierl & Stengel-Rutkowski, 1994	GS.52 differs from typical CBR systems, because cases are clustered into prototypes, which represent diagnoses, and the retrieval searches only among these prototypes.
ICONS - A suitable calculated antibiotics therapy advice for intensive care patients	Liang, Schmidt, 1993 Pollwein & Gierl, 1999	It gives an automatic interpretation of the renal state to elicit impairments of the kidney function on time.

Table 1. (*Continued*)

System	Author, Year	Description
BOLERO	Lopez & Plaza, 1997	The major aim is that of improving the performance of a rule-based diagnosis system by adapting its behavior using the most recent information available about a patient.
CARE-PARTNER - It supports the long-term follow-up care of stem-cell transplantation patients. The ideas it pioneered have been carried over into the ongoing Mémoire project	Bichindaritz, Kansu & Sullivan, 1998 Bichindaritz 2006, 2007	This system was built between 1996 and 2000 at the Fred Hutchinson Cancer Research Center, at the University of Washington, in Seattle. It assists clinicians with the long-term follow-up (LTFU) of stem cell transplant patients once they have returned to their home communities.
Hybrid case-based architecture	Hu & Hso, 2004	The architecture combines case-based reasoning (CBR), neural networks, fuzzy theory, induction, utility theory, and knowledge-based planning technology together to facilitate medical diagnosis.
Using CBR to diagnostic screening of children with developmental delay	Chang, 2005	Applied CBR to screening children with delayed development in order to detect their disorder early through analysis of their symptoms, thereby improving the chances of effective treatment.
4DSS - The 4 Diabetes Support System	Marling, Shubrook & Schwartz, 2009, 2011, 2012	Began in 2004, Frank Schwartz, an endocrinologist; and Jay Shubrook, a diabetologist are practicing clinicians, who have treated several hundred T1D patients, as well as faculty members of the Ohio University Heritage College of Osteopathic Medicine. 4DSS aims to assist patients with type 1 diabetes (T1D) and their professional caregivers.
RHENE - Retrieval of HEmodialysis in NEphrological Disorders	Bellazzi, 2004	Began in 2004, Roberto Bellazzi, a nephrologist at the Vigevano Hospital in Italy. Supports physicians working in the domain of end stage renal disease (ESRD).

Table 1. (*Continued*)

System	Author, Year	Description
ExcelicareCBR	Van den Branden, et.al, 2010	Excelicare integrates CBR to support clinical decision making by harnessing electronic patient records for clinical experience reuse.
Mälardalen - The Mälardalen Stress System (MSS) provides decision support for the diagnosis and treatment of stress	Ahmed, Begum & Funk, 2011, 2012	This system was built between 2002 and 2011 at Mälardalen University in Västerås, Sweden. The experts for this system were psychologists Bo von Schéele and Erik Olsson. Dr. von Schéele has over 30 years experience in clinical stress diagnosis and has pioneered new biosensor and biofeedback methods.

3   Concept of DePicT CBR System

DePicT (**De**tect and **P**redict diseases using **i**mage **c**lassification and **T**ext Information from patient health records) is a knowledge based system for the identification and diagnosis of diseases by utilization of graphical and textual data sources with a CBR recommendation approach. The system in practice employs image interpretation and text mining methods as well as considerations and suggestions of medical experts in the feature analysis process. The concept of DePicT is shown in Fig. 1. The main idea and advantage of DePicT is a combination of different data sources in the knowledge base of the system, which also leads to multi-dimensionality and flexibility of medical problem solving. In the process of patient's examination and continuous health monitoring data is collected in the knowledge base as Personal Health Record (PHR). The database of DePicT holds textual and graphical description of the patient's health condition and consists of three major components. The data of PHR, which includes patient images and predefined information such as personal and health condition as well as personal statements of the patient in a textual form, image interpretation of the records, and a current word association profile of the record from Knowledge Discovery from Text (KDT). All gathered patient records are stored in relational databases as structured or closed-format (e.g. parameters and statistics), or unstructured or open-format e.g. texts and images. For example, photographs of affected areas of a melanoma skin cancer can contribute and support early stage diagnosis. Also further information through answering questions or writing a statement about the patient's health condition, especially about the affected zone, can be added into the knowledge base and is beneficial to describe the current case. Domain Experts can validate and verify the collected information and also update or correct the data records of patients. Through knowledge extraction the patient data is pre-processed for a

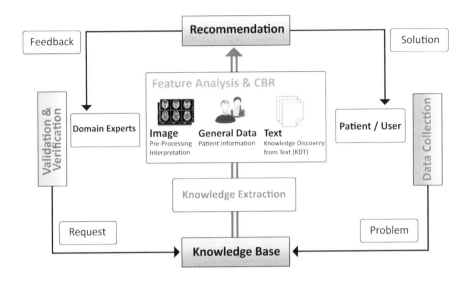

Fig. 1. Concept of DePicT CBR Recommender System

CBR recommendation methodology with feature selection and analysis. Taking into account a history-based and learnable approach for detection and prediction, therefore, a new image is pre-processed in comparison with other existing (clustered) images in the case base. This way the user disease can be classified and compared to the existing cases. DePicT utilizes statistical methods from data and text-mining i.e. knowledge discovery tools and algorithms to gather and extract the knowledge from recorded images, texts as well as information from other connected databases for the creation of the knowledge base. Moreover, to classify and analyze them in accordance with systematic and standard diagnostic methods, which are employed by the domain experts. And finally, to provide recommendations for users (i.e. patients or care-givers) depending on the type and severity of the detected case. DePicT collects and manages variety of patient data and records which are used for provision of recommendations, which is illustrated in Fig. 1. The recommendations are automatically generated based on the analysis of the patient and references records from evidence based consideration of domain experts. The recommendations also facilitate early diagnosis and homecare by care-givers and relatives, and ultimately guide the patient to visit a specialist.

The recommendations should include summary of findings through analysis of patient records and treatment plan based on consideration from domain expert. Recommendation techniques are typically based on knowledge sources [8]. These knowledge sources can be fed by "the knowledge of other users' preferences" or "ontological or inferential knowledge about the domain, added by a human knowledge engineer" [8]. DePicT utilizes the knowledge-based recommendation approach using CBR. In this type of recommendation, the system does not gather user ratings. The system provides recommendation and refers to specific domain knowledge about "how certain item features meet users needs and preferences" [9]. The procedure of recommendation within the DePicT System is described in the following section.

4   Recommendation Process

This section is focused on the recommendation process of DePicT and how graphical and textual information is used as feature within the CBR case matching and selection procedure. First, how these types of data can be used to enrich the knowledge base of DePicT is described. Afterwards, how the gathered data of the patient can contribute in the case matching and selection process is shown. To have a good recommendation, first we should make up the cases, and then find the appropriate similarity measure for the problem, after that matching the cases and finding the best similar case to recommend as a solution. Then the case is ready to be added in the knowledge base. Figure 2 presents the structure of cases within DePicT. Besides the image and text information also the general data from the patient is included in the case for problem description.

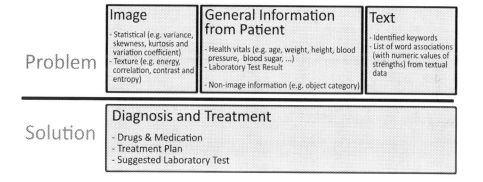

Fig. 2. DePicT Case Structure

The recommendation procedure is illustrated in Fig. 3. It starts on request. Image, text and patient general data is combined into a case in the case formation process. Since DePicT is working on personal data from patients, each patient needs to agree to the conditions of PHR. When patients fill the requests and want to have a recommendation, DePicT will anonymize their data in the case formation process to ensure a high security level. The structure of a case is predefined by the case base of DePicT. The problem description of the case summarizes the information about symptoms, references images and test results. When the new case is refined DePict checks the similarity of the current case and the existing cases in the case base in case matching and retrieval process step to find the best near case. After finding the case it is selected and the solution of the case is presented as recommendation to the user's request.

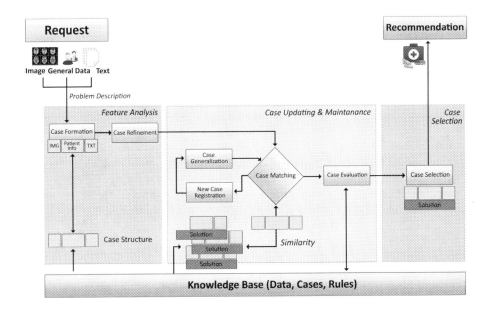

Fig. 3. DePicT Recommender Procedure

4.1   Visual Information from Image Representation

"Image interpretation is the process of mapping the numerical representation of an image into a logical representation such as is suitable for scene description. This is a complex process; these steps include image preprocessing, image segmentation, image analysis, and image interpretation. The image interpretation component identifies an object by finding the object to which it belongs (among the models of the object class). This is done by matching the symbolic description of the object to the model/concept of the object stored in the knowledge base. Most image-interpretation systems run on the basis of a bottom-up control structure" [10]. "The segmentation problem can be seen as a classification problem for which we have to learn the best classifier. Depending on the segmentation task, the output of the classifier can be the labels for the image regions, the segmentation algorithm selected as the most adequate, or the parameters for the selected segmentation algorithm. In any case, the final result is a segmented image" [11]. Different approaches involving medical imagery in CBR is surveyed that "cases can consist of papers, printed films, clinical patient information such as anamnesis, '1qdiagnosis, prescriptions in written text form, patient data, images, waveforms, and structured reports stored DICOM format, and SNOMED terms" [12].

Image and non-image (which is called general information in DePicT) features with different steps can classify the images in the case base, e.g. "18 RGB and 18 textural features extracted from each region of interest to form cases in two separate case bases" [12][13]. Similarity measures are defined regarding the image representation into 1-Pixel (iconic) matrix based, 2-featured based (numerical or symbolical or mixed type) and 3-structured similarity measures, but it is still proposed the new similarity measures for specific goals and different kinds of image interpretation [14].

According to DePicT features, similarity consists of three parts, image information, general information (non-image) and text information similarity. General information is defined based on application, e.g. in CT images, it includes patient-specific parameters (like age and sex), slice thickness, and the number of slices [14]. For finding the similarity of non-image information between a new case B and a case C_i that is in case base, we can use Teverskey's similarity measure [15]:

$$SIM_N = S(C_i, B) = \frac{|A_i|}{\alpha|A_i| + \beta|D_i| + \chi|E_i|}$$

$$\tag{1}$$

$$\alpha = 1, and\ \beta = \chi = 0.5$$

"Where A_i are the features common to both C_i and B, D_i are the features that belong to C_i but not to B, and E_i are the features that belong to B but not to C_i" [11].

For finding the similarity of image information between images A and B, we can use Frucci et al.'s image dissimilarity measure [11]:

$$dist_{AB} = \frac{1}{k} \sum_{i=1}^{K} w_i \left| \frac{C_{iA} - C_{imin}}{C_{imax} - C_{imin}} - \frac{C_{iB} - C_{imin}}{C_{imax} - C_{imin}} \right| \tag{2}$$

Where w_i is the weight for the ith feature with $w_1 + w_2 + ... + w_i + ... + w_k = 1$. That in each case, different value could be assigned to weights. C_{imax} and C_{imin} are the maximum and minimum value, respectively, of the ith feature of all images and C_{iA} and C_{iB} are the values of the ith feature of A and B [11].

"There is significant work in developing standardized vocabularies and medical ontologies for representation. In particular, for medical imagery, it is necessary to have a vocabulary that makes the meaning and the visual appearance of image features clear and universally understood among medical practitioners" [12]. In the following subsection these vocabularies (keywords), word association profiles and similarity measure of text information is discussed.

4.2   Strength of Word Associations Between Disease, Symptoms and Drugs

The frequent existence of word co-occurrences in different levels of texts is a good indication for a direct relation between selected words. For the estimation of semantic relationships of textual content we use the concept of imitation of the human ability of word association [16]. It is designed to imitate the Human Word Association (HWA) within a large collection of texts. The method is called CIMAWA and delivers the following equation as a final result for text mining application [16].

$$CIMAWA^{\zeta}_{ws}(x(y)) = \frac{Cooc_{ws}(x,y)}{(frequency(y))^{\alpha}} + \zeta \cdot \frac{Cooc_{ws}(x,y)}{(frequency(x))^{\alpha}} \qquad (3)$$

The hybrid character of (3) makes it possible to measure symmetric and asymmetric word associations with a damping factor ζ larger than 0. Co-occurrences ($Cooc_{ws}$) of two words x and y in a defined text window size ws are measured in a large document corpus. Best results were achieved in former studies with a text window size of 10 and a damping factor of 0.5 [17]. As a final result, all word associations from the selected word x within the document collection can be listed and ordered with the numeric value of CIMAWA.

We applied the methodology of word associations in medical context to evaluate the textual data from the knowledge base of DePicT. Substitutions of word x in (3) with medical expressions of diseases, symptoms and drugs results in lists of word associations and corresponding numeric strength values, which can be combined to build a semantic profile of the current textual data record of the case. The algorithm is visualized in Fig. 4.

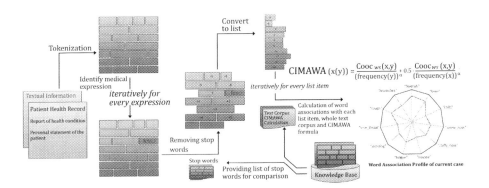

Fig. 4. Visualization of the CIMAWA algorithm to build word association profiles for cases in DePicT

In the first step all text information is tokenized, stop words are removed and all medical expressions are iteratively identified by a list of words from medical references within the knowledge base of DePicT. When a medical expression is found, a short list of text neighbor words, inside the text window (ws), is created. Again, iteratively, for each short list item the CIMAWA equation in (3) is applied. This leads to numerical values of all strengths of word association between medical expressions and other words (e.g. disease, symptoms, and drugs). In the final step, all word associations are summarized to create the association profile of the current case. This profile is stored in the case in the knowledge base. The system can compare the strengths of word associations of a new case to existing cases and their specific word associations in the knowledge base. A similarity is given if the profile of strengths of word associations related to disease, symptoms and drugs is similar in numeric values of CIMAWA to an existing profile. By adding the identified word associations as keywords to each case DePicT can use these medical expressions and tag the image that is related to the case. As a consequence, the system is able to search in the knowledge base for cases with similar text expressions and also simultaneously finding similar images.

4.3   Case Matching and Selection

In the case matching process within DePicT the criteria for comparison of existing and new cases are determined. For integrated medical systems to achieve significant success, it is necessary to provide a level of standardization for image diagnosis and retrieval besides the other information of patients. The matching procedure implemented in DePicT is a prioritization and weighting of features depending on the data source of the request. According to the concept of matching the features, critical and trivial features should be defined and set up [1]. While general data from the patient is present in every case, parts of the cases can be initially compared for similarity based on the non-image information shown in (1).

The case matching of word association profiles is illustrated in Fig. 5. After ordering and selection of cases from the knowledge base with similar patient information (e.g. age of the patient, gender, etc.), the image interpretation and textual problem descriptions within the case are considered in the second step.

With the implementation of the similarity measure of image interpretation, explained in (2) and the comparison of word association profiles created with (3), the current case is compared to existing cases in the knowledge base. The word association profile, which is added to each case, if textual data is existing in the problem description, can be compared to the existing profiles of cases. The matching of these profiles is visualized in Fig. 5.

The matching process can skip either the graphical or the textual similarity measure if none of the data is existing in the request. The case from the knowledge base that fits best to the current patient data, the textual problem description and, or the graphical interpretation of patient's image is selected and the solution from the case is presented as recommendation. If no case matches with the current case, a new case is created in the knowledge base of DePicT by the system, which should be validated by the domain experts.

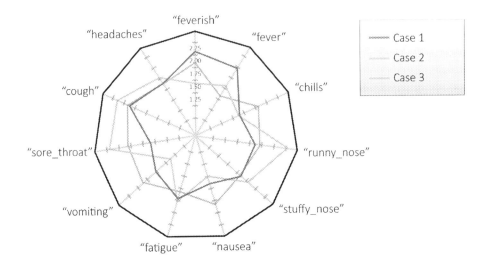

Fig. 5. Case Matching of Word Association Profiles

5   Conclusion and Future Work

As we have shown in our concept, a medical CBR system can help to improve disease diagnosis and recommendation of medication and patient treatment by searching in the knowledge base for existing cases which are similar to the current case. Therefore, different data types are useful. With the possibility to search for similar cases by evaluation of text, patient information and images in the case matching process DePicT is able to make recommendations. Even with some missing case components DePicT shows solutions for problem descriptions and is in this regard more flexible than existing systems.

In future work we want to deploy DePicT in the area of medical education. However, the recommendations should include summary of findings through analysis of patient records and treatment plan, with similarity methods and an innovative case matching approach, DePicT can recommend the solution regarding the combination of inputs. The characteristic of the system to be able to search for images based on word associations and keywords is interesting for trainees in medical vocational education. Also the vice versa approach to find textual problem descriptions based on an existing image is useful in many applications.

References

1. Huang, M.-J., Chen, M.-Y., Lee, S.-C.: Integrating data mining with case based reasoning for chronic diseases prognosis and diagnosis. Expert Systems with Applications **32**, 856–867 (2007)

2. Chang, C.L.: Using case-based reasoning to diagnostic screening of children with developmental delay. Expert Systems with Applications **28**(2), 237–247 (2005)
3. Hsu, C.C., Ho, C.S.: A new hybrid case-based architecture for medical diagnosis. Information Science **166**, 231–247 (2004)
4. Yang, H.-L., Wang, C.-S.: A recommender mechanism based on case-based reasoning. Expert Systems with Applications **39**, 4335–4343 (2012)
5. Carneiro, D., Novais, P., Costa, R., Neves, J.: Case-based reasoning decision making in ambient assisted living. In: Omatu, S., Rocha, M.P., Bravo, J., Fernández, F., Corchado, E., Bustillo, A., Corchado, J.M. (eds.) IWANN 2009, Part II. LNCS, vol. 5518, pp. 788–795. Springer, Heidelberg (2009)
6. Van den Branden, M., Wirtatunga, N., Burton, D., Craw, S.: Integrating case-based reasoning with an electronic patient record system. Artificial Intelligence in Medicine **51**, 117–123 (2007)
7. Marling, C., Montani, S., Bichindaritz, I., Funk, P.: Synergistic case-based reasoning in medical domains. Expert Systems with Applications **41**, 249–259 (2014)
8. Burke, R.: Hybrid web recommender systems. In: Brusilovsky, P., Kobsa, A., Nejdl, W. (eds.) Adaptive Web 2007. LNCS, vol. 4321, pp. 377–408. Springer, Heidelberg (2007)
9. Ricci, F., Rokach, L., Shapira, B.: Introduction to Recommender Systems. In: Ricci, F., et al. (eds.) Recommender Systems Handbook, pp. 1–35. Springer Science+Business Media, LLC (2011)
10. Perner, P.: Mining sparse and big data by case-based reasoning. Procedia Computer Science **35**, 19–33 (2014)
11. Frucci, M., Perner, P., Sanniti di Baja, G.: Case-Based Reasoning for Image Segmentation by Watershed Transformation, vol. 73. Springer-Verlag, Heidelberg (2008)
12. Wilson, D.C., O'Sullivan, D.: Medical imagery in case-based reasoning. In: Perner, P. (ed.) Case-Based Reasoning on Images and Signals. Studies in Computational Intelligence, vol. 73, pp. 389–418. Springer, Heidelberg (2008)
13. Galushka, M., Zheng, H., Patterson, D., and Bradley, L.: Case-based tissue classification for monitoring leg ulcer healing. In: Proceedings of the 18th IEEE Symposium on Computer-Based Medical Systems, pp. 353–358. IEEE Computer Society (2005)
14. Perner, P.: Introduction to case-based reasoning for signals and images. In: Perner, P. (ed.) Case-Based Reasoning on Images and Signals. Studies in Computational Intelligence, vol. 73, pp. 1–24. Springer, Heidelberg (2008)
15. Tversky, A.: Feature similarity. Psychological Review **84**(4), 327–350 (1977)
16. Uhr, P., Klahold, A., Fathi, M.: Imitation of the Human Ability of Word Association. International Journal of Soft Computing and Software Engineering **3**(3), 248–254 (2013)
17. Klahold, A., Uhr, P., Ansari, F., Fathi, M.: A Framework to utilize the Human Ability of Word Association for detecting Multi Topic Structures in Text Documents. IEEE Intelligent Systems **29**(5), 40–46 (2014)

Non Spontaneous Saccadic Movements Identification in Clinical Electrooculography Using Machine Learning

Roberto Becerra-García[1(✉)], Rodolfo García-Bermúdez[1,3],
Gonzalo Joya-Caparrós[2], Abel Fernández-Higuera[1],
Camilo Velázquez-Rodríguez[1], Michel Velázquez-Mariño[1],
Franger Cuevas-Beltrán[1], Francisco García-Lagos[2],
and Roberto Rodríguez-Labrada[4]

[1] Universidad de Holguín, Grupo de Procesamiento de Datos Biomédicos (GPDB),
Holguín, Cuba
{idertator,afernandezh,cvelazquezr,mvelazquez}@facinf.uho.edu.cu,
fcuevas@facii.uho.edu.cu, rodolfo.garcia@live.uleam.edu.ec
[2] Departamento de Tecnología Electrónica, Universidad de Málaga, Málaga,
Campus de Excelencia Internacional Andalucía Tech, Málaga, España
gjoya@uma.es, lagos@dte.uma.es
[3] Universidad Laica Eloy Alfaro de Manabí, Facultad de Informática,
Manta, Ecuador
[4] Centro para la Investigación y Rehabilitación de las Ataxias Hereditarias,
Holguín, Cuba
roberto@ataxia.hlg.sld.cu

Abstract. In this paper we evaluate the use of the machine learning algorithms Support Vector Machines (SVM), K-Nearest Neighbors (KNN), Classification and Regression Trees (CART) and Naive Bayes (NB) to identify non spontaneous saccades in clinical electrooculography tests. Our approach tries to solve problems like the use of manually established thresholds present in classical methods like identification by velocity threshold (I-VT) or identification by dispersion threshold (I-DT). We propose a modification to an adaptive threshold estimation algorithm for detecting signal impulses without the need of any user input. Also, a set of features were selected to take advantage of intrinsic characteristics of clinical electrooculography tests. The models were evaluated with signals recorded to subjects affected by Spinocerebellar Ataxia type 2 (SCA2). Results obtained by the algorithm show accuracies over 97%, recalls over 97% and precisions over 91% for the four models evaluated.

Keywords: Saccade identification · Clinical electrooculography · Classification

1 Introduction

The alteration of eye movements is one of the symptoms of many neurological diseases like Parkinsons syndrome, spinocerebellar ataxias or the Niemann-Pick

© Springer International Publishing Switzerland 2015
I. Rojas et al. (Eds.): IWANN 2015, Part II, LNCS 9095, pp. 56–68, 2015.
DOI: 10.1007/978-3-319-19222-2_5

disease [4]. Specifically in the Spinocerebellar Ataxia type 2 (SCA2) this alteration is an important clinical marker present in more than 90% of patients [29].

There are several kind of eye movements such as saccades, fixations and pursuits. Among them, saccades are critical to follow and evaluate subjects with SCA2. For instance, SCA2 patients have significantly slower saccades and with larger latencies than healthy subjects [29]. The analysis of this kind of movement is very often used in the researches conducted by the medical community, hence its importance.

A technique to measure eye movements called electrooculography consists in capturing the electrical potential of the eyes to calculate its magnitude and direction. This technique is widely used in electrophysiologic tests [16]. The resulting signals of this recording process are named electrooculograms [6].

There exists several methods and algorithms for identifying saccades in electrooculograms, the vast majority of them based on kinetic thresholds [11,14,26, 31], using suppervised learning [6,28], unsupervised learning [20] or other novel approachs [18,22] like particle filters [8]. These methods were designed to work in a not constrained scheme having advantages in a lot of scenarios. They are usually evaluated against data from healthy subjects where the differences between saccadic and non saccadic movements are very evident. However, in electrooculography clinical tests these methods try to detect as many saccades as posible, not distinguishing which of them are spontaneous and which not.

In a previous work [2], we proposed a method that identifies saccadic movements using a sample-to-sample approach. This method allows us to discriminate whether a sample belong to a saccadic movement or not. Now, in this work we have the task of identifying which of these movements are stimuli related using a feature-based approach.

Here we set out to evaluate the use of machine learning algorithms taking into account the strengths of clinical tests of electrooculography to solve the proposed task. Our approach have to use only horizontal movement signals and stimulus signals, and do not require the use of thresholds or any other user input. To do so, a new set of features were selected to train the models taking into account characteristics of valid saccadic movements.

To identify the ocurrence of saccadic movements we use an impulse detection method based on velocity thresholds. These thresholds are calculated adaptively with a modified version of the method proposed in [18]. Our algorithm uses a classification model to solve the presented task, so we evaluate four of them: Support Vector Machines (SVM) [7], K-Nearest Neighbors (KNN) [27], Classification and Regression Trees (CART) [5] and Naive Bayes (NB) [25]. The performance of the classification models were measured, obtaining very good results ($> 97\%$ accuracy) in all of them.

The rest of this paper is organized as follows: In section 2 we describe the designed experiments and available data. Section 3 is devoted to analize and comment the results. Finally, section 4 summarizes the main conclusions and future work lines.

2   Material and Methods

To test the selected algorithms an experiment was designed. The first step was detect potential impulses and annotating them to build a labeled dataset. Then, each classification method is evaluated with stratified k-fold cross validation. Finally, we compare the performance of the models using nonparametric statistical tests to select the fittest.

Clinical tests of electrooculography are setup as follows. Subjects with their head fixed are seated in front to a monitor at a previously known distance. Then, they are commanded to follow a visual stimuli which appears and disappears from one side to the other in the monitor. Capturing eye movements in these conditions using electrooculography allows to researchers the identification of which saccades respond to stimulus and which ones are spontaneous. Also allows to calculate important features of these movements like latency, duration, amplitude, deviation and maximal velocity.

The electrooculograms were recorded using the OtoScreen electronystamography device at a sampling rate of 200 Hz with a bandwith of 0.02 to 70 Hz. Records of 12 sick subjects with SCA2 were used to build a dataset with features extracted from signal impulses. Each one of the records have at least tests of $10°$, $20°$ and $30°$ of visual stimulation. Typically saccadic tests have at least one horizontal channel and one stimulus signal (Fig. 1).

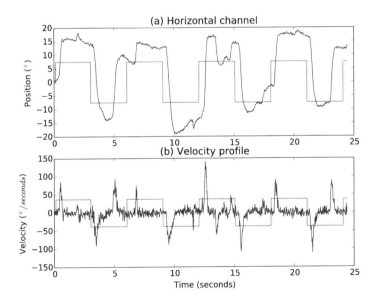

Fig. 1. Typical electrooculography signal with $30°$ stimulus angle of a subject suffering SCA2. Red signals are the scaled stimuli signals. Blue signals are the horizontal channel (a) and its velocity profile (b) respectively.

IPython notebooks [23] were used in conjunction with the Python language scientific facilities: NumPy [19], SciPy [12], Pandas [17], Matplotlib [10] and Scikit-Learn [21] for running the experiments. The intention behind using Python powered technologies is that the resulting algorithm (including trained models) will be used at NSEog, a processing platform developed by the authors.

2.1   Signals Preprocessing

Before the identification of potentially saccadic impulses, two common tasks need to be performed: denoising and differentiation. Noise removal is a very important matter in order to eliminate non desired spectral components produced by equipment malfunction, poor analog filtering or biological artifacts. Differentiation allows to obtain the velocity profile used later by the algorithm.

Median filter (Equation 1) has proven to be very robust in eliminating high frequency signal noise while preserving sharp edges. An study carried out in [13] demonstrated that this kind of filters is appropiate for eye movements signals. To eliminate non desired noise present in the signals used in the experiment, we use a median filter with a window size of 9 samples (approximately 45 milliseconds) obtaining very good results. This is accomplished using the **medfilt** function of SciPy.

$$y_i = median\{x_j | j = i - k, \ldots, j + k\} \tag{1}$$

Due to the discrete nature of these signals, numerical differentiation is employed to calculate the velocity profiles. According [3], Lanczos differentiators (Equation 2) with 11 points ($N = 11$) have good performance for signals with the same characteristics as the ones used in this experiment.

$$f'(x^*) \approx \frac{3}{h} \sum_{k=1}^{m} k \frac{f_k - f_{-k}}{m(m+1)(2m+1)}, \quad m = \frac{N-1}{2} \tag{2}$$

We implemented the rutine of a Lanczos 11 differentiator which have the following formula:

$$f'(x^*) \approx \frac{f_1 - f_{-1} + 2(f_2 - f_{-2}) + 3(f_3 - f_{-3}) + 4(f_4 - f_{-4}) + 5(f_5 - f_{-5})}{110h} \tag{3}$$

2.2   Detection of Impulses

Saccadic movements are represented as impulses in a velocity graph (Fig. 1b). Typically, this movements can be easily detected by its contrast in magnitude and shape with other movements such as fixations and microsaccades. However, for the same stimulus angle the range of values of true saccadic impulses vary from subject to subject. This situation is tied greatly on the degree of affectation present in the subject [24].

One of the critical parts of the algorithm is the detection of velocities impulses which can potentially be saccades. For that matter, a threshold is needed to know when the velocity has reached a certain value that can be considered as a saccade candidate. Due to the inter-subject variability explained before, this threshold should not be fixed a priori. Also should be large enough to ignore in most cases other movements like microsaccades and fixations, and not too large to miss valid saccadic movements.

To detect impulses we developed the algorithm described in Algorithm 1, which is a modification to the method introduced by Nystrm and Holmqvist in [18]. The algorithm uses the absolute values of the *velocities* samples to calculate the approximation of the initial threshold (*last threshold*). This initial threshold is calculated by adding σ times the standard deviation of the *velocities* to its mean. Then, iteratively it adjusts the *last threshold* with the same formula using only *selected samples* of velocities below the previous threshold. The stop condition happens when the difference between the current threshold and the last one is less or equal than one degree. The value of the resulting threshold is represented graphically by the red line in Fig. 2.

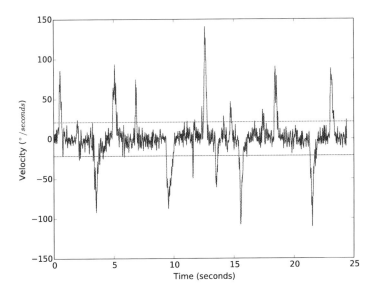

Fig. 2. Threshold estimated in a $30°$ stimulus angle test of a subject with SCA2

The original algorithm requires the initial threshold as input. This adds a little subjectivity to the main process, because to obtain good detection results this value must be variable and set by the user. The noise levels present on the signals and the degree of affectation of the subject have great influence on this issue. The proposed modification consists in calculate the initial threshold in a adaptive way using all velocity samples, so eliminating the subjectivity of the

Algorithm 1. Modified version of Nystrm and Holmqvist [18] threshold estimation algorithm

---

Input : velocity profile (Array of degree/seconds samples)
Input : σ (Safety margin)
Output: Threshold estimation
begin

    velocities $\longleftarrow$ Abs(velocity profile);
    last threshold $\longleftarrow$ Mean(velocities) $+ \sigma$ * Std(velocities);
    current threshold $\longleftarrow$ 0;
    while Abs(last threshold- current threshold) > 1 **do**
        selected samples $\longleftarrow$ samples from velocities below last threshold;
        current threshold $\longleftarrow$ last threshold;
        last threshold $\longleftarrow$ = Mean(selected samples) $+ \sigma$ * Std(selected samples);
    return last threshold;

---

original approach. Using the new approach on signals recorded to subjects with SCA2 in different stages seems to be adecuate to the task at hand.

The safety margin ($\sigma = 6$) employed by [18] ignores too many valid saccadic movements in lower angle tests for subjects with SCA2. A value of $\sigma = 3$ seems to be adequate for most cases at the expense of the detection of more non valid impulses. Even when has a penalty in runtime performance, the final accuracy of the method should not decrease significantly. Due the amplitude of this new impulses the classification model should avoid them.

Finally, we detect the impulses individually by finding a group of samples grouped together that exceeds the calculated threshold. The principle behind this algorithm is looping through the signal to find velocities above the threshold. When we encounter with one of these points, we move to the left and to the right until the velocity is zero or cross it. This approach allows further refinement of the saccade start and ending points because the impulses usually get more samples beyond the real saccade limits. If the length of a detected impulse is not greater than 10 samples, then it is discarded to avoid very small invalid movements. A typical output of this method is represented in Fig. 3.

2.3 Model Evaluation

Once we have the saccadic impulses candidates, we need to know if they are saccades and if they are related to the stimulus. For this reason, the strategy behind our approach uses human intuitive features to solve this task. To take advantage of the characteristics of the clinical tests, the following set of features was carefully selected:

Angle: Amplitude of the stimulus, it can take 3 values: 10, 20 or 30.
Absolute Latency: Time between the start of the stimulus transition and the maximal velocity point of the impulse in milliseconds (ms).

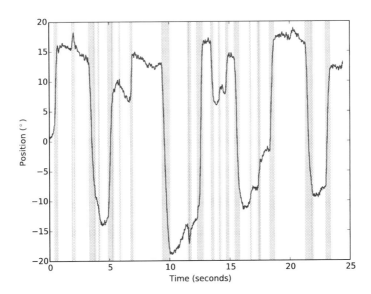

Fig. 3. Identified impulses in the same signal used in Figure 2

Normalized Latency: Normalized version of absolute latency with values between 0 and 1. The value 0 means that the maximal velocity is in the start of the fixation window, and the value 1 means that the maximal velocity is at the end of the fixation window.

Amplitude: Difference between the maximum position value and the minimum position value in the impulse.

Deviation: Difference between amplitude and the angle of the stimulus.

Maximum Velocity: Maximum velocity achieved during the impulse in $°/s$.

Maximum Acceleration: Maximum acceleration achieved during the impulse in $°/s^2$.

Maximum Jerk: Maximum jerk achieved during the impulse in $°/s^3$.

Direction: Take the value 1 if the movement follows the direction of the stimulus or -1 in other case.

End Relative Position: Values between 0 and 1, representing in which side of the stimulus the impulse ends. The value 0 represents the left side and the value 1 the right side.

Using the features previously selected, a dataset of signal impulses was created. To build this dataset, a human specialist aided by the NSEog classified the detected impulses in valid and non valid saccades (Figure 4). As results, 1797 valid saccades and 6809 not valid impulses were obtained, resulting in 8606 instances.

Because we are using Python technologies, Scikit-Learn was selected as machine learning library, hence we are constrained to a restricted set of models

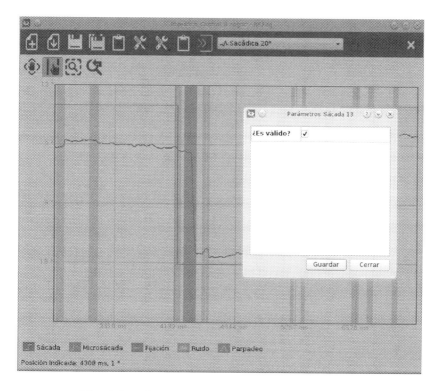

Fig. 4. Impulses annotation with the NSEog platform

implemented in it. The main policy of model selection was family representation, meaning that we try to choose methods with different working principles. So we evaluate four different models: Support Vector Machines, K-Nearest Neighbors, CART decision trees and the Gaussian version of Naive Bayes.

Support Vector Machines (SVMs) are a set of supervised learning methods very effective in high dimensional spaces [7]. There are also very versatile supporting a set of kernel functions. Scikit-Learn implements four kernel functions: linear $\langle x, x' \rangle$, polynomial $(\gamma \langle x, x' \rangle + r)^d$, rbf $e^{-\gamma |x-x'|^2}$ and sigmoid $tanh(\gamma \langle x, x' \rangle + r)$. Results from preliminary experiments showed that for the proposed task, the rbf kernel function have the best performance compared with the others. Further study are necessary to fine tune the parameter γ of this kernel.

K-Nearest Neighbors is a type of instance-based learning which can be used for supervised or unsupervised learning. Instead of creating a generalizing function, it stores all the data inside the models using different data structures like Ball Trees or KD Trees. The principle behind the algorithm is to find a number of training samples nearest to the analized point and predict the label from it [27]. To train our model we tried several numbers of neighbors starting from 2, giving the best results when this value is equal to 3. The data structure used is determined automatically by the Scikit-Learn implementation using optimization techniques.

Decision trees are nonparametric supervised learning techniques. This algorithm requires little preprocessing and its runtime performance is good enough to handle real time tasks. This method split the data trying to infere decision rules which can be used to clasify instances. Scikit-Learn uses an optimized version of the CART tree that support classification and regression [5]. The implementation used here do not require any parameter by default.

Naive Bayes classifiers are supervised methods based on Bayes theorem which assume independence between every pair of features [25]. We used a gaussian version of this classifier implemented in Scikit-Learn. Like for decision trees, the default implementation of this statistical classifier do not require any parameters.

As validation scheme we use an stratified 10-fold cross validation to evaluate internally the models. The metrics employed to measure the performance were accuracy (Equation 4), recall (Equation 5) and precision (Equation 6) [30]. The accuracy gives a general quality measure of the performance of the models, while the recall and the precision allow to know how well the model predict or miss predict valid saccadic movements. In the following equations, TP (true positives), TN (true negatives), FP (false positives) and FN (false negatives) are the items from the confusion matrix used to compute involved metrics.

$$Accuracy = \frac{TP + TN}{TP + FP + TN + FN} \tag{4}$$

$$Recall = \frac{TP}{TP + FN} \tag{5}$$

$$Precision = \frac{TP}{TP + FP} \tag{6}$$

The whole dataset was adjusted by removing the mean and scaled to unit variance. This technique is critical to obtain good results in the training of the RBF kernel version of Support Vector Machines. These scales was saved along with the model for further use by the algorithm.

To compare the real performance of the models, the Friedman's nonparametric statistical test was used as recommended in [9]. In this step we use records not used in the training phase. Each metric were analyzed by separate and the statistical calculations were performed using the Keel tool [1].

The resulting classification algorithm is very simple and flexible. It consists in the evaluation of the features calculated from impulses detected in the signal by the supervised model. This approach allows the parallelization of the algorithm and even swap the model if needed. Due the use of the proposed impulse detection algorithm, the need for parameters managed by the user is eliminated.

3   Results

The evaluated models were trained with 8606 impulses, 1797 valid saccades and 6809 invalid ones. Using 10-fold cross validation the internal performance of the trained process was measured with the metrics accuracy, recall and precision. Table 1 shows results above .97 of accuracy, .94 of recall and .90 of precision in all cases.

Table 1. 10-fold cross validation results

Model	Acc.	Rec.	Pre.
SVM	0.9833	0.9750	0.9467
KNN	0.9796	0.9666	0.9376
CART	0.9769	0.9449	0.9445
NaiveBayes	0.9747	0.9817	0.9056

To perform a more objective evaluation, the algorithm was tested against records obtained from five new subjects not used in the training phase. A total of 3797 impulses were evaluated this time, 704 real saccadic impulses and 3093 not saccadic.

Table 2. External validation results by stimulus amplitude

	SVM			KNN			CART			NaiveBayes		
Angle	Acc.	Rec.	Pre.	Acc.	Rec.	Pre.	Acc.	Rec.	Pre.	Acc.	Rec.	Pre.
10	.9765	.9703	.9051	.9659	.9449	.8745	.9636	.9237	.8790	.9575	.9661	.8261
20	.9858	.9837	.9377	.9844	.9837	.9305	.9822	.9633	.9365	.9780	.9837	.8993
30	.9720	.9686	.9038	.9646	.9686	.8745	.9674	.9462	.9017	.9543	.9686	.8372
Mean	.9780	.9742	.9155	.9716	.9657	.8932	.9711	.9444	.9058	.9633	.9728	.8542
Std	.0070	.0082	.0192	.0111	.0195	.0323	.0098	.0198	.0290	.0128	.0095	.0394

Results obtained analysing the performance individually by stimulus angle seems to favor slightly the SVM model (Table 2). However, doing the same analysis using independent subject records shows a more erratic behaviour (Table 3). Because of this situation, the Friedman's nonparametric statistical test was employed to compare the performance of the four models. Each record was considered as an individual dataset and each of the three performance metrics was analyzed independently using the data in Table 3. Results obtained by this method show that there are no significant differences in the performance of these models for a significance level of $p = 0.10$.

Literature about the task proposed in this work is scarce and no methods to specifically solve it were found. However, similar works reported a recall of .89 for 10° recordings on healthy subjects [22] and .80 of recall on subjects with Obstructive Sleep Apnea Syndrome (OSAS) [15]. Other related research conducted by Tigges et al. shows an accuracy of .92 [28]. Taking into account that we are dealing with signals recorded to subjects which suffers a very severe neurological disorder, results shown in Table 2 and Table 3 are better than the others presented in the literature.

Table 3. External validation results by subject record

	SVM			KNN			CART			NaiveBayes		
Subject	Acc.	Rec.	Pre.	Acc.	Rec.	Pre.	Acc.	Rec.	Pre.	Acc.	Rec.	Pre.
1	.9881	.9877	.9699	.9881	.9877	.9699	.9796	.9693	.9576	.9881	.9755	.9815
2	.9862	.9935	.9107	.9724	.9610	.8506	.9845	.9870	.9048	.9535	.9935	.7427
3	.9871	.9754	.9444	.9794	.9590	.9141	.9704	.8852	.9231	.9717	.9836	.8571
4	.9799	.9420	.9559	.9784	.9348	.9556	.9741	.9130	.9545	.9756	.9420	.9353
5	.9410	.9685	.8039	.9392	.9843	.7911	.9358	.9528	.7961	.9375	.9685	.7935
Mean	.9765	.9734	.9170	.9715	.9654	.8962	.9689	.9415	.9072	.9653	.9726	.8620
Std	.0201	.0201	.0669	.0189	.0215	.0748	.0193	.0417	.0659	.0199	.0195	.0982

4   Conclusions

In this work we have described a procedure to indentify spontaneous saccades from a set of detected impulses in electrooculography signals. To detect the impulses we made a modification to the algorithm proposed in [18], which consists in adaptively calculate the initial thresholds. This new algorithm avoids the need of thresholds or any other user input and works very well for noisy signals like the ones recorded to subjects with SCA2, which is a difficult task.

To clasify we used and compared four machine learning paradigms: Support Vector Machines, K-Nearest Neighbors, Classification and Regression Trees and Naive Bayes. The procedure has been applied to a database of eye movements recorded to subjects suffering spinocerebellar ataxias. The evaluation of the performance of the different paradigms were carried out using metrics such as Accuracy, Recall and Precision. The four used paradigms achieved an accuracy above 95%, a recall above 92% and a precision above 83% by external validation (using patterns not used for training). Specifically for Support Vector Machines the performance obtained was always above 97%, 96% and 90% for the three metrics respectively. These results exceed widely the reported by the literature in related works.

Acknowledgments. This work has been partially supported by the Universities of Holguín and Málaga through the joint project titled *"Mejora del equipamiento para la evaluación de la rehabilitación de enfermedades neurológicas de especial prevalencia en el oriente de Cuba"*. Also, we would like to thank the Agencia Española de Cooperación Internacional para el Desarrollo (AECID) and the Campus de Excelencia Internacional Andalucía Tech of University of Málaga because of the funding granted to this project. Finally, the feedback of the anonymous reviewers is gratefully acknowledged.

References

1. Alcalá-Fdez, J., Sánchez, L., García, S., del Jesús, M.J., Ventura, S., Garrell, J.M., Otero, J., Romero, C., Bacardit, J., Rivas, V.M., et al.: KEEL: a software tool to assess evolutionary algorithms for data mining problems. Soft Computing **13**(3), 307–318 (2009). http://link.springer.com/article/10.1007/s00500-008-0323-y

2. Becerra, R., Joya, G., Bermúdez, R.V.G., Velázquez, L., Rodríguez, R., Pino, C.: Saccadic Points Classification Using Multilayer Perceptron and Random Forest Classifiers in EOG Recordings of Patients with Ataxia SCA2. In: Rojas, I., Joya, G., Cabestany, J. (eds.) IWANN 2013, Part II. LNCS, vol. 7903, pp. 115–123. Springer, Heidelberg (2013)

3. Becerra Garca, R.A.: Plataforma de procesamiento de electrooculogramas. Caso de estudio: pacientes con Ataxia Espinocerebelosa Tipo 2. Master en matemtica aplicada e informtica para la administracin, Universidad de Holgun, Holgun (2013)

4. Bonnet, C., Hanuka, J., Rusz, J., Rivaud-Pchoux, S., Sieger, T., Majerov, V., Serranov, T., Gaymard, B., Rika, E.: Horizontal and vertical eye movement metrics: What is important?. Clinical Neurophysiology **124**(11), 2216–2229 (2013). http://www.clinph-journal.com/article/S1388245713006378/abstract

5. Breiman, L., Friedman, J., Stone, C.J., Olshen, R.A.: Classification and regression trees. CRC Press (1984)

6. Bulling, A., Ward, J., Gellersen, H., Troster, G.: Eye Movement Analysis for Activity Recognition Using Electrooculography. IEEE Transactions on Pattern Analysis and Machine Intelligence **33**(4), 741–753 (2011)

7. Cortes, C., Vapnik, V.: Support-vector networks. Machine Learning **20**(3), 273–297 (1995). http://link.springer.com/article/10.1007/BF00994018

8. Daye, P.M., Optican, L.M.: Saccade detection using a particle filter. Journal of Neuroscience Methods **235**, 157–168 (2014)

9. Demšar, J.: Statistical comparisons of classifiers over multiple data sets. The Journal of Machine Learning Research **7**, 1–30 (2006). http://dl.acm.org/citation.cfm?id=1248548

10. Hunter, J.D.: Matplotlib: A 2D Graphics Environment. Computing in Science and Engineering **9**(3), 90–95 (2007)

11. Inchingolo, P., Spanio, M.: On the Identification and Analysis of Saccadic Eye Movements-A Quantitative Study of the Processing Procedures. IEEE Transactions on Biomedical Engineering BME **32**(9), 683–695 (1985)

12. Jones, E., Oliphant, T., Peterson, P.: SciPy: Open source scientific tools for Python (2001). http://www.scipy.org/, http://www.citeulike.org/group/2018/article/2644428

13. Juhola, M.: Median filtering is appropriate to signals of saccadic eye movements. Computers in Biology and Medicine **21**(1–2), 43–49 (1991). http://www.sciencedirect.com/science/article/pii/0010482591900347

14. Juhola, M., Jäntti, V., Pyykkö, I., Magnusson, M., Schalén, L., Åkesson, M.: Detection of saccadic eye movements using a non-recursive adaptive digital filter. Computer Methods and Programs in Biomedicine **21**(2), 81–88 (1985). http://www.sciencedirect.com/science/article/pii/0169260785900665

15. Magosso, E., Provini, F., Montagna, P., Ursino, M.: A wavelet based method for automatic detection of slow eye movements: A pilot study. Medical Engineering & Physics **28**(9), 860–875 (2006). http://www.sciencedirect.com/science/article/pii/S1350453306000154

16. Marmor, M.F., Zrenner, E.: Standard for clinical electro-oculography. Documenta Ophthalmologica **85**(2), 115–124 (1993). http://dx.doi.org/10.1007/BF01371127

17. McKinney, W.: Data Structures for Statistical Computing in Python, pp. 51–56 (2010). http://conference.scipy.org/proceedings/scipy2010/mckinney.html

18. Nystrm, M., Holmqvist, K.: An adaptive algorithm for fixation, saccade, and glissade detection in eyetracking data. Behavior Research Methods **42**(1), 188–204 (2010). http://link.springer.com/article/10.3758/BRM.42.1.188

19. Oliphant, T.E.: Python for scientific computing. Computing in Science & Engineering **9**(3), 10–20 (2007). http://scitation.aip.org/content/aip/journal/cise/9/3/10.1109/MCSE.2007.58

20. Otero-Millan, J., Castro, J.L.A., Macknik, S.L., Martinez-Conde, S.: Unsupervised clustering method to detect microsaccades. Journal of Vision **14**(2), 18 (2014)

21. Pedregosa, F., Varoquaux, G., Gramfort, A., Michel, V., Thirion, B., Grisel, O., Blondel, M., Prettenhofer, P., Weiss, R., Dubourg, V., Vanderplas, J., Passos, A., Cournapeau, D., Brucher, M., Perrot, M., Duchesnay, E.: Scikit-learn: Machine Learning in Python. Journal of Machine Learning Research **12**, 2825–2830 (2011). http://jmlr.csail.mit.edu/papers/v12/pedregosa11a.html

22. Pettersson, K., Jagadeesan, S., Lukander, K., Henelius, A., Hæggström, E., Müller, K.: Algorithm for automatic analysis of electro-oculographic data. Biomedical Engineering Online **12**(1), 110 (2013)

23. Prez, F., Granger, B.E.: IPython: A System for Interactive Scientific Computing. Computing in Science & Engineering **9**(3), 21–29 (2007). http://scitation.aip.org/content/aip/journal/cise/9/3/10.1109/MCSE.2007.53

24. Rodrguez-Labrada, R., Velzquez-Prez, L.: Eye movement abnormalities in spinocerebellar ataxias. Spinocerebellar Ataxias, pp. 59–76. Intech, Rijeka (2012)

25. Russell, S., Norvig, P.: Artificial Intelligence: A Modern Approach, 3rd edn. Prentice Hall, Upper Saddle River (December (2009)

26. Salvucci, D.D., Goldberg, J.H.: Identifying fixations and saccades in eye-tracking protocols. In: ETRA 2000, pp. 71–78. ACM, New York (2000). http://doi.acm.org/10.1145/355017.355028

27. Silverman, B.W., Jones, M.C.: An Important Contribution to Nonparametric Discriminant Analysis and Density Estimation. International Statistical Review / Revue Internationale de Statistique **57**(3), 233–238 (1989). http://www.jstor.org/stable/1403796

28. Tigges, P., Kathmann, N., Engel, R.R.: Identification of input variables for feature based artificial neural networks-saccade detection in EOG recordings. International Journal of Medical Informatics **45**(3), 175–184 (1997). http://www.sciencedirect.com/science/article/pii/S1386505697000427

29. Velzquez-Prez, L., Rodrguez-Labrada, R., Garca-Rodrguez, J.C., Almaguer-Mederos, L.E., Cruz-Mario, T., Laffita-Mesa, J.M.: A comprehensive review of spinocerebellar ataxia type 2 in cuba. The Cerebellum **10**(2), 184–198 (2011). http://link.springer.com/article/10.1007/s12311-011-0265-2

30. Witten, I.H., Frank, E., Hall, M.A.: Data Mining. Practical Machine Learning Tools and Techniques, 3 edn. Morgan Kaufmann (2011)

31. Wyatt, H.J.: Detecting saccades with jerk. Vision Research **38**(14), 2147–2153 (1998). http://www.sciencedirect.com/science/article/pii/S0042698997004100

Applying a Hybrid Algorithm
to the Segmentation of the Spanish
Stock Market Index Time Series

Antonio Manuel Durán-Rosal[1]([✉]), Mónica de la Paz-Marín[2],
Pedro Antonio Gutiérrez[1], and César Hervás-Martínez[1]

[1] Department of Computer Science and Numerical Analysis, University of Córdoba,
Rabanales Campus, Albert Einstein building, 14071 Córdoba, Spain
{i92duroa,pagutierrez,chervas}@uco.es
[2] Department of Management and Quantitative Methods,
Loyola Andalucía University, Business Administration Faculty,
Escritor Castilla Aguayo 4, 14004 Córdoba, Spain
mpaz@uco.es

Abstract. Time-series segmentation can be approached by combining
a clustering technique and genetic algorithm (GA) with the purpose of
automatically finding segments and patterns of a time series. This is an
interesting data mining field, but its application to the optimal segmentation of financial time series is a very challenging task, so accurate algorithms are needed. In this sense, GAs are relatively poor at finding the
precise optimum solution in the region where the algorithm converges.
Thus, this work presents a hybrid GA algorithm including a local search
method, aimed to improve the quality of the final solution. The local
search algorithm is based on maximizing a likelihood ratio, assuming
normality for the series and the subseries in which the original one is
segmented. A real-world time series in the Spanish Stock Market field
was used to test this methodology.

Keywords: Time series segmentation · Hybrid algorithms · Clustering ·
Spanish stock market index

1 Introduction

Recently, the ubiquity of temporal data has initiated various research and development efforts in the field of data mining. In this sense, time series can be easily
obtained from financial and scientific applications, being one of the main sources
of temporal datasets. How to discover useful time series patterns is a very interesting field [1]. The continuous nature of time series data make them difficult

This work has been partially subsidised by the TIN2011-22794 project of the Spanish Ministry of Economy and Competitiveness (MINECO), FEDER funds and the
P2011-TIC-7508 project of the "Junta de Andalucía" (Spain).

I. Rojas et al. (Eds.): IWANN 2015, Part II, LNCS 9095, pp. 69–79, 2015.
DOI: 10.1007/978-3-319-19222-2_6

to be processed, analyzed and/or mined. Discretizing a continuous time series into significant symbols [2,3] is an option to alleviate this problem. The process, referred to as "numeric-to-symbolic" (N/S) conversion, is considered as one of the basic steps before mining the time series. Das *et al.* [2] use a fixed-length window to segment time series and represent it using the resulting simple patterns. Other approaches [4,5] suggest dividing the time series using previously identified change points and substituting the segments with suitable functions.

Evolutionary algorithms (EAs) are robust heuristics, since they perform a global multi-point search, quickly converging to high quality areas (even when the search space is very complex). On the contrary, they are relatively poor at finding the precise optimum solution in the region where the algorithm converges to [6]. The lack of precision of the EAs has been tackled, during the last few years, incorporating local optimization algorithms to improve the results. The idea is to combine local search procedures, which are good at finding local optima (local exploiter), and evolutionary algorithms (global explorer), resulting in what is usually known as hybrid algorithms or EA-LS methods.

There are several ways in which a LS procedure can be combined with an EAs. The way the combination is done is extremely important in terms of accuracy and computational efficiency. The best balance of local exploitation and global exploration has to be obtained. Some of the strategies previously used include the multistart approach, the Lamarckian learning, the Baldwinian learning, the partial Lamarckianism and or the process of random linkage [7–9].

In this paper, we propose a hybrid evolutionary algorithm for obtaining the most important cut points in time series, i.e. to divide the time series in segments. The methodology combines a genetic algorithm (global explorer), a clustering process and a local improvement procedures. The LS process maximizes the logarithmic likelihood-ratio between two likelihood values, the first one used to contrast the null hypothesis that the n observations of the time series are sampled from the same normal distribution, and the second one used to contrast the alternative hypothesis that the n observations are associated with two independent normal distributions, obtained by a potential cut point, assuming that the observations on the left and right hand sides are sampled from two different normal distributions.

The LS is applied to the final solution obtained by the GA, allowing the precise local optimum around the final solution to be found. We have compared the results of the algorithm when run with and without the LS process. To test the performance of the proposed hybrid algorithm, it is applied to a hard real-world time series in the Stock Market field (IBEX 35).

The rest of the paper is organized as follows. Section 2 presents the characteristics of the algorithm proposed, while Section 3 includes the description of the time series, the experiments performed and the discussion about the results. Finally, Section 4 establishes the conclusions.

2 Hybrid Segmentation Algorithm

2.1 Summary of the Algorithm

Given a time series $Y = \{y_n\}_{n=1}^N$, our objective is to divide the values of y_n into m consecutive subsets or segments, where the segments should be associated to a homogeneous behaviour of y_n. Time indexes $(n = 1, \ldots, N)$ are divided into segments: $s_1 = \{y_1, \ldots, y_{t_1}\}, s_2 = \{y_{t_1}, \ldots, y_{t_2}\}, \ldots, s_m = \{y_{t_{m-1}}, \ldots, y_N\}$, where the ts are the different cut points subscripted in ascending order ($t_1 < t_2 < t_{m-1}$). The cut points are the only points which belong to two segments (the one before and the one after). The number of segments m and the values of the cut points $t_i, i = 1, \ldots, m - 1$, have to be determined by the algorithm. Then, the methodology attempts to group the segments into k different clusters (where $k < m$ is a parameter to be defined by the user). Each segment will be associated to a class label, from k different possible labels, $\{\mathcal{C}_1, \ldots, \mathcal{C}_k\}$.

The main steps of the algorithm are summarized in Fig. 1. The algorithm proposed is an extension of that proposed in [10], where a final step of local search has been included. Each individual chromosome consists of an array of binary values, where the length of the chromosome is the time series length, N. Each position c_i stores whether the time index t_i of the time series represents a cut point for the evaluated solution[1].

Time series segmentation:
Input: Time series.
Output: Best segmentation of the time series.
  1: Generate a random population of t time series segmentations.
  2: Evaluate all segmentations of the initial population by using a predefined fitness
     function.
  3: **while not** Stop Condition **do**
  4:    Store a copy of the best segmentation.
  5:    Select parent segmentations from current population.
  6:    Generate offsping: apply crossover and mutation to construct new candidate
        segmentations.
  7:    Evaluate the fitness of the offsping segmentations.
  8:    Merge parent segmentations and offsping segmentations.
  9:    Replace current population with selected segmentations of the previous union.
 10: **end while**
 11: Improve the best segmentation from final population using the likelihood-based
     segmentation algorithm.
 12: **return** Resulting segmentation from hybridization.

Fig. 1. Main steps for the proposed hybrid algorithm

---

[1] Note that the first and last points of the chromosome are always considered cut points.

2.2   Genetic Algorithm

The characteristics of the GA are defined as follows.

Initial Population. The population of the GA is a set of binary vectors of length N, where 1s are the cut points (initialised randomly, in such a way that two consecutive points are separated by a value in $[s_{max} - s_{min}]$ positions) and the rest are 0s.

Fitness Evaluation. The evaluation of a segmentation consists of three steps: 1) obtaining the characteristics of the segments, 2) applying a clustering process, and 3) evaluating the quality of this clustering.

1. **Characteristics of the segments**: The segments of the chromosome can have different length, so all the segments are projected into the same five dimensional space. Five metrics measured for all segments of the chromosome:

   - Variance (S_s^2):
   $$S_s^2 = \frac{1}{t_s - t_{s-1} + 1} \sum_{i=t_{s-1}}^{t_s} (y_i - \overline{y_s})^2, \tag{1}$$

   where y_i are the time series values of the segment, and $\overline{y_s}$ is the average value of the segment.
   - Skewness (γ_{1s}):
   $$\gamma_{1s} = \frac{\frac{1}{t_s - t_{s-1} + 1} \sum_{i=t_{s-1}}^{t_s} (y_i - \overline{y_s})^3}{S_s^3}, \tag{2}$$

   where S_s is the standard deviation of the s-th segment.
   - Kurtosis (γ_{2s}):
   $$\gamma_{2s} = \frac{\frac{1}{t_s - t_{s-1} + 1} \sum_{i=t_{s-1}}^{t_s} (y_i - \overline{y_s})^4}{S_s^4} - 3. \tag{3}$$

   - Slope of a linear regression over the points of the segment (a_s):
   $$a_s = \frac{S_{s,yt}}{(S_{s,t})^2}, \tag{4}$$

   where, for the s-th segment, $S_{s,yt}$ is the covariance between the time indexes, t, and the time series values, y; and $S_{s,t}$ is the standard deviation of the time values. Covariance $S_{s,yt}$ is defined by:
   $$S_{s,yt} = \frac{1}{t_s - t_{s-1} + 1} \sum_{i=t_{s-1}}^{t_s} (i - \overline{t_s}) \cdot (y_i - \overline{y_s}). \tag{5}$$

   - Autocorrelation coefficient (AC_s):
   $$AC_s = \frac{\sum_{i=t_{s-1}}^{t_s} (y_i - \overline{y_s}) \cdot (y_{i+1} - \overline{y_s})}{S_s^2}. \tag{6}$$

2. **Clustering process**: After projecting all segments to the same five-dimensional space, a clustering process is applied to group them. A scaling of the metrics to the range $[0, 1]$ is considered. The algorithm chosen for the clustering step is the well-known k-means method. A deterministic process is used to select the initial centroids which ensures that a chromosome will have always the same fitness value. First, the characteristic with higher variability is selected. The first initial centroid will be the segment with the highest value in that characteristic. The second one will be the segment with the highest Euclidean distance from the first centroid previously selected. The third centroid will be that which is farthest from both, and so on.

3. **Evaluating the quality of the clustering process**: To evaluate the quality of the clustering the COP index [11, 12] is considered. COP is a ratio-type index with an estimate of the intra-cluster variance divided by an estimate of the inter-cluster variance:

$$COP(C) = \frac{1}{N} \sum_{c_k \in C} \frac{(1/|c_k|) \sum_{\mathbf{x}_i \in c_k} d(\mathbf{x}_i, \overline{\mathbf{c}_k})}{min_{\mathbf{x}_i \notin c_k} max_{\mathbf{x}_j \in c_k} d(\mathbf{x}_i, \mathbf{x}_j)}, \tag{7}$$

where k is the number of clusters, $|c_k|$ is the number of segments of cluster k, $\overline{\mathbf{c}_k}$ is the centroid of cluster k, and $d(\mathbf{x}, \mathbf{y})$ is the Euclidean distance between vector $\mathbf{x}$ and vector $\mathbf{y}$. As this index has to be minimised, the fitness will be defined as $f = \frac{1}{1+COP}$. The cohesion is estimated by the distance from the points in a cluster to its centroids and the separation is based of the furthest neighbour distance.

Selection and Replacement Processes. All the individuals are considered for reproduction and generation of offspring. However, a replacement process is applied to the joint offspring and parent populations by roulette wheel selection. The selection process promotes diversity, but the replacement process is promoting elitism.

Mutation and Crossover Operators. Two kinds of operators are included in the algorithm to reduce the dependency with respect to the initial population and escaping from local optima[2]:

– Mutation operator: the probability p_m of performing any mutation is decided by the user. The kind of mutation applied to the individual is randomly selected from the following two: 1) add or remove (with the same probability) a given number of cut points of the segmentation; and 2) move a given number of cut points of the segmentation to the left or the right (with the same probability).

– Croosover operator: For each parent individual, the crossover operator is applied with a given probability p_c. The operator randomly selects the other parent and a time index, interchanging the left and right parts of the chromosomes selected with respect to the time index.

---

[2] For more information about these operators see [10].

2.3   Likelihood-Based Segmentation Algorithm

To find the $m-1$ unknown cut points t_l (separating segment l and $l+1$, where m is the number of segments) a recursive segmentation scheme could be considered as done in previous works [13–15]. This method is based on the likelihood-ratio test under an *i.i.d.* Gaussian distribution (assuming that each segment is sampled from a Gaussian distribution with different mean and variance) and a joint distribution consisting of two different Gaussian models for the complete time series.

Suppose a segment $\mathbf{s}_s$ with a number of elements n_s, where $\mathbf{s}_s = (x_{t_l}, \ldots, x_{t_{l+1}})$ and $s = 1, \ldots, m$, following a Gaussian distribution with parameters μ_s and σ_s^2. Then, let us denote a potential cut point for the observations $\mathbf{s}_s$ as u, in such a way that the observations on the left hand side are assumed to be sampled from a normal distribution, $N(\mu_{sL}, \sigma_{sL}^2)$, and the ones on the right hand side from another normal distribution, $N(\mu_{sR}, \sigma_{sR}^2)$, with different parameters. In this case, we define the likelihood-ratio between L_1 and $L_2(u)$ to contrast the null hypothesis that the n_s observations are sampled from the same normal distribution (the alternative hypothesis being that the n_s observations are associated with two independent normal distributions). L_1 and $L_2(u)$ can be defined as:

$$L_1 = \prod_{i=t_l}^{t_{l+1}} f(x_i; \mu_s, \sigma_s^2),$$

$$L_2(u) = \prod_{i=t_l}^{t_l+u} f(x_i; \mu_{sL}, \sigma_{sL}^2) \prod_{i=t_l+u+1}^{t_{l+1}} f(x_i; \mu_{sR}, \sigma_{sR}^2).$$

The logarithmic likelihood-ratio between L_1 and $L_2(u)$ (i.e., $\log L(u)$) can be defined in the form:

$$\log L(u) = \log \frac{L_2(u)}{L_1} = \log L_2(u) - \log L_1 =$$

$$= \sum_{i=t_l}^{t_l+u} f(x_i; \mu_{sL}, \sigma_{sL}^2) + \sum_{i=t_l+u+1}^{t_{l+1}} f(x_i; \mu_{sR}, \sigma_{sR}^2) - \sum_{i=t_l}^{t_{l+1}} f(x_i; \mu_s, \sigma_s^2),$$

where for the normal distributions hypothesis it holds that:

$$\log L(u) = n_s \log \sigma_s - u \log \sigma_{sL} - (n_s - u) \log \sigma_{sR},$$

where σ_s, σ_{sL} and σ_{sR} are approximated as the maximum likelihood estimators, based on a sufficiently large value of n_s, u and $(n_s - u)$ (in order to warranty that these estimators have good properties, such as consistency, efficiency and asymptotic normality).

To find the optimal cut point u for a segment $\mathbf{s}_s$, the logarithmic likelihood-ratio between L_1 and $L_2(u)$ that has been defined previously can be used an indicator. More specifically, an adequate way to separate the observations is to choose u so that $\log L(u)$ takes the maximum value. In other words, an adequate segmentation should be done at $u^* = \arg\max_u (\log L(u))$.

Note that, a predefined threshold p can be selected in order to restrict the divisions to be done. That is, if $\max(\log L(u))/n_s \log \sigma_s$ is less than p, then the segment is not divided. This is used as the stopping condition for the recursive segmentation procedure. As we can see, it is a relative stopping condition. Then, the decision rule to divide a segment $\mathbf{s}_s$ is:

IF $\max(\log L(u))/n_s \log \sigma_s > p$

THEN the initial segment $\mathbf{s}_s$ must be split at u^*. Continue dividing the resultant left subsegment $\mathbf{s}_{sL}$ provided that $u > 2s_{min}$, and/or the resultant right subsegment $\mathbf{s}_{sR}$ if $(n_s - u) > 2s_{min}$.

ELSE Stop the division procedure.

The defined segmentation procedure is a Top-Down technique, since the time series is recursively partitioned until the stopping criterion is met. Other alternative approaches in the literature are the bottom-up approach or the sliding window method [16].

Note that, once that the hybrid segmentation step is finished, the last step is to evaluate the segmentation with the clustering process defined in section 2.2.

3   Experimental Results and Discussion

The experiments performed and the results obtained are analysed in this section.

3.1   Spanish Stock Market Index Dataset

As said, we analysed one of the official indexes of the Madrid Stock Market: the Ibex-35, an index composed of the 35 most liquid values listed in the Computer Assisted Trading System. For our study. we considered the daily closing prices of the Spanish Ibex-35 stock index from 14th January 1992 to 26th September 2014, presenting thus a total of 5730 observations. The complete time series used in the experiments can be seen in Fig. 2, where the most relevant financial phases have been included as vertical lines, which were extracted from the literature.

3.2   Experimental Setting

The experimental design for the stock index under study is presented in this subsection.

The GA was configured with the following parameter values obtained by a *trial and error* procedure. The number of individuals of the population is fixed to $P = 100$. The crossover probability is taken as $p_c = 0.8$ and the mutation probability as $p_c = 0.2$. The percentage of cut points to be mutated is set to 20% of the current number of cut points. The maximum number of generations is configured as $g = 100$, and the k-means clustering process is allowed a maximum of 20 iterations. Experts established three decisive parameters: the initial minimum and maximum size of the segments, $s_{min} = 20$ and $s_{max} = 120$, respectively; and the number of clusters, $k = 5$. The likelihood-based segmentation algorithm was configured with a $p = 0.01$.

Given the stochastic nature of GAs, our algorithm was run 30 times with different seeds, and we selected the best one in terms of the fitness function.

Table 1. IBEX35 centroids of clusters

Cluster	Variance	Asymmetry	Kurtosis	Slope	Autocorrelation
1	24146.468331	0.799105	0.383196	2.397703	15.824783
2	541224.303442	-0.282932	-0.130286	-98.559788	20.027348
3	91390.874077	0.099044	-0.597932	-0.344728	56.589522
4	51136.677812	-0.019688	-1.594107	-89.333291	5.249476
5	39063.510562	-0.177806	-0.630814	4.852070	18.763463

3.3   Discussion

Because the optimum number of segments can not be known for real world time series, this section aims to evaluate the effectiveness of the proposed segmentation algorithm analysing the original IBEX35 Index time series and to present the main results of the segmentation algorithm. The segmentation returned by the hybrid algorithm in the last generation is analysed. The best segmentation, with a total of 215 segments, can be seen in Fig. 2. The relevant financial phases phases (taken from the literature) are represented as vertical lines and each segment is coloured according to its assigned cluster in the clustering process. Note that, given the nature of the used statistics for the clustering process, the shape of patterns belonging to the same cluster might differ (as patterns presenting different shapes can present similar statistical properties), but, instead, the segmentation shows grouped patterns with similar trends and homogeneous characteristics. Table 1 shows the characteristics of the centroids of each cluster. Using this information in combination with the fifteenth phases in Fig. 2, the clusters can be described as follows:

- Cluster 1 (red colour) groups the segments with the lowest variance. The distribution is the most asymmetric one (to the right) and segments in this cluster presents the only one positive kurtosis. The slope is positive, which

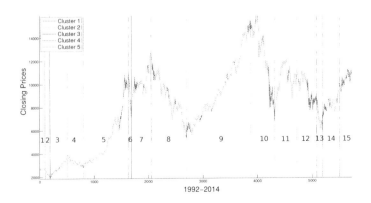

Fig. 2. IBEX35 Segmentation (Online version in colour)

indicates the presence of an increasing linear trend in the segments, with a low autocorrelation. It could be said that red cluster are changing trends, although sometimes very small, but always are.

– Cluster 2 (green colour) groups the segments with the highest variance. The distribution is asymmetric (to the left) and the kurtosis is slightly oriented to the left too. The slope is the most negative one, so the segments in this cluster present present values in time t which are positive correlated with values in time $t - 1$. In stock market terms, they are assimilated to crashes or spikes before a fall.

– The segments with the second highest variance are included in cluster 3 (dark blue colour), with a symmetric distribution (the values near zero); kurtosis value is negative and the slope presents a moderate negative trend (negatives values but near zero), representing an absence of a linear trend. The autocorrelation is the highest one. They are more intense oscillations that others.

– Cluster 4 (pink colour) presents a variance below the average, with a symmetric distribution as the previous cluster, and the highest negative kurtosis of all clusters. The slope presents a big decreasing linear trend as segments in cluster two, but in this case with a low autocorrelation. They don't have a special meaning; they could be interpreted as interferences.

– Cluster 5 (light blue colour) is composed of segments with the second lowest variance and with a slightly asymmetric distribution (to the left); kurtosis is negative and the slope presents the highest positive value, but it is not very big. The autocorrelation is similar to that of cluster one. The segments in this cluster represent the main trend of the Spanish Stock Market Index.

In general, the obtained segments in the same cluster are similar in both shape and trend. Additional interesting discussions about phases and clusters can be pointed out as general rules (see Table 1):

– The period 1992–2000 showed relatively small fluctuations, while the shape of the graphics mutates since 2000 year until nowadays.
– Segments in cluster 1 (red) modulate the segments in cluster 5 (light blue) because they appear inside of this cluster and they help segments in cluster 5 to change its inertia.
– Segments in cluster 3 (dark blue) are interesting because they are intense turbulences inside of periods of a pronounced trend.
– Segments in cluster 1 (red) show the beginning of a great change in the graphic (they can be interpreted as a signal of change of trend).
– We note that pink (cluster 4) is necessary to correctly segment and group the rest of the time series.

4   Conclusions

This paper presents a hybrid optimisation time series segmentation methodology. The segments obtained are used for a posterior clustering-based analysis (where these segments are mapped to a 5-dimensional space representing their statistical properties and grouped according to their similarity). The characteristics of the resultant clusters and their relation to well-known financial patterns and phases (cycles) are analysed in the discussion section. The segmentation obtained appears to be consistent with the key milestones in the IBEX35 and describes this time series acceptably. A genetic algorithm (GA) from the field of time series segmentation has been applied to stock market data to identify common patterns that would act as early signals for trend change. The common patterns can be visualized in a straightforward manner by looking at their segment class label. Such capability is extremely useful for analysing stock data in a novel fashion.

The trend of the Spanish Stock market is easily found in the resulting segmentation and the clustering of the obtained segments allows us to detect important incidents that influence the stock market, for example the 2008 financial crisis, the bursting of the real state bubble, the technological boom, and so on.

A future line of work could correspond to the prediction of complete time periods, using the temporal patterns identified in the segmentation/clustering phase. This prediction will allow us to determine the shape of the next period, which is especially useful in financial applications to identify future market behaviours ('bear', 'bull' and 'sluggish' periods, or any other financial patterns that indicates a change of trend).

References

1. Chung, F.L., Fu, T.C., Ng, V., Luk, R.W.: An evolutionary approach to pattern-based time series segmentation. IEEE Transactions on Evolutionary Computation 8(5), 471–489 (2004)
2. Das, G., ip Lin, K., Mannila, H., Renganathan, G., Smyth, P.: Rule discovery from time series, pp. 16–22. AAAI Press (1998)
3. Ou-Yang, K., Jia, W., Zhou, P., Meng, X.: A new approach to transforming time series into symbolic sequences. In: [Engineering in Medicine and Biology, 1999. 21st Annual Conference and the 1999 Annual Fall Meetring of the Biomedical Engineering Society] BMES/EMBS Conference, Proceedings of the First Joint. vol. 2, 974, October 1999
4. Oliver, J., Forbes, C.: Bayesian approaches to segmenting a simple time series. Technical Report 14/97, Monash University, Department of Econometrics and Business Statistics (1997)
5. Oliver, J.J., Baxter, R.A., Wallace, C.S.: Minimum message length segmentation. In: Wu, X., Kotagiri, R., Korb, K. (eds.) Research and Development in Knowledge Discovery and Data Mining. LNCS, vol. 1394, pp. 222–233. Springer, Berlin Heidelberg (1998)
6. Houck, C.R., Joines, J.A., Kay, M.G., Wilson, J.R.: Empirical investigation of the benefits of partial lamarckianism. Evol. Comput. 5(1), 31–60 (1997)
7. Kolen, A., Pesch, E.: Genetic local search in combinatorial optimization. Discrete Applied Mathematics 48(3), 273–284 (1994)
8. Joines, J.A., Kay, M.G.: Utilizing hybrid genetic algorithms. In: Optimization, E. (ed.) International Series in Operations Research & Management Science, vol. 48, pp. 199–228. Springer, US (2002)

9. Ulder, N.L.J., Aarts, E.H.L., Bandelt, H.J., Laarhoven, P.J.M.v., Pesch, E.: Genetic local search algorithms for the travelling salesman problem. In: Schwefel, H.-P., Männer, R. (eds.) PPSN I 1990. LNCS, vol. 496, pp. 109–116. Springer, Heidelberg (1991)

10. Nikolaou, A., Gutiérrez, P., Durán, A., Dicaire, I., Fernández-Navarro, F., Hervás-Martínez, C.: Detection of early warning signals in paleoclimate data using a genetic time series segmentation algorithm. Climate Dynamics, 1–15 (2014)

11. Gurrutxaga, I., Albisua, I., Arbelaitz, O., Martín, J.I., Muguerza, J., Pérez, J.M., Perona, I.: Sep/cop: An efficient method to find the best partition in hierarchical clustering based on a new cluster validity index. Pattern Recognition 43(10), 3364–3373 (2010)

12. Arbelaitz, O., Gurrutxaga, I., Muguerza, J., Pérez, J.M., Perona, I.: An extensive comparative study of cluster validity indices. Pattern Recognition 46(1), 243–256 (2013)

13. Cheong, S.A., Fornia, R.P., Lee, G.H.T., Kok, J.L., Yim, W.S., Xu, D.Y., Zhang, Y.: The japanese economy in crises: A time series segmentation study. Economics: The Open-Access. Open-Assessment E-Journal 6 (2012)

14. Bernaola-Galván, P., Román-Roldán, R., Oliver, J.L.: Compositional segmentation and long-range fractal correlations in dna sequences. Phys. Rev. E 53, 5181–5189 (1996)

15. Sato, A.H.: A comprehensive analysis of time series segmentation on japanese stock prices. Procedia Computer Science 24, 307–314 (2013); 17th Asia Pacific Symposium on Intelligent and Evolutionary Systems. In: IES 2013 (2013)

16. Keogh, E., Chu, S., Hart, D., Pazzani, M.: An online algorithm for segmenting time series. In: Proceedings IEEE International Conference on Data Mining, ICDM 2001, pp. 289–296 (2001)

Nonlinear Ordinal Logistic Regression Using Covariates Obtained by Radial Basis Function Neural Networks Models

Manuel Dorado-Moreno$^{(\boxtimes)}$, Pedro Antonio Gutiérrez,
Javier Sánchez-Monedero, and César Hervás-Martínez

Department of Computer Science and Numerical Analysis, University of Cordoba,
Campus de Rabanales, C2 building, 14071 Cordoba, Spain
{i92domom,pagutierrez,jsanchezm,chervas}@uco.es

Abstract. This paper proposes a nonlinear ordinal logistic regression method based on the hybridization of a linear model and radial basis function (RBF) neural network models for ordinal regression. The process for obtaining the coefficients is carried out in several steps. In the first step we use an evolutionary algorithm to determine the structure of the RBF neural network model, in a second step we transform the initial feature space (covariate space) adding the nonlinear transformations of the input variables given by the RBFs of the best individual in the final generation of the evolutionary algorithm. Finally, we apply an ordinal logistic regression in the new feature space. This methodology is tested using 8 benchmark problems from the UCI repository. The hybrid model outperforms both the linear and the nonlinear part obtaining a good compromise between them and better results in terms of accuracy and ordinal classification error.

Keywords: Artificial neural networks · Radial basis function · Proportional odds model · Evolutionary algorithms · Ordinal classification · Ordinal regression

1 Introduction

There are many fields of study such as medicine, social sciences and others, where it is needed to classify items into naturally ordered classes. These problems are traditionally handled by conventional methods intended for classification of nominal classes, where the order relation is ignored. This kind of supervised learning problems are referred to as ordinal classification or ordinal regression, where an ordinal scale ($Class_1 > Class_2 > ...Class_J$) [12] is used to label the examples. Therefore, in ordinal classification problems, the goal is to learn how

This work has been partially subsidised by the TIN2011-22794 project of the Spanish Ministry of Economy and Competitiveness (MINECO), FEDER funds and the P2011-TIC-7508 project of the "Junta de Andalucía" (Spain).

© Springer International Publishing Switzerland 2015
I. Rojas et al. (Eds.): IWANN 2015, Part II, LNCS 9095, pp. 80–91, 2015.
DOI: 10.1007/978-3-319-19222-2_7

to classify examples in the correct class. But one should take into account that the higher distance between predicted and real labels are (with respect to the ordinal scale), the more the misclassification error should be penalised.

Logistic Regression (LR) models have been widely used in statistics for many years. This traditional statistical tool arises from the desire to model the posterior probabilities of the class level via linear functions of the predictor variables. In this way, the LR model serves the purpose of predicting the class where a pattern belongs to. In general, LR is a simple and useful procedure, although it poses problems when applied to real-problems, where, frequently, we cannot make the assumption of additive and purely linear effects of the covariates. As suggested by [9], an obvious way to generalise the linear logistic regression is to replace the linear predictors with structured nonparametric models such as an additive model of basis function. In this paper, we extend the ideas introduced in [8], where a combination of LR and Neural Networks (NN) models was used to solve nominal classification problems. We present an adaptation of the corresponding algorithm to tackle ordinal classification.

One of the first models specifically designed for ordinal classification, and the one our work is based on, is the Proportional Odds Model (POM) [14]. This model is based on the assumption of stochastic ordering in the input space, and the use of thresholds to split the input space into different ordered classes. Our hybrid model is based on the idea of augmenting/replacing the vector of inputs with nonlinear covariates obtained by radial basis functions (RBFs), which are transformations of the linear input variables, and then use POM in this new space of derived input features.

The estimation of the coefficients is carried out in several steps. In a first step, an evolutionary algorithm [19] (EA) is applied to design the structure and train the weights of an RBF neural network [5,11] (RBFNN). Evolutionary computation has been widely used in the late years to evolve NN architectures and weights. There have been many applications for parametric learning [17] and for both parametric and structural learning [2,13,15,18]. EAs are global search methods, which incorporate the semantics of natural evolution to optimisation processes, becoming blind stochastic search methods for optimal solutions. In order to work, they maintain a population of individuals, in our case, a set of NNs (solutions to the problem). These are subject to a series of transformations to obtain new solutions (NNs) and a selection process aimed at favoring the best models. Thus, the networks will evolve to keep improving and get a good solution to the problem, although it does not have to be the best one. This evolutionary process determines the number of RBFs in the model and the corresponding variables, which will be the new covariates in the nonlinear LR model. The best model in the last generation is used for that purpose. In a second step, we transform the input space in two different ways: the first one will transform the whole input space replacing the linear covariates with the RBFs, and the second one will augment the input space adding the RBF nonlinear covariates to the linear ones. It is known that the EAs perform well at exploring the search surface, but they are not that good at finding the optimum solution. That led us to add a third step, where we perform a local optimization algorithm using the new input covariates, with a maximum likelihood method for ordinal LR, based on the structure of the POM.

The rest of the paper is organized as follows. Section 2 introduces the POM. In section 3, the neural network model for ordinal regression is explained. Section 4 presents the algorithm developed in order to obtain the coefficients for the hybrid model. Section 5 includes the experiments: experimental design, information about the datasets and results of the experiments. Finally, in section 6, we present the conclusions of the paper.

2   Proportional Odds Model (POM)

As previously mentioned, we are developing a model to overcome the problems of liner LR when applied to real world datasets, where the covariates cannot be assumed to be linear. Moreover, we focus on ordinal classification, so we consider an ordinal regression model.

Let formally define the ordinal classification problem. Given an input vector $\mathbf{x}_i$, we have to predict the label y_i, where $\mathbf{x}_i \in \mathcal{X} \subseteq \mathbb{R}^k$ and $y_i \in \mathcal{Y} \in \{\mathcal{C}_1, \mathcal{C}_2, \ldots, \mathcal{C}_J\}$ (J is the number of classes). A classification rule or function has to be estimated, $R : \mathcal{X} \to \mathcal{Y}$, able to predict the categories of new patterns. Because we are in a supervised setting, a training set of N points is given, $D = \{(\mathbf{x}_i, y_i), 1 \leq i \leq N\}$. What makes the difference between nominal classification and ordinal regression is that, in ordinal regression, the following constraint appears: $\mathcal{C}_1 \prec \mathcal{C}_2 \prec \ldots \prec \mathcal{C}_Q$. The symbol $\prec$ expresses that a label is before another in the ordinal scale.

The Proportional Odds Model (POM) is a direct extension of LR, and one of the first specifically designed for ordinal regression. It can be grouped under the Cumulative Link Models (CLMs) family [1], which predict probabilities of adjacent categories, taking the ordinality into account. CLM models estimate cumulative probabilities as follows:

$$P(y \preceq \mathcal{C}_j|\mathbf{x}) = P(y = \mathcal{C}_1|\mathbf{x}) + \ldots + P(y = \mathcal{C}_j|\mathbf{x}),$$
$$P(y = \mathcal{C}_j|\mathbf{x}) = P(y \preceq \mathcal{C}_j|\mathbf{x}) - P(y \preceq \mathcal{C}_{j-1}|\mathbf{x}),$$

for $j \in \{1, \ldots, J\}$, assuming, by definition, that $P(y \preceq \mathcal{C}_J|\mathbf{x}) = 1$ and $P(y = \mathcal{C}_1|\mathbf{x}) = P(y \preceq \mathcal{C}_1|\mathbf{x})$. CLMs link a linear model of the input variables to these cumulative probabilities:

$$f(\mathbf{x}) = g^{-1}\left(P(y \preceq \mathcal{C}_j|\mathbf{x})\right) = \beta_0^j - \mathbf{w}^T\mathbf{x},$$

where $g^{-1} : [0, 1] \to (-\infty, +\infty)$ is a monotonic transformation (the inverse link function), β_0^j is the threshold defined for class $\mathcal{C}_j$, and $\mathbf{w}$ is the coefficient vector of the linear model. The most common choice for the link function is the logistic function. The logit link function is the inverse of the standard logistic cumulative distribution function (CDF), with the following expression:

$$g^{-1}\left(P(y \preceq \mathcal{C}_j|\mathbf{x})\right) = \ln\left(\frac{P(y \preceq \mathcal{C}_j|\mathbf{x})}{(1 - P(y \preceq \mathcal{C}_j|\mathbf{x}))}\right). \tag{1}$$

Considering that $f(\mathbf{x})$ follows a logistic CDF and following the idea of the POM model [14], the cumulative likelihood of a pattern being associated with a class less than or equal to class C_j is defined as:

$$P(y \preceq C_j|\mathbf{x}) = \frac{1}{1 + \exp(f(\mathbf{x}) - \beta_0^j)},$$

where $j = 1, \ldots, J$, and, by definition, $P(y \preceq C_J|\mathbf{x}) = 1$. Therefore, this model approximates the posterior probability of a class j as:

$$P(y = C_j|\mathbf{x}) = P(y \preceq C_j|\mathbf{x}) - P(y \preceq C_{j-1}|\mathbf{x}) =$$
$$= \frac{1}{1 + \exp(f(\mathbf{x}) - \beta_0^j)} - \frac{1}{1 + \exp(f(\mathbf{x}) - \beta_0^{j-1})}. \tag{2}$$

where thresholds must satisfy $\beta_0^1 < \beta_0^2 < \cdots < \beta_0^{J-1}$ in order to split the real line into J contiguous ordered intervals.

3   Artificial Neural Network (ANN) Model Used

The POM model [14], as the majority of existing ordinal regression models, can be represented in the following general form:

$$C(\mathbf{x}) = \begin{cases} C_1, & \text{if } f(\mathbf{x}, \boldsymbol{\theta}) \leq \beta_0^1 \\ C_2, & \text{if } \beta_0^1 < f(\mathbf{x}, \boldsymbol{\theta}) \leq \beta_0^2 \\ \cdots \\ C_J, & \text{if } f(\mathbf{x}, \boldsymbol{\theta}) > \beta_0^{J-1} \end{cases}, \tag{3}$$

where $\beta_0^1 < \beta_0^2 < \cdots < \beta_0^{J-1}$ (this will be the most important constraint in order to adapt the nominal classification model to ordinal classification), J is the number of classes, $\mathbf{x}$ is the input pattern to be classified, $f(\mathbf{x}, \boldsymbol{\theta})$ is a ranking function and $\boldsymbol{\theta}$ is the vector of parameters of the model. Indeed, the analysis of Eq. (3) uncovers the general idea previously presented: patterns, $\mathbf{x}$, are projected to a real line by using the ranking function, $f(\mathbf{x}, \boldsymbol{\theta})$, and the biases or thresholds, β_0^j, are separating the ordered classes.

We are using an adaptation of the POM to artificial neural networks. This adaptation is based on two elements: the first one is a second hidden linear layer with only one node whose inputs are the non-linear transformations of the first hidden layer. The task of this node is to project the values into a line, to make them have an order. After this one node linear layer, an output layer is included with one bias for each class, whose objective is to set the optimum thresholds to classify the patterns in the class they belong to.

The structure of our model is presented in Fig. 1 which has two main parts. The lower one shows the RBFNN model, where $\mathbf{x} = (x_1, \ldots, x_k)$, is the vector of input variables and k is the number of variables that describe a pattern. $\boldsymbol{w}_j = (w_{j0}, \ldots, w_{jk})$ is the vector of weights of the connections from the input

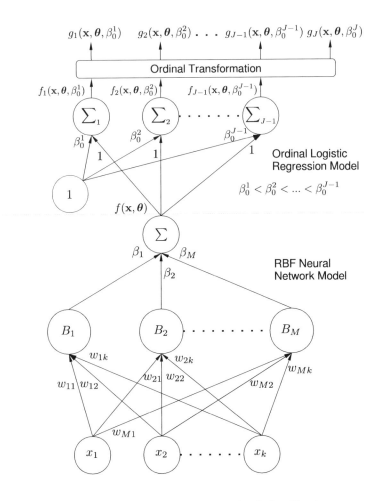

Fig. 1. Neural Network model for ordinal classification

nodes to the m-th hidden node ($m \in \{1, \ldots, M\}$), B are the RBF nodes and M is the number of nodes in the hidden layer. Finally, "1" is the bias of the layer.

The upper part of the figure shows a lonely node in the second hidden layer of the model, which is the one that performs the linear transformation of the POM model. Its result, $f(\mathbf{x}, \boldsymbol{\theta})$, is connected, together with a second bias, to the output layer, where J is the number of classes, and $\beta_0^0, \ldots, \beta_0^{J-1}$ are the thresholds for the different classes. These $J-1$ thresholds are able to separate the J classes, but they have to fulfil the order constraint shown in the figure. Finally, the output layer obtains the outputs of the model, $f_j(\mathbf{x}, \boldsymbol{\theta}, \beta_0^j)$, for $j \in \{1, \ldots, J-1\}$. These outputs are transformed using the function of the POM model, Eq. (2), which transforms them into a probability ($g_j(\mathbf{x}, \boldsymbol{\theta}, \beta_0^j)$). This is the probability that each pattern has to belong to the different classes, and the class with the greatest probability is the one selected by the NN to be the class of the pattern.

4   Estimation of the Coefficients

The methodology proposed is based on the combination of an EA and an ordinal maximum likelihood optimization method. In a first step, the EA is applied to determine the number of RBF neurons in the hidden layer (architecture of the NN) and the corresponding matrix of RBF parameters $\mathbf{W} = (\mathbf{w}_1, \mathbf{w}_1, \ldots, \mathbf{w}_m)$, where $\mathbf{w}_j = (\mathbf{c}_j, r_j)$, $\mathbf{c}_j = (c_{j1}, c_{j2}, \ldots, c_{jk})$ is the center of the jth Gaussian RBF transformation [6], r_j is the radius and $c_{ji}, r_j \in \mathbb{R}$.

Once the basis functions have been determined by the EA, we proceed with the second step and perform two different transformations in the initial covariate space. One of the transformations replaces the initial covariates by their nonlinear transformations given by the RBFs of the best individual in the final generation of the EA. The other transformation keeps the initial covariates and add the nonlinear covariates to the covariates space. The model is linear in these new variables. Finally, in the last step, we use a gradient descent algorithm in order to optimise the likelihood of the new model. Fig. 2 represents the different steps of the algorithm and the different models obtained for the experiments.

The different steps of the algorithm are now explained:

Step 1: We apply and EA to find the basis functions of the RBFs:

$$\mathbf{B}(\mathbf{x}, \mathbf{W}) = \{B_1(\mathbf{x}, \mathbf{w}_1), B_2(\mathbf{x}, \mathbf{w}_2), \ldots, B_m(\mathbf{x}, \mathbf{w}_m)\}$$

corresponding to the nonlinear part of the hybrid logistic regression model presented in this paper. The NN model for the EA is presented in Fig. 1. The EA begins with a random initial population, and each iteration the population is updated using a population-update algorithm [7]. The population is subject to operations of mutation and replication with ordinal constraints. Crossover is not used because of its disadvantages in evolving NNs [2].

Step 2: We perform a transformation of the input space, including the nonlinear transformations of the inputs obtained by the EA in Step 1:

$$\mathbb{H} : \mathbb{R}^k \to \mathbb{R}^{k+m}$$

$$(x_1, x_2, \ldots, x_k) \to (x_1, x_2, x_k, \ldots, z_1, z_2, \ldots, z_m),$$

where $z_1 = B_1(\mathbf{x}, \mathbf{w}_1)$, $z_2 = B_2(\mathbf{x}, \mathbf{w}_2), \ldots, z_m = B_m(\mathbf{x}, \mathbf{w}_m)$.

Step 3: We apply an ordinal maximum likelihood optimization method in the new input space obtained in step 2. The optimization of the maximum likelihood is performed using a gradient descent algorithm called iRProp+ [10], which optimises the nonlinear ordinal logistic regression for a defined number of epochs.

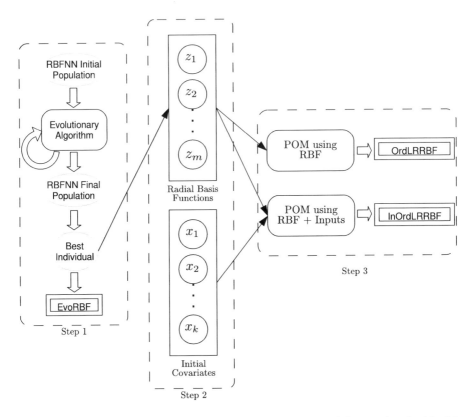

Fig. 2. Steps in the proposed methodology. The different models associated with this methodology are presented in a double squared box

5 Experiments

In order to analyse the performance of the proposed models, eight datasets have been tested, their characteristics being shown in Table 5. The collection of datasets is taken from the UCI [3] and the `mldata.org` [16] repositories. The experimental design was conducted using 10 random holdout procedures with 3 repetitions (to evaluate the randomness of both the training/test split and the EA), with $3n/4$ instances for the training set and $n/4$ instances for the generalization set (where n is the size of the dataset).

Three different versions of the models optimised by the EA were compared: EvoRBF is the RBF neural network obtained by the EA, OrdLRRBF is the same model after applying maximum likelihood optimisation and InOrdLRRBF augments OrdLRBF including the input variables as additional information. We compare also the results against the original POM model, which, being a deterministic method, was run using a 30 random holdout procedure with only one repetition per training/test split.

All the parameters of the algorithm are common to these eight problems. The main parameters of the algorithm are:

- Number of Generations: 50.
- Population Size: 250.
- Mutation Percentage: 10%.
- Minimum number of hidden nodes: 10.
- Maximum number of hidden nodes: 20.

In order to set up the minimum number of hidden neurons, a preliminary experiment was done with one partition of each dataset. A 5-fold cross-validation (using only the training split) was done and repeated with the following minimum number of hidden nodes, $\{1, 2, 4, ..., 20\}$. We conclude that the optimum minimum number of hidden nodes was 10 and we added 10 nodes for the maximum, in order to give the EA some freedom to optimise the NN.

The idea of having a little population size and a small number of generations is to give less importance to the evolutionary algorithm because its computational time is much higher than the optimiation of the nonlinear ordinal logistic regression. After finishing the EA, we took the hidden layer RBFs from the best model in the last generation to augment/replace the input space and applied a gradient descent algorithm with 500 epochs to optimise the nonlinear ordinal logistic regression model.

We will also compare the results obtained by the models presented in this paper with some standard methods for nominal classification, such as, Support Vector Machines (LibSVM), Multilayer Perceptron (MLP) and Radial Basis Function Neural Networks (RBFNN). That comparison will allow us to see the improvements of the results when using ordinal models for this type of datasets.

Table 1. Characteristics of the eight datasets used for the experiments: number of instances (Size), inputs (#In.), classes (#Out.) and patterns per-class (#PPC)

Dataset	Size	#In.	#Out.	#PPC
automobile	205	71	6	(3,22,67,54,32,27)
balance	625	4	3	(288,49,288)
car	1728	21	4	(1210,384,69,65)
ESL	488	4	9	(2,12,38,100,116,135,62,19,4)
LEV	1000	4	5	(93,280,403,197,27)
newthyroid	215	5	3	(30,150,35)
SWD	1000	10	4	(32,352,399,217)
toy	300	2	5	(35,87,79,68,31)

The following two measures have been used for comparing the models:

- CCR: The Correct Classification Rate (CCR) is the rate of correctly classified patterns:

$$CCR = \frac{1}{n} \sum_{i=1}^{N} [\![y_i^* = y_i]\!],$$

where y_i is the true label, y_i^* is the predicted label and $[\![\cdot]\!]$ is a Boolean test which is 1 if the inner condition is true and 0 otherwise. CCR values range from 0 to 1. It represents a global performance on the classification task. This measure is not taking into account category order.

– MAE: The Mean Absolute Error (MAE) is the average deviation (number of categories) in absolute value of the predicted class from the true class [4]:

$$MAE = \frac{1}{n} \sum_{i=1}^{N} e(\mathbf{x}_i),$$

where $e(\mathbf{x}_i) = |\mathcal{O}(y_i) - \mathcal{O}(y_i^*)|$ is the distance between the true and the predicted ranks, $\mathcal{O}(\mathcal{C}_j) = j$ and MAE values range from 0 to $J - 1$. This a way of evaluating the ordering performance of the classifier.

Table 2 shows the mean test value and standard deviation of the correct classified rate (CCR) and the mean absolute error (MAE) over the 30 models obtained (10 holdout procedures $\times$3 repetitions or 30 holdout procedures).

Table 2. Generalization results obtained for benchmark datasets

Dataset	Classification	CCR(%) Func.	MAE Mean ± SD	Mean ± SD
automobile	Ordinal	InOrdLRRBF	52.5000 ± 5.9480	0.6153 ± 0.0883
		OrdLRRBF	52.1153 ± 5.9756	0.6153 ± 0.0864
		EvoRBF	44.4230 ± 5.0839	0.7865 ± 0.0922
		POM	46.6667 ± 1.9415	0.9532 ± 0.6868
	Nominal	LibSVM	*68.8461 ± 8.0473*	*0.4796 ± 0.1181*
		MLP	**70.5769 ± 9.2027**	**0.4384 ± 0.1424**
		RBFNN	63.6538 ± 8.2316	0.5115 ± 0.1158
balance	Ordinal	InOrdLRRBF	*95.0955 ± 1.8763*	*0.0535 ± 0.0225*
		OrdLRRBF	94.6496 ± 1.8799	0.0579 ± 0.0217
		EvoRBF	86.6878 ± 1.5731	0.1834 ± 0.0314
		POM	90.5520 ± 1.8557	0.1067 ± 0.0208
	Nominal	LibSVM	**98.6624 ± 8.1955**	**0.0133 ± 0.0081**
		MLP	91.2739 ± 1.4725	0.0987 ± 0.0188
		RBFNN	86.2420 ± 2.5157	0.1847 ± 0.0551
car	Ordinal	InOrdLRRBF	*88.3796 ± 2.6093*	*0.1254 ± 0.0307*
		OrdLRRBF	88.0787 ± 2.7242	0.1312 ± 0.0335
		EvoRBF	81.9675 ± 1.7200	0.2194 ± 0.0200
		POM	15.7485 ± 3.0633	1.4505 ± 0.5482
	Nominal	LibSVM	88.3025 ± 2.2459	0.1755 ± 0.0132
		MLP	**98.7731 ± 1.1290**	**0.0129 ± 0.0110**
		RBFNN	87.5880 ± 2.6099	0.1333 ± 0.0373
ESL	Ordinal	InOrdLRRBF	**72.2131 ± 2.3959**	**0.2893 ± 0.0276**
		OrdLRRBF	*72.2131 ± 2.7445*	*0.2893 ± 0.0304*
		EvoRBF	66.8852 ± 2.3248	0.3680 ± 0.0344
		POM	70.5464 ± 3.3620	0.3103 ± 0.0380
	Nominal	LibSVM	71.4754 ± 2.3756	0.3024 ± 0.0209

Continued on next page

Table 2 – *Continued from previous page*

		MLP	69.5902 ± 3.4000	0.3213 ± 0.0298
		RBFNN	69.6721 ± 2.7594	0.3459 ± 0.0446
LEV	Ordinal	InOrdLRRBF	**64.4000 ± 2.1664**	0.4260 ± 0.0295
		OrdLRRBF	60.6400 ± 2.2167	0.4280 ± 0.0309
		EvoRBF	58.0800 ± 3.7371	0.4540 ± 0.0428
		POM	62.3333 ± 2.7992	*0.4093 ± 0.03039*
	Nominal	LibSVM	*62.4400 ± 2.3735*	**0.4016 ± 0.0227**
		MLP	62.0800 ± 2.6985	0.4280 ± 0.0288
		RBFNN	59.8400 ± 3.5034	0.4336 ± 0.0395
newthyroid	Ordinal	InOrdLRRBF	*95.3703 ± 1.3094*	*0.0462 ± 0.0130*
		OrdLRRBF	95.3703 ± 1.5737	0.0462 ± 0.1573
		EvoRBF	88.8888 ± 2.7605	0.1111 ± 0.0276
		POM	**97.2222 ± 2.2153**	**0.0277 ± 0.0221**
	Nominal	LibSVM	94.4444 ± 3.0240	0.0611 ± 0.0462
		MLP	92.5926 ± 6.0481	0.0740 ± 0.0604
		RBFNN	94.8150 ± 2.0380	0.0485 ± 0.0238
SWD	Ordinal	InOrdLRRBF	**58.3600 ± 1.4041**	**0.4344 ± 0.0154**
		OrdLRRBF	*58.3200 ± 1.2336*	*0.4356 ± 0.0145*
		EvoRBF	56.2400 ± 1.800	0.4516 ± 0.0167
		POM	56.7867 ± 2.9571	0.4501 ± 0.0304
	Nominal	LibSVM	57.4400 ± 2.6613	0.4476 ± 0.0339
		MLP	55.4000 ± 2.0677	0.4820 ± 0.0316
		RBFNN	56.5200 ± 2.1503	0.4696 ± 0.0264
toy	Ordinal	InOrdLRRBF	*95.3333 ± 1.8053*	*0.0466 ± 0.0180*
		OrdLRRBF	**95.6000 ± 1.8908**	**0.0440 ± 0.0189**
		EvoRBF	81.0666 ± 6.5561	0.1906 ± 0.0671
		POM	28.9333 ± 2.5527	0.9808 ± 0.0389
	Nominal	LibSVM	93.6000 ± 3.1926	0.0640 ± 0.0319
		MLP	55.3333 ± 7.9567	0.5333 ± 0.1064
		RBFNN	91.8667 ± 2.7720	0.0826 ± 0.0306

The best result is shown in bold and the second best in italics

Observing the results, it is clear that hybrid methodologies proposed in this paper overcome both EvoRBF and POM models, obtaining higher CCR and lower MAE for almost all the different benchmark datasets. They also overcome nominal standard methods in most of the datasets, except automobile, which presents a clear nominal distribution, and balance, where LibSVM obtains very good CCR and MAE, followed by our best model. Differences between the augmented input space (InOrdLRRBF) and the replaced input space (OrdLRRBF) are not that significant, but it can be observed that the model combining input linear covariates with nonlinear RBFs presents better results in more datasets than the one that only uses the RBFs for the ordinal logistic regression.

6   Conclusions

This work proposes to transform an ordinal linear logistic regresion model into a nonlinear one. To this end, the linear model is added nonlinear covariates using the outputs of the hidden layer neurons in a RBFNN. These neural network models are trained using an evolutionary algorithm that optimizes its architecture.

Moreover, the coefficients of the ordinal logistic regression model, consisting of the initial covariates and RBF basis functions, are estimated by a gradient descent algorithm that tries to optimize the maximum likelihood.

Initial experiments show that this hybrid approach is promising and generally improve accuracy and order quality, performing better than the corresponding base models (evolutionary neural network and POM).

References

1. Agresti, A.: Analysis of ordinal categorical data, vol. 656. John Wiley & Sons (2010)
2. Angeline, P.J., Sauders, G.M., Pollack, J.B.: An evolutionary algorithm that constructs recurrent neural networks. IEEE Transactions on Neural Networks 5(1), 54–65 (1994)
3. Asuncion, A., Newman, D.: UCI machine learning repository (2007). http://www.ics.uci.edu/mlearn/MLRepository.html
4. Baccianella, S., Esuli, A., Sebastiani, F.: Evaluation measures for ordinal regression. In: Proceedings of the Ninth International Conference on Intelligent Systems Design and Applications (ISDA 2009), Pisa, Italy, December 2009
5. Buchtala, O., Klimek, M., Sick, B.: Evolutionary optimization of radial basis function classifiers for data mining applications. IEEE Transactions on Neural Networks Part B 35(5), 928–947 (2005)
6. Chu, W., Ghahramani, Z.: Gaussian processes for ordinal regression. Journal of Machine Learning Research 6, 1019–1041 (2005)
7. Dorado-Moreno, M., Gutiérrez, P.A., Hervás-Martínez, C.: Ordinal Classification Using Hybrid Artificial Neural Networks with Projection and Kernel Basis Functions. In: Corchado, E., Snášel, V., Abraham, A., Woźniak, M., Graña, M., Cho, S.-B. (eds.) HAIS 2012, Part II. LNCS, vol. 7209, pp. 319–330. Springer, Heidelberg (2012)
8. Gutiérrez, P.A., Hervás-Martínez, C., Martínez-Estudillo, F.J.: Logistic regression by means of evolutionary radial basis function neural networks. IEEE Transactions on Neural Networks 22(2), 246–263 (2011)
9. Hastie, T., Tibshirani, R.: Generalized additive models. Chapman & Hall, London (1990)
10. Igel, C., Hsken, M.: Empirical evaluation of the improved rprop learning algorithms. Neurocomputing 50(6), 105–123 (2003)
11. Lee, S.H., Hou, C.L.: An art-based construction of RBF networks. IEEE Transactions on Neural Networks 13(6), 1308–1321 (2002)
12. Lippmann, R.P.: Pattern classification using neural networks. IEEE Transactions on Neural Networks 27, 47–64 (1989)
13. Maniezzo, V.: Genetic evolution of the topology and weight distribution of neural networks. IEEE Transactions on Neural Networks 5, 39–53 (1994)
14. McCullagh, P.: Regression models for ordinal data (with discussion). Journal of the Royal Statistical Society 42(2), 109–142 (1980)
15. Odri, S.V., Petrovacki, D.P., Krstonosic, G.A.: Evolutional development of a multilevel neural network. Neural Networks 6, 583–595 (1993)

16. PASCAL: Pascal (pattern analysis, statistical modelling and computational learning) machine learning benchmarks repository (2011). http://mldata.org/
17. van Rooij, A.J.F., Jain, L.C., Johnson, R.P.: Neural Networks Training Using Genetic Algorithms, Series in Machine Perception and Artificial Intelligence, vol. 26. World Scientific, Singapore (1996)
18. Yao, X., Liu, Y.: A new evolutionary system for evolving artificial neural networks. IEEE Transactions on Neural Networks **8**, 694–713 (1997)
19. Yao, X.: Evolving artificial neural networks. Proceedings of the IEEE 87(9) (1999)

Energy Flux Range Classification by Using a Dynamic Window Autoregressive Model

Pedro Antonio Gutiérrez[1]([✉]), Juan Carlos Fernández[1], Mária Pérez-Ortiz[1],
Laura Cornejo-Bueno[2], Enrique Alexandre-Cortizo[2], Sancho Salcedo-Sanz[2],
and César Hervás-Martínez[1]

[1] Department of Computer Science and Numerical Analysis, University of Cordoba,
Campus de Rabanales, C2 building, 14071 Cordoba, Spain
{pagutierrez,jfcaballero,i82perom,chervas}@uco.es
[2] Department of Signal Processing and Communications, Universidad de Alcalá,
Madrid, Spain
{laura.cornejo,enrique.alexandre,sancho.salcedo}@uah.es

Abstract. This paper tackles marine energy prediction from the classification point of view, by previously discretising the real objective variable into a set of consecutive categories or ranges. Given that the range of energy flux is enough to obtain an approximation of the amount of energy produced, the purpose of this discretisation is to simplify the prediction task. A special kind of autoregressive models are considered, where the category to be predicted depends on both the previous values of energy flux and a set of meteorological variables estimated by numerical models. Apart from this, this paper introduces two different ways of adjusting the order of the autoregressive models, one based on nested cross-validation and the other one based on a dynamic window. The results show that these kind of models are able to predict the time series in an acceptable way, and that the dynamic window procedure leads to the best accuracy without needing the additional computational cost of adjusting the order of the model.

Keywords: Wave energy prediction · Multi-class classification · Flux of energy · Autoregressive models

1 Introduction

Oceans are being increasingly considered in many countries as a promising source of clean and sustainable energy. Off-shore wind energy, ocean thermal and tidal and wave energy conversion [1] are some of the marine energy technologies which are currently under exploitation. Although off-shore wind energy is currently the most exploited one, wave energy conversion is being paid attention because of its good balance between cost and efficiency.

This work has been partially subsidised by the TIN2014-54583-C2-1-R project of the Spanish Ministry of Economy and Competitiveness (MINECO), FEDER funds, the P2011-TIC-7508 project of the "Junta de Andalucía" (Spain) and by the "Comunidad de Madrid" (Spain), under project number S2013/ICE-2933.

© Springer International Publishing Switzerland 2015
I. Rojas et al. (Eds.): IWANN 2015, Part II, LNCS 9095, pp. 92–102, 2015.
DOI: 10.1007/978-3-319-19222-2_8

Wave Energy Converters (WECs) are able to transform the energy of waves into electricity using either the vertical oscillation or the linear motion of the waves [2], i.e. WECs convert potential and kinetic energy of waves into electricity. These WECs are being used in areas with great wave power density located near populated regions demanding energy, e. g. Norway, UK, Ireland or Portugal, to name a few.

Wave resource prediction becomes a crucial topic for the design, deployment, and control of WECs [3,4], that require a proper characterization of waves in a given area. Wave energy resource prediction is difficult because of their stochastic nature and the large amount of factors which influence, e. g. changes in the wind, sea depth, closeness to the coast, etc. This prediction is, however, very important to ensure a correct management of WECs facilities, in renewable and sustainable systems for energy supply (the demand of energy in certain regions fluctuates throughout the day and even depending on the season of year). Alternative applications of wave height prediction are decision making in operational works at sea, risk evaluation of marine energy facilities, etc.

The data for characterising waves (and subsequent prediction) can be basically obtained from radars and buoys arrays, which generate time series. Using these time series, the corresponding wave spectrum, $S(f)$, can be computed. In turn, based on its spectral moments, the most useful wave parameter to estimate the *wave power density* at a given location is the wave's *flux of energy* (F_e), which will be used by the WEC in order to generate electricity [2,5]. Basically, the conversion to electrical power [6] is subject to hugely varying energy flux, which can be obtained from meteorological data as the *significant wave height* (H_s) and the *wave energy period* (T_e).

This paper deals with a problem of marine wave energy flux prediction (F_e). Contrary to the previous approaches, the problem is tackled as a multi-class classification problem by discretising the time series, F_e, in different categories. Given that a predefined number of categories or ranges in terms of F_e is enough for obtaining practical information, the main advantage of using this discretisation is the corresponding simplification of the prediction problem. For this purpose, real data of a sea buoy located at the southeast coast of the USA are used for evaluationg the proposed methodologies. F_e is obtained from two standard meteorological data: H_s and T_e, collected hourly by the sensors in the buoy. This information can be obtained from the National Data Buoy Center (NDBC) [7], belonging to the National Data Buoy Service of the United States.

For tackling this classification problem, we consider a longitudinal study by using AutoRegressive (AR) models [8]. The motivation for using AR models comes from the fact that many observed time series exhibit serial autocorrelation, i.e. linear association between lagged observations. This kind of models are built considering the past lagged values of the time series as inputs (or independent variables) and the next value of the time series as output (or dependent variable). However, the order of AR models (i.e. the number of lagged values to be included as inputs, usually denoted as p) decisively influences its final performance.

It has to be defined prior to fitting the AR model and it usually depends on the dynamics of the time series.

For the case of discrete time series, autoregressive logistic regression predicts a discrete response for time t based on the p previous discrete responses [9]. Although the same approach could be tackled for F_e range prediction, we propose to use the actual values of F_e instead of the discretised ones, which would ideally increase the quality of the predictions thanks to a finer-grain input information. Finally, in order to deal with the optimum determination of the order p for the AR model, we propose to consider a dynamic window instead of a fixed order one. This dynamic window procedure is based on the selection of those past values of the time series where no change of class is observed with respect to the current class value. Then, different statistics are calculated over the values of this window (in our case, the average value, the standard deviation and the amplitude of the window).

In addition to consider the past lagged values of the time series as inputs, four meteorological predictive variables in four different points around the buoy studied in this work have been added as independent variables. The four meteorological variables have been obtained from the NCEP/NCAR Reanalysis Project of the Earth System Research Laboratory (ESRL) Physical Sciences Division [10]: air temperature, sea level pressure, the zonal component of the wind and the meridional component of the wind. To carry out the use of variables of numerical models, a matching procedure every six hours (the minimum temporal resolution of the Reanalysis data used) between the meteorological variables and the F_e values has been necessary. Therefore, the independent variables of the models obtained are formed by the past values of F_e and the values of the four meteorological variables in those instants.

The results obtained show that the models proposed achieve promising accuracy, with over 67% of correctly predicted ranges when using 4 possible class values. Moreover, the dynamic window procedure alleviates the computational cost needed for selecting a proper window size for this kind of AR models. The rest of this paper is structured as follows: Firstly, the next section presents a brief background about machine learning techniques applied to solve the prediction of data for characterising waves. In Section 3, the building of the dataset used in this paper is detailed. The specific problem modelling considered is stated in Section 4. Section 5 presents the experimental part of the paper and the results obtained. Finally, Section 6 gives some concluding remarks for closing the paper.

2   Brief Background

Taking into account wave energy prediction from the machine learning point of view [11], the most common techniques consider Artificial Neural Networks (ANNs). Deo et al. [12] proposed the use of ANNs for predicting H_s, and his proposal was improved later in [13]. In [14], H_s and T_e are predicted from the observed wave records using time series neural networks. In [15], a hybrid genetic algorithm-adaptive network-based fuzzy inference system model was developed to forecast H_s and the peak spectral period in Lake Michigan. In this methodology, both clustering and rule base parameters are simultaneously optimised using genetic algorithms and ANNs. Recently, in [16], ANNs have been applied to estimate the wave energy resource in the northern coast of Spain.

Other works as [17] apply efficient regression methods based on machine learning and soft computing. In [18], support vector regression is used, and in [19] and [20] genetic programming and fuzzy logic are used respectively. Alternative approaches using numerical models of atmosphere and Ocean, usually hybridised with time series prediction, can also be found in the literature in [21,22].

3   Data Source

In this paper, the values of F_e over a time range have been obtained from a buoy located at the East Coast of the USA. The buoy collects standard meteorological data hourly using the sensors installed on it. These data are stored and they can be obtained by downloadable annual text files in the National Oceanic and Atmospheric Administration (NOAA), specifically in the National Data Buoy Center (NDBC) [7], that maintains a network of data collecting for buoys and coastal stations. To check the robustness of the models used in this work (see Section 4) the standard meteorological data of the buoy studied from January 1st (00:00) to December 31st (23:00), for the years 2012 and 2013 have been considered.

The selected buoy is the Station 41013 (LLNR 815) [23] - Frying Pan Shoals, NC Buoy. This buoy is geographically located at coordinates 33.436N 77.743W (33° 26′ 11″ N 77° 44′ 35″ W). This location is a region of interest to exploit wave energy conversion, because it is relatively close to massive population zones. The considered buoy is a near-shore location with sufficient range of values for F_e to carry out this first study (a good location will have an annual average typically in the range $20 - 70$ kilowatts per meter (kW/m) [1]).

It is possible to obtain the value of F_e considering H_s, measured in meters, and T_e, measured in seconds, as shown below:

$$F_e = 0.49 \cdot H_s^2 \cdot T_e, \tag{1}$$

where F_e is the energy flux generated by the waves measured in kW/m. Note that F_e is defined in Equation (1) as an average energy flux (H_s is a kind of average wave height), though for simplicity it will be referred just as energy flux.

F_e values obtained from the model used must be discretised in order to tackle the problem as a classification one. To check out how the change in the discretisation of F_e can affect the prediction task, datasets with different number of classes according to different thresholds used for each one have been built. Considering the expert knowledge about the problem, the threshold values were adjusted trying to cover the most important kind of waves from the point of view of their energy production in that geographical location. It is important to note that this discretisation may be adjustable depending on each specific buoy location and even the WEC type. Moreover, note that the inclusion of many different classes could compromise the performance of the classification task. Table 1 shows the thresholds for 4, 5 and 6 classes, respectively, for F_e and for the buoy studied in this work.

Table 1. Discretisation in $Q = 4$, $Q = 5$ and $Q = 6$ classes for F_e in the Station 41013, Southeast of United States. For each class, the range of the class and the corresponding number of patterns are specified.

Discretization	C_1	C_2	C_3	C_4	C_5	C_6
$Q = 4$	$[0, 2)$	$[2, 4)$	$[4, 10)$	$[10, \infty)$		
	858	863	762	437		
$Q = 5$	$[0, 1)$	$[1, 2)$	$[2, 4)$	$[4, 10)$	$[10, \infty)$	
	233	625	863	762	437	
$Q = 6$	$[0, 1)$	$[1, 2)$	$[2, 3)$	$[3, 4)$	$[4, 10)$	$[10, \infty)$
	233	625	521	342	762	437

Regarding the meteorological variables used as inputs, they were taken from the NCEP/NCAR Reanalysis Project web page [10], that maintains sea surface level data around the world in a global grid of resolution $2.5° \times 2.5°$. The four points closest to the buoy (north, south, east and west) have been taken into account in a 6-hours time horizon resolution, since it is the minimum resolution provided by the ESRL. In each point four representative meteorological variables were used for the prediction: air temperature, sea level pressure, the zonal component of the wind and the meridional component of the wind [24]. Then, a matching procedure was carried out every six hours between these meteorological variables obtained from NCEP/NCAR Reanalysis Project and between the F_e hourly collected by the buoy, obtained from the NDBC. After this matching step, there were only four missing values, corresponding to four dates when no data were recorded in the buoy. These values were approximated by considering the average of the previous and next point of the missing value. The total number of data values is 2924, including the four ones which were approximated.

4 Models Used for Prediction

The problem of F_e can be described as a time series forecasting problem. Let $\{y_t\}_{t=1}^{N+p}$ be a time series of $N + p$ points. The objective is to obtain a function $f : \mathbb{R}^p \rightarrow \mathbb{R}$ which is derived from a dataset of N patterns, $\mathbf{D} = (\mathbf{X}, \mathbf{Y}) = \{(\mathbf{x}_t, y_t)\}_{t=1}^{N}$ where $\mathbf{x}_t = \{y_{t-p}, y_{t-p+1}, \ldots, y_{t-1}\}$ is the vector with the values of the independent variables (i.e., the p past values of the time series), and the dependent variable, y_t, is the value of the time series for time t. A standard AR model of order p is defined in the following way:

$$y_t = \beta_0 + \sum_{i=1}^{p} \beta_i y_{t-i} + \epsilon_t,$$

where $\beta_0, \beta_1, \ldots, \beta_p$ are the coefficients of the model and ϵ_t is white noise.

The values of the time series can be discretised in a set of classes, given that the range of F_e provides enough information for taking decisions, the exact value not being needed for obtaining a general idea about the energy produced. In this way, we considered the ranges presented in Table 1 to map the time series of real values, $\{y_t\}_{t=1}^{N+p}$, to a time series of symbols or classes, $\{c_t\}_{t=1}^{N+p}$, where $c_i \in \{C_1, \ldots, C_Q\}$. In order to predict discrete responses, the AR logistic regression [9] resembles the following structure:

$$\ln \frac{P(c_t = C_q | c_{t-1}, \ldots, c_{t-p})}{P(c_t = C_Q | c_{t-1}, \ldots, c_{t-p})} = \beta_{q0} + \sum_{i=1}^{p} \beta_{qi} \mathcal{O}(c_{t-i}),$$

where $q \in \{1, \ldots, Q-1\}$, $\mathcal{O}(C_q) = q$, and $\beta_{q0}, \beta_{q1}, \ldots, \beta_{qp}$ are the parameters of class q. One of the downsides of this model is that it assumes that distances between different ranges are the same for all pairs of labels, which, as can be checked in Table 1, is not the case for our problem. Moreover, given that we still have the original values of F_e, i.e. y_t, we hypothesise that these more accurate inputs can provide better results for the prediction of the discrete responses. Finally, the analysis is extended by including the meteorological variables (exogenous variables) previously described, which will be referred to as $\mathbf{x} = \{x_1, \ldots, x_D\}$. Summarising, the following model is considered:

$$\ln \frac{P(c_t = C_q | y_{t-1}, \ldots, y_{t-p}, \mathbf{x})}{P(c_t = C_Q | y_{t-1}, \ldots, y_{t-p}, \mathbf{x})} = \beta_{q0} + \sum_{i=1}^{p} \beta_{qi} y_{t-i} + \sum_{i=1}^{D} \beta_{q(p+i)} x_i,$$

where $q \in \{1, \ldots, Q-1\}$, $\beta_{q0}, \beta_{q1}, \ldots, \beta_{qp}, \beta_{q(p+1)}, \ldots, \beta_{q(p+D)}$ are the parameters of class q. With these exogenous variables, the AR model becomes an ARX model (AR with eXogenous variables). The final probability can predicted as:

$$P(c_t = C_q | y_{t-1}, \ldots, y_{t-p}, \mathbf{x}) = \tag{2}$$
$$\frac{\exp \left(\beta_{q0} + \sum_{i=1}^{p} \beta_{qi} y_{t-i} + \sum_{i=1}^{D} \beta_{q(p+i)} x_i \right)}{\sum_{j=0}^{Q} \exp \left(\beta_{j0} + \sum_{i=1}^{p} \beta_{ji} y_{t-i} + \sum_{i=1}^{D} \beta_{j(p+i)} x_i \right)},$$

for $q \in \{1, \ldots, Q-1\}$.

However, it is still necessary to determine the order p for this ARX model. It depends on the regime of the time series, thus it has to be adjusted for each time series. A possible solution is to consider a nested cross-validation procedure to decide the value of p resulting in the maximum estimated accuracy.

This paper explores the possibility of extending an AR model with order $p = 1$ with different statistics for a dynamic window. This dynamic window is formed by those past values of the time series which belong to a range equal to the last one:

$$W_t = \{y_{t-1}, y_{t-2}, \ldots, y_{t-s} | c_{t-1} = c_t \wedge \ldots \wedge c_{t-s} = c_t \wedge c_{t-s-1} \neq c_t\},$$

where the width of the window s depends on the point being examined. The window W_t is formed by all those adjacent past values which have the same label than the last value. The behaviour of this window is shown in Fig. 1, where the evaluated points are shadowed. As can be observed, the static window used in AR models always consider the same number of precedent values (in this case, two values), while the dynamic window considers all those previous time indexes whose class is the same than the last one. This can provide the model with dynamic information of the time series, more values being used when the time series is stable (so the range is constant) and less points being considered for those cases in which the labels have changed.

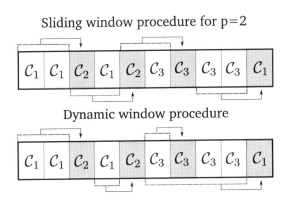

Fig. 1. Static window with $p = 2$ and dynamic window

Using this window, the final model extends an ARX model with $p = 1$ and has the following structure:

$$P(c_t = C_q | W_t, \mathbf{x}) = \tag{3}$$

$$\frac{\exp\left(\beta_{q0} + \beta_{q1}y_{t-1} + \beta_{q2}\overline{W_t} + \beta_{q3}S_{W_t} + \beta_{q4}R_{W_t} + \sum_{i=1}^{D}\beta_{q(4+i)}x_i\right)}{\sum_{j=0}^{Q}\exp\left(\beta_{j0} + \beta_{j1}y_{t-1} + \beta_{j2}\overline{W_t} + \beta_{j3}S_{W_t} + \beta_{j4}R_{W_t} + \sum_{i=1}^{D}\beta_{q(4+i)}x_i\right)},$$

where $q \in \{1, \dots, Q - 1\}$, and $\overline{W_t}$, S_{W_t} and R_{W_t} are the average, the standard deviation and the range of the window W_t, respectively:

$$\overline{W_t} = \frac{1}{s}\sum_{y \in W_s} y, \quad S_{W_t} = \sqrt{\sum_{y \in W_s} \frac{(y - \overline{W_t})}{s - 1}}, \quad R_{W_t} = \max_{y \in W_t} y - \min_{y \in W_t} y.$$

5   Experiments

As previously mentioned, the final number of F_e values is 2924 including data from 2012 and 2013. Regarding the experimental design, we have applied a 10-fold cross-validation, where the time series was divided into 10 disjoint folds without randomizing the data. Then, the training process was repeated 10 times, each one considering a different fold for test and the remaining ones for training.

The values of F_e were discretised using the intervals presented in Table 1, and we considered three different versions of a logistic regression model (with different input variables) for predicting the corresponding categories or labels:

- ARX_1, which is the ARX model presented in Eq. (2) with only one previous real value of F_e used as predictor, i.e. $p = 1$.
- ARX_p, i.e. an ARX model using a range of values for p ranging from 1 to 10. In order to decide the best value, a nested cross-validation process with 10 folds was applied using only the training data (again, without data randomization). This nested cross-validation was repeated for $p \in \{1, \ldots, 10\}$ and the value of p resulting in the maximum cross-validation accuracy was selected. Then, the training was repeated for this value of p but this time using the whole training set.
- ARX_{W_t}, which is the dynamic window ARX model defined in Eq. (3), where there is no need to specify the order of the model.

To perform the experiments, we have used `scikit-learn` [25,26], an open source machine learning library for Python. Given that some of the models involve a quite large number of features, we have included a regularisation term to control over-fitting. Specifically, the logistic regression implementation considered is that included in `liblinear` [27,28], and we have applied L2 regularisation. This implies adjusting a regularisation parameter C, which represents the inverse of the regularisation strength. For this, a nested 10-fold cross-validation has been applied (similar to the one previously discussed), using the following values $C \in \{10^{-5}, 10^{-4}, \ldots, 10^2\}$. In order to apply the AR models, the first p data points had to be discarded and used only as data input. We discarded the first 10 points for all the models (independently of the order considered) in order to do a fair comparison.

We have evaluated two different aspects of the results, the percentage of labels correctly classified (accuracy) and the time needed for obtaining the corresponding models, including the cross-validation process for adjusting all the parameters (C and, in the case of ARX_p, the order p). Table 2 includes the mean and standard deviation of the 10 accuracy results, while Table 3 is focused on the computational time. In general, it can be observed how the accuracy of the models decreases as the number of classes (Q) increases, which is due to the increased complexity of the classification problem. However, the level of accuracy obtained is acceptable, if we take into account that they are quite higher than the accuracy obtained by random guessing ($CCR = 33.3\%$ for 3 classes, $CCR = 25.0\%$ for 4 classes and $CCR = 16.6\%$ for 6 classes).

Although the ARX_1 is able to achieve very good results, they are consistently worse results than those obtained by ARX_p and ARX_{W_t}. When comparing the performances of ARX_p and ARX_{W_t}, they tend to be quite similar, but the computational cost of the ARX_p method is much higher, because 10 different values of p are tested for each fold (recall that the value of p is independently optimised for each fold).

Table 2. Mean and standard deviation of the accuracy (%) obtained by the different methods considered

	$Q = 4$ (Mean ± SD)	$Q = 5$ (Mean ± SD)	$Q = 6$ (Mean ± SD)
ARX_1	66.679 ± 2.434	58.754 ± 3.555	54.636 ± 4.793
ARX_p	66.679 ± 2.129	58.960 ± 3.896	55.084 ± 4.880
ARX_{W_t}	**67.500 ± 2.548**	**59.542 ± 3.224**	**55.871 ± 4.547**

Table 3. Mean and standard deviation of the total cross-validation time (seconds) needed for obtaining the models

	$Q = 4$ (Mean ± SD)	$Q = 5$ (Mean ± SD)	$Q = 6$ (Mean ± SD)
ARX_1	**0.6502 ± 0.0064**	**0.8131 ± 0.0094**	**0.9736 ± 0.0129**
ARX_p	44.6132 ± 24.5891	56.8958 ± 31.1296	67.9879 ± 37.4131
ARX_{W_t}	0.8238 ± 0.0419	1.0429 ± 0.0367	1.2648 ± 0.0390

The labels of F_e obtained by the ARX_{W_t} model when considering $Q = 4$ classes are shown in Fig. 2. As can be seen, the predictions follow the general tendency of the real values, resulting in an enough accurate idea of the quantity of energy produced.

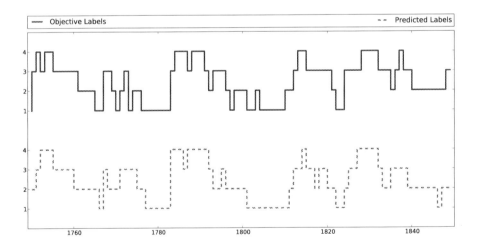

Fig. 2. Observed labels for a range of the F_e time series and label predicted by the ARX_{W_t} ($Q = 4$)

6   Conclusions

This paper evaluates different AutoRegressive models with eXogenous variables (ARX) based on logistic regression in a problem of wave energy flux prediction (F_e). Specifically, F_e prediction in a 6 hours time-horizon is considered, and classification techniques for tackling prediction are introduced. Classification is accomplished by considering consecutive intervals of values (ranges), which is enough information for energy prediction purposes. The input variables include both the past values of F_e (instead of the labels) and meteorological variables from numerical meteorological models from four grid points surrounding the considered buoy. Three models are used: an ARX model using one past value of the time series, an ARX model where the number of past values (order of the model, p) is decided by cross-validation, and a dynamic window ARX model, all of them based on logistic regression. The results shows that the dynamic window avoids the computational burden needed for adjusting the order of the model, while increasing the accuracy with respect to the ARX model of order $p = 1$. The way of performing F_e prediction is novel for marine energy prediction, and allows to apply well-tuned machine learning classifiers. As future work, ordinal logistic regression [29] could be considered, instead of standard nominal logistic regression, given that the different ranges of F_e are categories showing a predefined order, which can improve the quality of the classifiers.

References

1. Bahaj, A.S.: Generating electricity from the oceans. Renewable and Sustainable Energy Reviews **15**, 3399–3416 (2011)
2. de O. Falcao, A.F.: Wave energy utilization: a review of the technologies. Renewable and Sustainable Energy Reviews 14, 899–918 (2010)
3. Fusco, F., Ringwood, J.V.: Short-term wave prediction for real-time control of Wave Energy Converters. IEEE Transactions on Sustainable Energy **1**(2), 99–106 (2010)
4. Richter, M., Magaña, M.E., Sawodny, O., Brekken, T.K.: Nonlinear Model Predictive Control of a Point Absorber Wave Energy Converter. IEEE Transactions on Sustainable Energy **4**(1), 118–126 (2013)
5. Waters, R., Engström, J., Isberg, J., Leijon, M.: Wave climate off the Swedish west coast. Renewable Energy **34**(6), 1600–1606 (2009)
6. Falnes, J.: A review of wave-energy extraction. Marine Structures **20**, 185–201 (2007)
7. NOAA/NDBC: National Oceanic and Atmospheric Administration (NOAA), National Data Buoy Center (NDBC). http://www.ndbc.noaa.gov
8. Brockwell, P.J., Davis, R.A.: Time series: theory and methods. Springer Science & Business Media (2009)
9. Guanche, Y., Mínguez, R., Méndez, F.J.: Autoregressive logistic regression applied to atmospheric circulation patterns. Climate Dynamics **42**(1–2), 537–552 (2014)
10. NCEP/NCAR: The NCEP/NCAR Reanalysis Project, NOAA/ESRL Physical Sciences Division. http://www.esrl.noaa.gov/psd/data/reanalysis/reanalysis.shtml
11. Bishop, C.: Pattern Recognition and Machine Learning. Springer (2006)

12. Deo, M.C., Naidu, C.S.: Real time wave prediction using neural networks. cean. Engineering **26**(3), 191–203 (1998)
13. Agrawal, J.D., Deo, M.C.: Wave parameter estimation using neural networks. Marine Structures **17**, 536–550 (2004)
14. Tsai, C.P., Lin, C., Shen, J.N.: Neural network for wave forecasting among multi-stations. Ocean Engineering **29**(13), 1683–1695 (2002)
15. Zanaganeh, M., Jamshid-Mousavi, S., Etemad-Shahidi, A.F.: A hybrid genetic algorithm-adaptive network-based fuzzy inference system in prediction of wave parameters. Engineering Applications of Artificial Intelligence **22**(8), 1194–1202 (2009)
16. Castro, A., Carballo, R., Iglesias, G., Rabuñal, J.R.: Performance of artificial neural networks in nearshore wave power prediction. Applied Soft Computing **23**, 194–201 (2014)
17. Mahjoobi, J., Etemad-Shahidi, A.F., Kazeminezhad, M.H.: Hindcasting of wave parameters using different soft computing methods. Applied Ocean Research **30**(1), 28–36 (2008)
18. Mahjoobi, J., Mosabbeb, E.A.: Prediction of significant wave height using regressive support vector machines. Ocean Engineering **36**, 339–347 (2009)
19. Nitsure, S.P., Londhe, S.N., Khare, K.C.: Wave forecasts using wind information and genetic programming. Ocean Engineering **54**, 61–69 (2012)
20. Ozger, M.: Prediction of ocean wave energy from meteorological variables by fuzzy logic modeling. Expert Systems with Applications **38**(5), 6269–6274 (2011)
21. Reikard, G., Pinson, P., Bidlot, J.R.: Forecasting ocean wave energy: The ECMWF wave model and time series methods. Ocean Engineering **38**, 1089–1099 (2011)
22. Akpinar, A., Komurcu, M.I.: Assessment of wave energy resource of the Black Sea based on 15-year numerical hindcast data. Applied Energy **101**, 502–512 (2013)
23. NDBC: National Data Buoy Center (NDBC) - Station 41013 (LLNR 815) - Frying Pan Shoals, NC Buoy. http://www.ndbc.noaa.gov/station_history.php?station=41013
24. NCEP/NCAR: The NCEP/NCAR Reanalysis Project, Sea Surface Level Variables 6-hourly. http://www.esrl.noaa.gov/psd/data/gridded/data.ncep.reanalysis.surface.html
25. Pedregosa, F., Varoquaux, G., Gramfort, A., Michel, V., Thirion, B., Grisel, O., Blondel, M., Prettenhofer, P., Weiss, R., Dubourg, V., Vanderplas, J., Passos, A., Cournapeau, D., Brucher, M., Perrot, M., Duchesnay, E.: Scikit-learn: Machine Learning in Python. Journal of Machine Learning Research **12**, 2825–2830 (2011)
26. Scikit-Learn: Scikit-Learn, Machine Learning in Python. http://scikit-learn.org/stable/
27. Fan, R.E., Chang, K.W., Hsieh, C.J., Wang, X.R., Lin, C.J.: LIBLINEAR: A library for large linear classification. The Journal of Machine Learning Research **9**, 1871–1874 (2008)
28. Liblinear: Liblinear - A Library for Large Linear Classification. http://www.csie.ntu.edu.tw/cjlin/liblinear
29. McCullagh, P.: Regression Models for Ordinal Data. Journal of the Royal Statistical Society. Series B (Methodological) **42**(2), 109–142 (1980)

Automatic Eye Blink Detection Using Consumer Web Cameras

Beatriz Remeseiro[1]([✉]), Alba Fernández[1], and Madalena Lira[2]

[1] Departamento de Computación, Universidade da Coruña, A Coruña, Spain
{bremeseiro,alba.fernandez}@udc.es
[2] Centro de Física, Universidade do Minho, Guimarães, Portugal
mlira@fisica.uminho.pt

Abstract. This research aims to advance blinking detection in the context of work activity. Rather than patients having to attend a clinic, blinking videos can be acquired in a work environment, and further automatically analyzed. Therefore, this paper presents a methodology to perform the automatic detection of eye blink using consumer videos acquired with low-cost web cameras. This methodology includes the detection of the face and eyes of the recorded person, and then it analyzes the low-level features of the eye region to create a quantitative vector. Finally, this vector is classified into one of the two categories considered —open and closed eyes— by using machine learning algorithms. The effectiveness of the proposed methodology was demonstrated since it provides unbiased results with classification errors under 5%.

Keywords: Image processing · Feature extraction · Pattern recognition · Classification · Eye blink · Clinical application

1 Introduction

Blinking is vital to maintain optical performance and ocular surface health. Blinks can be spontaneous, reflex and voluntary, and eye blink rate (EBR) depends of environmental factors, type of activity and individual characteristics [1]. Blinking is a protective mechanism for the eye, serving to maintain a stable tear film over the ocular surface that is necessary for corneal health and optical performance [2].

The blinking process is influenced by a number of work-related factors [3]. Eye irritation symptoms and dry eyes are common and abundant symptoms reported in office-like environments. Work with visual display units (VDU), inadequate lighting and the position of the VDU may destabilize the tear film by changing eye blink frequency [4]. A review of studies that examined EBR in human subjects, indicated sample-average blink rates that ranged between 10 and 22 blinks per minute [5]. The disparities between studies demonstrate how EBR is dependent on varying experimental conditions for measurement since during computer use, EBR was found to be 4 blinks per minute, only 20% of the rate recorded for

© Springer International Publishing Switzerland 2015
I. Rojas et al. (Eds.): IWANN 2015, Part II, LNCS 9095, pp. 103–114, 2015.
DOI: 10.1007/978-3-319-19222-2_9

the same subjects during a period of general conversation. Blink frequency and duration changed significantly over time during the task [6]. After a blink, the gradual increase in optical aberration associated with the increasingly irregular tear film may cause a progressive reduction in the optical quality of the eye [7].

The most commonly employed and, up to recently, only available measure for startle eye-blink is the electrophysiological recording of the orbicularis oculi muscle via electromyography (EMG) [8]. Although EMG is a reliable measure of the eye blink, it is also limited in its range of applications as eye-blink response per se might be compromised by the measurement since this technique requires the attachment of electrodes directly underneath the eye.

The automatic assessment of eye blinks has still not been satisfactorily addressed. It is not easy to measure a movement as fast as a spontaneous blink. Depending on its amplitude, a blink can be completed in less than 100 ms. Many different techniques have been employed to detect and register blinks. The evolution of these methods reflects advances in image processing that have emerged in recent decades. The direct assessment of the eye blink can be a valuable enhancement of the available methodology. The eye blink startle response serves to protect the eye against physical impact, therefore a physical stimulus in the proximity of the eye might have an influence on startle induced blinks.

1.1   Related Work

There has now been extensive work focus on eye blink, but for many different purposes. Grauman et al. [9] proposed a method to automatically detect eye blinks and measure their duration, whose main target was to allow people with severe disabilities to access computers. However, the development of the system was done with a Sony EVI-D30 color video CCD camera, and two monitors were helpful when running it. Similarly, Morris et al. [10] proposed a method for eye tracking aimed at people with motor difficulties, which used spatio-temporal filtering and maps to locate the head and eyes, and a modified version of the Lucas-Kanade algorithm to track the succeeding frames. Other common purpose is the drowsiness detection, very useful to monitor drivers and prevent car accidents. In this sense, Park et al. [11] proposed an illumination compensation algorithm to measure eyelids movements by using a CCD camera whose lens had an infrared band-pass filter. More recently, an edge-based method was presented in [12] to classify open and closed eyes by using a NIR camera and dual NIR illuminations; whilst a system focus on color and texture segmentation was proposed in [13], and validated on CCD and CMOS cameras.

Other approaches for blink detection include: eye state recognition and closed-eye photo correction [14]; eye blink detection using split-interlaced images acquired with NTSC video cameras [15]; an analytical model of the eye blink including lid movement and ocular retraction by means of a slit lamp and an external fast camera with a zoom lens [16]; and a semiautomatic model-based contour detection and tracking algorithm for the analysis of high-speed video records [17]. To the best of our knowledge no attempts have been made so far, in the literature or commercially, to provide eye blink assessment using low-cost web cameras.

Our framework: The framework proposed herein takes advantage of using web cameras to carry out eye blink assessment in a work environment, instead of attending a clinic or using more invasive methods. The idea lies in analyzing the frames of a video, and classifying them in open or closed eyes in order to automatically detect the blinks. Moreover, the use of this framework does not imply the availability of any specific camera or expensive instrument, which often require medical supervision. Therefore, our framework makes three important contributions: (1) it is able to tackle eye blink detection in work environments; (2) it does not require medical supervision; and (3) it provides reliable results from consumer videos acquired with low-cost cameras.

The outline of the paper is as follows: Section 2 describes a methodology to automatically detect eye blinks from webcam videos, Section 3 presents the dataset and the experimental results, and Section 4 includes the conclusions and future lines of research.

2 Research Methodology

The proposed methodology is composed of four main steps (see Figure 1), which are applied to each frame of an input video. Therefore, for each input frame acquired with a web camera, its face and eyes are located and some low-level features are extracted, and finally the frame is classified into one of the two categories considered (open eyes, or closed eyes). These four steps will be subsequently presented in depth.

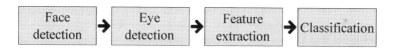

Fig. 1. Steps of the research methodology for eye blink detection, applied to each frame of a webcam video

2.1 Face Detection

The location of the face is the first step of the methodology. Proper face location will allow us to narrow the subsequent search areas. This way, this initial location reduces the computational cost of the next steps and makes them less error prone. Although the location of the face is a natural process for a human being, it becomes a challenging task in computer vision. In addition to the inherent complexity of defining a face for a computer, the variations in scale, orientation, pose, facial expression, lighting conditions, and background, increase the complexity of the problem.

In this particular case, since the domain is very stable in terms of location (the subject will be seated in front of the screen, and thus, in front of the web camera), it can be ensured that the face will be recorded in frontal position.

The certainty of having a frontal position of the patient's face allows to apply the Viola and Jones approach [18]. The Viola-Jones detector is a general object detection framework which provides competitive object detection rates in real-time. It can be trained to detect a variety of different objects; however, its initial motivation was to provide a solution for the face detection task. As consequence of this, an optimized classifier for the face detection was obtained. Particularly, an implementation of the Viola-Jones algorithm for the detection of frontal faces is available in the OpenCV library [19], which is the solution applied at this point. Figure 2 illustrates the face detection on two different images.

Fig. 2. Face detection over frames from two different subjects located at different work environments

2.2   Eye Detection

Once the face area has been properly delimited, the next step entails the location of the eye region. For the location of the eye region the Viola and Jones object detection framework was considered again by applying a cascade specifically trained for the detection of eyes. Since the eye blink is expected to be symmetric, only the detection of left eye is considered in order to simplify the further analysis. Note that this model is composed of weak classifiers, based on a decision stump.

Eyes are detected based on the hypothesis that they are darker than other parts of the face, so the detection framework search for small patches in the input image that are roughly as large as an eye and are darker than their neighborhoods. Then, anthropological characteristics of human eyes are applied in order to discard potential eye regions and maintain the true positive ones. Figure 3 depicts the eye detection over two different faces.

2.3   Feature Extraction

After detecting the left eye, the next step entails analyzing its low-level features to obtain a discriminant descriptor which allows to distinguish open eyes from closed eyes. Since the proportion of skin in a closed eye is greater than in a open eye, the TSL color space has been considered due to its effectiveness in skin analysis. On the other hand, the distribution of the pixel intensity values

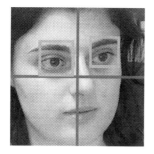

Fig. 3. Eye detection over the two different faces previously located

is also considered by means of image histograms and wavelets. All the methods involved in this step are following explained.

TSL Color Space. TSL is a perceptual color space which defines color as tint (T), saturation (S), and lightness (L). It was proposed by Terrillon et al. [20], and was primarily developed for the purpose of face detection. One of its main advantages is that it allows for an easy detection of different skin tones. The eyes can be perceived with the highest contrast, with respect to the skin area, in the L component. Thus, the input frame in RGB is transformed to the TSL color space, and only its lightness channel is considered in this stage. Figure 4 illustrates some examples of open and closed eyes from two subjects, and shows their visible differences in terms of the distribution of the pixel intensity values. Consequently, the two following methods for intensity analysis are applied to the L channel of the input frames.

Uniform Histograms. The histogram represents the distribution of the pixel intensity values of an image, i.e. it shows the number of pixels at each different intensity value found in the image. Analyzing the histograms computed from the input frames, it can be seen that they concentrated most of the information in the intermediate bins, which made their comparison difficult. For this reason, histograms with equiprobable bins, i.e., with non-equidistant bins, are computed instead of the traditional ones. The process to obtain uniform histograms is described as follows: given all the images, the limits of the histogram are defined so that each bin contains a maximum of $\frac{N}{N_{bins}}$ pixels, where N is the number of pixels in the corresponding frequency and N_{bins} the number of histogram bins.

Therefore, using n-bin histograms, the descriptor of an input frame is composed of n features.

Discrete Wavelet Transform. The discrete wavelet transform [21] generates a set of wavelets by scaling and translating a *mother wavelet*, which is a function

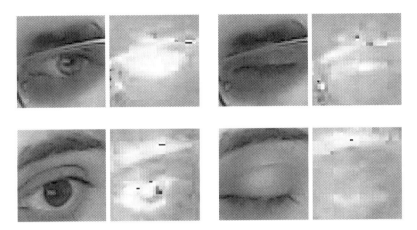

Fig. 4. Open and closed eyes of two different subjects, in the RGB color space and using the L component of the TSL color space

defined in the spatial and frequency domains, that can be represented in 2D as:

$$\phi^{a,b}(x,y) = \frac{1}{\sqrt{a_x a_y}} \phi \left(\frac{x - b_x}{a_x}, \frac{y - b_y}{a_y} \right) \tag{1}$$

where $a = (a_x, a_y)$ governs the scale, and $b = (b_x, b_y)$ the translation of the function. The values of a and b control the band-pass of the filter in order to generate high-pass (H) or low-pass (L) filters.

The wavelet decomposition of an image consists in applying wavelets horizontally and vertically in order to generate 4 subimages (LL, LH, HL and HH), which are then subsampled by a factor of 2. Some statistical measures are used in order to create the descriptor from an input image: mean, absolute average deviation and energy. These measures are respectively defined as:

$$\mu = \frac{1}{N} \sum_{i=1}^{N} p(i) \tag{2}$$

$$aad = \frac{1}{N} \sum_{i=1}^{N} |p(i) - \mu| \tag{3}$$

$$e = \frac{1}{N^2} \sum_{i=1}^{N} p(i)^2 \tag{4}$$

where $p(i)$ is the ith entry in the image, and N represents its number of pixels.

Thus, the descriptor of an input image is composed of 7 features: the μ and the aad of the input and LL images, and the e of the LH, HL and HH images.

Different mother wavelets can be considered, and the most popular ones are Haar and Daubechies [22]. Haar is the simplest nontrivial wavelet and Daubechies is one representative type of basis for wavelets. Daub$i$ represents the Daubechies orthonormal wavelet, where the number of vanishing moments is equal to half the coefficient i. Notice that the Haar wavelet is equivalent to Daub2.

2.4 Classification

Supervised learning entails learning a mapping between a set of input features and output labels, and applying this mapping to predict the outputs for new data [23]. The resulting classifier is then used to assign class labels to the new instances where the values of the features are known, but the value of the class label is unknown [24].

For this step, five popular machine learning algorithms were selected aiming to provide different approaches of the learning process [24]:

- **Naive Bayes** (NB). It is an statistical learning algorithm based on the Bayesian theorem that can predict class membership probabilities.
- **Random tree** (RT). It is a tree drawn at random from a set of possible trees, i.e. each tree in the set of trees has an equal chance of being sampled.
- **Random forest** (RF). It is a combination of tree predictors, in which each tree depends on the values of a random vector sampled independently and with the same distribution for all trees in the forest.
- **Support vector machine** (SVM). It is based on the statistical learning theory and revolves around a hyperplane that separates two classes.
- **Multilayer perceptron** (MLP). It is a feedforwad artificial neural network which consists of a set of units, joined together in a pattern of connections.

3   Experimental Results

In this section, the dataset used for validation and the results obtained with the proposed methodology are presented. Notice that the results are shown in terms of predictive accuracy for all the combinations of methods and classifiers. Additionally, a complete set of metrics is used for the best configuration of parameters in order to estimate the performance of the best approach in detail. These metrics are defined as follows [25]:

- The *accuracy* is the percentage of correctly classified instances:

$$Accuracy = \frac{TN + TP}{TP + FP + FN + TN} \tag{5}$$

- The *precision* is the proportion of the true positives against all the positives:

$$Precision = \frac{TP}{TP + FP} \tag{6}$$

- The *sensitivity* is the proportion of positives which are correctly classified:

$$Sensitivity = \frac{TP}{TP + FN} \tag{7}$$

– The *specificity* is the proportion of negatives which are correctly classified:

$$Specificity = \frac{TN}{TN + FP} \qquad (8)$$

3.1   Dataset

The acquisition of the input videos was carried out with a *c920 Logitech HD Pro Webcam*, a web camera used at consumer level. The configuration of the camera was set at Full HD and 30 fps, and the videos were stored in AVI format. Note that the frame rate achievable with a web camera is enough for eye blink detection. The recorded videos were analyzed frame by frame, and the spatial resolution per frame is 1280×720 pixels.

The dataset obtained for evaluation consists of videos acquired from seven different people at work environments, including spectacle wearers which may constitute an additional difficulty in the automatic analysis. The subjects were asked to move as they would normally when using a personal computer at work. Among the whole set of frames obtained from those videos, a total of 392 images were selected to validate the proposed methodology. Note that this selection was necessary to create a balanced dataset of images since the number of frames corresponding to open eyes is much greater than the number of closed eyes. In this manner, the 50% of the images correspond to each class.

3.2   Results

The objective here is to find which low-level features describe better the characteristics of the eye blink, and determine which classifier performs better for the problem at hand. Firstly, the different methods for feature extraction were applied to each frame selected from the video dataset. Next, the five classifiers were trained using a 10-fold cross-validation. Finally, the effectiveness of the methods were evaluated in terms of the predictive accuracy of the classifiers. Note that experimentation was performed on an Intel®Core™i5 CPU 760 @ 2.80GHz with RAM 16 GB.

The first experiment was performed using the uniform histograms, and the five different classifiers. Different number of bins were considered to create the histograms: 2, 4, 8, 16, 32, 64, 128, and 256. See Table 1, which shows the results in terms of accuracy for all the configurations.

Analyzing these results it can be seen that the greatest numbers of bins are more discriminative than the lowest and intermediates ones, since they provide the best results for all classifiers but RT, for which the best results are obtained when using the intermediate numbers of bins. In general terms, the results obtained are over 75% of correct classifications regardless of the machine learning algorithm. The best combination provides a classification rate over 95%, and is obtained when using the MLP and 128 bins. Note that this result is closely followed by the best combination achieved with the SVM, which provides an accuracy over 94% when using 256 bins. Thus, the MLP and the SVM are the

Table 1. Accuracy of the different classifiers when using the uniform histogram

Classifier	No. of bins							
	2	4	8	16	32	64	128	256
NB	50.00%	70.41%	71.17%	68.11%	69.90%	71.94%	71.43%	**75.00%**
RT	61.99%	81.38%	83.67%	**84.69%**	**84.69%**	83.42%	76.28%	79.34%
RF	62.5%	82.91%	85.71%	89.54%	89.80%	91.07%	91.33%	**92.09%**
SVM	63.27%	83.16%	89.80%	92.86%	92.09%	93.37%	94.39%	**94.64%**
MLP	50.51%	72.45%	83.16%	88.27%	88.27%	88.52%	**95.15%**	90.82%

most competitive classifiers, with no significant differences, when analyzing the low-level features by means of histograms. Notice that the results are quite stable since the highest numbers of bins (128 and 256) provide accuracies over 90% for three out the five classifiers considered (RF, SVM and MLP).

The second experiment was performed with the discrete wavelet transform and aimed at analyzing the behavior of each mother wavelet, in addition to the five classifiers. Note that the mother wavelets considered were Haar and Daubechies (Daub4, Daub6 and Daub8), and that the Haar wavelet is equivalent to Daub2. See Table 2, which depicts the results in terms of percentage accuracy for all the mother wavelets and classifiers.

Table 2. Accuracy of the different classifiers when using the discrete wavelet transform

Classifier	Mother wavelet			
	Haar	Daub4	Daub6	Daub 8
NB	71.68%	**73.47%**	71.94%	72.45%
RT	88.52%	90.82%	90.31%	**92.09%**
RF	92.35%	**92.60%**	92.35%	91.84%
SVM	94.64%	**95.15%**	93.62%	93.88%
MLP	85.20%	**86.74%**	85.20%	85.97%

Analyzing these results it can be seen that the Daub4 is the most appropriate mother wavelet for the problem at hand, since it provides the best results for all classifiers but RT, for which the best result is achieved when using Daub8. Regarding the behavior of the classifiers themselves, the SVM is the most competitive classifier with an accuracy over 95%. In this case, the performance of the MLP is worse since it does not reach the 90% in any case. Additionally, the results are quite stable since there is a set of mother wavelets (Daubechies) which provides accuracies over 90% for three out the five classifiers considered (RT, RF and SVM).

Table 3 depicts a confusion matrix in order to analyze how good is the detection of each class. Thus, the approaches which provide a classification rate of 95.15% have been considered in this case. As can be seen, there are no significant differences among the two classes.

Table 3. Confusion matrix for the best configuration

	Predicted class	
Real class	Open eye	Closed eye
Open eye	187	9
Closed eye	10	186

Finally, the performance of the best configuration is evaluated by means of a complete set of measures computed from the confusion matrix (see Table 4). As can be expected with such a well-balanced confusion matrix, all the performance measures are around 95% which demonstrates the validity of the method, and the reliability of the automatic detection of eye blinks.

Table 4. Performance measures for the best configuration

Accuracy	Precision	Sensitivity	Specificity
95.15%	94.92%	95.40%	94.90%

4   Conclusions

The portability and real-time response of consumer web cameras bring new challenges in the field of eye blink assessment. Millions of these cameras have already been sold or combined with laptops and PCs. Consequently, millions of people have already the ability to record themselves e.g. at work. In this sense, this paper presents a methodology for the automatic assessment of the eye blink which is able to analyze videos acquired with web cameras. For each frame of an input video, it detects the face of the person recorded and one of its eyes; and then it analyzes the low-level features in order to determine if the eye is open or closed, i.e. in order to automatically detect eye blinks.

The proposed methodology was tested with a video dataset, which covers diverse people and includes contact lens wearers. Results demonstrate the adequacy of the methodology, which provides maximum accuracy over 95% when using the TSL color space and regardless the method used to measure the intensity distribution of the pixels. Other performance measures considered are also around 95%, which states the reliability of our framework. This finding can be very useful in the field of eye blink assessment, since it will allow people to assess themselves at work, i.e. without attending a hospital and with no clinical supervision.

While the initial findings are promising, and demonstrate that the automatic detection of eye blink is feasible, further research is necessary. In this sense, it is planned to develop a complete system which will allow patients to assess eye blinks at work, and specialists could monitored them via remote viewing/displaying of the obtained data. This clinician-patient interaction will allow professionals of vision and psychological science to remotely assess eye blinks and, with many users, it is possible to compare blink assessments, accompanying treatments and convalescences in an objective and large-scaled way.

Acknowledgments. This research has been partially funded by the Secretaría de Estado de Investigación of the Spanish Government and FEDER funds of the European Union through the research project PI14/02161, and by the Consellería de Cultura, Educación e Ordenación Universitaria of the Xunta de Galicia through the research project GPC2013/065. Beatriz Remeseiro acknowledges the support of the European Grouping for Territorial Cooperation Galicia-Norte de Portugal under the IACOBUS Program.

References

1. Declerck, C.H., DeBrabander, B., Boone, C.: Spontaneous eye blink rates vary according to individual differences in generalized control perception. Perceptual and Motor Skills **102**, 721–735 (2006)
2. Wolkoff, P., Nojgaard, J., Troiano, P., Piccoli, B.: Eye complaints in the office environment: precorneal tear film integrity influenced by eye blinking efficiency. Occupational and Environmental Medicine **62**, 4–12 (2005)
3. Wolkoff, P., Skov, P., Franck, C., Petersen, L.N.: Eye irritation and environmental factors in the office environment-hypotheses, causes and a physiological model. Scandinavian Journal of Work, Environment & Health **29**, 411–430 (2003)
4. Wolkoff, P.: "Healthy" eye in office-like environments. Environment International **34**, 1204–1214 (2008)
5. Doughty, M.J.: Consideration of three types of spontaneous eyeblink activity in normal humans: during reading and video display terminal use, in primary gaze, and while in conversation. Optometry & Vision Science **78**, 712–725 (2001)
6. McIntire, L.K., McKinley, R.A., Goodyear, C., McIntire, J.P.: Detection of vigilance performance using eye blinks. Applied Ergonomics **45**, 345–362 (2014)
7. Koh, S., Maeda, N., Hori, Y., Inoue, T., Watanabe, H., Hirohara, Y., Mihashi, T., Fujikado, T., Tano, Y.: Effects of suppression of blinking on quality of vision in borderline cases of evaporative dry eye. Cornea **27**, 275–278 (2008)
8. Blumenthal, T.D., Cuthbert, B.N., Filion, D.L., Hackley, S., Lipp, O.V., vanBoxtel, A.: Committee report: Guidelines for human startle eyeblink electromyographic studies. Psychophysiology **42**, 1–15 (2005)
9. Grauman, K., Betke, M., Gips, J., Bradski, G.R.: Communication via eye blinks - detection and duration analysis in real time. In: IEEE Computer Society Conference on Computer Vision and Pattern Recognition (CVPR 2001), vol. 1, pp. I-1010–I-1017 (2001)
10. Morris, T., Blenkhorn, P., Zaidi, F.: Blink detection for real-time eye tracking. Journal of Network and Computer Applications **25**(2), 129–143 (2002)
11. Park, I., Ahn, J., Byun, H.: Efficient Measurement of Eye Blinking under Various Illumination Conditions for Drowsiness Detection Systems. In: 18th International Conference on Pattern Recognition (ICPR 2006), vol. 1, pp. 383–386 (2006)
12. Jo, J., Lee, S.J., Lee, Y.J., Jung, H.G., Park, K.R., Kim, J.: An Edge-based Method to Classify Open and Closed Eyes for Monitoring Driver's Drowsiness. In: International Conference on Electronics, Informations and Communications (ICEIC 2010), pp. 510–513 (2010)

13. Lenskiy, A.A., Lee, J.: Driver's eye blinking detection using novel color and texture segmentation algorithms. International Journal of Control, Automation and Systems **10**(2), 317–327 (2012)
14. Liu, Z., Ai, H.: Automatic eye state recognition and closed-eye photo correction. In: 19th International Conference on Pattern Recognition (ICPR 2008), pp. 1–4 (2008)
15. Abe, K., Ohi, S., Ohyama, M.: Automatic Method for Measuring Eye Blinks Using Split-Interlaced Images. In: Jacko, J.A. (ed.) HCI International 2009, Part I. LNCS, vol. 5610, pp. 3–11. Springer, Heidelberg (2009)
16. Perez, J., Espinosa, J., Domenech, B., Mas, D., Illueca, C.: Blinking kinematics description through non-invasive measurement. Journal of Modern Optics **58**, 1857–1863 (2011)
17. Bernard, F., Deuter, C.E., Gemmar, P., Schachinger, H.: Eyelid contour detection and tracking for startle research related eye-blink measurements from high-speed video records. Computer Methods and Programs in Biomedicine **112**(1), 22–37 (2013)
18. Viola, P., Jones, M.: Robust real-time object detection. International Journal of Computer Vision (2001)
19. Bradski, G.: The OpenCV Library. Dr. Dobb's Journal of Software Tools (2000)
20. Terrillon, J., David, M., Akamatsu, S.: Automatic Detection of Human Faces in Natural Scene Images by Use of a Skin Color Model and of Invariant Moments. In: Proceedings of Third IEEE International Conference on Automatic Face and Gesture Recognition, pp. 112–117 (1998)
21. Mallat, S.G.: A theory for multiresolution signal decomposition: the wavelet representation. IEEE Transactions on Pattern Analysis and Machine Intelligence **11**, 674–693 (1989)
22. Daubechies, I.: Ten Lectures on Wavelets. SIAM, CBMS series (1992)
23. Mitchell, T.M.: Machine Learning. McGraw-Hill (1997)
24. Kotsiantis, S.B.: Supervised Machine Learning: A Review of Classification Techniques. Informatica **31**, 249–268 (2007)
25. Sokolova, M., Lapalme, G.: A systematic analysis of performance measures for classification tasks. Information Processing & Management **45**(4), 427–437 (2009)

Insights on the Use of Convolutional Neural Networks for Document Image Binarization

J. Pastor-Pellicer[1]([✉]), S. España-Boquera[1], F. Zamora-Martínez[2],
M. Zeshan Afzal[3], and Maria Jose Castro-Bleda[1]

[1] Departamento de Sistemas Informáticos y Computación,
Universitat Politècnica de València, Valencia, Spain
jpastor@dsic.upv.es
[2] Departamento de Ciencias Físicas, Matemáticas y de la Computación,
Universidad CEU Cardenal Herrera, Alfara del Patriarca (Valencia), Valencia, Spain
[3] German Research Center for Artificial Intelligence (DFKI),
Kaiserslautern, Germany

Abstract. Convolutional Neural Networks have systematically shown good performance in Computer Vision and in Handwritten Text Recognition tasks. This paper proposes the use of these models for document image binarization. The main idea is to classify each pixel of the image into foreground and background from a sliding window centered at the pixel to be classified. An experimental analysis on the effect of sensitive parameters and some working topologies are proposed using two different corpora, of very different properties: DIBCO and Santgall.

1 Introduction

Document analysis combines image analysis and pattern recognition techniques to process and extract information from documents. Documents include images of paper documents after scanning or camera capture. This active area of research includes document layout analysis, text recognition, graphics recognition, text extraction from images, and many others. This work will focus on preprocessing, more precisely, in document image binarization. Neural Networks (NNs) have been widely applied for image processing due to their ability to learn very complex nonlinear input/output relationships from examples (see [6, 15, 23] for a review on image processing with neural networks). In particular, Convolutional Neural Networks (CNNs) are hierarchical NNs which have demonstrated a vast representational capacity, specially for image tasks. CNNs have been successfully applied to many problems such as visual object recognition or handwritten text recognition [11, 13, 25], but, to the best of our knowledge, they have never been applied to document image binarization. This paper presents our work in binarization by using CNNs and shows how this approach has been successfully accomplished on two different corpora, of very different properties, DIBCO and Santgall. The first corpus comes from several editions of well known binarization contests, which comprises a very heterogeneous corpus, and the Historical IAM

© Springer International Publishing Switzerland 2015
I. Rojas et al. (Eds.): IWANN 2015, Part II, LNCS 9095, pp. 115–126, 2015.
DOI: 10.1007/978-3-319-19222-2_10

Santgall Database, which are images captured from the Manuscript images of the *Codex Sangallensis 562*s.

The paper is structured as follows. Section 2 reviews the related work in binarization. The following section is devoted to the use of Convolutional Neural Networks for image binarization, and Section 4 describes the performed experiments on the two databases. In the last section, we briefly discuss our conclusions and we propose some research directions.

2   Document Image Binarization

Document image binarization is one of the first stages of document preprocessing, analysis and the further recognition pipeline. It consists in classifying the pixels of the image as background or foreground. At this stage, binarization also implies cleaning and enhancement, since detecting the foreground pixel will remove, for instance, ink spots, degradations or bleed through problems, and it will recover lost strokes or other foreground information. Undoubtedly, the performance of this stage will greatly affect the outcome of other document analysis steps as document layout and text line extraction and also the final recognized text [17].

Binarization methods can be split in three main groups: local thresholding, global thresholding and mixing strategies. Global thresholding stablishs an optimal threshold value for all pixels of the image, whereas local thresholding applies different thresholds for each pixel, depending on their local features. Hybrid methods try to compose both strategies. Global thresholding techniques need few computation and can work well in simple cases, though fail in complex images. The most popular global image threshold is Otsu's method [19] and other refinements [3,12]. Examples of local thresholding include Niblack [18] and Sauvola [24] filters. It is also interesting to combine several methods to improve the overall performance as in [1,26].

In addition, NNs have been applied for image processing due to their ability to learn very complex nonlinear input/output relationships from examples. In particular, Multilayer Perceptrons (MLPs) have been used for binarization and preprocessing in [2,4,10,15,16]. For instance, in [10], the MLP is fed with the pixels of a fixed size moving window as a regression model to enhance the image document. In [4], a MLP is trained as a classificator for binarization, also considering a moving window at the input. What they have in common is that they require a supervised set of data in order to train the classifier or the regression model. In this case, the training data quality has a direct impact in the overall system performance.

3   Convolutional Neural Networks

CNNs [14] are a kind of deep neural networks inspired biologically by the brain visual cortex and how the cells are arranged in layers. The cells in one layer are sensitive to a small region of the input, called receptive field, and these regions are tiled to cover the whole input space. Different layers are connected sequentially in order to extract useful information from the input.

In mathematical terms, this process is described as a convolution between the whole input space and a kernel matrix. The size of the kernel corresponds to the receptive field of the cell. Convolutions with different kernels are computed together to produce several output maps, which are different transformations of the input. This convolution allows the extraction of local features which are invariant to translation in the input space. The dimensionality of the input space constrains the dimensionality of the convolution, being a 1D convolution in case of time-series data, 2D convolution in case of grayscale image, 3D convolution for color image or grayscale video, and 4D convolution for color video. More complex convolutions are possible, for example for multiple video streams. Convolutions of subsequent layers may traverse the set of maps previously computed.

In order to reduce the dimensionality of hidden layers, pooling operations are applied to the output maps of every convolution layer. Different pooling strategies have been proposed in the literature [13, 25, 29]. Regarding the advantages of more complex pooling layers, max-pooling is widely used in CNN systems. A nonlinear activation function, such as the hyperbolic tangent, the rectified linear or the logistic function, is applied between layers of convolution and pooling.

Normally, a CNN sequentially combines two or three layers of convolution-activation-pooling, acting as a deep extractor for high level features. The output of this convolutional part is fed into a standard MLP, being the output of this MLP linear for regression, or logistic/softmax for classification tasks.

3.1 Convolutional Neural Networks for Image Binarization

The use of CNNs for image image processing tasks, as described in the literature, is usually based on applying a set of convolutions and max-pooling transforms to the input image. Convolutions are able to learn useful features and several kernels are used to obtain a set of maps. Relating the input, it can be composed by one or by several maps depending on the use of grayscale or color images: when using grayscale images the input is only one map with the brightness of each pixel. When processing RGB color images, the input is formed by three maps. In this work, grayscale images are used, performing a grayscale conversion when necessary. In this way, the input of the model is only one map.

Since the binarization can be seen as a classification performed at the pixel level, a suitable set of features should be extracted for each pixel in order to feed this classifier. This work relies on the use of automatic machine learning techniques to allow the system to select the best features for this classification task. Indeed, previous work based on MLPs [4, 10] has already shown that it is possible to use supervised learning techniques successfully.

A priori, there are several advantages when using CNNs instead of MLPs: in the first place, the kernels operate on a smaller scale and each one shares its weights at different positions on the input window, which reduces the number of parameters decreasing the possibilities of overfitting and improving generalization. As with MLPs, the features extracted from CNN are computed using a

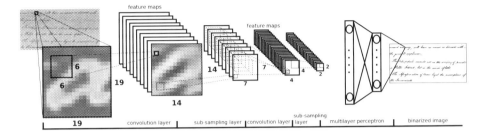

Fig. 1. CNN for document image binarization

window centered at the pixel to be classified so that nearby pixels will have a large amount of features in common.

Like in other local binarization approaches, the optimal window size must be provided. We believe that a contribution of this work is not only to validate empirically the suitability of CNNs for document image binarization but also to provide some insight relating the most adequate optimal window sizes and CNN topologies in order to obtain the best trade-off between binarization quality and computational cost.

Relating the most appropriate window sizes, previous work [10,15] has experimentally shown that small windows do not provide enough information whereas large windows add too much variability and leads to problems due to the curse of dimensionality. We can take profit of prior experience with MLPs trying window sizes similar to earlier MLP experiments. The results, compared with the MLPs counterpart, indicates that it is a good point of departure.

Besides the window size, the configuration of the CNNs must be provided. When using MLPs, the usual topologies comprise one or more hidden layers. Now, we have to take into account a functional set of convolutional and pooling layers. Given a quite reasonable range of input sizes, it seems sensible to limit the number of layers composed by convolutional and max-pooling layers. By limiting the scanned topologies to up to two sets of convolutional followed by max-pooling layers, and up to two hidden layers, the range of configuration parameters suitable to describe the CNN topologies are:

- The window input size.
- Size of the kernel of the first convolution.
- Number of kernels in the first convolution.
- Size of the first sub-sampling layer.
- Size of the kernel of the second convolution.
- Number of kernels in the second convolution.
- Size of the second sub-sampling layer.
- Topology of the final hidden layers.

As shown in Figure 1, the final setup of the CNN is formed by an input window of the image, two sets of convolution and sub-sampling layers, and a MLP with two hidden layers followed by a single output neuron. The model estimates

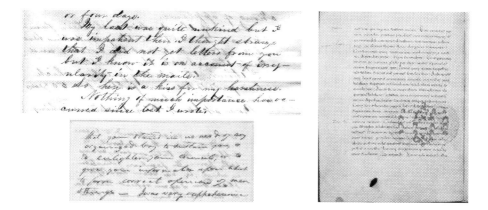

Fig. 2. Two documents from DIBCO (top-left, bottom-left), and one page from Sant-gall (right)

the value of the image after binarizing (0 background and 1 foreground). During training, the output is the groundtruth value of the pixel.

4   Experimentation

4.1   Datasets: DIBCO and Santgall

In order to test the perfomance of our binarization methods, we have used the benchmarking dataset from the well known Document Binarization Contest (DIBCO) [9,22] and the Handwritten Document Binarization Contest (H-DIBCO) [20,21], hold in the context of the International Conference on Frontiers in Handwriting Recognition (ICFHR) and International Conference on Document Analysis and Recognition (ICDAR) since 2009. The document images present a great variability of styles (see Figure 2 for an example). We have used the data from DIBCO 2009, H-DIBCO 2010, DIBCO 2011, H-DIBCO 2012, for training and development sets, whereas DIBCO 2013 as been used as the final test set.

The proposed approach has been also applied on the noisy images from the Historical IAM Santgall Database. This corpus is composed of images from *Codex Sangallensis 562* [7,8]. These images have homogeneous text and data along the different pages (see Figure 2).

The chosen training and evaluation partition are the standard ones in the corpus description of the IAM Santgall Database:

- Training: 20 pages corresponding to manuscript *Codex Sangallensis 562*, pages 3 to 20.
- Development: 10 pages of manuscript *Codex Sangallensis 562*, pages 24, 28, 32, 36, 49, 44, 48, 55, 59 and 63.
- Test: 30 pages of manuscript *Codex Sangallensis 562*, pages 23, 25-27, 29-31, 33-35, 37-39, 41-43, 45-47, 49, 50, 54, 56-58, 60-62, 64 and 65.

4.2   Metrics and Evaluation

The F-Measure (FM) is the usual metric to evaluate document image binarization performance. FM is the harmonic mean of *precision* and *recall*, and is defined as FM $= 2 \cdot precision \cdot recall/(precision + recall)$, where precision (also called positive predictive value) is the fraction of well classified foreground predicted pixels, and recall (also known as sensitivity) is the fraction of ground-truth foreground correctly.

We have also computed two figures of merit which do not take in to account the relevancy of the pixel: the Mean Squared Error (MSE), and the signal/noise ratio Peak Signal to Noise Ratio (PSNR), based on the MSE and computed as $10 \cdot \log_{10}(\frac{1}{\sqrt{\text{MSE}}})$.

4.3   Topology and Parameters Setup

The architecture of the CNN was described in Section 3.1. Our main interest in the experimental evaluation was to analyze the effect of the number of features extracted from the CNN before using the MLP classifier. This value depends on the number of kernels and on the size of max-pooling layers.

Since each kernel learns a different characteristic from the input, this number is a relevant parameter, although increasing this value too much has a direct impact on the computational cost. On the other side, max-pooling layers can be used to reduce the dimensionality of the CNN.

In order to limit the number of configurations of this experimentation, and from our previous experience with MLPs, as well as from some preliminary experiments with CNNs, some parameters have been fixed:

- The window input size to compute the value of each pixel has been fixed to a 9 neighbors squared window centered at each pixel leading to a 19×19 window.
- Size of the kernel of the first convolution is set to 6×6 leading to maps of 14×14 real values.
- Size of the first sub-sampling layer is fixed to 2×2 reducing the maps to 7×7, one map for each kernel of the first convolution.
- Size of the kernel of the second convolution 4×4, leading to maps of 4×4 before applying the last max-pooling sub-sampling layer.

The number of kernels in the first convolution has been varied from 10 up to 160, whereas the number of kernels of the second convolution is twice as much the number of the first convolution. Two different settings have been tested for the second sub-sampling layers: 2×2 and 4×4. In the last case, the number of extracted features is drastically reduced, 16 values on the last convolution are reduced to 1 for each different map.

The set of features computed by the last layer of the CNN is fed to a fully connected MLP, composed by two hidden layers of sizes 32 and 16, respectively. Several experiments have shown that this topology works well.

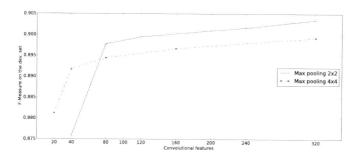

Fig. 3. FM on the DIBCO development set given the number of features generated by different CNNs

The CNN has been trained with the APRIL-ANN toolkit [27] using the backpropagation algorithm with the ADADELTA [28] adaptive per-dimension learning rate method combined with regularization methods such as weight decay penalty and gaussian noise on the CNN input.

Figure 3 shows the FM score for the DIBCO development set (H-DIBCO 2012) for different number of features computed by different CNNs. Since each map of the second convolution is 4×4, the number of features when using 2×2 max-pooling in the second sub-sampling layer is 4 times the number of kernels of the second convolution layer. When using 4×4 max-pooling, the number of features is the number of maps. Thus, given a set of F features, the number of kernels in each convolution layer is $F/8$ and $F/4$, respectively, for 2×2 max-pooling configurations. In the case of 4×4 max-pooling layers, these values are $F/2$ and F, respectively. For instance, a CNN with 20 kernels in the second convolution layer generates 80 features when using 2×2 max-pooling layer or 20 features when using the 4×4 max-pooling configuration.

As expected, the more extracted features the better, although the performance slows down around 80 features. As shown in Figure 3, comparing the two CNN configurations, the 4×4 max-pooling layer has better performance when using less than 80 features, since more kernels are applied. When more than 80 features are used, the 2×2 max-pooling networks perform better probably due to a less drastic sub-sampling.

Figure 4 visualizes the kernel weights learned for different nets. Brighter values shows higher values. The kernels are directly applied to the image in order to extract features like edges or corners. The figure shows maps for 5, 10, 20 and 60 kernels. Note that, while increasing the number of kernels, the CNN extracts more features and shows better behavior.

For the sake of a better comprehension of the CNN, Figure 5 shows the activation values for each layer. The first convolution extracts several maps of the input image and the max-pooling reduces the dimensionality. The second convolution combines the extracted features and, finally, two hidden layers compute the output value. As in previous figure, brighter values means higher activations.

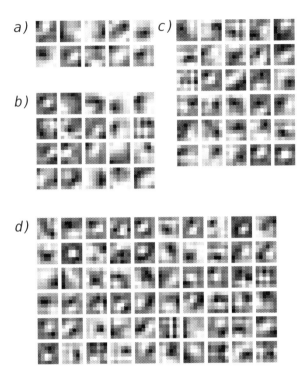

Fig. 4. Kernels learned from the first convolution of 6×6 for: (a) 10 maps, (b) 20 maps, (c) 30 maps, (d) 60 maps

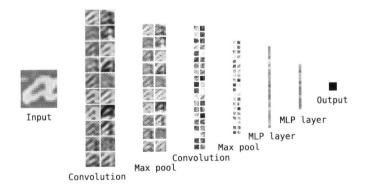

Fig. 5. Example illustrating the neuron activation of a CNN for an input sample

Table 1. Comparison of binarization performance for CNNs and other methods for the test sets of DIBOC and Santgall databases

Method	DIBCO 2013			Santgall		
	FM	MSE	PSNR	FM	MSE	PSNR
Otsu	83.94	0.056	16.94	80.71	0.020	17.09
Sauvola	85.02	0.047	16.63	88.68	0.010	19.86
MLP	82.31	0.029	16.89	93.94	0.005	22.80
MLP+Features	85.82	0.021	18.18	94.75	0.005	23.39
CNN	87.74	0.020	18.91	97.02	0.002	27.22

4.4 Overall Performance

After analyzing several nets and their performance on the development set as shown in the previous section, we have selected for the final evaluation a CNN showing a good compromise between FM and computational cost. The chosen CNN has 10 kernels for the first convolution layer and 20 kernels in the second one. Both max-pooling layers are of size 2×2, leading to 80 extracted features.

Table 1 shows the evaluation of the CNNs over the test sets of DIBCO and Santgall databases, compared with other binarization algorithms. Besides Otsu and Sauvola, two methods based on MLPs are considered. The fist MLP evaluated method uses the same 19×19 input windows and two hidden layers of 256 and 64, respectively. The MLP+Features is the same but adding to the input to the MLP the following features: a median filter and a vertical and horizontal histograms computed over the input window. As can be observed, the CNN outperforms the baselines and the other connectionist methods for both datasets. Some examples together with the result of these binarization techniques are illustrated in Figure 6.

Comparing the performance of the CNN approach on the DIBCO dataset with respect to the results reported in the DIBCO 2013 contest [22], the best FM measure on this competition was 92.70. Although the obtained result (87.74) is far from this position, it still remains competitive with respect to other competitors.

5   Summary and Conclusions

This work describes a practical use of CNNs applied to document image binarization tasks. This approach has some resemblances with previous works based on MLPs but, now, a CNN is used to classify each pixel value from a sliding window centered at the pixel to be binarized.

Experimental results on two different datasets show that CNNs systematically outperform MLPs for this task. The improvement obtained for Santgall database seems more prominent than the results obtained on DIBCO test dataset. This difference may be due to the fact that Santgall is a more homogeneous collection than DIBCO. Indeed, it seems that methods based on supervised learning techniques excels in this kind of documents where the font and size of

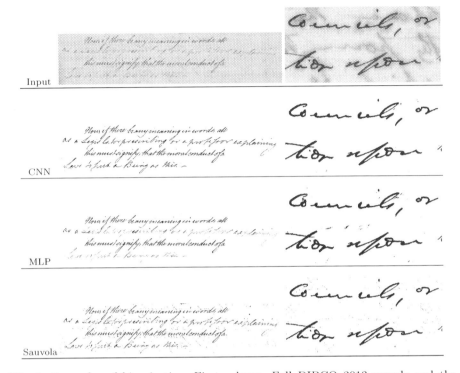

Fig. 6. Examples of binarization. First column: Full DIBCO 2013 sample and the obtained binarizations. Second column: DIBCO 2013 image and its binarizations.

the text as well as the kind of noise is more homogeneous along the full collection. The practical interest of the proposed technique is supported by the existence of collections composed by thousands of similar documents.

Due to the huge number of topologies and setting parameters of these models, the study of the influence of the most sensitive parameters and the proposal of working and practical topologies provides a useful insight on the use of CNNs for this binarization task.

It is mandatory to obtain a good compromise between binarization quality and computational cost. We have also shown that small kernels are enough to obtain competitive results and we have empirically validated the effect of the number of such kernels on the overall binarization performance. In fact, increasing this number over a given threshold does not improve significantly the FM.

As a future work, we consider the use of several input layers in order to include other features and the possibility of using multi-column deep CNNs [5] to allow the simultaneous use of different resolutions of the image.

Acknowledgments. This work has been partially supported by the Spanish Government TIN2010-18958.

References

1. Badekas, E., Papamarkos, N.: Optimal combination of document binarization techniques using a self-organizing map neural network. Engineering Applications of Artificial Intelligence **20**(1), 11–24 (2007)
2. Banerjee, J., Namboodiri, A.M., Jawahar, C.: Contextual restoration of severely degraded document images. In: IEEE Conference on Computer Vision and Pattern Recognition, 2009. CVPR 2009, pp. 517–524. IEEE (2009)
3. Brink, A.: Thresholding of digital images using two-dimensional entropies. Pattern recognition **25**(8), 803–808 (1992)
4. Chi, Z., Wong, K.: A two-stage binarization approach for document images. In: Proceedings of 2001 International Symposium on Intelligent Multimedia, Video and Speech Processing, pp. 275–278 (2001)
5. Ciresan, D.C., Meier, U., Schmidhuber, J.: Multi-column deep neural networks for image classification. CoRR abs/1202.2745 (2012)
6. Egmont-Petersen, M., de Ridder, D., Handels, H.: Image processing with neural networks - a review. Pattern Recognition **35**(10), 2279–2301 (2002)
7. Fischer, A., Frinken, V., Fornés, A., Bunke, H.: Transcription alignment of latin manuscripts using hidden markov models. In: Proceedings of the 2011 Workshop on Historical Document Imaging and Processing, pp. 29–36. ACM (2011)
8. Fischer, A., Indermühle, E., Bunke, H., Viehhauser, G., Stolz, M.: Ground truth creation for handwriting recognition in historical documents. In: Proceedings of the 9th IAPR International Workshop on Document Analysis Systems, pp. 3–10. ACM (2010)
9. Gatos, B., Ntirogiannis, K., Pratikakis, I.: Icdar 2009 document image binarization contest (dibco 2009). ICDAR **9**, 1375–1382 (2009)
10. Hidalgo, J.L., España, S., Castro, M.J., Pérez, J.A.: Enhancement and cleaning of handwritten data by using neural networks. In: Marques, J.S., Pérez de la Blanca, N., Pina, P. (eds.) IbPRIA 2005. LNCS, vol. 3522, pp. 376–383. Springer, Heidelberg (2005)
11. Kang, L., Kumar, J., Ye, P., Li, Y., Doermann, D.: Convolutional neural networks for document image classification. In: Intern. Conf. on Pattern Recognition, pp. 3168–3172. IEEE (2014)
12. Kittler, J., Illingworth, J.: On threshold selection using clustering criteria. IEEE Transactions on Systems, Man and Cybernetics **5**, 652–655 (1985)
13. Krizhevsky, A., Sutskever, I., Hinton, G.E.: Imagenet classification with deep convolutional neural networks. In: Pereira, F., Burges, C., Bottou, L., Weinberger, K. (eds.) Advances in Neural Information Processing Systems 25, pp. 1097–1105. Curran Associates, Inc. (2012). http://papers.nips.cc/paper/4824-imagenet-classification-with-deep-convolutional-neural-networks.pdf
14. Lecun, Y., Bottou, L., Bengio, Y., Haffner, P.: Gradient-based learning applied to document recognition. Proceedings of the IEEE **86**(11), 2278–2324 (1998)
15. Marinai, S., Gori, M., Soda, G.: Artificial neural networks for document analysis and recognition. IEEE Transactions on Pattern Analysis and Machine Intelligence **27**(1), 23–35 (2005)
16. Mehrara, H., Zahedinejad, M., Pourmohammad, A.: Novel edge detection using bp neural network based on threshold binarization. In: Second International Conference on Computer and Electrical Engineering, 2009. ICCEE 2009, vol. 2, pp. 408–412. IEEE (2009)

17. Nagy, G.: Twenty years of document image analysis in pami. IEEE Transactions on Pattern Analysis and Machine Intelligence **22**(1), 38–62 (2000)
18. Niblack, W.: An introduction to digital image processing. Strandberg Publishing Company (1985)
19. Otsu, N.: A threshold selection method from gray-level histograms. Automatica **11**(285–296), 23–27 (1975)
20. Pratikakis, I., Gatos, B., Ntirogiannis, K.: H-dibco 2010-handwritten document image binarization competition. In: 2010 International Conference on Frontiers in Handwriting Recognition (ICFHR), pp. 727–732. IEEE (2010)
21. Pratikakis, I., Gatos, B., Ntirogiannis, K.: Icfhr 2012 competition on handwritten document image binarization (h-dibco 2012). ICFHR **12**, 18–20 (2012)
22. Pratikakis, I., Gatos, B., Ntirogiannis, K.: Icdar 2013 document image binarization contest (dibco 2013). In: 2013 12th International Conference on Document Analysis and Recognition (ICDAR), pp. 1471–1476. IEEE (2013)
23. Rehman, A., Saba, T.: Neural networks for document image preprocessing: state of the art. Artificial Intelligence Review **42**(2), 253–273 (2014)
24. Sauvola, J., Pietikäinen, M.: Adaptive document image binarization. Pattern Recognition **33**(2), 225–236 (2000)
25. Sermanet, P., Chintala, S., LeCun, Y.: Convolutional neural networks applied to house numbers digit classification. In: 2012 21st International Conference on Pattern Recognition (ICPR), pp. 3288–3291 (2012)
26. Su, B., Lu, S., Tan, C.L.: Combination of document image binarization techniques. In: 2011 International Conference on Document Analysis and Recognition (ICDAR), pp. 22–26. IEEE (2011)
27. Zamora-Martínez, F., España-Boquera, S., Gorbe-Moya, J., Pastor-Pellicer, J., Palacios-Corella, A.: APRIL-ANN toolkit, A Pattern Recognizer In Lua with Artificial Neural Networks (2013). https://github.com/pakozm/april-ann
28. Zeiler, M.D.: ADADELTA: an adaptive learning rate method. CoRR abs/1212.5701 (2012). http://arxiv.org/abs/1212.5701
29. Zeiler, M.D., Fergus, R.: Stochastic pooling for regularization of deep convolutional neural networks. CoRR abs/1301.3557 (2013). http://arxiv.org/abs/1301.3557

A Genetic Algorithms-Based LSSVM Classifier for Fixed-Size Set of Support Vectors

Danilo Avilar Silva and Ajalmar R. Rocha Neto[(✉)]

Department of Teleinformatics, Federal Institute of Ceará, Fortaleza, Ceará, Brazil
{daniloavilar,ajalmar}@gmail.com

Abstract. Least Square Support Vector Machines (LSSVMs) are an alternative to SVMs because the training process of LSSVM classifiers only requires to solve a linear equation system instead of solving a quadratic programming optimization problem. Nevertheless, the absence of sparseness in the solution (i.e. the Lagrange multipliers vector) obtained is a significant drawback which must be overcome. This work presents a new approach to building Sparse Least Square Support Vector Machines with fixed-size of support vectors for classification tasks. Our proposal named FSGAS-LSSVM relies on a binary-encoding single-objective genetic algorithms, in which the standard reproduction and mutation operators must be modified. The main idea is to leave a few support vectors out of the solution without affecting the classifier's accuracy and even improving it. In our proposal, GAs are used to select a suitable fixed-size set of support vectors by removing non-relevant patterns or those ones, which can be corrupted with noise and thus prevent classifiers to achieve higher accuracies.

Keywords: LSSVM · Sparse LSSVM · Pruning methods · Genetic algorithms

1 Introduction

Different real-world research areas such as bioinformatics, medicine, economics, chemistry have problems which can be handled by Evolutionary Computation (EC) tools such as Genetic Algorithms (GAs). GAs as optimization tools aim at generating useful solutions to search problems. Due to the underlying features of GAs, some optimization problems can be solved without supposing linearity, differentiability, continuity or convexity of the objective function. Unfortunately, these desired properties are not found in several classical mathematical methods found in literature.

GAs have been used with several types of machine learning methods and neural networks in classification tasks only to tune their parameters (see an example [11]). Similarly, other meta-heuristic methods have also been used to tune classifier parameters(see [1]). As example of machine learning methods, we highlight large margin classifiers such as Support Vector Machines (SVMs) [17] and Least Square Support Vector Machines (LSSVMs) [15].

© Springer International Publishing Switzerland 2015
I. Rojas et al. (Eds.): IWANN 2015, Part II, LNCS 9095, pp. 127–141, 2015.
DOI: 10.1007/978-3-319-19222-2_11

A theoretical advantage of large margin classifiers as those aforementioned concerns the empirical and structural risk minimization which balances the complexity of the model against its success at fitting the training data, along with the production of sparse solutions [13]. By sparseness we mean that the decision surface, i.e, a hyperplane in the feature space built by the induced classifier can be written in terms of a relatively small number of input examples, the so-called support vectors (SVs), which usually lie close to the decision borders of two classes. In practice, however, it is observed that the application of different training approaches over identical training sets yield distinct sparseness [5], i.e., produce solutions (support vector set) with a greater number of elements (SVs) than are strictly necessary.

As it is well known by the research community, the LSSVM is an alternative to the standard SVM formulation [17]. A solution for the LSSVM is achieved by solving the linear systems resulting from the optimality conditions that appear from minimizing the primal optimization problem in a least square sense. Therefore, the solution follows directly from solving a linear equation system, instead of solving a QP optimization problem. On the one hand, it is in general less computationally intensive to solve a linear system than a QP problem, as well as the way of solving the problem is easier so that we only need to be able to compute the inverse of a matrix. On the other hand, the resulting solution is too far from sparse, in the sense that it is common to have all training samples being used as SVs.

To handle the lack of sparseness in large margin classifiers, several *reduced set* (RS) and pruning methods have been proposed, respectively. These methods comprise a bunch of techniques which aims at simplifying the internal structure of SVM and LSSVM classifiers, while keeping the decision boundaries as similar as possible to the original ones. The basic idea behind these kind of methods was firstly introduced by Burges [3].

Reduced set and pruning methods are very useful to reduce the computational complexity of the original models, because they speed up the decision process by reducing the number of SVs. They are particularly important to handle large datasets, when a great number of data samples may be selected as support vectors, either by pruning less important SVs [7,8] or by constructing a smaller set of training examples [4,9], often with minimal impact on performance. A smaller number of SVs is also useful to have better understanding of the internal structure of SVM and LSSVM classifiers by means of more succinct prototype-based rules [2].

In order to combine the advantages of LSSVM and GAs, this work aims to put both of them to work together in order to reduce the number of support vectors into a fixed-size set and also to achieve equivalent or even superior performances than standard LSSVM classifiers. Our genetic algorithms-based proposal searches for sparse classifiers based on a single optimization genetic algorithms in which the optimization process is guided by using the accuracy (of classifier over the training or validation dataset) as the fitness function.

We must point out that a related method called Multi-Objective Genetic Algorithm for Sparse Least Square Support Vector Machines (MOGAS-LSSVM) was recently proposed [12]. Even though the resulting classifiers in MOGAS-LSSVM are based on Genetic Algorithms, such method does not achieve sparse classifiers with a fixed-size of support vectors as our novel proposal presented in this work. Thus, since they are not so similar, we decide not to compare our novel proposal presented in this work with our previous one (MOGAS-LSSVM).

The remaining part of this paper is organized as follows. In Section 2 we review the fundamentals of the LSSVM classifiers. In Section 3 we briefly present some methods for obtaining sparse LSSVM classifiers, such as the Pruning LSSVM and IP-LSSVM. In Section 4, we introduce Genetic Algorithms which are necessary to understanding of our proposal presented in Section 5 and then, in Section 6, we present our simulations. After all, the paper is concluded in Section 7.

2   LSSVM Classifiers

The primal problem formulation for LSSVM classifiers [15] is given by

$$\min_{\mathbf{w},\xi_i,b} \left\{ \frac{1}{2}\mathbf{w}^T\mathbf{w} + \gamma\frac{1}{2}\sum_{i=1}^{L}\xi_i^2 \right\}, \tag{1}$$

$$\text{subject to} \quad y_i[(\mathbf{w}^T\mathbf{x}_i)+b] = 1 - \xi_i, i = 1,\ldots,L$$

where γ is a positive cost parameter, $\{\xi_i\}_{i=1}^{L}$ are the slack variables and b is the bias. Rearranging the Eq. (1), the following Lagrangian function for LSSVM classifiers is achieved

$$L(\mathbf{w},b,\boldsymbol{\xi},\boldsymbol{\alpha}) = \frac{1}{2}\mathbf{w}^T\mathbf{w} + \gamma\frac{1}{2}\sum_{i=1}^{L}\xi_i^2 - \sum_{i=1}^{L}\alpha_i(y_i(\mathbf{x}_i^T\mathbf{w}+b)-1+\xi_i), \tag{2}$$

where $\{\alpha_i\}_{i=1}^{L}$ are the Lagrange multipliers . The conditions for optimality are given by the partial derivatives

$$\frac{\partial L(\mathbf{w},b,\boldsymbol{\xi},\boldsymbol{\alpha})}{\partial \mathbf{w}} = \mathbf{0} \Rightarrow \mathbf{w} = \sum_{i-1}^{L}\alpha_i y_i \mathbf{x}_i,$$

$$\frac{\partial L(\mathbf{w},b,\boldsymbol{\xi},\boldsymbol{\alpha})}{\partial b} = 0 \Rightarrow \sum_{i=1}^{L}\alpha_i y_i = 0, \tag{3}$$

$$\frac{\partial L(\mathbf{w},b,\boldsymbol{\xi},\boldsymbol{\alpha})}{\partial \alpha_i} = 0 \Rightarrow y_i(\mathbf{x}_i^T\mathbf{w}+b)-1+\xi_i = 0,$$

$$\frac{\partial L(\mathbf{w},b,\boldsymbol{\xi},\boldsymbol{\alpha})}{\partial \xi_i} = 0 \Rightarrow \alpha_i = \gamma\xi_i.$$

Thus, based on Eq. (3), one can formulate a linear system in order to represent this problem as

$$\mathbf{Dz} = \mathbf{1}, \tag{4}$$

where

$$D = \left[\begin{array}{c|c} 0 & \mathbf{y}^T \\ \hline \mathbf{y} & \Omega + \gamma^{-1}\mathbf{I} \end{array} \right] \quad , \quad \mathbf{z} = \left[\begin{array}{c} b \\ \alpha \end{array} \right] \quad , \quad \mathbf{1} = \left[\begin{array}{c} 0 \\ \mathbf{1} \end{array} \right], \tag{5}$$

$\Omega \in \mathbb{R}^{L \times L}$ is a matrix whose entries are given by $\Omega_{i,j} = y_i y_j \mathbf{x}_i^T \mathbf{x}_j$, $i,j = 1, \ldots, L$. In addition, $\mathbf{y} = [y_1 \cdots y_L]^T$ and the symbol $\mathbf{1}$ denotes a vector of ones with dimension L. The solution of this linear system can be computed by direct inversion of matrix $\mathbf{D}$ as follows

$$\mathbf{z} = \mathbf{D}^{-1}\mathbf{1} = \left[\begin{array}{c|c} 0 & \mathbf{y}^T \\ \hline \mathbf{y} & \Omega + \gamma^{-1}\mathbf{I} \end{array} \right]^{-1} \left[\begin{array}{c} 0 \\ \mathbf{1} \end{array} \right]. \tag{6}$$

Once we have the values of the Lagrange multipliers and the bias, the output can be calculated based on the classification function described as

$$f(\mathbf{x}) = \text{sign} \left(\sum_{i=1}^{l} \alpha_i y_i \mathbf{x}^T \mathbf{x}_i + b \right). \tag{7}$$

It is easy to see in Eq. (7) that whenever a Lagrange multiplier α_i is zero, we do not have to keep the associated input vector $\mathbf{x}_i$ on hand for future usage. It is also straightforward the usage of the *kernel trick*, which is applied to generate non-linear versions of the standard linear SVM classifier. This procedure works by replacing the dot product $\mathbf{x}^T \mathbf{x}_i$ with the kernel function $k(\mathbf{x}, \mathbf{x}_i)$.

3   Sparse Classifiers

In this section, we present two of the most important methods used to obtain sparse classifiers: Pruning LSSVM and IP-LSSVM. Both of them reduce the number of support vectors based on the values of Lagrange multipliers. These methods are compared with our proposal in the sections results and discussions.

3.1   Pruning LSSVM

Pruning LSSVM (P-LSSVM) was proposed by Suykens in 2000 [14]. In this method, support vectors and then their respective input vectors are eliminated according to the absolute value of their Lagrange multipliers. The process is carried out recursively, with gradual vector elimination at each iteration, until a stop criterion is reached, which is usually associated with decrease in performance on a validation set. Vectors are eliminated by setting the corresponding Lagrange multipliers to zero, without any change in matrix dimensions. The resolution of the current linear system, for each new reduced set, is needed at each iteration, and the reduced set is selected from the best iteration. This is a multistep method, since the linear system needs to be solved many times until the convergence criterion is reached. We point out that the P-LSSVM was modified in order to reach only the desired reduction. Such modification was accomplished in order to have a fair assessment with our proposal.

3.2   IP-LSSVM

IP-LSSVM [4] uses a criterion in which patterns close to the separating surface and far from the support hyperplanes are very likely to become support vectors. In fact, since patterns with $\alpha_i \gg 0$ are likely to become support vectors, the margin limits are located closer to the separating surface than the support hyperplanes. As we can see, that idea applied in IP-LSSVM training is based on SVM classifiers working. According to these arguments, the new relevance criteria proposed in IP-LSSVM work can be described as:

- $\mathbf{x}_i$ with $\alpha_i \gg 0$ is a support vector. In this case, $\mathbf{x}_i$ is placed on the border between the two classes or on the opposite class area, corresponding to vectors associated with non-zero $\alpha_i \gg 0$ when compared to SVM classifiers solved by Quadratic Programming (QP) algorithms.
- $\mathbf{x}_i$ with $\alpha_i \geq 0$ is removed. In this situation, $\mathbf{x}_i$ is correctly classified, close to the support hyperplane, corresponding to vectors associated with $\alpha_i = 0$ in SVM classifier solved by QP Algorithms.
- $\mathbf{x}_i$ with $\alpha_i < 0$ or $\alpha_i \ll 0$ is eliminated. In this case, $\mathbf{x}_i$ is correctly classified, far from the decision surface, corresponding to vectors associated with $\alpha_i = 0$ in QP SVM classifiers.

The proposed criteria is applied to IP-LSSVM in order to eliminate non-relevant columns of the original matrix $\mathbf{D}$ and to build a non-squared reduced matrix $\mathbf{D}_2$ to be used a posteriori. The eliminated columns correspond to the least relevant vectors for the classification problem, selected according to their Lagrange multiplier values. The rows of $\mathbf{D}$ are not removed, because its elimination would lead to a loss of labeling information and in performance [16].

The first step is accomplished by using the inverse function to solve the system of linear equations represented by Eq. (4). The solution of this system is $\mathbf{z} = \mathbf{D}^{-1}\mathbf{l}$. The first element of $\mathbf{z}$ is discarded, due to it corresponds to the bias value and only α values are aimed in this vector elimination phase. In second step, a system $\mathbf{D}_2\mathbf{z}_2 = \mathbf{l}_2$ is solved using the pseudo-inverse function, whose solution $\mathbf{z}_2$. The training process of the proposed sparse classifier can be described as:

1. The system of linear equations presented in Eq. (4) can be solved, with all training vectors, using $\mathbf{z} = \mathbf{D}^{-1}\mathbf{l}$, since $\mathbf{D}$ is a square matrix.
2. The parameter $\tau \in [0, 1]$ defines the fraction of training vectors that will be considered support vectors.
3. The training vectors are ordered by their α values.
4. The fraction $1 - \tau$ of training data that corresponds to the smaller values is selected.
5. The non-squared matrix $\mathbf{D}_2$ is generated by removing from $\mathbf{D}$ the columns associated to the selected elements of a.
6. The new system of linear equations, represented by $\mathbf{D}_2\mathbf{z}_2 = \mathbf{l}_2$, is solved as $\mathbf{z}_2 = \mathbf{D}_2^*\mathbf{l}_2$ where

$$\mathbf{D}^* = (\mathbf{D}_2^T\mathbf{D}_2)^{-1}\mathbf{D}_2^T. \tag{8}$$

7. The training points of $\mathbf{D}_2$ are the support vectors.
8. The α's and b values are obtained from the solution $\mathbf{z}_2$.

4   Genetic Algorithms

Genetic algorithm is a search meta-heuristic method inspired by natural evolution, such as inheritance, mutation, natural selection, and crossover. In a genetic algorithm, a population of candidate individuals (or solutions) to an optimization problem is evolved toward better solutions by natural selection, i.e., a fitness function. In this population, each individual has a set of genes (gene vector), named chromosome, which can be changed by mutation or combined generation-by-generation with other one to build new individuals by reproduction processes which use crossover. The most common way of representing solutions is in a binary format, i.e., strings of 0s and 1s. Despite that other kind of encodings are also possible.

The first population - set of solutions - is generated in a pseudo-randomized mode and it evolves in cycles, usually named generations. A value of the fitness function (its accuracy) for each individual ranks how 'good' the solution is. One can calculate the fitness value after decoding the chromosome and the best solution of the problem to be solved has the best fitness value. This value is used in order to guide the reproduction process where individuals with high fitness value have more chance to spread out their genes over the population.

5   Proposal: Fixed-Size Genetic Algorithm for Sparse LSSVM (FSGAS-LSSVM)

Our proposal called FSGAS-LSSVM relies on a single objective genetic algorithms which aims at improving (maximizing) the classifier's accuracy over the training (or validation) dataset. The individual or chromosome in our simulations is represented by a binary vector of genes where each one is set to either "one" (true) whether a certain pattern will be used as a support vector in the training process or "zero" (false) otherwise. Our fitness function is the value of the resulting accuracy of a classifier by using the pseudo-inverse method to solve the non-square linear system (as presented in Eq. 8). To do so, we build a data matrix for the linear system taking into account only the genes that were set to "one". In this approach, each individual has as many support vectors as the number of genes set up to "one"[1].

FSGAS-LSSVM is guided by accuracy, however, in parallel this approach achieves sparse classifiers due to the flexibility added, i.e., the ability of having or not a certain pattern as a support vector. Thus, each individual is a solution in terms of which patterns do belong to the set of support vectors. In this way, some patterns will not belong to the set of support vectors and then ones which are outliers, corrupted with noise or even non-relevant can fortunately be eliminated.

---

[1] Except if the solution of linear system owns zero values as components.

For this reason, our population can evolve to the best solution or at least to a better one.

It is important to be aware that, in a similar way to IP-LSSVM, our proposal only remove non-relevant columns related to the patterns we are intend to throw out of the support vectors set. Therefore, we maintain all of the rows which means to maintain the restrictions in order to avoid a loss of labeling information and in performance.

5.1 Individuals or Chromosomes

In order to clarify our ideas concerning the individuals, we present simple examples of how to construct individuals using our proposal, so-called FSGAS-LSSVM. Let us consider the matrix $\mathbf{D}$ of LSSVM linear system for a very small training set only with four patterns as presented in Fig. (1). In our proposal, for all the patterns in the support vectors set, we have to set up a gene vector to $[1\ 1\ 1\ 1]$.

```
D =

        0     1.0000   -1.0000   -1.0000    1.0000
    1.0000    0.4218    0.6557    0.6787    0.6555
   -1.0000    0.9157    0.0357    0.7577    0.1712
   -1.0000    0.7922    0.8491    0.7431    0.7060
    1.0000    0.9595    0.9340    0.3922    0.0318
```

Fig. 1. Matrix $\mathbf{D}$ without leaving any patterns out of the support vectors set. This means a individual stood for a gene vector equals to $[1\ 1\ 1\ 1]$.

For a individual without the second pattern in the SV set, we need to set up a gene vector to $[1\ 0\ 1\ 1]$. This means the fifth column have to be eliminated as presented in Fig. (2).

```
D1 =

        0     1.0000   -1.0000    1.0000
    1.0000    0.4218    0.6787    0.6555
   -1.0000    0.9157    0.7577    0.1712
   -1.0000    0.7922    0.7431    0.7060
    1.0000    0.9595    0.3922    0.0318
```

Fig. 2. Matrix $\mathbf{D_2}$ leaving only the second pattern out of the support vectors set. This means a individual stood for a gene vector equals to $[1\ 0\ 1\ 1]$. In this situation, we have to remove the third column.

Generally speaking, in order to remove a certain pattern from the support vector set which is described by the i-th gene (g_i) in the vector $\mathbf{g} = [g_1\ g_2\ g_3\ g_4]$, it is necessary to eliminate the $i + 1$ column. It is also worth emphasizing that we must not remove the first column so that to avoid lost of performance.

5.2   Reproduction

The main issue concerning the individuals in our proposal (FSGAS-LSSVM) is to keep the number of "one"s in each individual constant whenever applying the reproduction operators. This is required because a direct swap[2] between individuals genes does not work well, since invalid individuals will be produced. Therefore, we propose to use a indirect coding of individuals so that we transform binary-encoding individuals into integer-encoding ones to which we are able to apply permutation operators, such as partially matched (PMX), cycle (CX) and order crossover (OX) operators [6].

In the Figure 3 is depicted the way the transformation between binary and integer encodings happens and then how the order crossover operator works over the integer-encoding parents. After such process, each offspring is turned back to the binary-encoding so that each child will be suitable to belong to next population.

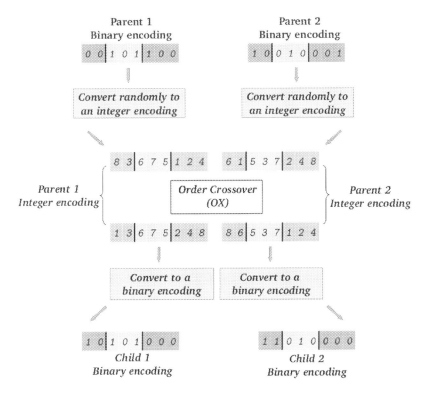

Fig. 3. Transformation process between binary-integer encoding for application of OX crossover operator followed by the transformation between integer-binary encoding

---

[2] As it is usually done by the default operators: one-point, two-point or even uniform crossover.

5.3   Mutation

As stated before, in our modeling, the individual is represented by a binary vector of genes, where each one is set to either "one" (true) whenever a certain pattern is a support vector or "zero" (false) otherwise. In addition to that, let m be the number of support vector to be pruned. Our mutation operator is based on a pair of genes that exchange their genetic material. In order to keep the number of genes with "one" at most m exchanges of gene values must be performed, where each gene value might be changed once. To do so, the position of the first individual gene is obtained from those genes with "one" and the position of the second is chosen at random between valid values, i.e., positions from those genes with "zero". After that, their values are exchanged. This is done for each and every gene with "one". With this considerations, the number of "one"s in the individual remains constant even after applying the mutation operator. In fact, we model such mutation operator to change 10% of those genes with "one".

5.4   Fitness Function for FSGAS-LSSVM

Our single objective fitness function is the classifier's accuracy over the training (or validation) set obtained by using the pseudo-inverse method to solve the LSSVM linear system. As stated before, we use pseudo-inverse method due to the resulting non-square matrix achieved by eliminating those patterns which have gene value equals zero.

5.5   FSGAS-LSSVM Algorithm

FSGAS-LSSVM algorithm for training a classifier can be described as follows.

1. Initiate $t = 0$, where t stands for the current generation;
2. Generate initial population $P(t)$ of individuals, i.e., gene sets with a fixed quantity of "one"s and build their related matrices $\{\mathbf{D}_i\}_{i=1}^s$, where s is the number of individuals at the generation t, randomly;
3. For each individual i in $P(t)$
4.   Solve the LSSVM linear $(\mathbf{z}_i = (\mathbf{D}_i^T \mathbf{D}_i)^{-1} \mathbf{D}_i^T \mathbf{l}_i)$;
5.   Evaluate fitness function, i.e., the training (or validation) set accuracy;
6. While $t \leq t_{max}$, where t_{max} is the maximum number of generations
7.   Select individuals i (and their matrices $\mathbf{D}_i$) by the roulette wheel method;
8.   Apply the order crossover operation over the previous selected individuals to generate new ones as presented in Subsection 5.2;
9.   Apply the mutation operation over those selected individuals as described in Subsection 5.3;
10.   Compute $t = t + 1$;
11.   Build a matrix $\mathbf{D}_i$ for each individual in $P(t + 1)$.
12.   Evaluate training (or validation) set accuracies for each $\mathbf{D}_i$;
13. Select the best individual or solution.

6 Simulations and Discussion

The results for simulations carried out in this work are presented in this section. From the total, 80% of the data examples were randomly selected for training purposes and so the remaining 20% of the examples were used for assessing the classifiers' generalization performances. Tests with real-world benchmarking datasets were also evaluated in this work. We used three UCI datasets; namely, Diabetes (PID) with 768 patterns, Haberman (HAB) with 306 patterns and Breast Cancer (BCW) with 683 patterns. We also used the vertebral column pathologies dataset (VCP) with 310 patterns as described in [10]. To do so, we transformed the original three-class VCP problem into a binary one by aggregating the two classes of pathologies, the disc hernia and spondylolisthesis, into a single one; however, the normal class remained unchanged. In addition to those datasets, we evaluated our proposal over other two well-known datasets: Ripley (RIP) with 1250 patterns and Banana (BAN) with 1001 patterns.

Initially, the GA randomly creates a population of feasible solutions. Each solution is a string of binary values in which each bit represents the presence (1) or absence (0) of a pattern in the support vector set (i.e., whether the Lagrange multiplier related to a certain pattern is zero or not). For the next generation, we take into account the fact that 10% of the best individuals are selected due to a elitist selection scheme, 80% of new individuals were generated as result of applying the crossover operator and then the remaining individuals were obtained by mutation.

In Table 1, we report performance metrics (mean value and standard deviation of the accuracy) for aforementioned methodology $80\% - 20\%$ on testing set averaged over 30 independent runs. We also show the average number of SVs ($\#SVs$), the number of training patterns ($\#TP$), the values of the parameter γ (LS-SVM) and the reduction obtained (Red.). In Table 1, we show results for LSSVM, Pruning LSSSVM (P-LSSVM)[3] and FSGAS-LSSVM.

In Table 2, we present some results concerning LSSVM, Pruning LSSSVM (P-LSSVM) and FSGAS-LSSVM. In order to compare our novel approach with other proposed methods (such as Pruning LSSVM) and to be in accordance with their assessment methodology, we carried out simulations splitting our data set into three other ones. In this configuration 60% of the patterns were in the first one (training set), 20% of examples for validation purposes and, as usual, the remaining 20% of the patterns were used for testing the classifiers' generalization performances.

By analyzing these tables, one can conclude that the performances of the fixed-size reduced-set classifiers FSGAS-LSSVM were equivalent or even superior to those achieved by the full-set classifiers. Similarly, as shown in these tables, the performances of the FSGAS-LSSVM classifiers were in general better than the IP-LSSVM or P-LSSVM.

---

[3] P-LSSVM was modified in order to have a fair assessment with FSGAS-LSSVM. See the Subsection 3.1.

Table 1. Results for the LSSVM, IP-LSSVM and FSGAS-LSSVM classifiers with 80% (20%) of the full dataset for training (testing)

Dataset	Model	γ	Accuracy	# TP	# SVs	Red.
VCP	LSSVM	0.05	81.2 ± 4.9	248	248.0	−
VCP	IP-LSSVM	0.05	75.1 ± 6.6	248	198.0	20.0%
			54.1 ± 17.8		124.0	50.0%
			36.4 ± 7.7		50.0	80.0%
VCP	FSGAS-LSSVM	0.05	83.7 ± 3.7	248	198.0	20.0%
			83.6 ± 4.1		124.0	50.0%
			81.2 ± 3.9		50.0	80.0%
HAB	LSSVM	0.04	73.8 ± 5.0	245	245.0	−
HAB	IP-LSSVM	0.04	57.6 ± 9.4	245	196.0	20.0%
			52.0 ± 12.5		122.0	50.0%
			40.7 ± 17.6		49.0	80.0%
HAB	FSGAS-LSSVM	0.04	74.3 ± 4.5	245	196.0	20.0%
			75.4 ± 6.2		122.0	50.0%
			74.4 ± 4.8		49.0	80.0%
BCW	LSSVM	0.04	96.7 ± 1.1	546	546.0	−
BCW	IP-LSSVM	0.04	96.9 ± 1.1	546	437.0	20.0%
			96.9 ± 1.3		273.0	50.0%
			97.4 ± 1.0		109.0	80.0%
BCW	FSGAS-LSSVM	0.04	95.9 ± 1.6	546	437.0	20.0%
			95.9 ± 1.8		273.0	50.0%
			95.9 ± 1.6		109.0	80.0%
PID	LSSVM	0.04	75.8 ± 1.1	614	614.0	−
PID	IP-LSSVM	0.04	71.6 ± 8.4	614	491.0	20.0%
			30.6 ± 4.6		307.0	50.0%
			28.9 ± 4.0		123.0	80.0%
PID	FSGAS-LSSVM	0.04	77.5 ± 3.7	614	491.0	20.0%
			76.4 ± 3.3		307.0	50.0%
			76.6 ± 3.1		123.0	80.0%
RIP	LSSVM	0.04	87.7 ± 1.6	1000	1000.0	−
RIP	IP-LSSVM	0.04	87.7 ± 1.7	1000	800.0	20.0%
			86.1 ± 2.9		500.0	50.0%
			15.2 ± 5.4		200.0	80.0%
RIP	FSGAS-LSSVM	0.04	87.8 ± 2.2	1000	800.0	20.0%
			87.7 ± 1.9		500.0	50.0%
			87.9 ± 2.0		200.0	80.0%
BNA	LSSVM	0.04	96.9 ± 1.3	801	801.0	−
BNA	IP-LSSVM	0.04	96.7 ± 1.2	801	641.0	20.0%
			96.7 ± 1.3		400.0	50.0%
			90.3 ± 5.8		160.0	80.0%
BNA	FSGAS-LSSVM	0.04	96.8 ± 1.0	801	641.0	20.0%
			96.9 ± 1.0		400.0	50.0%
			96.6 ± 0.9		160.0	80.0%

Table 2. Results for the LSSVM, P-LSSVM and FSGAS-LSSVM and classifiers with 60%, 20% and 20% of the full dataset for training, validating and testing, respectively

Dataset	Model	γ	Accuracy	# TS	# SVs	Red.
VCP	LSSVM	0.05	80.1 ± 3.2	186	186.0	–
VCP	P-LSSVM	0.05	69.1 ± 14.1	186	149.0	20.0%
			54.7 ± 14.2		93.0	50.0%
			50.9 ± 14.9		37.0	80.0%
VCP	FSGAS-LSSVM	0.05	81.7 ± 4.4	186	149.0	20.0%
			81.1 ± 4.8		93.0	50.0%
			81.1 ± 4.4		37.0	80.0%
HAB	LSSVM	0.04	73.7 ± 3.6	184	184.0	–
HAB	P-LSSVM	0.04	64.6 ± 6.9	184	147.0	20.0%
			63.1 ± 8.4		92.0	50.0%
			70.9 ± 6.7		37.0	80.0%
HAB	FSGAS-LSSVM	0.04	72.5 ± 5.8	184	147.0	20.0%
			75.0 ± 7.4		92.0	50.0%
			73.7 ± 5.0		37.0	80.0%
BCW	LSSVM	0.04	96.9 ± 0.8	410	410.0	–
BCW	P-LSSVM	0.04	96.2 ± 1.5	410	328.0	20.0%
			96.4 ± 1.6		205.0	50.0%
			95.6 ± 5.3		82.0	80.0%
BCW	FSGAS-LSSVM	0.04	96.4 ± 1.6	410	328.0	20.0%
			96.1 ± 1.8		205.0	50.0%
			96.1 ± 1.0		82.0	80.0%
PID	LSSVM	0.04	75.8 ± 1.5	461	461.0	–
PID	P-LSSVM	0.04	68.9 ± 9.8	461	369.0	20.0%
			33.3 ± 3.3		230.0	50.0%
			30.2 ± 3.3		92.0	80.0%
PID	FSGAS-LSSVM	0.04	76.8 ± 2.4	461	369.0	20.0%
			77.1 ± 3.0		230.0	50.0%
			76.9 ± 3.0		92.0	80.0%
RIP	LSSVM	0.04	87.9 ± 0.9	750	750.0	–
RIP	P-LSSVM	0.04	87.8 ± 2.0	750	600.0	20.0%
			83.3 ± 8.3		375.0	50.0%
			16.7 ± 8.0		150.0	80.0%
RIP	FSGAS-LSSVM	0.04	88.2 ± 1.5	750	600.0	20.0%
			87.9 ± 1.9		375.0	50.0%
			87.7 ± 1.7		150.0	80.0%
BNA	LSSVM	0.04	96.8 ± 0.6	601	601.0	–
BNA	P-LSSVM	0.04	96.8 ± 1.2	601	481.0	20.0%
			96.4 ± 1.5		300.0	50.0%
			89.6 ± 6.1		120.0	80.0%
BNA	FSGAS-LSSVM	0.04	97.1 ± 0.9	601	481.0	20.0%
			96.8 ± 1.2		300.0	50.0%
			97.1 ± 0.8		120.0	80.0%

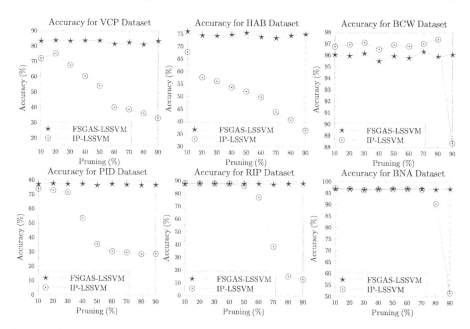

Fig. 4. Comparative accuracy between the FSGAS-LSSVM and IP-LSSVM methods according to the pruning percentage for each problem obtained evaluated with 80% and 20% of data for training and testing, respectively

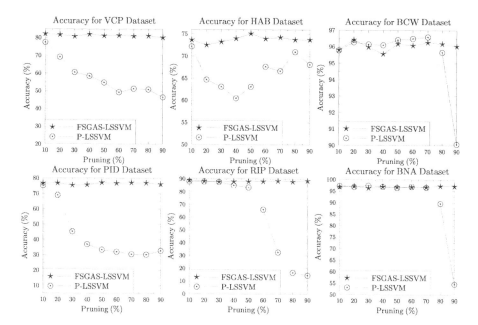

Fig. 5. Comparative accuracy between the FSGAS-LSSVM and P-LSSVM methods according to the pruning percentage for each problem evaluated with 60%, 20% and 20% of data for training, validation and testing, respectively

In addition to the previous results, we show in Figure 4 the FSGAS-LSSVM and IP-LSSVM accuracies for $10\%, 20\%, \ldots, 90\%$ of support vectors over 30 independent runs for each aforementioned dataset. In Figure 5, we present the same analysis for FSGAS-LSSVM and P-LSSVM classifiers.

We notice in those figures that the FSGAS-LSSVM accuracies are higher than IP-LSSVM (and P-LSSVM) for each percentage of support vectors, except for the BCW dataset. Even for such dataset the FSGAS-LSSVM worked well when the percentage is 90%; the same is not true for IP-LSSVM and P-LSSVM classifiers.

7   Conclusions

Our proposal called Fixed-Size Genetic Algorithm for Sparse LSSVM (FSGAS-LSSVM) is based on a single objective genetic algorithms, in which the desired number of support vectors is fixed in advance. Our proposal is guided by the training (or validation) accuracy, so that a suitable set of support vector might be found. In this work, it is proposed a new way of keeping a fixed number of support vectors by a reproduction operator. Moreover, the achieved results indicate that the FSGAS-LSSVM works very well, providing a reduced number of support vectors while maintaining or increasing classifier's accuracy.

References

1. Aydin, I., Karakose, M., Akin, E.: A multi-objective artificial immune algorithm for parameter optimization in svm. Applied Soft Computing **11**(1), 120–129 (2011)
2. Blachnik, M., Kordos, M.: Simplifying SVM with weighted LVQ algorithm. In: Yin, H., Wang, W., Rayward-Smith, V. (eds.) IDEAL 2011. LNCS, vol. 6936, pp. 212–219. Springer, Heidelberg (2011)
3. Burges, C.J.C.: Simplified support vector decision rules. In: Proceedings of the 13th (ICML 1996). pp. 71–77. Morgan Kaufmann (1996)
4. Carvalho, B.P.R., Braga, A.P.: IP-LSSVM: A two-step sparse classifier. Pattern Recognition Letters **30**, 1507–1515 (2009)
5. D'Amato, L., Moreno, J.A., Mujica, R.: Reducing the complexity of kernel machines with neural growing gas in feature space. In: Lemaître, C., Reyes, C.A., González, J.A. (eds.) IBERAMIA 2004. LNCS (LNAI), vol. 3315, pp. 799–808. Springer, Heidelberg (2004)
6. Eiben, A., Smith, J.: Introduction to Evolutionary Computation. Springer (2003)
7. Geebelen, D., Suykens, J.A.K., Vandewalle, J.: Reducing the number of support vectors of SVM classifiers using the smoothed separable case approximation. IEEE Transactions on Neural Network and Learning Systems **23**(4), 682–688 (2012)
8. Li, Y., Lin, C., Zhang, W.: Improved sparse least-squares support vector machine classifiers. Neurocomputing **69**, 1655–1658 (2006)
9. Peres, R., Pedreira, C.E.: Generalized risk zone: Selecting observations for classification. IEEE Transactions on Pattern Analysis and Machine Intelligence **31**(7), 1331–1337 (2009)

10. Neto, A.R.R., Barreto, G.A.: Opposite maps: Vector quantization algorithms for building reduced-set SVM and LSSVM classifiers. Neural Processing Letters **37**(1), 3–19 (2013)
11. Samadzadegan, F., Soleymani, A., Abbaspour, R.: Evaluation of genetic algorithms for tuning svm parameters in multi-class problems. In: 11th International Symposium on Computational Intelligence and Informatics (CINTI), pp. 323–328 (2010)
12. Silva, D.A., Rocha Neto, A.R.: Multi-objective genetic algorithms for sparse least square support vector machines. In: Corchado, E., Lozano, J.A., Quintián, H., Yin, H. (eds.) IDEAL 2014. LNCS, vol. 8669, pp. 158–166. Springer, Heidelberg (2014)
13. Steinwart, I.: Sparseness of support vector machines. Journal of Machine Learning Research **4**, 1071–1105 (2003)
14. Suykens, J.A.K., Lukas, L., Vandewalle, J.: Sparse least squares support vector machine classifiers. In: Proceedings of the 8th European Symposium on Artificial Neural Networks (ESANN 2000), pp. 37–42 (2000)
15. Suykens, J.A.K., Vandewalle, J.: Least squares support vector machine classifiers. Neural Processing Letters **9**(3), 293–300 (1999)
16. Valyon, J., Horvath, G.: A sparse least squares support vector machine classifier. In: IEEE IJCNN, 2004. vol. 1, p. 548 (2004)
17. Vapnik, V.N.: Statistical Learning Theory. Wiley-Interscience (1998)

Ensemble of Minimal Learning Machines for Pattern Classification

Diego Parente Paiva Mesquita[1], João Paulo Pordeus Gomes[1],
and Amauri Holanda Souza Junior[2]([⊠])

[1] Department of Computer Science, Federal University of Ceará,
Fortaleza, Ceará, Brazil
diego@diegoparente.com, jpaulo@lia.ufc.br
[2] Department of Computer Science, Federal Institute of Ceará,
Maracanaú, Ceará, Brazil
amauriholanda@ifce.edu.br

Abstract. The use of ensemble methods for pattern classification have gained attention in recent years mainly due to its improvements on classification rates. This paper evaluates ensemble learning methods using the Minimal Learning Machines (MLM), a recently proposed supervised learning algorithm. Additionally, we introduce an alternative output estimation procedure to reduce the complexity of the standard MLM. The proposed methods are evaluated on real datasets and compared to several state-of-the-art classification algorithms.

1 Introduction

Recently there has been a lot of interest in ensemble methods for classification. This is mainly due to their ability to achieve high performance in a variety of classification tasks, such as face recognition [1], recognition of spontaneous face expressions [2], hyperspectral remote sensing [3] and character recognition [4]. According to [6] and [7], a necessary and sufficient condition for an ensemble of classifiers to be more accurate than any of its individual components is to use classifiers that are accurate and diverse. In this context, a classifier is said to be accurate if it achieves an error classification rate smaller than what is achieved using random guesses. Two classifiers are called diverse if they produce different classification errors. In order to achieve a suitable balance between accuracy and diversity, various strategies have been developed to extend standard classification algorithms to the ensemble framework. Manipulations of training examples [8], selection of subsets of features [9], and injection of randomness in the initialization step [5] are some of the most popular strategies. Ensemble strategies based on combinations of classifiers' results using voting, weighted voting, summation, mean- and median-based averaging schemes are also commonly used [10].

Among the recently proposed supervised learning algorithms, the Minimal Learning Machine (MLM, [11]) has gained attention for its simple and easy implementation, additionally requiring the adjustment of only a single hyperparameter (K, the number of reference points). Learning in MLM consists in

© Springer International Publishing Switzerland 2015
I. Rojas et al. (Eds.): IWANN 2015, Part II, LNCS 9095, pp. 142–152, 2015.
DOI: 10.1007/978-3-319-19222-2_12

building a linear mapping between input and output distance matrices. In the generalization phase, the learned distance map is used to provide an estimate of the distance from K output reference points to the target output value. Then, the output point estimation is formulated as multilateration problem based on the predicted output distance and the locations of the reference points.

This work aims to evaluate the MLM on a ensemble framework using well known methods for ensemble generation and classifier combination. In addition, we propose a simplification of the output estimation step in MLM that speeds up its computational time complexity. Ensembles of MLMs are benchmarked against a number of to state-of-the-art classifiers using UCI classification datasets. The results show that the proposed methods are competitive, achieving higher classification rates than the reference methods in most of the selected problems.

The remainder of the paper is organized as follows. Section 2 introduces the Minimal Learning Machine and discusses the novel proposal for the output estimation step. Section 3 presents methods for ensemble learning using MLMs. The experiments are reported in Section 4. Conclusions are given in Section 5.

2   Minimal Learning Machine

We are given a set of N input points $X = \{\mathbf{x}_i\}_{i=1}^N$, with $\mathbf{x}_i \in \mathbb{R}^D$, and the set of corresponding outputs $Y = \{\mathbf{y}_i\}_{i=1}^N$, with $\mathbf{y}_i \in \mathbb{R}^S$. Assuming the existence of a continuous mapping $f : \mathcal{X} \to \mathcal{Y}$ between the input and the output space, we want to estimate f from data with the multiresponse model

$$\mathbf{Y} = f(\mathbf{X}) + \mathbf{R}.$$

The columns of the matrices $\mathbf{X}$ and $\mathbf{Y}$ correspond to the D inputs and S outputs respectively, and the rows to the N observations. The columns of the $N \times S$ matrix $\mathbf{R}$ correspond to the residuals.

The MLM is a two-step method designed to

1. reconstruct the mapping existing between input and output distances;
2. estimating the response from the configuration of the output points.

In the following, the two steps are discussed.

2.1   Distance Regression

For a selection of reference input points $R = \{\mathbf{m}_k\}_{k=1}^K$ with $R \subseteq X$ and corresponding outputs $T = \{\mathbf{t}_k\}_{k=1}^K$ with $T \subseteq Y$, define $\mathbf{D}_x \in \mathbb{R}^{N \times K}$ in such a way that its kth column $\mathbf{d}(X, \mathbf{m}_k)$ contains the distances $d(\mathbf{x}_i, \mathbf{m}_k)$ between the N input points $\mathbf{x}_i$ and the kth reference point $\mathbf{m}_k$. Analogously, define $\boldsymbol{\Delta}_y \in \mathbb{R}^{N \times K}$ in such a way that its kth column $\boldsymbol{\delta}(Y, \mathbf{t}_k)$ contains the distances $\delta(\mathbf{y}_i, \mathbf{t}_k)$ between the N output points $\mathbf{y}_i$ and the output $\mathbf{t}_k$ of the kth reference point.

We assume that there exists a mapping g between the input distance matrix $\mathbf{D}_x$ and the corresponding output distance matrix $\boldsymbol{\Delta}_y$ that can be reconstructed using the multiresponse regression model

$$\boldsymbol{\Delta}_y = g(\mathbf{D}_x) + \mathbf{E}.$$

The columns of the matrix $\mathbf{D}_x$ correspond to the K input vectors and columns of the matrix $\boldsymbol{\Delta}_y$ correspond to the K response vectors, the N rows correspond to the observations. The columns of matrix $\mathbf{E} \in \mathbb{R}^{N \times K}$ correspond to the K residuals.

Assuming that mapping g between input and output distance matrices has a linear structure for each response, the regression model has the form

$$\boldsymbol{\Delta}_y = \mathbf{D}_x \mathbf{B} + \mathbf{E}. \tag{1}$$

The columns of the $K \times K$ regression matrix $\mathbf{B}$ correspond to the coefficients for the K responses. Under the normal conditions where the number of selected reference points is smaller than the number of available points available (i.e., $K < N$), the matrix $\mathbf{B}$ can be approximated by the usual least squares estimate:

$$\hat{\mathbf{B}} = (\mathbf{D}_x' \mathbf{D}_x)^{-1} \mathbf{D}_x' \boldsymbol{\Delta}_y. \tag{2}$$

For an input test point $\mathbf{x} \in \mathbb{R}^D$ whose distances from the K reference input points $\{\mathbf{m}_k\}_{k=1}^K$ are collected in the vector $\mathbf{d}(\mathbf{x}, R) = [d(\mathbf{x}, \mathbf{m}_1) \dots d(\mathbf{x}, \mathbf{m}_K)]$, the corresponding estimated distances between its unknown output $\mathbf{y}$ and the known outputs $\{\mathbf{t}_k\}_{k=1}^K$ of the reference points are

$$\hat{\boldsymbol{\delta}}(\mathbf{y}, T) = \mathbf{d}(\mathbf{x}, R)\hat{\mathbf{B}}. \tag{3}$$

The vector $\hat{\boldsymbol{\delta}}(\mathbf{y}, T) = [\hat{\delta}(\mathbf{y}, \mathbf{t}_1) \dots \hat{\delta}(\mathbf{y}, \mathbf{t}_K)]$ provides an estimate of the geometrical configuration of $\mathbf{y}$ and the reference set T, in the $\mathcal{Y}$-space.

2.2   Output Estimation

The problem of estimating the output $\mathbf{y}$, given the outputs $\{\mathbf{t}_k\}_{k=1}^K$ of all the reference points and estimates $\hat{\boldsymbol{\delta}}(\mathbf{y}, T)$ of their mutual distances, can be understood as a multilateration problem [12] to estimate its location in $\mathcal{Y}$.

Numerous strategies can be used to solve a multilateration problem [13]. From a geometric point of view, locating $\mathbf{y} \in \mathbb{R}^S$ is equivalent to solve the overdetermined set of K nonlinear equations corresponding to S-dimensional hyper-spheres centered in $\mathbf{t}_k$ and passing through $\mathbf{y}$. Figure 1 graphically depicts the problem for $S = 2$.

Given the set of $k = 1, \dots, K$ spheres each with radius equal to $\hat{\delta}(\mathbf{y}, \mathbf{t}_k)$

$$(\mathbf{y} - \mathbf{t}_k)'(\mathbf{y} - \mathbf{t}_k) = \hat{\delta}^2(\mathbf{y}, \mathbf{t}_k), \tag{4}$$

the location of $\mathbf{y}$ is estimated from the minimization of the objective function

$$J(\mathbf{y}) = \sum_{k=1}^K \left((\mathbf{y} - \mathbf{t}_k)'(\mathbf{y} - \mathbf{t}_k) - \hat{\delta}^2(\mathbf{y}, \mathbf{t}_k) \right)^2. \tag{5}$$

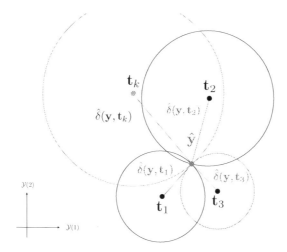

Fig. 1. Output estimation

The cost function has a minimum equal to 0 that can be achieved if and only if **y** is the solution of (4). If it exists, such a solution is thus global and unique. Due to the uncertainty introduced by the estimates $\hat{\delta}(\mathbf{y}, \mathbf{t}_k)$, an optimal solution to (5) can be achieved by any minimizer $\hat{\mathbf{y}} = \arg\min_{\mathbf{y}} J(\mathbf{y})$ like the nonlinear least square estimates from standard gradient descent methods. The original MLM proposal applies the Levenberg-Marquardt (LM) method [14] to solve the output estimation step.

2.3 Extension to Classification

An important class of problems is classification, where we are concerned with the prediction of categorical variables or class labels. For the task, we are still given N input points $X = \{\mathbf{x}_i\}_{i=1}^{N}$, with $\mathbf{x}_i \in \mathbb{R}^D$, and corresponding class labels $L = \{l_i\}_{i=1}^{N}$, with $l_i \in \{C_1, \ldots, C_S\}$, where C_j denotes the j-the class. For $S = 2$, we have binary classification, whereas for $S > 2$ we have multi-class classification.

The MLM can be extended to classification in a straightforward manner by representing the S class labels in a vectorial fashion through an 1-of-S encoding scheme [15]. In such approach, a S-level qualitative variable is represented by a vector of S binary variables or bits, only one of which is *on* at a time. In the classification of a test observation $\mathbf{x}$ of unknown class label $l \in \{C_1, \ldots, C_S\}$, the estimated class $\hat{l}$ associated to the output estimate $\hat{\mathbf{y}}$ is $\hat{l} = C_{s^*}$, where

$$s^* = \underset{s=1,\ldots,S}{\arg\max} \{\hat{y}^{(s)}\}. \tag{6}$$

Given this formulation, the Minimal Learning Machine provides a general framework that can be used for regression, binary and multi-class classification.

Complexity Analysis. The training procedure of the Minimal Learning Machine can be roughly decomposed into two parts: i) calculation of the pairwise distance matrices in the output and input space; ii) calculation of the least-square solution for the multiresponse linear regression problem on distance matrices. The first part takes $\Theta(KN)$ time. The computational cost of the second part is driven by the calculation of the Moore-Penrose pseudoinverse matrix, which runs in $\Theta(K^2N)$ time if we consider the SVD algorithm.

In order to establish a comparison, the MLM training computational cost is similar to what is presented by an Extreme Learning Machine when the number of hidden neurons is equal to the number of reference points. It is worthy to notice that the ELM is considered one of the fastest methods for nonlinear regression and classification tasks.

Concerning the computational analysis of the generalization (output estimation) step in MLM, we consider the Levenberg-Marquardt method due to its fast and stable convergence, even though any gradient descent method can be used on the minimization step in Eq. 5. For each iteration, the LM method involves the computation of the Jacobian $\mathbf{J} \in \mathbb{R}^{K \times S}$ and the inverse of $\mathbf{J}^T\mathbf{J}$. The computational complexity of the LM algorithm is approximately $\Theta(I(KS^2 + S^3))$, where S is the dimensionality of $\mathbf{y}$ and I denotes the number of iterations.

Speeding up the Output Estimation. The MLM was proposed as a general supervised learning method, capable of dealing with classification and regression tasks. This resulted in a general formulation that does not take advantage of the particularities of each task. Considering the classification case, one can notice that the finite and discrete set of possible outputs may help finding a solution to the output estimation without solving an optimization problem. In fact, the output estimation step is the computationally critical part in MLM in comparison to standard classification methods. Thus, reducing the computational complexity of the output estimation step is particularly useful in the context of ensemble methods, since such an step is extensively computed for each ensemble member.

The output estimation part aims to find the best estimation for the label (numerical output vector) $\hat{\mathbf{y}}$ of a given input test vector $\mathbf{x}$ from estimated distances to the reference points $\hat{\boldsymbol{\delta}}(\mathbf{y}, T)$ in the output space. In the classification task, since the set of feasible values for $\hat{\mathbf{y}}$ is limited to the number of classes, a few trials should be computed to select an estimate to the output. Moreover, an alternative to trying all possible values is to take the output (label) associated with the nearest reference point in the output space, whose distance is likely to be about 0 — given that all the classes are represented in the set of reference points.

Consider a classification problem with S classes. In this situation, the reference output points $\mathbf{t}_k$ assume S possible values and the objective function (5) can be rewritten as:

$$J(\mathbf{y}) = \sum_{s=1}^{S} N_s \left((\mathbf{y} - \mathbf{t}^s)'(\mathbf{y} - \mathbf{t}^s) - \hat{\delta}^2(\mathbf{y}, \mathbf{t}^s) \right)^2 \tag{7}$$

where N_s is the number of reference points belonging to class C_s, and $\mathbf{t}^s$ denotes the vectorial representation (1-of-S encoding) of the class C_s.

For the sake of simplicity, let us assume that all classes are equally represented in the set of reference points. Therefore, the factor N_s can be neglected from the objective function. The argument $(\mathbf{y} - \mathbf{t}^s)'(\mathbf{y} - \mathbf{t}^s)$ may assume only two possible values: zero, when $\mathbf{y} = \mathbf{t}^s$, or a positive number, otherwise. Similarly, the estimated distance $\hat{\delta}^2(\mathbf{y}, \mathbf{t}^s)$ return S distinct values: the estimate for the distance to the correct class (supposed to be 0 in a perfect reconstruction of the distance), and estimates for the distances to the other classes (positive numbers). We are interested in minimizing Eq. 7. It is accomplished when we set $\mathbf{y} = \mathbf{t}^{s^*}$ to the class s^* with smallest distance estimate, i.e., $s^* = \underset{s=1,...,S}{\operatorname{argmin}} \hat{\delta}^2(\mathbf{y}, \mathbf{t}^s)$. It means that the classification step can be carried out based on the label of the reference point associated to the smallest output distance estimate. In practice, we use the estimated distances $\hat{\delta}(\mathbf{y}, \mathbf{t}_k)$ given from Eq. 3 in such a way that the estimated class $\hat{l}$ for an input pattern $\mathbf{x}$ is given by $\hat{l} = l_{k^*}$, where

$$k^* = \underset{k=1,...,K}{\operatorname{argmin}} \hat{\delta}^2(\mathbf{y}, \mathbf{t}_k), \tag{8}$$

and l_{k^*} represents the label of the k^*th reference point.

This strategy corresponds to carrying out a nearest neighbors classifier based on the estimated distances in the output space. It turns out the computational complexity of the output estimation step to $O(K)$ with small constant factor. One may notice that only S distinct values need to be evaluated in order to find the minimum estimated distance.

3 Ensemble Strategies

The proposed ensemble methods consists of two strategies for manipulating the training examples, and two output combination schemes. Manipulating the training examples is one of the simplest strategies to generate different classifiers. For example, it can be achieved by resampling the training set or by using different samples in the model training. These strategies aim to increase the classifiers variability and consequently the overall ensemble performance. On the other hand, a combination strategy consists of combining the output of each ensemble member to generate a final consensus output. Many strategies have been proposed and the most commonly used one is based on a majority voting scheme.

The strategies used in this work are detailed in the following.

3.1 Selecting Training Examples and Reference Points

The standard procedure for selecting the number of reference points K in MLM is based on resampling methods, e.g., cross-validation. Then, K randomly selected points from the learning points comprise the set of reference points. The first ensemble strategy consists of using M classifiers with the standard procedure to select K. By doing so, we expect the classifiers to differ with respect to the randomly chosen reference points. Henceforth this procedure will be referred as *sampling procedure 1*.

The second strategy consists of inserting uncertainty by using a randomly selected fraction P of the total available training data as learning points to each of the M classifiers. Then, all the learning points are used as reference points ($K = P$). This procedure will be referred to as *sampling procedure 2*. One may notice that, in the procedure 2, the fraction P (and consequently K) is defined beforehand whereas that in the first procedure it is selected through cross-validation. Also, the number of samples used for learning is different in the two strategies.

3.2 Combination Strategies

The combination of the classifiers' outputs is based on two approaches. The first method uses the majority voting scheme; in the case, an out-of-sample point is assigned to the most voted class among all the classifiers. It corresponds to the standard voting method.

The second approach consists of a weighted majority voting; in this case, each classifier vote is associated with a weighting factor. Such a factor is related to the classifier prediction confidence. Assuming the 1-of-S encoding scheme for the classifiers, the weight associated with each classifier is given by

$$w = \frac{\max_{s=1,\ldots,S} \hat{y}^{(s)}}{\sum_{s=1}^{S} \hat{y}^{(s)}}, \tag{9}$$

where $\hat{y}^{(s)}$ is the sth component of the output estimate vector $\hat{\mathbf{y}}$. It is straightforward noticing that the larger is the maximum component value of $\hat{\mathbf{y}}$ in comparison to the other components, the higher is the prediction confidence and, consequently, the associated weight w. After calculating the weights for each classifier, the class is chosen according to the weighted majority voting.

4 Performance Evaluation

Using the described procedures for ensemble generation, it is possible to create 4 different MLM ensembles. Combining *procedure 1* with the voting and the weighted voting scheme generates, respectively, the voting based MLM (V-MLM) and the weighted voting based MLM (WV-MLM). The combination of *procedure 2* with the voting and the weighted voting scheme generates, respectively, the random sampling voting based MLM (RSV-MLM) and the random sampling weighted voting based MLM (RSWV-MLM).

The performances of the proposed MLM ensemble strategies are compared to standard MLM and some state of the art methods under real-world datasets. Simulations using V-ELM, SVM, OP-ELM, BP and and KNN are conducted on 10 UCI datasets. Datasets used are described in Table 1.

Table 1. Datasets description

Datasets	Attributes #	Classes #	Training data #	Testing data #
Balance	4	3	400	225
Breast	30	2	300	269
Diabetes	8	2	576	192
Glass	10	2	100	114
Heart	13	2	100	170
Sonar	60	2	100	108
Wine	13	3	100	78
Monk 1	6	2	124	432
Monk 2	6	2	169	432
Monk 3	6	2	122	432

For the first 7 datasets in Table 1, at each trial we deal the data at random between the training and testing set. For the remaining ones, training and testing data are fixed for all trials. Each value in Table 2 and Table 3 reflects the outcome of 50 similar trials.

4.1   Comparisons with MLM

Table 2 shows the performance of the proposed methods compared to the MLM. For WV-MLM, V-MLM and MLM the number of reference point was selected using 10-fold cross validation and the reference point were chosen randomly from the training set. For RSV-MLM and RSWV-MLM the value adopted for P was 0.8. All MLM ensembles in the experiments are comprised of 7 MLMs.

As expected, ensemble MLM strategies achieved higher accuracies when compared to the standard MLM for most datasets. Particularly, WV-MLM achieved the best results in 4 of the datasets. It is also important to notice that for most datasets, the ensemble methods achieved a low standard deviation.

4.2   Comparisons with SVM, OP-ELM, BP, KNN and V-ELM

Table 3 compares the performance of MLM ensembles against SVM [17], OP-ELM [18], BP [19], KNN [20] and V-ELM [5]. MAX-MLM denotes the results from the MLM variant with higher average accuracy on Table 2. For SVM, the Gaussian RBF is used as the kernel function, the cost parameter C and the kernel parameter γ are searched in a grid formed by $C = [2^{12}, 2^{11}, \ldots, 2^{-2}]$ and $\gamma = [2^4, 2^3, \ldots, 2^{-10}]$. For OP-ELM, the three possible kernels, linear, sigmoid and Gaussian are used as a combination. For BP, Levenberg-Marquardt algorithm is used to train the neural network. For KNN, 7 nearest neighbors are used and the Euclidean norm is adopted to calculate the distance. For V-ELM, 7 independent ELMs are adopted for training and majority voting, the number of hidden node is gradually increased up to 50 and chosen using cross validation.

Table 2. Performance comparison with MLM

Datasets	MLM		V-MLM		RSV-MLM		WV-MLM		RSWV-MLM	
	Acc.	Dev.	Acc.	Dev.	Acc.	Dev.	Acc.	Dev.	Acc.	Dev.
Balance	89.85	1.43	89.97	1.54	89.31	1.45	**90.20**	1.48	89.75	1.45
Breast	97.19	0.65	97.33	0.68	**97.40**	0.68	97.33	0.68	97.33	0.67
Diabetes	75.49	2.50	**76.03**	2.46	74.07	2.57	75.92	2.57	74.36	2.46
Glass	95.71	3.16	96.04	2.96	96.00	2.80	**96.05**	2.92	95.81	2.96
Heart	82.01	2.28	82.52	2.19	81.25	245	**82.56**	2.33	81.49	2.54
Sonar	**82.03**	4.11	81.92	4.20	82.00	3.79	81.79	4.10	81.46	3.92
Wine	98.41	1.23	98.41	1.23	98.43	1.27	98.41	1.23	**98.48**	1.28
Monk 1	82.04	2.17	83.98	0.78	**84.28**	0.56	83.25	0.66	83.46	0.78
Monk 2	**82.17**	0.00	**82.17**	0.00	82.02	0.61	**82.17**	0.00	81.86	0.68
Monk 3	93.32	0.06	93.51	0.00	93.37	0.36	93.54	0.07	**93.73**	0.28

Table 3. Performance comparison between MLM variants, SVM, OP-ELM, BP, KNN and V-ELM

	MAX-MLM		SVM		OP-ELM		BP		KNN		V-ELM	
	Acc.	Dev	Acc.	Dev	Acc.	Dev	Acc.	Dev	Acc.	Dev	Acc.	Dev
Balance	90.2	1.48	**95.88**	1.31	92.31	1.83	90.92	2.14	87.00	1.8	91.24	1.49
Breast	**97.40**	0.68	95.55	0.82	95.33	1.29	95.01	1.66	96.32	1.03	96.75	0.94
Diabetes	76.03	2.46	77.31	2.73	77.34	3.17	77.23	2.81	74.09	2.73	**78.56**	2.46
Glass	**96.05**	2.92	91.84	2.78	91.65	3.23	91.30	3.09	90.18	2.60	93.33	2.21
Heart	**82.56**	2.33	76.10	3.46	81.05	2.96	71.25	8.54	80.79	2.57	82.53	2.27
Sonar	82.03	4.11	**83.48**	3.88	71.70	4.79	70.31	5.40	66.30	4.93	79.11	3.51
Wine	**98.48**	1.28	97.48	1.57	98.18	1.72	94.10	3.12	96.23	2.01	98.31	1.61
Monk1	84.28	0.56	**94.44**	0.01	74.79	3.91	69.99	13.82	80.56	0.01	85.75	1.41
Monk2	82.17	0.00	84.72	0.01	70.35	3.58	72.84	2.92	71.53	0.01	**85.84**	1.27
Monk3	**93.73**	0.28	90.04	0.01	88.77	2.31	80.41	6.07	80.79	0.01	90.44	1.00

Detailed procedure for all methods, except for the MLM approaches, can be found in [5].

From the table, one can see that, regarding accuracy, the MLM variants outperform the other methods in 5 out of 10 cases, while SVM wins in 3 cases and V-ELM wins in 2. As for the standard deviation, the MLM variants are usually comparable to SVM and V-ELM, while the other methods present significantly higher deviation. It is worth highlighting the performance of the MLM ensembles on the datasets with the highest number of features. The proposed methods achieved better classification rates in 4 out of 5 datasets of highest dimensionality. Another important result can observed when comparing the proposed methods with V-ELM. The MLM ensembles outperformed V-ELM in 6 out of 10 datasets.

5   Conclusion

This work evaluates Minimal Learning Machine ensembles for classification tasks. Four MLM ensembles were proposed based on sampling and voting strategies. Additionally, we introduced a fast alternative for computing the output estimation step in the MLM, thus allowing its usage for ensemble learning.

Despite the simple ensemble generation strategies used in this work, MLM ensembles showed promising results, outperforming some state-of-the-art algorithms in many of the evaluated datasets. Future works may investigate different strategies for ensemble generation using boosting approaches.

Acknowledgments. The authors acknowledge the support of CNPq (Grant 456837/2014-0).

References

1. Lu, J., Plataniotis, K.N., Venetsanopoulos, A.N., Li, S.Z.: Ensemble-based discriminant learning with boosting for face recognition. IEEE Transactions on Neural Networks **17**, 166–178 (2006)
2. El Abd Meguid, M.K., Levine, M.D.: Fully automated recognition of spontaneous facial expressions in videos using random forest classifiers. IEEE Transactions on Affective Computing **5**, 141–154 (2014)
3. Chen, Y., Zhao, X., Lin, Z.: Optimizing Subspace SVM Ensemble for Hyperspectral Imagery Classification. IEEE Journal of Selected Topics in Applied Earth Observations and Remote Sensing **7**, 1295–1305 (2014)
4. Asadi, N., Mirzaei, A., Haghshenas, E.: Multiple Observations HMM Learning by Aggregating Ensemble Models. IEEE Transactions on Signal Processing **61**, 5767–5776 (2013)
5. Cao, J., Lin, Z., Huang, G.-B., Liu, N.: Voting based extreme learning machine. Information Sciences **185**, 66–77 (2012)
6. Hansen, L.K., Salamon, P.: Neural network ensembles. IEEE Transactions on Pattern Analysis and Machine Intelligence **12**, 993–1001 (1990)
7. Dietterich, T.G.: Ensemble methods in machine learning. In: Kittler, J., Roli, F. (eds.) MCS 2000. LNCS, vol. 1857, p. 1. Springer, Heidelberg (2000)
8. Yoav, F., Schapire, R.E.: Experiments with a New Boosting Algorithm. Proceedings of the International Conference on Machine Learning **1**, 148–156 (1996)
9. Tsymbal, A., Pechenizkiy, M., Cunningham, P.: Diversity in search strategies for ensemble feature selection. Information Fusion **6**, 83–98 (2005)
10. Kittler, J., Hatef, M., Duin, R.P.W., Matas, J.: On combining classifiers. IEEE Transactions on Pattern Analysis and Machine Intelligence **20**, 226–239 (1998)
11. de Souza Junior, A.H., Corona, F., Miche, Y., Lendasse, A., Barreto, G.A., Simula, O.: Minimal learning machine: a new distance-based method for supervised learning. In: Rojas, I., Joya, G., Gabestany, J. (eds.) IWANN 2013, Part I. LNCS, vol. 7902, pp. 408–416. Springer, Heidelberg (2013)
12. Niewiadomska-Szynkiewicz, E., Marks, M.: Optimization Schemes For Wireless Sensor Network Localization. International Journal of Applied Mathematics and Computer Science **19**, 291–302 (2009)
13. Navidi, W., Murphy Jr., W.S., Hereman, W.: Statistical methods in surveying by trilateration. Computational Statistics & Data Analysis **27**, 209–227 (1998)

14. Marquardt, D.W.: An Algorithm for Least-Squares Estimation of Nonlinear Parameters. Journal of the Society for Industrial and Applied Mathematics. **11**, 431–441 (1963)
15. Souza Junior, A.H., Corona F., Miché Y., Lendasse, A., Barreto, G.: Extending the minimal learning machine for pattern classification. In: Proceedings of the 1st BRICS countries conference on computational intelligence, vol. 1, pp. 1–8 (2013)
16. Frank, A., Asuncion, A.: UCI Machine Learning Repository University of California. Irvine, School of Information and Computer Sciences (2010)
17. Hsu, C.W., Lin, C.J.: A comparison of methods for multiclass support vector machines. IEEE Transactions on Neural Networks **13**, 415–425 (2002)
18. Miche, Y., Sorjamaa, A., Bas, P., Simula, O., Jutten, C., Lendasse, A.: OP-ELM: Optimally Pruned Extreme Learning Machine. IEEE Transactions on Neural Networks **21**, 158–162 (2010)
19. Haykin, S.: Neural Networks, A Comprehensive Foundation 2nd ed., Pearson education Press (2001)
20. Cover, T.M., Hart, P.E.: Nearest neighbor pattern classification. IEEE Transactions on Information Theory **13**, 21–27 (1967)

Extreme Learning Machines for Multiclass Classification: Refining Predictions with Gaussian Mixture Models

Emil Eirola[1], Andrey Gritsenko[2], Anton Akusok[1,2], Kaj-Mikael Björk[1],
Yoan Miche[3,4], Dušan Sovilj[1,2,3], Rui Nian[5], Bo He[5],
and Amaury Lendasse[1,2(✉)]

[1] Arcada University of Applied Sciences, Helsinki, Finland
emil.eirola@arcada.fi,
amaury-lendasse@uiowa.edu
[2] Department of Mechanical and Industrial Engineering and the Iowa
Informatics Initiative, The University of Iowa, Iowa City, USA
[3] Department of Information and Computer Science,
Aalto University School of Science, 00076 Aalto, Finland
[4] Nokia Solutions and Networks Group, Espoo, Finland
[5] College of Information Science and Engineering,
Ocean University of China, 266003 Qingdao, China

Abstract. This paper presents an extension of the well-known Extreme Learning Machines (ELMs). The main goal is to provide probabilities as outputs for Multiclass Classification problems. Such information is more useful in practice than traditional crisp classification outputs. In summary, Gaussian Mixture Models are used as post-processing of ELMs. In that context, the proposed global methodology is keeping the advantages of ELMs (low computational time and state of the art performances) and the ability of Gaussian Mixture Models to deal with probabilities. The methodology is tested on 3 toy examples and 3 real datasets. As a result, the global performances of ELMs are slightly improved and the probability outputs are seen to be accurate and useful in practice.

Keywords: Classification · Machine learning · Neural network · Extreme learning machines · Gaussian mixture models · Multiclass classification · Leave-one-out cross-validation · PRESS statistics · Parental control · Internet security

1 Introduction

The Extreme Learning Machines and other neural networks have a successful history of being used to solve classification problems. The standard procedure is to convert the class labels into numerical $0/1$ binary variables (or equivalently, $+1/-1$), effectively transforming the situation into a regression task. When a new sample is fed through the network to produce a result, the class is assigned based on which numerical value it is closest to. While this leads to good performance

© Springer International Publishing Switzerland 2015
I. Rojas et al. (Eds.): IWANN 2015, Part II, LNCS 9095, pp. 153–164, 2015.
DOI: 10.1007/978-3-319-19222-2_13

in terms of classification accuracy and precision, the network outputs as such are not very meaningful. This paper presents a method which converts the outputs into more interpretable probabilities by using Gaussian Mixture Models (GMM).

Most classifiers based on neural networks provide results which can not directly be interpreted as probabilities. Probabilities are useful for understanding the confidence in classification, and evaluating the possibility of misclassification. In a multiclass problem, for instance, certain misclassification results may be considerably more harmful or expensive than others.

One example is in website filtering based on user-defined categories, where neural networks are used to classify previously uncategorized sites [1,2]. More reliable estimates of the risks involved are necessary for cloud security service provider to make informed (but automated) filtering decisions. Other such cases where the penalty for choosing the wrong class may vary greatly, include detecting malicious software activity [3–5], bankruptcy prediction [6] and nuclear accident prediction [7].

It is true that the optimal least-squares estimator is equivalent to the conditional probability:

$$\hat{y}(x) = \mathrm{E}[Y \mid x] = p(Y = 1 \mid x).$$

In practice, however, the results can be outside the range 0–1, and this interpretation is not very easy or useful.

Gaussian Mixture Models can be used to transform the values in the output layer to more interpretable probabilities. Specifically, this is accomplished by fitting the model to the training data and using it to calculate the probability of a sample belonging to a class, conditional on the output of the ELM. This procedure of refining the classification result of the ELM also leads to better classification accuracy and precision in some cases, as illustrated in the Experiments (section 3.2).

In related work, the Sparse Bayesian Extreme Learning Machine [8] presents another approach to use an ELM and obtain estimates of the posterior probability for each class. In the SBELM, the parameters of the ELM and the Bayesian inference mechanism are linked, and must be learned together through an iterative optimization scheme. This contrasts the currently proposed method, where the ELM and GMM layers are entirely decoupled, and can be trained separately.

The remainder of this paper is structured as follows: Section 2 reviews the Extreme Learning Machines and Gaussian Mixture Models before introducing two variants of the proposed refinement procedure. An experimental comparison on a variety of datasets is provided in Section 3. Section 4 presents conclusions and further works.

2   Global Methodologies

2.1   Extreme Learning Machines

Extreme Learning Machines (ELMs) [9] are single hidden-layer feed-forward neural networks where *only* the output weights are optimised, and all the weights between the input and hidden layer are assigned randomly (see Figure 1). Due to its fast computational speed and theoretical guarantees [10], the method recently received an active development both theoretically [11–13], including optimally pruned modification of ELM [14,15], and in applications [16], in particular: finding mislabeled samples using ELM [17], ELM for time series prediction [18,19], identification of evolving fuzzy systems using OP-ELM [20], accelerating ELM using GPU [21], ELM for regression with missing data [22], solving feature selection problem using ELM [23], ELM for nominal data classification [24], etc.

Training this model is simple, as the optimal output weights $\boldsymbol{\beta}$ can be calculated by ordinary least squares or various regularised alternatives.

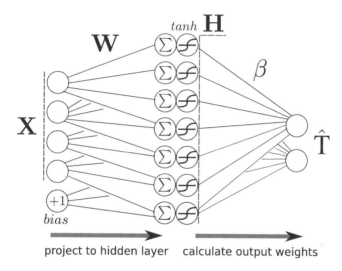

Fig. 1. Extreme learning machine with multiple outputs. Bias is conveniently included as an additional constant +1 input. Hidden layer weights $\mathbf{W}$ are fixed, only output layer weights $\boldsymbol{\beta}$ are calculated.

In the following, a multi-class classification task is assumed. The data is a set of N distinct samples $\{\boldsymbol{x}_i, y_i\}$ with $\boldsymbol{x}_i \in \mathbb{R}^d$ and $y_i \in \{1, \ldots, c\}$ where c is the number of distinct classes. Encode classification targets as one binary variable for each class (one-hot encoding). $\mathbf{T}$ is the matrix of targets such that $T_{ij} = 1$ if and only if $y_i = j$, i.e., sample i belongs to class j. Otherwise, $T_{ij} = 0$. In the case of two classes, a single output variable is sufficient.

A single (hidden) layer feedforward neural network (SLFN) with d input nodes, c output nodes, and M neurons in the hidden layer can be written as

$$f(\boldsymbol{x}) = \sum_{k=1}^{M} \boldsymbol{\beta}_k h\left(\boldsymbol{w}_k \cdot \boldsymbol{x}\right), \tag{1}$$

where $\boldsymbol{w}_k$ are randomly assigned d-dimensional weight vectors, the output layer weights $\boldsymbol{\beta}_k$ are c-dimensional vectors, and $h(\cdot)$ an appropriate nonlinear activation function, e.g., the sigmoid function. The output of f is a c-dimensional vector, and class assignment is determined by which component is the largest.

In terms of matrices, the training of the network can be re-written as finding the least-squares solution to the matrix equation.

$$\mathbf{H}\boldsymbol{\beta} = \mathbf{T}, \quad \text{where} \quad H_{ik} = h(\boldsymbol{w}_k \cdot \boldsymbol{x}_i). \tag{2}$$

Constant bias terms are commonly included by appending a 1 to each $\boldsymbol{x}_i$ and concatenating a column of 1s to $\mathbf{H}$.

2.2   PRESS Statistics for Selecting the Optimal Number of Neurons

The number of hidden neurons is the only tunable hyperparameter in an ELM model. It is selected using a Leave-One-Out (LOO) Cross-Validation error. The LOO method is usually a costly approach to optimize a parameter since it requires to train the model on the whole dataset but one sample, and evaluate on this sample repeatedly for all the samples of the dataset. However, the output layer is linear for the ELM model, and the LOO error has a closed form given by Allen's Prediction Sum of Squares (PRESS) [25]. This closed form allows for fast computation of the LOO Mean Square Error, which gives an estimate of the generalization error of ELM. The optimal number of hidden neurons is found as the minimum of that Meas Squared Error.

The Allen's PRESS formula written with the multi-output notations of the paper is

$$\text{MSE}_{\text{LOO}}^{\text{PRESS}} = \frac{1}{Nc} \sum_{n=1}^{N} \sum_{k=1}^{c} \left(\frac{\mathbf{T} - \mathbf{H}\mathbf{H}^{\dagger}\mathbf{T}}{[\mathbf{1}_N - \text{diag}(\mathbf{H}\mathbf{H}^{\dagger})]\,\mathbf{1}_c^T} \right)_{ik}^2, \tag{3}$$

where $\mathbf{H}^{\dagger}$ denotes the Moore-Penrose pseudo-inverse [26] of $\mathbf{H}$, and the division and square operations are applied element-wise.

2.3   Gaussian Mixture Models

Mixtures of Gaussians can be used for a variety of applications by estimating the density of data samples [27,28]. A Gaussian Mixture Model can approximate any distribution by fitting a number of components, each representing a multivariate normal distribution. See Figure 2 as an example.

The model is defined by its parameters, which consist of the mixing coefficients π_k, the means $\boldsymbol{\mu}_k$, and covariance matrices $\boldsymbol{\Sigma}_k$ for each component k ($1 \leq k \leq K$) in a mixture of K components. The combination of parameters is represented as $\boldsymbol{\theta} = \{\pi_k, \boldsymbol{\mu}_k, \boldsymbol{\Sigma}_k\}_{k=1}^{K}$.

The model specifies a distribution in $\mathbb{R}^d$, given by the probability density function

$$p(\boldsymbol{x} \mid \boldsymbol{\theta}) = \sum_{k=1}^{K} \pi_k \mathcal{N}(\boldsymbol{x} \mid \boldsymbol{\mu}_k, \boldsymbol{\Sigma}_k), \tag{4}$$

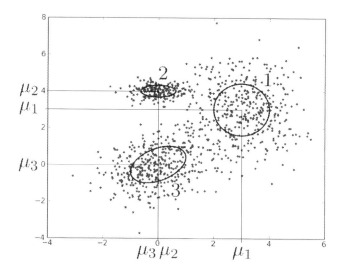

Fig. 2. An example of 2D data with 3 Gaussian components after the convergence of GMM

where $\mathcal{N}(x \mid \boldsymbol{\mu}, \boldsymbol{\Sigma})$ is the probability density function of the multivariate normal distribution

$$\mathcal{N}(x \mid \boldsymbol{\mu}, \boldsymbol{\Sigma}) = \frac{1}{\sqrt{(2\pi)^d \det(\boldsymbol{\Sigma})}} \exp\left(-\frac{1}{2}(x - \boldsymbol{\mu})^T \boldsymbol{\Sigma}^{-1}(x - \boldsymbol{\mu})\right). \quad (5)$$

The standard procedure for fitting a Gaussian Mixture Model to a dataset is maximum likelihood estimation by the Expectation-Maximisation (EM) algorithm [28–30]. The E-step and M-step are alternated until convergence is observed in the log-likelihood. Initialisation before the first E-step is arbitrary, but a common choice is to use the clustering algorithm K-means to find a reasonable initialisation [27].

The only parameter to tune select is the number of components K. This can be done by separately fitting several models with different values for K, and using the BIC criterion [31] to select the best model. In the proposed methodology, we are using the BIC criterion to select the value of K. Several further criteria are discussed in [32, Ch. 6].

2.4   ELM-GMM

The main idea of the proposed method is to first train a standard ELM for classification, and then use a GMM to refine the results into more interpretable probabilities. This is accomplished by building a separate GMM for each class, on the ELM outputs of the samples from that class. If Y is the output of the

ELM, the GMM is a model for the conditional distribution $p(Y \mid C)$ for each class. This leads to c separate GMMs.

Given a new sample, prediction is conducted as follows: calculate the ELM output Y, and apply Bayes' theorem to find the posterior probability of each class C:

$$p(C \mid Y) = p(Y \mid C)\frac{p(C)}{p(Y)}.$$

Specifically:

$$p(C \mid Y) \propto p(Y \mid C)p(C),$$

where the proportionality constant is determined by the condition of adding up to 1. The class priors $p(C)$ are given by the proportions in the training set (i.e., the maximum likelihood estimate).

The end result is now interpretable as a probability. A summary of the training and testing algorithms is presented in Algorithm 1.

Algorithm 1. Training the model and finding the conditional class probabilities for unseen data.

▷ **Training step**
Require: Input data $\mathbf{X}$, targets $\mathbf{T}$
1: Randomly assign input vectors $\boldsymbol{w}_k$ and form $\mathbf{H}$
2: Calculate $\boldsymbol{\beta}$ as the least squares solution to eq. (2)
3: Calculate outputs on training data: $\mathbf{Y} = \mathbf{H}\boldsymbol{\beta}$
4: **For each** class C **do**
5:     Fit a GMM$_C$ to the rows of $\mathbf{Y}$ corresponding to the class C
6: **End for**
7: Calculate $p(C)$ based on proportions of each class
8: **Return** $\boldsymbol{w}_k$, $\boldsymbol{\beta}$, GMM$_C$, $p(C)$

▷ **Testing step**
Require: Test data $\mathbf{X}_t$, weights $\boldsymbol{w}_k$, $\boldsymbol{\beta}$, GMM$_C$ and $p(C)$ for each class C
1: Form $\mathbf{H}_t$ by using the weights $\boldsymbol{w}_k$
2: Calculate outputs: $\mathbf{Y}_t = \mathbf{H}_t\boldsymbol{\beta}$
3: **For each** class C **do**
4:     Use GMM$_C$ to calculate $p(Y_t \mid C)$ for each sample
5: **End for**
6: Calculate $p(C \mid Y_t) \propto p(Y_t \mid C)p(C)$ for each sample
7: **Return** Conditional probabilities $p(C \mid Y_t)$ for each class for each sample

---

To evaluate performances of this model for each sample, we consider the class with the highest conditional probability as the result of classification. A second criterion is presented and used in the Experiments Section 3.4 in order to evaluate the quality and the applicability of the predicted probabilities.

2.5   Refine the Training for GMM

It is obvious that GMM built of the ELM outputs would inherit the error of the ELM model. To avoid this error accumulation, we are proposing to build GMM using only the correct classifications of the ELM. This training approach will be denoted by suffix 'r' added to the corresponding GMMs. Compared to the algorithm presented in Algorithm 1, the only change is an additional step between steps 3 and 4 of the training phase: delete the rows of $\mathbf{Y}$ corresponding to misclassified samples. In the Experiments Section 3.4, it is shown that this second approach is especially relevant when the original multiclass classification task is challenging.

3   Experiments

In the following subsections, three methodologies are compared using several classification tasks. These compared methods are the original ELM, and the two variants of the proposed combination of ELM and GMM: ELM-GMM and ELM-GMMr.

3.1   Datasets

Six different datasets have been chosen for the experiments: three small datasets and three large ones. Datasets are collected from the University of California at Irvine (UCI) Machine Learning Repository [33] and they have been chosen by the overall heterogeneity in terms of number of samples, variables, and classes for classification problems. Furthermore, the large datasets have high number of variables and a large number of classes. This is done in order to validate the quality of the predicted probabilities.

Table 1. Information about the selected datasets

Dataset	Variables	Classes	Samples Train	Test
Wisconsin Breast Cancer	30	2	379	190
Pima Indians Diabetes	8	2	512	256
Wine	13	3	118	60
Image Segmentation	18	7	1540	770
First-Order Theorem Proving	51	6	4078	2040
Cardiotocography	21	10	1417	709

Table 1 summarizes the different attributes for the six datasets. All datasets have been preprocessed in the same way. Two thirds of the points are used to create the training set and the remaining third is used as the test set. The First-Order Theorem Proving dataset has predefined training, validation and testing sets in proportion of 2:1:1. We have have performed random permutation for the validation set. Afterwards, the result of permutation for the validation set was split in two parts to be added to the test and train sets in such a way that the resulting ratio between these sets becomes 2:1. Then for all datasets, the training set is standardized to zero mean and unit variance, and the test set is also standardized using the same mean and variance calculated and used for the training set. Because the test set is standardized using the same parameters as for the training set, it is most likely not exactly zero mean and unit variance.

It should also be noted that the proportions of the classes have been kept balanced: each class is represented in an equal proportion, in both training and test sets. This is important in order to have relevant test results.

3.2   Experimental Procedure

All experiments have been run on the same Windows machine with 16 GB of memory (no swapping for any of the experiments) and 3.6 GHz processor, single-threaded execution on one single core, for the sake of comparisons.

Because ELM is a single hidden-layer feed-forward neural network with randomly assigned weights w_k, we run each method 1000 times and average its performance. We also compute the optimal value of neurons for ELM on each step using the PRESS Leave-One-Out Cross-Validation technique [25,34] with a maximum number of neurons equal to 300 based on the performance results obtained by [11].

3.3   Results

Table 2 shows the test results for the three models and six datasets. In this Table 2, each GMM is built of the ELM outputs for a certain dataset. In that table, we have removed ELMs from the names of the global methodologies for the sake of clarity.

Comparing the accuracies of ELMs to the ones of the GMM variants, some datasets (Wine, Cardiotocography) are showing that the GMM is providing a clear improvement. In the other cases, the results are not notably different, but never statistically worse. The First-Order Theorem Proving dataset is the only situation where ELM-GMM performs clearly worse, but ELM-GMMr is again better than the original ELM. For all datasets, ELM-GMMr provides similar or better results than ELM-GMM.

3.4   Reevaluate Performance of Probability Classification Methods

When calculating the performance of a probability-based classification method by just picking the class with the highest probability and treating it as a result of classification, we lose the advantage of the probability itself.

There are several possible solutions to take into account the predicted probabilities. One of the most simple solutions is to consider a classification to be correct if one of the two highest probabilities is for the correct class. If the predicted probabilities were not meaningful, the increase of performance measured

Table 2. Correct classification rates (and standard deviation in brackets) for all six datasets obtained using 3 different methods. "Wisc. B.C." for Wisconsin Breast Cancer dataset, "Pima I.D." for Pima Indians Diabetes dataset, "Image Seg." for Image Segmentation dataset, "F.-O. T.P." for First-Order Theorem Proving dataset and "Card." for Cardiotocography dataset.

	Wisc. B.C.	Pima I.D.	Wine	Image Seg.	F.-O. T.P.	Card.
ELM	95.05 (1.49)	70.90 (1.69)	93.00 (2.92)	93.67 (0.66)	52.34 (0.77)	73.63 (1.34)
GMM	95.00 (1.84)	70.40 (2.22)	94.00 (3.16)	93.96 (0.68)	50.42 (0.90)	76.62 (1.27)
GMMr	95.00 (1.49)	70.98 (1.51)	96.67 (2.81)	93.92 (0.64)	52.45 (0.75)	76.59 (1.29)

by this second criterion would be limited. For example, in the Cardiotocography dataset with a total of 10 classes the improvement is close to 15%. The standard deviation is decreased. The correct class is nearly certainly one of the two most probable predicted classes. Eight classes are then certainly discarded. Similar considerations can be made for the First-Order Theorem Proving dataset, for which the improvement is even more significant, and the other examples.

Table 3 shows the resulting improvement in accuracy for those datasets with more than two classes. This second criterion is imperfect, and will be replaced in further works. For example, probabilistic classification will be investigated in order to provide a probability distribution of the performances of the proposed methodologies.

Table 3. Comparing the improvement in classification accuracy when considering top 2 labels. "Image Seg." for Image Segmentation dataset, "F.-O. T.P." for First-Order Theorem Proving dataset and "Card." for Cardiotocography dataset.

	Wine	Image Seg.	F.-O. T.P.	Card.
GMM	94.00 (3.16)	93.96 (0.68)	50.42 (0.90)	76.62 (1.27)
GMM 2	99.50 (0.81)	97.89 (0.43)	68.48 (0.85)	91.42 (0.86)
GMMr	96.67 (2.81)	93.92 (0.64)	52.45 (0.75)	76.59 (1.29)
GMMr 2	99.50 (0.81)	97.83 (0.39)	68.79 (0.89)	91.03 (0.91)

4   Conclusions and Further Works

The proposed methodology is based on the well-known ELMs that has been shown to provide accurate classification results.

Including GMM as postprocessing preserves the qualities of ELMs. Based on the results obtained on six datasets, it has been shown that the provided predicted probabilities are accurate, useful and robust.

The drawback of the given methodology is an increase of the overfitting risk based on the fact that both ELM and GMM are trained on the same training sets. Furthermore, the optimal number of neurons for the original ELM is probably not optimal when the GMMs are added. In the future, selecting the optimal number of neuron for the proposed global methodology will be rigorously investigated.

Comparison with Sparse Bayesian Extreme Learning Machines [8] will also be done in the future, and computational times will be compared.

As described in the Experiments Section, there are needs to develop a better criterion to evaluate the quality and the advantages of dealing with probability outputs.

In the future, the proposed methodology will be tested on very large datasets, including more than one million samples, several hundreds of input variables and ten to twenty classes. For example, to perform website classification [35] where the number of given output classes is very large, and the number of samples is nearly unlimited.

References

1. Qi, X., Davison, B.D.: Web page classification: Features and algorithms. ACM Comput. Surv. **41**(2), 12:1–12:31 (2009)
2. Patil, A.S., Pawar, B.: Automated classification of web sites using naive baycsian algorithm. In: Proceedings of the International MultiConference of Engineers and Computer Scientists. vol. 1 (2012)
3. Dahl, G., Stokes, J.W., Deng, L., Yu, D.: Large-scale malware classification using random projections and neural networks. In: Proceedings IEEE Conference on Acoustics, Speech, and Signal Processing, IEEE SPS, May 2013
4. Rieck, K., Trinius, P., Willems, C., Holz, T.: Automatic analysis of malware behavior using machine learning. J. Comput. Secur. **19**(4), 639–668 (2011)
5. Miche, Y., Akusok, A., Hegedus, J., Nian, R.: A Two-Stage Methodology using K-NN and False Positive Minimizing ELM for Nominal Data Classification. Cognitive Computation, pp. 1–26 (2014)
6. Akusok, A., Veganzones, D., Björk, K.M., Séverin, E., du Jardin, P., Lendasse, A., Miche, Y.: ELM clustering-application to bankruptcy prediction-. In: International Work Conference on TIme SEries, pp. 711–723 (2014)
7. Sirola, M., Talonen, J., Lampi, G.: SOM based methods in early fault detection of nuclear industry. In: ESANN (2009)
8. Luo, J., Vong, C.M., Wong, P.K.: Sparse bayesian extreme learning machine for multi-classification. IEEE Transactions on Neural Networks and Learning Systems **25**(4), 836–843 (2014)
9. Huang, G.B., Zhu, Q.Y., Siew, C.K.: Extreme learning machine: Theory and applications. Neurocomputing **70**(1–3), 489–501 (2006)
10. Huang, G.B., Chen, L., Siew, C.K.: Universal approximation using incremental constructive feedforward networks with random hidden nodes. IEEE Transactions on Neural Networks **17**(4), 879–892 (2006)
11. Miche, Y., van Heeswijk, M., Bas, P., Simula, O., Lendasse, A.: TROP-ELM: A double-regularized ELM using LARS and Tikhonov regularization. Neurocomputing **74**(16), 2413–2421 (2011)

12. Lendasse, A., Akusok, A., Simula, O., Corona, F., van Heeswijk, M., Eirola, E., Miche, Y.: Extreme learning machine: a robust modeling technique? yes!. In: Rojas, I., Joya, G., Gabestany, J. (eds.) IWANN 2013, Part I. LNCS, vol. 7902, pp. 17–35. Springer, Heidelberg (2013)

13. Yu, Q., van Heeswijk, M., Miche, Y., Nian, R., He, B., Séverin, E., Lendasse, A.: Ensemble delta test-extreme learning machine (dt-elm) for regression. Neurocomputing **129**, 153–158 (2014). cited By 2

14. Miche, Y., Sorjamaa, A., Bas, P., Simula, O., Jutten, C., Lendasse, A.: Op-elm: Optimally pruned extreme learning machine. IEEE Transactions on Neural Networks **21**(1), 158–162 (2010)

15. Miche, Y., Sorjamaa, A., Lendasse, A.: OP-ELM: theory, experiments and a toolbox. In: Kůrková, V., Neruda, R., Koutník, J. (eds.) ICANN 2008, Part I. LNCS, vol. 5163, pp. 145–154. Springer, Heidelberg (2008)

16. Cambria, E., Huang, G.B., Kasun, L.L.C., Zhou, H., Vong, C.M., Lin, J., Yin, J., Cai, Z., Liu, Q., Li, K., Leung, V.C., Feng, L., Ong, Y.S., Lim, M.H., Akusok, A., Lendasse, A., Corona, F., Nian, R., Miche, Y., Gastaldo, P., Zunino, R., Decherchi, S., Yang, X., Mao, K., Oh, B.S., Jeon, J., Toh, K.A., Teoh, A.B.J., Kim, J., Yu, H., Chen, Y., Liu, J.: Extreme Learning Machines. IEEE Intelligent Systems **28**(6), 30–59 (2013)

17. Akusok, A., Veganzones, D., Miche, Y., Severin, E., Lendasse, A.: Finding originally mislabels with MD-ELM. In: Proc. of the 22th European Symposium on Artificial Neural Networks, Computational Intelligence and Machine Learning (ESANN 2014), pp. 689–694 (2014)

18. van Heeswijk, M., Miche, Y., Lindh-Knuutila, T., Hilbers, P.A.J., Honkela, T., Oja, E., Lendasse, A.: Adaptive ensemble models of extreme learning machines for time series prediction. In: Alippi, C., Polycarpou, M., Panayiotou, C., Ellinas, G. (eds.) ICANN 2009, Part II. LNCS, vol. 5769, pp. 305–314. Springer, Heidelberg (2009)

19. Grigorievskiy, A., Miche, Y., Ventelä, A.M., Séverin, E., Lendasse, A.: Long-term time series prediction using op-elm. Neural Networks **51**, 50–56 (2014). cited By 4

20. Pouzols, F., Lendasse, A.: Evolving fuzzy optimally pruned extreme learning machine for regression problems. Evolving Systems **1**(1), 43–58 (2010)

21. van Heeswijk, M., Miche, Y., Oja, E., Lendasse, A.: Gpu-accelerated and parallelized ELM ensembles for large-scale regression. Neurocomputing **74**(16), 2430–2437 (2011). Advances in Extreme Learning Machine: Theory and Applications Biological Inspired Systems. Computational and Ambient Intelligence Selected papers of the 10th International Work-Conference on Artificial Neural Networks (IWANN2009)

22. Yu, Q., Miche, Y., Eirola, E., van Heeswijk, M., Séverin, E., Lendasse, A.: Regularized extreme learning machine for regression with missing data. Neurocomputing **102**, 45–51 (2013). cited By 9

23. Benoît, F., van Heeswijk, M., Miche, Y., Verleysen, M., Lendasse, A.: Feature selection for nonlinear models with extreme learning machines. Neurocomputing **102**, 111–124 (2013). cited By 8

24. Akusok, A., Miche, Y., Hegedus, J., Nian, R., Lendasse, A.: A two-stage methodology using k-nn and false-positive minimizing elm for nominal data classification. Cognitive Computation **6**(3), 432–445 (2014). cited By 0

25. Allen, D.M.: The relationship between variable selection and data agumentation and a method for prediction. Technometrics **16**(1), 125–127 (1974)

26. Rao, C.R., Mitra, S.K.: Generalized Inverse of Matrices and Its Applications. John Wiley & Sons Inc (1971)

27. Bishop, C.M.: Pattern Recognition and Machine Learning. Springer (2006)
28. Eirola, E., Lendasse, A., Vandewalle, V., Biernacki, C.: Mixture of gaussians for distance estimation with missing data. Neurocomputing **131**, 32–42 (2014)
29. Dempster, A.P., Laird, N.M., Rubin, D.B.: Maximum likelihood from incomplete data via the EM algorithm. Journal of the Royal Statistical Society **39**(1), 1–38 (1977). Series B (Methodological)
30. McLachlan, G., Krishnan, T.: The EM Algorithm and Extensions. Wiley Series in Probability and Statistics. John Wiley & Sons, New York (1997)
31. Schwarz, G.: Estimating the dimension of a model. The annals of statistics **6**(2), 461–464 (1978)
32. McLachlan, G.J., Peel, D.: Finite Mixture Models. Wiley Series in Probability and Statistics. John Wiley & Sons, New York (2000)
33. Lichman, M.: UCI Machine Learning Repository (2013). http://archive.ics.uci.edu/ml
34. Myers, R.: Classical and Modern Regression with Applications. Bookware Companion Series, PWS-KENT (1990)
35. Akusok, A., Grigorievskiy, A., Lendasse, A., Miche, Y.: Image-based classification of websites. In: Villmann, T., Schleif, F.M. (eds.) Machine Learning Reports 02/2013. Volume ISSN: 1865–3960 of Machine Learning Reports., Saarbrücken, Germany, Workshop of the GI-Fachgruppe Neuronale Netze and the German Neural Networks Society in connection to GCPR 2013, Proceedings of the Workshop - New Challenges in Neural Computation 2013, pp. 25–34 September 2013

Modeling the EUR/USD Index Using LS-SVM and Performing Variable Selection

Luis-Javier Herrera[1]([✉]), Alberto Guillén[1], Rubén Martínez[2], Carlos García[2], Hector Pomares[1], Oresti Baños[1], and Ignacio Rojas[1]

[1] Department of Computer Architecture and Technology,
Universidad de Granada, Granada, Spain
jherrera@ugr.es
[2] CoTrading S. L., Barcelona, Spain

Abstract. As machine learning becomes more popular in all fields, its use is well known in finance and economics. The growing number of people using models to predict the market's behaviour can modify the market itself so it is more predictable. In this context, the key element is to find out which variables are used to build the model in a macroeconomic environment. This paper presents an application of kernel methods to predict the EUR/USD relationship performing variable selection. The results show how after applying a proper variable selection, very accurate predictions can be achieved and smaller historical data is needed to train the model.

1 Introduction

The foreign exchange market is one of the most liquid markets together with the derivate financial market. The EUR/USD relationship is one of the most negotiated as they represent the two most world powerful economies, Europe and the United States of America. The price of the foreign exchange is the main point of interest of various Economists and experts that are wondering about those prices and whether they can be a picture of their economies or not [2]. It is important to highlight that this asset is very complex and it can be affected by several circumstances, as there are several agents on both sides, offering and demanding.

The role of the central banks, the Federal Reserve as well as from the European Central Bank, have been meaningful in the last few years for the fluctuation of foreign exchanges [7]. The importance of the changes related to interests has been decisive for their economies and for the appreciation and depreciation of the relationship EUR/USD along their history. The improvement of the currency mass has been reflected in the stock markets like in the relationship analysed in this paper. These and other aspects can be detected studying how the macroeconomic variables affect the final price EUR/USD.

This study will focus on the creation of a model that is able to check the impact that variables have on the quotation. In order to detect those facts in the

© Springer International Publishing Switzerland 2015
I. Rojas et al. (Eds.): IWANN 2015, Part II, LNCS 9095, pp. 165–172, 2015.
DOI: 10.1007/978-3-319-19222-2_14

fluctuation of foreign exchanges, we will choose as a reference the macroeconomic data from Europe and the USA, stock markets and raw materials like gold and oil.

2   Data Set Definition

There are plenty of reference data that could be chosen, however, this paper presents a selection based on the experience of a trading company. The variables have been shown trough the time that are meaningful and important to each economy, making posible to create a solid model. As the paper is focused in the EUR/USD relationship, the macroeconomic variables are taken from both, European and Northamerican sources.

The EUR/USD Foreign Exchange Rate is the output variable to be predicted. It is a type of negotiated change in financial markets, and it is the most liquid asset.The data used in the following section can be downloaded from http://research.stlouisfed.org/fred2/series/EXUSEU.

The subset of macroeconomic variables and their sources are listed below:

1. **Consumer Price Index for All Urban Consumers (CPIAUCNS)** The Consumer Price Index is a statistical measurement of the development, of all prices of goods and services consumed. A high Price Index Consumer means a significant loss of purchasing power. Source: http://research.stlouisfed.org/fred2/series/CPIAUCNS/downloaddata?cid=9
2. **Civilian Unemployment Rate (UNRATE)**. Unemployment rate is the percentage of the total work force that is unemployed actively. The lower the better for the currency value. Source: http://research.stlouisfed.org/fred2/series/UNRATE/downloaddata.
3. **10-Year Treasury Constant Maturity Rate (GS10)**. It represents the interest rate the U.S government would pay on top of principal to the bond holder once ten years have passed. The market value of an existing bond will move in the opposite direction of the change in market interests. The bond yields actually serve as an excellent indicator of the strength of a nations stock market, which increases the value of currencies. Source: http://research.stlouisfed.org/fred2/series/GS10.
4. **Effective Federal Funds Rate (FEDFUNDS)**. It is the interest ratio at which depositary institutions exchange money. The Reserve Federal regulating banks and other important financial institutions to ensure the safety and soundness between them. Monetary policy decisions involve setting the interest rate. Source: http://research.stlouisfed.org/fred2/series/FEDFUNDS/.
5. **Real Effective Exchange Rates Based on Manufacturing Consumer Price Index for the United States (CCRETT01USM661N)**. Wellbeing or utility depends on consumption. This indicator measures the health of the economy over consumption. Source: http://research.stlouisfed.org/fred2/series/CCRETT01USM661N/downloaddata.
6. **All Employees: Total nonfarm (PAYEMS)**. This measure provides useful insights into the current economic situation. It measures the number of U.S

workers in the economy that excludes proprietors, private household employees and unpaid volunteers. The stability currency depends on the labor market. Source: https://research.stlouisfed.org/fred2/series/PAYEMS/.

7. **M2 Money Stock (M2NS)**. It is a monetary indicator sign that shows
   us the amount of money that there is in a country or region. In this fact, it
   is the amount of money that the houses of the USA have. M2 is formed by
   saving deposits, the small-denomination time deposits less than 100.000$ and
   the balances in retail money market mutual funds in amount retail trade. A
   good situation of M2 makes us understand an upward trend of the currency,
   because saving is a fundamental matter for the good working of the country.
   The more savings we have, the more capacity to affront a financial crisis we
   will have. The currency values significantly at a bigger saving in the reference
   country. Source: http://research.stlouisfed.org/fred2/series/M2NS.

8. **Trade Balance: Goods and Services, Balance of Payments Basis
   (BOPGSTB)**. The trade balance in a country shows the exportations minus
   the importations. If the balance were positive, we would be talking about a
   positive trade balance that reflects income in the balance of payments. A negative trade balance would damage the economy each time the impact in the
   trade balance is more relevant because of the globalization. A strong country
   should have a positive trade balance and it would also have an appreciation in
   its currency. Source: http://research.stlouisfed.org/fred2/series/BOPGSTB.

9. **WTI Crude Oil Spot Price Cushing**. The price of oil affects significantly
   to the price of one currency for its relationship with other economic indices.
   This variable shows the price of a barrel in dollars. Oil is directly related to
   prices, which is one of the variables studied, the level of household savings
   and exports. Source: http://www.eia.gov/dnav/pet/hist/LeafHandler.ashx?
   n=PET&s=RWTC&f=D.

10. **S&P 500 Index**. The S&P500 is an American stock index based on the cap-
    italization of five hundred large companies with shares traded on the NYSE
    or NASDAQ. Source: https://www.quandl.com/data/YAHOO/INDEX_
    GSPC-S-P-500-Index.

11. **DAX Index (Germany)**. It is the benchmark of Germany and possibly
    Europe. It contains thirty companies over capitalization and it is a clear
    indicator of reference when assessing the euro. Source: https://www.quandl.
    com/data/YAHOO/INDEX_GDAXI-DAX-Index-Germany.

12. **Gold**. It is one of the most traded commodities worldwide. Its price is rel-
    evant because the gold is used to protect the part of investors in times of
    crisis. Therefore it is a good sign of the state of the economy. Source: http://
    research.stlouisfed.org/fred2/series/GOLDAMGBD228NLBM.

3   Model Design

To perform the prediction of the future values, Least Squares Support Vector
Machines (LS-SVMs) [10], have been used. They are kernel-based methods so
they are also known as Kernel Ridge Regression method (KRR) [9].

These models are well suited for function approximation and they have some advantages over classical Support Vector Regression (SVR):

- easier mathematical resolution
- the parameter ε used in SVR is not needed
- the number of Lagrange multipliers is reduced to half.

Nonetheless, one of the main problems with LS-SVMs is that they do not generate sparse models so risk of overfitting has to be controlled.

In case we consider Gaussian kernels, σ is the width of the kernel, that together with the regularization parameter γ, are the hyper-parameters of the problem. Note that in the case in which Gaussian kernels are used, the models obtained resemble Radial Basis Function Networks (RBFN); with the particularities that there is an RBF node per data point, and that overfitting is controlled by a regularization parameter instead of by reducing the number of kernels [8].

In LS-SVM, the hyper-parameters of the model can be optimized by cross-validation. Nevertheless, in order to speed-up the optimization, a special formulation for a reduced cost evaluation of the cross-validation error of order l (l-fold CV) taken from the work [1] was used. With this formulation, the error evaluation cost of cross-derivation does not depend on the order l, but on the number of data points of the problem, since in fact the computational cost is dominated by the inversion of the kernels K activation matrix. Such inversion is performed through a Cholesky decomposition; the most efficient exact algorithm for this case is $O(N^3)$ where N is the number of samples.

In order to perform the evaluation of the performance of the stopping criteria in the forward selection strategy, it is necessary to learn a number of LS-SVMs, each one considering the eventual state of the selected subset in the iterative process X_G of the variable selection process. This requires therefore the training of a considerable number of LS-SVM, depending on the problem. In this work, this process was distributed in a computer cluster, so that each training process of a LS-SVM was sent to a different node. This way the computational time was reduced in a factor of N (considering a computer with N nodes, ignoring communication delays), supposing that every execution takes the same amount of computational time.

4   Feature Selection

In this work, a Mutual Information (MI) -based feature selection algoritm has been used with the objective of finding the most relevant factors needed to predict the EUR/USD exchange rate. Mutual information comes from Shannon's Information Theory, and can be expressed as

$$I(X,Y) = H(Y) - H(Y|X), \tag{1}$$

To estimate the mutual information, only the estimate of the joint probability density function (PDF) between X and Y is needed [8]. For continuous variables, this estimation is complex. However in recent years, a k-nearest neighbours-based mutual information estimator technique [6] has opened the door to more robust MI estimations among groups of variables [3].

The feature selection algorithm used in this work [4] makes use of the Markov Blanket concept [5]. Markov blankets are groups of variables M_i that subsume all the information that a single variable x_i has with respect to a different variable (or group of variables) Y; in practice and for our purposes, with respect to the objective variable. The algorithm consists of a backwards variable selection method which starts with the complete set of variables, and iteratively discards those which are detected to have a Markov Blanket in the remaining set X_G of variables, i.e. those whose information with respect to Y is already present in the remaining set X_G of variables [11].

The algorithm states the following steps:

1. Calculate the MI between every pair of input variables $I(x_i, x_j)$
2. Starting from the complete set of input variables $X_G = X$, iterate:
   a) For each variable x_i, let the candidate Markov blanket M_i be the set of p variables in X_G for which $I(x_i, x_j)$ is highest.
   b) Compute for each x_i
   c) Choose the x_i for which $Loss_i$ is lowest and eliminate x_i from X_G.
3. Continue with step 2 until no variables remain.

This way, a ranking of relevance of variables (in reverse order) is obtained. Under this operation, it is to be noted that variables that have low influence with respect to the output variable (irrelevant variables) will be soon discarded, as $Loss_i$ value should tend to 0. Similarly, redundant variables will be iteratively discarded at earlier stages. Relevant variables with low redundancy will be the last ones in being chosen.

The p parameter of the algorithm (in step 2.a of the algorithm) will take the value $p = 1$, as recommended in previous works [4] [11].

5   Experiments

This section presents the dataset and how the series were defined considering different subsets of variables. Afterwards, the models are designed showing the approximation errors and a final comment on the behaviour of the models considering different variables is made.

5.1   Defining Regressors and Data Sets

Taking the data from the sources specified in previously, the data sets is built considering a monthly based. More concretely, all the measurements in the previous month, three months ago and a year ago. Output variable was differentiated so that difference between the current month and previous month is the objective to be estimated. Given $X(t) = \{x_1(t), x_2(t), x_3(t)..., x_{12}(t)\}$ as the value of the independent variables at month t, the initial regressors considered for the modeling problem are

$$\hat{Y}(t) = y(t) - y(t-1) = F(X(t), X(t-1), X(t-3, X(t-12))) \qquad (2)$$

as suggested by the trading experts, and being $y(t)$ the EUR/USD change at time step t. After arranging all the variables, the final dataset consist of 139 samples corresponding to the 11 years and 7 months of data available.

5.2 Variable Selection and Regression Results

The variable selection was performed building the corresponding LSSVM for each subset of variables obtaining several values for Root Mean Squared Error for test that are represented graphically in Figure 1. It is easy to see that the information provided by most of the variables does not improve the accuracy although there are some of them that are critical to obtain proper results.

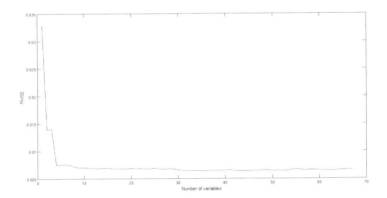

Fig. 1. Evolution of the Root Mean Squared Test Error as the number of variables increases

In Figure 2 are depicted the real output for the EUR/USD asset and the approximations obtained by the LSSVM using 3,10 and all variables. Three variables correspond to the first local minimum in the optimization process (LS-SVM attain a 0.94 of R2, meaning the model explains 94% of the variance of the output); ten variables corresponds to the second local minimum observed in the process(LS-SVM attain a 0.95% of R2, meaning the model explains 94% of the variance of the output). Choosing one or another would depend on the tradeoff between interpretability and accuracy desired. Optimal performance is obtained using in any case 10 variables. Table 1 shows the approximation errors of LSSVM and Radial Basis Function Neural Networks (RBFNNs) showing that, the variable selection is adequate and the modeling can be performed by several paradigms correctly.

Table 1. Approximation errors (using RMSE) comparison between LSSVM and RBFNN with 5 neurons

	LSSVM	RBFNN
3 variables		
Train	0.0138 (1.1e-4)	0.0313 (7.01e-3)
Test	0.0105 (9.3e-3)	0.0116(1.01e-2)
10 variables		
Train	0.0068(2.8e-4)	0.0102 (3.5e-3)
Test	0.0055(4.4e-3)	0.0067(8.4e-3)
All variables		
Train	0.0099(2.7e-4)	0.0230(3.8e-3)
Test	0.00554(3.9e-3)	0.0130(9.9e-3)

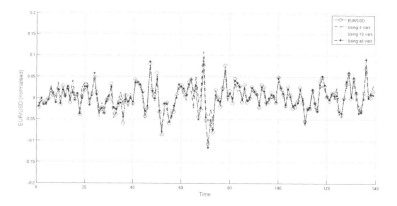

Fig. 2. This figure plots the real output (blue line) versus the approximations made by LSSVM using 3, 10 and all variables, according to equation 2

6   Conclusions and Further Work

The prediction of the relationship between the euro and the US dollar remains as a difficult task due to the macroeconomic environment and the operations on this assets that condition its own value. This paper has presented the application a ranking of macroeconomic variables based on experts' advice and numerical results. The subset of variables were feeded to an LSSVM in order to predict the final value. The reduced subset of variables were able to provide enough information to model the relationship, making the trading easier as traders could consider less variables.

Acknowledgments. This work has been supported by the GENIL-PYR-2014-12 project from the GENIL Program of the CEI BioTic, Granada, and the Junta de Andalucia Excellence Project P12-TIC-2082.

References

1. An, S., Liu, W., Venkatesh, S.: Fast cross-validation algorithms for least squares support vector machine and kernel ridge regression. Pattern Recogn. **40**(8), 2154–2162 (2007)
2. Ettredge, M., Gerdes Jr., J., Karuga, G.G.: Using web-based search data to predict macroeconomic statistics. Commun. ACM **48**(11), 87–92 (2005)
3. François, D., Rossi, F., Wertz, V., Verleysen, M.: Resampling methods for parameter-free and robust feature selection with mutual information. CoRR, abs/0709.3640 (2007)
4. Herrera, L.J., Pomares, H., Rojas, I., Verleysen, M., Guilén, A.: Effective input variable selection for function approximation. In: Kollias, S.D., Stafylopatis, A., Duch, W., Oja, E. (eds.) ICANN 2006. LNCS, vol. 4131, pp. 41–50. Springer, Heidelberg (2006)
5. Koller, D., Sahami, M.: Toward optimal feature selection. In: Saitta, L. (ed.) Proceedings of the Thirteenth International Conference on Machine Learning (ICML), pp. 284–292. Morgan Kaufmann Publishers (1996)
6. Kraskov, A., Stogbauer, H., Grassberger, P.: Estimating mutual information. Phys. Rev. E **69**, 066138 (2004)
7. Martinsen, K., Ravazzolo, F., Wulfsberg, F.: Forecasting macroeconomic variables using disaggregate survey data. International Journal of Forecasting **30**(1), 65–77 (2014)
8. Rossi, F., Lendasse, A., Franois, D., Wertz, V., Verleysen, M.: Mutual information for the selection of relevant variables in spectrometric nonlinear modelling. Chem. and Int. Lab. Syst. **80**, 215–226 (2006)
9. Saunders, C., Gammerman, A., Vovk, V.: Ridge regression learning algorithm in dual variables. In: Proceedings of the 15th International Conference on Machine Learning, pp. 515–521. Morgan Kaufmann (1998)
10. Suykens, J.A.K., Van Gestel, T., De Brabanter, J., De Moor, J., Vandewalle, B.: Least Squares Support Vector Machines. World Scientific, Singapore (2002)
11. Del Mar Perez, M., Val, J., Negueruela, I., Lafuente, V., Herrera, L.J.: Firmness prediction in prunus persica calrico peaches by visible/short-wave near infrared spectroscopy and acoustic measurements using optimised linear and non-linear chemometric models. J. Sci. Food Agric., 15, September 2014

Embedded intelligent systems

Modeling Retina Adaptation
with Multiobjective Parameter Fitting

Pablo Martínez-Cañada$^{(\boxtimes)}$, Christian Morillas, Samuel Romero,
and Francisco Pelayo

CITIC and Department of Computer Architecture and Technology,
University of Granada, Granada, Spain
{pablomc,cmg,sromero,fpelayo}@ugr.es

Abstract. The retina continually adapts its kinetics, average response
and sensitivity to the conditions of the environment. Retinal neurons
adapt essentially to the mean light intensity and its temporal fluctuations over the mean, also called temporal contrast. Contrast adaptation has two distinct temporal expressions with fast and slow components. Here, we present a configurable retina simulation environment that
accurately reproduces both contrast components. A contrast increase in
the visual input accelerates kinetics of the filter, reduces sensitivity and
depolarizes the membrane potential. Slow adaptation does not affect
the temporal response but produces a progressive hyperpolarization of
membrane potential. The implemented model for contrast adaptation
provides a neural basis of each retinal stage, from photoreceptors up to
ganglion cells, to explain the observed retina behavior. Both forms of
contrast adaptation, fast and slow, are captured by a combined model of
shunting feedback of bipolar cells and short-term plasticity (STP) at the
bipolar-to-ganglion synapse. Biological accuracy of the model is evaluated by comparison of the measured neural response with the simulated
response fitted to published physiological data. One problem with the
simulated model is finding its optimal parameter settings, since the model
response is described by a complex system of different retina stages with
linear, nonlinear and feedback connections. We propose to use a multiobjective genetic optimization to automatically search the parameter space
and easily find a feasible configuration solution.

Keywords: Visual adaptation · Multiobjective genetic optimization ·
Retina simulator · Shunting inhibition · Short-term plasticity

1 Introduction

The visual system quickly adapts its dynamic range to encode more efficiently
changes in the environment [1–3]. Neural sensitivity is increased when input signals are weak to improve the signal-to-noise ratio. However, when input signals
are strong the neural amplification factor is reduced and the time course accelerated to prevent the response from saturation and to anticipate faster temporal
patterns. At the earliest stages of the visual system neurons adapt primarily to

© Springer International Publishing Switzerland 2015
I. Rojas et al. (Eds.): IWANN 2015, Part II, LNCS 9095, pp. 175–184, 2015.
DOI: 10.1007/978-3-319-19222-2_15

the mean light intensity and its standard deviation relative to the mean over time, known as temporal contrast. Our work focused on reproducing contrast adaptation mechanisms using a retina model automatically configured by a multiobjective genetic algorithm.

Two different temporal components have been observed for contrast adaptation: a fast change that occurs within the first 100 ms and a slow change over 10 s [4–8]. When contrast of the visual input changes from low to high values, temporal filtering quickly accelerates, sensitivity decreases, and the average response increases. If a high variance is maintained over time, the temporal response is not modified but the ganglion membrane potential shows a slow decay. Upon a decrease in contrast, all these changes reverse direction but with asymmetric time constants for slow adaptation [4,7,9].

Several models have been proposed for contrast adaptation [4,10–13]. However they focused on only a few aspects of adaptation or do not fully characterize all neural stages in the retina. Moreover, some of them used functional modules that do not clearly connect with the underlying biophysical mechanisms. We present a retina model that reproduces both slow and fast components of contrast adaptation and provides a neural mechanism at each retinal stage, from photoreceptors up to ganglion cells, to explain the observed retina behavior. To evaluate the biological accuracy of the model, the physiological experiment described by Ozuysal and Baccus [4] was reproduced using a configurable retina simulation environment that can approximate different retina models. Further details of this platform are provided in previous publications [14]. A multiobjective genetic algorithm automatically fits responses of the simulated multistage model to neural responses measured in the experiment. Since the simulated response and its objective function cannot be easily described by a mathematical model, because of the multiple complex interactions among the different retina stages (e.g., feedback connections), a genetic algorithm is used as an efficient approach to find an optimal solution.

The rest of the paper is organized as follows. In section 2 we detail the neural model implemented to reproduce contrast adaptation. Parameter fitting and simulation results of the physiological experiment are described in section 3. Finally, in section 4, we discuss the conclusions.

2   Retinal Circuitry for Contrast Adaptation

Contrast adaptation originates in bipolar cells and neither photoreceptors nor horizontal cells are involved in the process [5,6]. Recent experiments have shown that contrast adaptation effects are still present under physiological blockade of amacrine synapses, ruling out a critical role for amacrine cells in driving contrast adaptation [2,5,15]. Slow adaptation mechanisms are apparently driven by prolonged depression of glutamate release at bipolar cell synapses [4,9,16–19], whereas inactivation of voltage-dependent Na+ channels in ganglion cells [7,20] and calcium-related mechanism in bipolar cells [15] may be responsible for the fast component. In addition, a large fraction of adaptation has been observed at the bipolar-to-ganglion synapse [5,21].

A well-known mathematical tool, the linear-nonlinear analysis (LN) [5–8,22], is usually used to fully characterize contrast adaptation. A LN analysis generates two output plots, represented by a linear filter and a static nonlinearity, that describe the filtering properties of a neuron. The LN analysis separates the temporal behavior of the cell from nonlinear response components (e.g., synaptic rectification or membrane depolarization). The neural response is first correlated with the input pattern to obtain the temporal filter. This filter is convolved afterwards with the stimulus to generate a linear model of the response. Then, the fixed nonlinearity is calculated by plotting the response against this linear model of the response [2,6].

Our model of contrast adaptation places temporal adaptation and changes of the static nonlinearity at different retina stages. Both forms of contrast adaptation, fast and slow, are captured by a combined model of shunting feedback [10,11] and short-term plasticity (STP) at the bipolar-to-ganglion synapse [4,9,16–19] (Figure 1). A whole retina architecture is described by this model. Every retinal layer of the model is represented by a series of biophysical mechanisms that explain some specific aspect of the signal processing, such as membrane potential integration or synaptic rectification. These mechanisms, provided by a simulation environment [14], are based on well-known retina models recurrently used in the literature to characterize different physiological experiments.

Visual input is first processed by the photoreceptor layer through a double-stage process that includes a temporal linear filter and a static nonlinearity. Since contrast adaptation is not present at this retina stage, a linear approximation of the neural response, $L(t)$, is defined based on the linear kernel $K(x, y, \tau)$ [23,24]:

$$L(t) = \int_0^\infty d\tau \int_{(x,y)\epsilon RF} K(x, y, \tau) s(x_0 - x, y_0 - y, t - \tau) dx dy \qquad (1)$$

where $s(x, y, t)$ is the visual stimulus and RF the receptive field of the cell. The neural response depends linearly on all past values of the input stimulus located in the cells receptive field RF. This integral corresponds to the well-defined convolution operation:

$$L(t) = (s * K)(x_0, y_0, t) \qquad (2)$$

$K(x, y, t)$ can be broken down as a product of two functions, one that accounts for the spatial receptive field and the other one for the temporal receptive field:

$$K(x, y, t) = K_s(x, y) K_t(t) \qquad (3)$$

The spatial receptive field, $K_s(x, y)$, is modeled as a Gaussian function, similarly to kernels used in the receptive field model proposed by Rodieck [25] and Enroth-Cugell and Robson [26]. An exponential cascade function, $E_t(t)$, based on the implementation of Virtual Retina [10], was adapted to model the temporal filter, $K_t(t)$. This type of filters, with multiple low-pass stages, has been commonly used to characterize processes such as the phototransduction cascade in cones [27].

$$E_t(t) = \frac{(nt)^n \exp(-nt/\tau)}{(n-1)! \tau^{n+1}} \qquad (4)$$

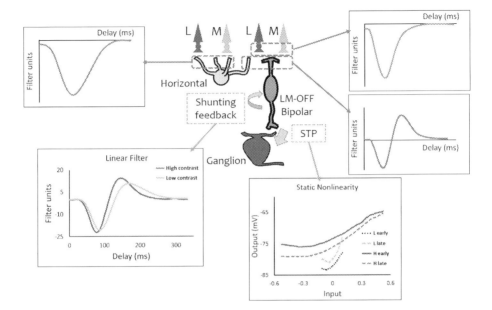

Fig. 1. Schematic view of the retina circuit proposed to reproduce contrast adaptation, where L and M correspond to L- and M-cones, respectively. Temporal kernels of photoreceptors (top-right) and horizontal cells (top-left) are represented by low-pass kernels implemented as exponential cascade filters with different time constants. Subtraction of the signal from photoreceptors and horizontal cells at the Outer Plexiform Layer, which is the layer of neural synapses that connects photoreceptors, horizontal and bipolar cells, produces the typical biphasic shape observed in bipolar and subsequent neural layers. The proposed model places temporal adaptation and changes of the static nonlinearity at different retina stages. Bipolar shunting feedback mechanism is responsible for adaptation of the linear filter, whereas the STP module at bipolar-to-ganglion synapse for polarization and hyperpolarization offsets of the nonlinearity. It is shown at the bottom of the figure a representation of the type of results that are obtained for the contrast experiment (shunting feedback and STP module). Further details of this experiment and the nomenclature used are included in the text and in Figure 2.

The exponential cascade filter peaks at time τ and the filter shape is controlled by the number of low-pass stages, n. The synaptic output of photoreceptors is again delayed by a similar low-pass scheme implemented at horizontal cells (Figure 1). Linear subtraction of the signal from photoreceptors and horizontal cells at the Outer Plexiform Layer, which is the layer of neural synapses that connects photoreceptors, horizontal and bipolar cells, produces the typical biphasic shape observed in bipolar and subsequent neural layers [5,6] (Figure 1).

Membrane potential of bipolar cells is described by a single-compartment model. The basic equation that explains the temporal evolution of a single-compartment model is [23]:

$$C_m \frac{dV(t)}{dt} = \sum_i I_i(t) + \sum_j g_j(E_j(t) - V(t)) \tag{5}$$

where the index j indicates the input ionic channel, C_m is the membrane capacitance, V the membrane potential, g_j is the conductance of the channel, E_j the reversal potential of the channel and the term $\sum_i I_i$ denotes the sum of external input currents. Channel conductances are modified by delayed and rectified feedback from bipolar output to reproduce the shunting inhibition effect. Shunting inhibition has been used to reproduce nonlinear mechanisms of the retina, such as contrast and luminance gain control [10,11], directional selectivity to motion [28,29], and normalization of the linear response in the primary visual cortex [30]. Temporal adaptations in the linear filter of the LN analysis are produced by shunting inhibition (Figure 1).

Polarization and hyperpolarization offsets of the nonlinearity are implemented by a model of short-term plasticity. It was suggested that opposing mechanisms of plasticity (i.e., depression and facilitation) could be combined together to compensate the mutual information loss [31]. Following this idea, the model includes a short-term plasticity module that correlates synaptic weight with the neural input to simulate a depolarizing offset of the ganglion membrane for high contrast steps [4,6]. On the other hand, synaptic depression occurs for maintained values of contrast with the synaptic offset decaying exponentially back to its resting value. This module is defined by:

$$P = P + k_f(k_m(t)abs(input) - P) \tag{6}$$

where P is the offset of the synapse, the parameter k_f controls the degree of facilitation, and the factor $(k_m(t)abs(input) - P)$ prevents the offset from growing indefinitely. A rectification of the input is applied by the term of absolute value. A normalization of the input would be required, and the term would become $abs(input - E_V)$, if the bipolar input had an offset E_V. The variable k_m is responsible for the slow depression of the synapse. Its exponential decay is approximated by:

$$k_m(t+1) = k_{mInf} + (k_m(t) - k_{mInf})\exp(-step/tau) \tag{7}$$

with a temporal constant defined by the quotient of the simulation $step$ and the parameter tau. k_{mInf} fixes the resting value and is inversely proportional to the input using a depression factor k_d:

$$k_{mInf} = \frac{k_d}{abs(input)} \tag{8}$$

3   Parameter Fitting and Simulated Neural Response

The model for contrast adaptation is described by a complex multistage system
with different retinal layers interconnected by linear, nonlinear and feedback
synaptic mechanisms. Thus, its simulated neural response and consequently the
optimization error function, understood as the difference between the simulated
and the measured responses, cannot be easily described by mathematical functions. Moreover, the lack of smoothness in the error function and the existence
of multiple local minima are another critical aspects to be considered when
selecting the optimization method. A multiobjective genetic algorithm can be
an efficient and easy solution to simultaneously fit the model to the different
measured neural responses.

We fit the model using a multiobjective genetic algorithm provided by the
Python library DEAP [32]. DEAP is a very flexible and intuitive evolutionary
computation framework that allows a rapid prototyping of different optimization
algorithms. Besides, the retina simulation platform is interfaced with NEST [33]
to generate spiking activity of ganglion cells. An evolutionary computation tool
that is executed in Python, such as DEAP, facilitates its integration with the
NEST simulation script, defined in PyNEST (its Python interface), and sim-
plifies the code by combining both simulation and optimization into a single
script.

A general evolutionary search was configured whereby the random initial
population of solutions is evolved by applying crossover (two-point crossover)

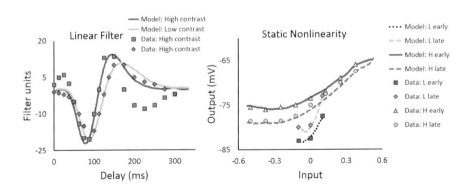

Fig. 2. Comparison of the LN analysis obtained from simulation results (solid line)
and the LN analysis of physiological data (markers). Four different contrast intervals
are considered in the measurements: 'L early' corresponds to the first 10 seconds after
a low contrast step and 'L late' to the period from 10 to 20 seconds after a low contrast
step. 'H early' and 'H late' are defined similarly for a high contrast step. A contrast
increase in the visual input accelerates kinetics of the filter and its response becomes
more differentiating. At the same time, the static nonlinearity shows an increase of
the offset and a decrease of the average slope. Slow adaptation does not affect the
temporal response, thus only two contrast periods are considered for the linear filter,
but produces a progressive hyperpolarization of membrane potential.

and mutation (gaussian mutation) operators in combination with a selection mechanism (tournament selection). Individuals are evaluated every new generation by a multiobjective fitness function with the same minimization weights for all objectives. The fitness function simulates a forty-second sequence of the retina model with parameters set by values of the individual evaluated. It computes then the LN analysis of simulation results and generates an error metric between simulated and measured data. By simplicity and effectiveness, a mean squared error (MSE) was used as estimator of this error function:

$$MSE = \sum_i (r_{measured}(i) - r_{simulated}(i))^2 \qquad (9)$$

where $r_{measured}$ are sampled values from the LN analysis of physiological results and $r_{simulated}$ are generated by simulation. Target physiological data were obtained by sampling curves of the LN analysis published by Ozuysal and Baccus [4]. Nearly total independence of the shunting feedback mechanism, affecting only the linear filter, and the STP module, responsible for variations of the offset, allows a double-stage minimization process that reduces the parameter search space and hence the computation time.

Twelve parameters of the retina model, up to bipolar cells, were first fitted simultaneously to the high and low contrast curves of the linear filter. These parameters control the shape of temporal kernels, such as τ and n in equation 4,

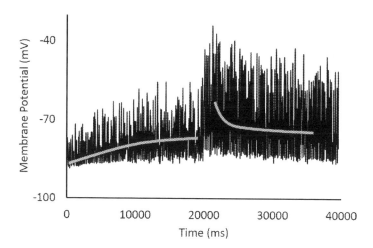

Fig. 3. Simulated neural response of ganglion cells over time. A low contrast stimulus is presented from 0 to 20000 ms, after a high contrast period before 0 ms. At 20000 ms there is a high contrast step. The response to a high contrast step is characterized by two distinctive temporal behaviors: first a fast hyperpolarization of the membrane potential is produced, followed by a slow decay with a higher time constant. A low contrast step is described by a single time constant that produces a slow increase in the offset, prolonged over the whole period.

and the shunting feedback mechanism described in equation 5. Then, by fixing the optimized parameters of the first retina stages we fitted seven parameters of the STP module to the four curves observed for the static nonlinearity (Figure 2). Optimal parameter values are within the biological range reported in different physiological studies (e.g., values of τ for photoreceptors and the slow mechanism of the STP module are 0.085 and 12 s, respectively).

For a high step contrast, the retina model captures both the decrease in the time to peak and a more differentiating response of the linear filter, and the decrease of the average slope of the nonlinearity. Similarly to the LNK model [4] our simulation cannot reproduce oscillations of the high contrast curve at the beginning (< 50 ms) and the end (> 150 ms) of the filter time course. These oscillations of the high contrast curve do not represent the neural response of all cells measured [6,21]. However, they tend to appear in cells that strongly adapt to contrast and further research is required to model this behavior.

A high contrast step also produces a fast depolarization of the membrane that is represented by the increase of the offset in the static nonlinearity (Figure 2). When a high variance is maintained over time the temporal response is not modified but the membrane potential shows a slow decay. Upon a decrease in contrast, all these changes reverse direction but with asymmetric time constants for slow adaptation [4,7] (Figure 3).

4   Discussion

We described a neural model that reproduces contrast adaptation in the retina. Both forms of contrast adaptation, fast and slow, are captured by a combined model of shunting feedback [10,11] and short-term plasticity at the bipolar-to-ganglion synapse [4,9,16–19] (Figure 1). Unlike other models, a whole retina architecture is proposed, which provides a neural basis of each retinal stage, from photoreceptors up to ganglion cells, to explain the observed retina behavior. We used the neural modules provided by a simulation environment [14] that reproduce widely studied properties of the retina processing, such as membrane potential integration or synaptic rectification.

We fit the model using a multiobjective genetic approach since it provides an easy an efficient solution for dealing with the complexity of the simulated response and its objective function. Target physiological data were obtained by the LN analysis published by Ozuysal and Baccus [4]. Independence of the shunting feedback mechanism, affecting only the linear filter, and the STP module, responsible for variations of the offset, allows a double-stage minimization process that reduces the computational load of searching the whole parameter space. However, we think that a one-step minimization process may improve the fitting of some intervals, such as the oscillations observed in some parts of the high contrast curve. Another computation architectures should be then studied (e.g., high-performance computing) to speed up the optimization process. Different optimization strategies can be also considered to fully exploit the potential of this model.

Acknowledgments. This work has been supported by the Human Brain Project (SP11 - Future Neuroscience), project P11-TIC-7983, Junta of Andalucia (Spain), Spanish National Grant TIN2012-32039, co-financed by the European Regional Development Fund (ERDF), and the Spanish Government PhD scholarship FPU13/01487.

References

1. Rieke, F., Rudd, M.E.: The challenges natural images pose for visual adaptation. Neuron **64**(5), 605–616 (2009)
2. Demb, J.B.: Functional circuitry of visual adaptation in the retina. The Journal of Physiology **586**(18), 4377–4384 (2008)
3. Kohn, A.: Visual adaptation: physiology, mechanisms, and functional benefits. Journal of Neurophysiology **97**(5), 3155–3164 (2007)
4. Ozuysal, Y., Baccus, S.A.: Linking the computational structure of variance adaptation to biophysical mechanisms. Neuron **73**(5), 1002–1015 (2012)
5. Beaudoin, D.L., Borghuis, B.G., Demb, J.B.: Cellular basis for contrast gain control over the receptive field center of mammalian retinal ganglion cells. The Journal of Neuroscience **27**(10), 2636–2645 (2007)
6. Baccus, S.A., Meister, M.: Fast and slow contrast adaptation in retinal circuitry. Neuron **36**(5), 909–919 (2002)
7. Kim, K.J., Rieke, F.: Temporal contrast adaptation in the input and output signals of salamander retinal ganglion cells. The Journal of Neuroscience **21**(1), 287–299 (2001)
8. Chander, D., Chichilnisky, E.: Adaptation to temporal contrast in primate and salamander retina. The Journal of Neuroscience **21**(24), 9904–9916 (2001)
9. Manookin, M.B., Demb, J.B.: Presynaptic mechanism for slow contrast adaptation in mammalian retinal ganglion cells. Neuron **50**(3), 453–464 (2006)
10. Wohrer, A., Kornprobst, P.: Virtual retina: a biological retina model and simulator, with contrast gain control. Journal of Computational Neuroscience **26**(2), 219–249 (2009)
11. Mante, V., Bonin, V., Carandini, M.: Functional mechanisms shaping lateral geniculate responses to artificial and natural stimuli. Neuron **58**(4), 625–638 (2008)
12. van Hateren, J.V., Rüttiger, L., Sun, H., Lee, B.: Processing of natural temporal stimuli by macaque retinal ganglion cells. The Journal of Neuroscience **22**(22), 9945–9960 (2002)
13. Victor, J.D.: The dynamics of the cat retinal x cell centre. The Journal of Physiology **386**(1), 219–246 (1987)
14. Martínez-Cañada, P., Morillas, C., Nieves, J.L., Pino, B., Pelayo, F.: First stage of a human visual system simulator: the retina. In: Trémeau, A., Schettini, R., Tominaga, S. (eds.) CCIW 2015. LNCS, vol. 9016, pp. 118–127. Springer, Heidelberg (2015)
15. Rieke, F.: Temporal contrast adaptation in salamander bipolar cells. The Journal of Neuroscience **21**(23), 9445–9454 (2001)
16. Euler, T., Haverkamp, S., Schubert, T., Baden, T.: Retinal bipolar cells: elementary building blocks of vision. Nature Reviews Neuroscience **15**(8), 507–519 (2014)
17. Jarsky, T., Cembrowski, M., Logan, S.M., Kath, W.L., Riecke, H., Demb, J.B., Singer, J.H.: A synaptic mechanism for retinal adaptation to luminance and contrast. The Journal of Neuroscience **31**(30), 11003–11015 (2011)
18. Dunn, F.A., Rieke, F.: Single-photon absorptions evoke synaptic depression in the retina to extend the operational range of rod vision. Neuron **57**(6), 894–904 (2008)

19. Singer, J.H., Diamond, J.S.: Vesicle depletion and synaptic depression at a mammalian ribbon synapse. Journal of Neurophysiology **95**(5), 3191–3198 (2006)
20. Kim, K.J., Rieke, F.: Slow na+ inactivation and variance adaptation in salamander retinal ganglion cells. The Journal of Neuroscience **23**(4), 1506–1516 (2003)
21. Zaghloul, K.A., Boahen, K., Demb, J.B.: Contrast adaptation in subthreshold and spiking responses of mammalian y-type retinal ganglion cells. The Journal of Neuroscience **25**(4), 860–868 (2005)
22. Zaghloul, K.A., Boahen, K., Demb, J.B.: Different circuits for on and off retinal ganglion cells cause different contrast sensitivities. The Journal of Neuroscience **23**(7), 2645–2654 (2003)
23. Dayan, P., Abbott, L.: Theoretical neuroscience: computational and mathematical modeling of neural systems. Journal of Cognitive Neuroscience **15**(1), 154–155 (2003)
24. Wohrer, A.: Model and large-scale simulator of a biological retina, with contrast gain control. PhD thesis, Nice (2008)
25. Rodieck, R.W.: Quantitative analysis of cat retinal ganglion cell response to visual stimuli. Vision Research **5**(12), 583–601 (1965)
26. Enroth-Cugell, C., Robson, J.G.: The contrast sensitivity of retinal ganglion cells of the cat. The Journal of Physiology **187**(3), 517–552 (1966)
27. Smith, V.C., Pokorny, J., Lee, B.B., Dacey, D.M.: Primate horizontal cell dynamics: an analysis of sensitivity regulation in the outer retina. Journal of Neurophysiology **85**(2), 545–558 (2001)
28. Torre, V., Poggio, T.: A synaptic mechanism possibly underlying directional selectivity to motion. Proceedings of the Royal Society of London. Series B. Biological Sciences **202**(1148), 409–416 (1978)
29. Amthor, F.R., Grzywacz, N.M.: Nonlinearity of the inhibition underlying retinal directional selectivity. Visual Neuroscience **6**(03), 197–206 (1991)
30. Carandini, M., Heeger, D.J., Movshon, J.A.: Linearity and normalization in simple cells of the macaque primary visual cortex. The Journal of Neuroscience **17**(21), 8621–8644 (1997)
31. Kastner, D.B., Baccus, S.A.: Coordinated dynamic encoding in the retina using opposing forms of plasticity. Nature Neuroscience **14**(10), 1317–1322 (2011)
32. Fortin, F.-A., De Rainville, F.-M., Gardner, M.-A., Parizeau, M., Gagné, C.: DEAP: Evolutionary algorithms made easy. Journal of Machine Learning Research **13**, 2171–2175 (2012)
33. Gewaltig, M.-O., Diesmann, M.: Nest (neural simulation tool). Scholarpedia **2**(4), 1430 (2007)

Stochastic-Based Implementation
of Reservoir Computers

Miquel L. Alomar[✉], Vincent Canals, Víctor Martínez-Moll, and Josep L. Rosselló

Physics Department, University of Balearic Islands, Palma De Mallorca, Spain
miquellleo.alomar@uib.es

Abstract. Hardware implementations of Artificial Neural Networks (ANNs) allow to exploit the inherent parallelism of these architectures. Nevertheless, ANN hardware implementation requires a large amount of hardware resources. Recently, Reservoir computing (RC) has arisen as an advantageous technique to implement Recurrent Neural Networks RNNs). In this work, we present an efficient approach to implement RC systems. The proposed methodology employs probabilistic logic to reduce the hardware area required to implement the arithmetic operations present in neural networks and conventional binary logic for the nonlinear activation function. We show the functionality and low hardware resources used by the proposed methodology.

Keywords: Field-Programmable Gate Array (FPGA) · Hardware implementation · Reservoir Computing (RC) · Recurrent Neural Networks (RNNs) · Probabilistic logic

1   Introduction

Reservoir Computing (RC) [1]–[5] is a recently introduced paradigm of understanding and training Recurrent Neural Networks (RNNs). It resulted from the observation that generic properties of RNNs can be maintained if all interconnection weights are kept fixed and only the output layer is trained (see Fig. 1). The RNN is called a reservoir in this context. As RNNs are notoriously difficult to train, they were not widely used until the advent of RC.

Basically, the RC architecture consists of three distinct parts: an input layer, the reservoir and an output layer as illustrated in Fig. 1. The input layer feeds the input signals $u(t)=(u_1(t),..., u_k(t))$ to the reservoir via fixed random weight connections W_{in}. The reservoir consists of a large number N of randomly interconnected neurons with states $x(t)=(x_1(t), ..., x_N(t))$ and internal weights W, constituting a RNN, that is to say, a network with internal feedback loops. Under the influence of input signals, the network exhibits transient responses which are read out at the output layer $y(n)=(y_1(n), ..., y_L(n))$ by means of a linear weighted sum of the individual node states. As the only part of the system which is trained (assessment of the output weights W_{out}) is the output layer, the training does not affect the dynamics of the reservoir itself unless a recurrence exists between the reservoir and the readout (recurrence weights given by W_{back}).

© Springer International Publishing Switzerland 2015
I. Rojas et al. (Eds.): IWANN 2015, Part II, LNCS 9095, pp. 185–196, 2015.
DOI: 10.1007/978-3-319-19222-2_16

The individual neuron states are updated according to

$$x(t + 1) = f(W^{in}u(t + 1) + Wx(t) + W^{back}y(t)) \tag{1}$$

where $f=(f_1, ..., f_N)$ are the internal unit's output functions (typically sigmoid functions). In the simplest case that avoids feedback connections at the output layer and connections from the input to the output layer, each output is computed according to

$$y_{nn}(t + 1) = \sum_{i=1}^{N} W_i^{out} \cdot x_i(t + 1) \tag{2}$$

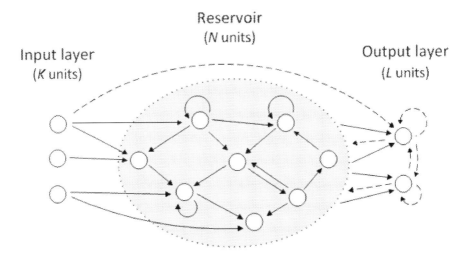

Fig. 1. Classical RC architecture. Dashed arrows indicate connections that are possible but not required. All connections are randomly chosen and kept fixed except the ones that couple the reservoir on the output layer.

A recent study [6] shows that a simple deterministically constructed cycle reservoir is comparable to the standard RC methodology. The simplified reservoir topology is composed of units organized in a cycle as illustrated in Fig. 2. All the connections between internal units have the same weight value r. The input layer is fully connected to the reservoir. All the input connections have the same absolute weight value $v > 0$, the sign of each input weight is determined randomly with equal probability.

The simple cycle reservoir structure presents only a slightly worse performance than the classical topology [6]. Nonetheless, it presents some advantages that make it particularly useful for the hardware implementation. The greatest benefit is that the number of connections within the reservoir is constant independently of the number of neurons N (while it increases with N in the case of the classical structure). This fact allows a great reduction of the number of required multipliers in the case of implementations using a high number of neurons. On the other hand, the design of the networks can be more easily automated since all neurons have the same number of connections (one connection from a neighboring neuron and one connection for each input unit) and this number does not change with the number of neurons.

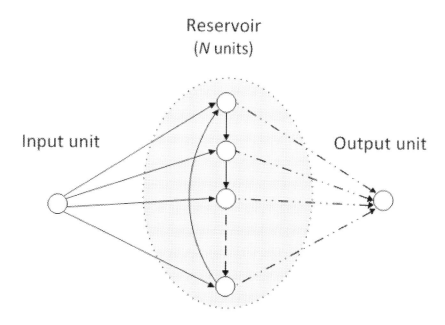

Fig. 2. Simple cycle reservoir (SCR) topology. Units are organized in a cycle.

RC systems and, in general, Artificial Neural Networks (ANNs) are often implemented through simulation on sequential computers. Nevertheless, hardware implementations take advantage of the inherent parallelism of ANNs which allows a much greater processing speed compared with software solutions.

The RC principle can be used to implement computations on generic dynamical systems treating them as reservoirs. For example, it has been used to perform computation on hardware platforms such as analog electronics [7], opto-electronic [7], [8] and optical [9] systems. Digital implementations of RC systems are limited to the use of spiking neurons (Liquid State Machine approach) [10]. To the best of our knowledge, this paper is the first hardware implementation example of an RC system using classical sigmoid neurons (Echo State Network approach).

The circuits involved in the digital implementation of the ANN's complex computations (addition, multiplication and assessment of the hyperbolic tangent function) consume many hardware resources. This fact hinders the implementation of massive neural networks in a single chip.

Stochastic computing is a feasible alternative to implement complex computations due to the simplicity of the involved circuitry. It is based on the result of applying probabilistic laws to logic cells where variables are represented by random pulse streams [11], [12]. Stochastic logic pulsed signals codify the information with the bit switching probability. Although the evaluation time of an operation's result is greater, the stochastic logic considerably reduces the use of hardware resources if compared to traditional digital implementations. As an illustrative example, the multiplication operation is performed by a simple AND gate in the stochastic logic (Figure 4). Therefore, stochastic computing is very useful for applications that require parallel-processing techniques.

Several works have used the stochastic logic to implement feed forward neural networks [13]–[16]. In addition, there were some attempts to implement the RC framework using stochastic bit-stream neurons [17], [18]. But these examples were only performed by simulations and not using actual hardware.

In a previous study [19], we presented an stochastic-based FPGA implementation of a Reservoir Computer using sigmoid neurons. The feasibility to implement the classical RC structure (Fig. 1) by means of the stochastic computing methodology was demonstrated for a regression application.

The present work focuses on the implementation of the Simple Cycle Reservoir topology (Fig. 2), which allows an improved performance of the system in terms of hardware resources. Another major contribution of this paper is an efficient methodology to implement RC systems that combines the use of stochastic and conventional deterministic computing. In particular, the proposed approach employs probabilistic logic to reduce the hardware area required to implement the arithmetic operations present in the neural network and conventional binary logic for the nonlinear activation function.

2   Stochastic Computing

In stochastic-based computations a global clock signal provides the time interval during which all stochastic signals are stable (settled to 0 or 1, LOW or HIGH). During a clock cycle, each node of the circuit has a probability p of being in the HIGH state (see Fig. 3). This probabilistic-based coding provides a natural way of operating with analog quantities (since probabilities are defined between 0 and 1) using digital circuitry.

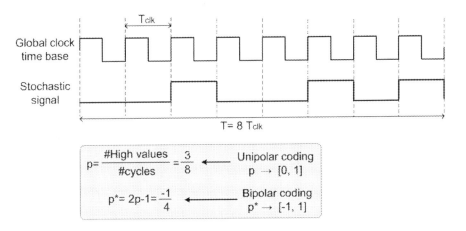

Fig. 3. Basic concept of the stochastic codification. Information is coded as the probability 'p' of the pulsed signal being in the high level.

2.1   Basic Operations

Pulsed signals follow probabilistic laws when they are evaluated through logic gates. For instance, the AND gate provides at the output the product of their inputs (that is, the collision probability between signals) as it is illustrated in Fig.4. Notice that the pulsed signals do not follow any particular pattern. Furthermore, they must be uncorrelated so that the operations can be performed properly. The basic stochastic arithmetic circuits are depicted in Fig. 5. A NOT gate converts the probability p at the input to the complementary 1-p at the output. The weighted addition of two switching signals (mean value of the two inputs) is implemented using a multiplexer and a modulus 2 binary counter. The counter supplies the selection signal to the multiplexer so that its output provides one of the input signals (either p or q) alternately at each clock cycle.

Negative numbers cannot be represented directly using this probabilistic scheme since probabilities are defined in the interval $[0, +1]$. To obtain negative numbers we must perform a variable change $p^*=2p$-1 (where p is the switching probability of the signal). In this case we obtain the bipolar coding (in contrast to the simpler unipolar coding based on the direct use of probabilities). Therefore, since p is delimited between 0 and 1, p^* is bounded in the interval $[-1, +1]$, and the zero value is located at $p=1/2$. Using this notation the product is implemented by a single XNOR gate, the negation is obtained with a NOT gate and the addition is implemented as in the unipolar codification (Fig. 5).

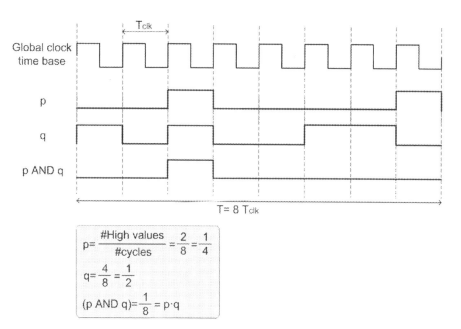

Fig. 4. Product operation of two stochastic signals with switching activities $p=0.25$ and $q=0.5$ performed by means of an AND gate

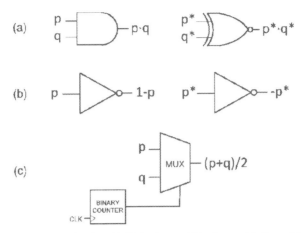

Fig. 5. Stochastic arithmetic circuits. (a) Unipolar and bipolar multipliers. (b) Unipolar complementary operation and bipolar negation. (c) Adder.

2.2   Data Conversion

A stochastic computing system requires converting any binary magnitude P to its equivalent stochastic signal p before it can perform the necessary probabilistic computations. Similarly, the resulting pulsed signals must be finally converted into their equivalent binary values. Binary numbers are converted to pulsed signals using a B2P block (Fig. 6b), which consists of a comparator and a linear feedback shift register circuit (LFSR) used as pseudorandom number generator. On the other hand, pulsed signals are converted to binary numbers using counters (Fig. 6a). We define a pulse to binary converter of order N (a $P2B(N)$) as the digital circuit that evaluates the number of HIGH values provided by a stochastic signal throughout N clock cycles. The output of a $P2B(N)$ block is an n-bit number that changes every N cycles so that the evaluation time is $T_{EVAL}=N \cdot T$ (where T is the clock cycle).

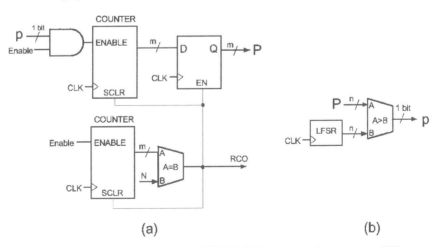

(a)   (b)

Fig. 6. (a) Pulse to binary converter $P2B(N)$. (b) Binary to pulse converter $B2P$.

A probabilistic error is always present during conversions. This inherent error of the stochastic logic can be reduced by increasing the evaluation time although it decreases the processing speed of the system.

2.3   Neural Network Implementation

We have shown that the multiplication and addition operations can be implemented by means of the bipolar codification using an XOR gate and a multiplexer, respectively. Another crucial issue for the neural implementation is to compute the nonlinear activation function. The stochastic computing literature [15], [16] offers several probabilistic approaches to reproduce the hyperbolic tangent function. Nonetheless, the present research adopts a classical approximation of the nonlinear function [20], which has proved to be an effective strategy in terms of accuracy, speed, and area resources.

We have improved the accuracy of the purely combinational sigmoid function approximation proposed in [20] by performing linear interpolation. In addition, this approach has been adapted so that it can be used in the stochastic computing framework. Basically, the stochastic signals are converted to deterministic binary values before they can be processed by means of the classical approach. The resulting binary values are finally transformed to pulsed signals again. Fig. 7 shows the configuration of a two-input neuron performing the weighting, the addition and the evaluation of the inputs through the activation function.

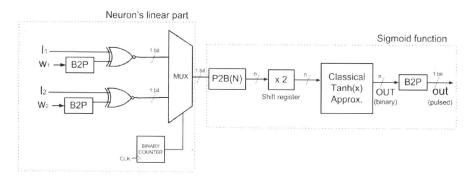

Fig. 7. Two-input sigmoid neuron. The linear part uses probabilistic logic whereas the nonlinear activation function is implemented classically. The output of the linear part is multiplied by 2 (shifting the binary word one position to left) to compensate the scaled sum performed by the multiplexer. A value $N=2^{16}$ was used for the $P2B(N)$ converter to approximately generate the $Tanh[I_1 \cdot w_1 + I_2 \cdot w_2]$ function.

Therefore, what we propose in this work is to combine the use of stochastic and conventional deterministic computing. Stochastic arithmetic allows reducing the computation hardware area required to implement the arithmetic operations present in neural networks while conventional binary logic can be used to efficiently implement the nonlinear activation function.

3   Results

As an example of functionality of the proposed methodology, a small reservoir computer was synthesized on a Cyclone III (EP3C16F484C6N) low cost FPGA (see Table 1 for the spent hardware resources). The implemented reservoir network consisted of 10 sigmoid neurons organized in a cycle as shown in Fig. 2. All the weights within the reservoir were set to the same value $r=0.95$ whereas the input weights were set to values of -0.95 and $+0.95$ with equal probability. No feedback connections were set between the output layer and the reservoir units.

Table 1. Spent hardware resources of the low cost Cyclone III (EP3C16F484C6N) FPGA for the 10-unit reservoir network

TOTAL LOGIC ELEMENTS (LE)	1,888 /15,408 (12%)
TOTAL COMBINATIONAL FUNCTIONS	1,639/15,408 (11%)
DEDICATED LOGIC REGISTERS	1,216 /15,408 (8%)
TOTAL MEMORY BITS	0/516,096 (0%)
EMBEDDED MULTIPLIER 9-BIT ELEMENT	0/112 (0%)

A single-channel sinusoid input $u(t)=Sin[2\pi \cdot t/P]$ with 20 points per period ($P=20$) was used to drive the system. An internal memory supplied the input values to the reservoir every time step. The network was trained to produce a non-linear transformation of the input: $y(t)_{teach}=3/4\ Sin^3[2\pi \cdot t/P]\ =3/4\ u(t)^3$. The training (assessment of the output layer optimal weights) was carried out using the experimental outputs of the individual neurons and consisted of computing a linear regression of the teacher output y_{teach} on the reservoir states.

Concretely, we first let the network run for $t=0$ to $t_{max}=250$, secondly we dismissed an initial transient of 50 time steps, then we collected the network states (reservoir individual neurons' outputs) for $t=51$ to $t=150$, and we computed the output layer weights (W_i^{out}) offline from the collected states. Finally, the trained network was tested with the output results obtained from $t=151$ to $t_{max}=250$.

The final network output y_{nn} (eq. 2) was calculated by software using the experimental reservoir states. However, it could be effectively implemented in hardware together with the reservoir neurons using either stochastic logic or the conventional FPGA embedded multipliers. A test error $mse_{test}=4.7 \cdot 10^{-4}$ was obtained.

Fig. 8 displays the dynamics of some selected nodes. The Simple Cycle Reservoir topology ensures a rich reservoir echo state network. The experimental results are plotted along with the expected numerical ones. A good agreement can be observed.

Fig. 9 shows how the linear combination of the reservoir states achieves a good approximation to the desired output.

16-bit counters ($P2B(N)$ blocks with $N=2^{16}$) were used to integrate the output stochastic signals. Therefore a time of 1.3 ms was spent in each time step when using a

50MHz clock. Since the computation time is mainly due to the conversion of the sto-chastic streams to binary values, higher order networks would present a similar speed.

It should be noticed that the use of the Simple Cycle Reservoir structure of Fig. 2 instead of the classical topology of Fig. 1 together with an efficient implementation approach for the sigmoid activation function have allowed an improved performance of the system in terms of hardware resources. In particular, a 22% reduction of the required logic elements has been achieved compared to a previous implementation [19]. Furthermore, the higher accuracy of the sigmoid function approximation has enabled a diminution of the test error by one order of magnitude.

Regarding the comparison of the present stochastic-based implementation with conventional ones, it is difficult to evaluate since there are no available studies in the literature using exactly the same type of device and the same network design. Never-theless, it appears that the stochastic methodology makes possible a significant reduc-tion of the resource requirements. For instance, a conventional implementation of a feed-forward network's layer with five sigmoid neurons [21] required more than half of the hardware resources of an FPGA with similar characteristics to the one used for the present research.

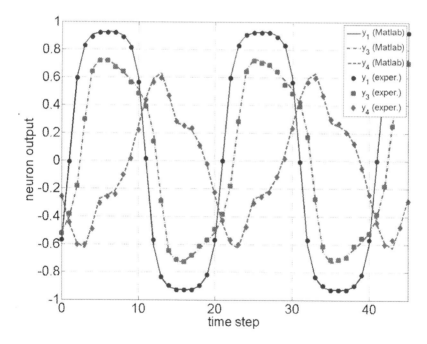

Fig. 8. Traces of three selected units of the 10-neuron implemented reservoir computer driven by a sinusoid input. Experimental values (symbols) are plotted together with the numerically obtained results (lines).

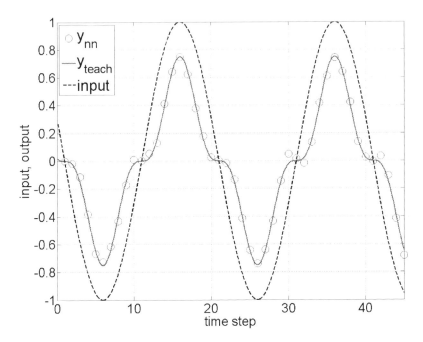

Fig. 9. Input signal ($Sin[2\pi \cdot t/20]$) along with the desired output signal $y_{teach} = 3/4\ Sin^3[2\pi \cdot t/20]$, and the experimental FPGA neural network output y_{nn}

4   Discussion

With a basic mathematical regression application we have demonstrated the feasibility of the stochastic logic to implement a reservoir computer. Specifically, we have proposed an area-efficient mixed implementation that employs probabilistic logic for the arithmetic operations and conventional binary logic for the nonlinear activation function.

We have focused on the implementation of a particular Reservoir Computer topology (the Simple Cycle Reservoir) which makes possible an improved performance of the system in terms of hardware resources due to the reduced number of connections.

The use of few hardware resources has shown that the proposed methodology is appropriate to implement massive reservoir networks. This fact paves the way to the stochastic-based hardware implementation of RC systems to perform complex tasks such as temporal pattern classification, temporal pattern generation, time series prediction or memorizing. The next step in our research is to implement reservoir networks with a high number of neurons and to apply them to real world tasks.

Reservoir networks have some advantages over conventional Recurrent Neural Networks that enable a more efficient hardware implementation. The fact that the RC architecture does not include intermediate layers (but only the "reservoir" layer between the input and output layers) makes possible a complete parallel processing of the input signals. Another benefit of RC networks is their sparse connectivity. This characteristic allows easy wireability that matches the intrinsic FPGA wiring capabilities well. Finally, the simple training process can be performed offline using the experimental neuron outputs.

The use of the stochastic logic implies certain constraints. The greatest shortcomings are the long evaluation time and the imprecision of the implementations due to the random nature of the signals. Nevertheless, these drawbacks are compensated for by the much simpler architecture and by the stochastic logic's inherent noise immunity that, all in all, allow a massive, parallel and reliable implementation.

Acknowledgements. This work has been partially supported by the Spanish Ministry of Economy and Competitiveness (MINECO), the Regional European Development Funds (FEDER), the Comunitat Autònoma de les Illes Balears under grant contracts TEC2011-23113, AA/EE018/2012 and a fellowship (FPI/1513/2012) financed by the European Social Fund (ESF) and the Govern de les Illes Balears (Conselleria d'Educació, Cultura i Universitats).

References

1. Jaeger, H.: The ' echo state ' approach to analysing and training recurrent neural networks – with an Erratum note. In: GMD Report 148, German National Research Center for Information Technology (2010)
2. Lukoševičius, M., Jaeger, H., Schrauwen, B.: Reservoir computing trends. KI - Künstliche Intelligenz **26**(4), 365–371 (2012)
3. Maass, W., Natschläger, T., Markram, H.: Real-time computing without stable states: a new framework for neural computation based on perturbations. Neural Comput. **14**(11), 2531–2560 (2002)
4. Jaeger, H., Haas, H.: Harnessing nonlinearity: predicting chaotic systems and saving energy in wireless communication. Science **304**(5667), 78–80 (2004)
5. Verstraeten, D., Schrauwen, B., D'Haene, M., Stroobandt, D.: An experimental unification of reservoir computing methods. Neural Networks **20**, 391–403 (2007)
6. Rodan, A., Tiňo, P.: Minimum complexity echo state network. IEEE Trans. Neural Networks **22**(1), 131–144 (2011)
7. Larger, L., Soriano, M.C., Brunner, D., Appeltant, L., Gutierrez, J.M., Pesquera, L., Mirasso, C.R., Fischer, I.: Photonic information processing beyond turing: an optoelectronic implementation of reservoir computing. Opt. Express **20**(3), 3241–3249 (2012)
8. Paquot, Y., Duport, F., Smerieri, A., Dambre, J., Schrauwen, B., Haelterman, M., Massar, S.: Optoelectronic reservoir computing. Sci. Rep. **2**, 1–6 (2012)
9. Vandoorne, K., Dierckx, W., Schrauwen, B., Verstraeten, D., Baets, R., Bienstman, P., Van Campenhout, J.: Toward optical signal processing using photonic reservoir computing. Opt. Express **16**(15), 11182–11192 (2008)
10. Schrauwen, B., 'Haene, M.D., Verstraeten, D., Van Campenhout, J.: Compact hardware liquid state machines on FPGA for real-time speech recognition. Neural Networks **21**, 511–523 (2008)
11. Gaines, B.R.: R68-18 random pulse machines. IEEE Trans. Comput. **C–17**(4), 410 (1968)

12. Toral, S.L., Quero, J.M., Franquelo, L.G.: Stochastic pluse coded arithmetic. In: International Symposium on Circuits and Systems, pp. I–599–I–602 (2000)
13. Kondo, Y., Sawada, Y.: Functional abilities of a stochastic logic neural network. IEEE Trans. Neural Networks 3(3), 434–443 (1992)
14. Bade, S.L., Hutchings, B.L.: FPGA-based stochastic neural networks-implementation. In: IEEE Workshop on FPGAs for Custom Computing Machines (1994)
15. Brown, B.D., Card, H.C.: Stochastic neural computation I: Computational elements. IEEE Trans. Comput. 50(9), 891–905 (2001)
16. Rosselló, J.L., Canals, V., Morro, A.: Hardware implementation of stochastic-based neural networks. In: Proceedings of the International Joint Conference on Neural Networks (2010)
17. Verstraeten, D.: Stochastic bitstream-based reservoir computing with feedback. In: Fifth FirW Ph.D. Symposium (2005)
18. Verstraeten, D., Schrauwen, B., Stroobandt, D.: Reservoir computing with stochastic bitstream neurons. In: Proceedings of the 16th Annual ProRISC Workshop, pp. 454–459 (2005)
19. Alomar, M.L., Canals, V., Martinez-Moll, V., Rossello, J.L.: Low-cost hardware implementation of reservoir computers. In: 2014 24th International Workshop on Power and Timing Modeling, Optimization and Simulation (PATMOS), pp. 1–5 (2014)
20. Tommiska, M.T.: Efficient digital implementation of the sigmoid function for reprogrammable logic. IEE Proceedings - Computers and Digital Techniques 150(6), 403–411 (2003)
21. Himavathi, S., Anitha, D., Muthuramalingam, A.: Feedforward neural network implementation in FPGA using layer multiplexing for effective resource utilization. IEEE Trans. Neural Networks 18(3), 880–888 (2007)

FPGA Implementation Comparison Between C-Mantec and Back-Propagation Neural Network Algorithms

Francisco Ortega-Zamorano[1], José M. Jerez[1], Gustavo Juárez[2], and Leonardo Franco[1](✉)

[1] Department of Computer Science, ETSI Informática, Universidad de Málaga, Málaga, Spain
{fortega,jja,lfranco}@lcc.uma.es
[2] Universidad Nacional de Tucumán, San Miguel de Tucumán, Argentina
gjuarez@herrera.unt.edu.ar

Abstract. Recent advances in FPGA technology have permitted the implementation of neurocomputational models, making them an interesting alternative to standard PCs in order to speed up the computations involved taking advantage of the intrinsic FPGA parallelism. In this work, we analyse and compare the FPGA implementation of two neural network learning algorithms: the standard Back-Propagation algorithm and C-Mantec, a constructive neural network algorithm that generates compact one hidden layer architectures. One of the main differences between both algorithms is the fact that while Back-Propagation needs a predefined architecture, C-Mantec constructs its network while learning the input patterns. Several aspects of the FPGA implementation of both algorithms are analysed, focusing in features like logic and memory resources needed, transfer function implementation, computation time, etc. Advantages and disadvantages of both methods are discussed in the context of their application to benchmark problems.

Keywords: Constructive neural networks · FPGA · Hardware implementation

1 Introduction

Artificial Neural Networks (ANN) [1] are mathematical models inspired in the functioning of the brain that can be utilized in clustering and classification problems, and that have been successfully applied in several fields, including pattern recognition, stock market prediction, control tasks, medical diagnosis and prognosis, etc. The implementation of ANN in digital circuits (standard PCs, embedded systems, etc.) has been limited by the computational power needed, mainly given its intrinsic parallelism. In this sense, the capacity and performance of current FPGAs are a realistic alternative for the real time implementation of ANN. FPGAs [2] are reprogrammable silicon chips, using prebuilt logic blocks and

© Springer International Publishing Switzerland 2015
I. Rojas et al. (Eds.): IWANN 2015, Part II, LNCS 9095, pp. 197–208, 2015.
DOI: 10.1007/978-3-319-19222-2_17

programmable routing resources, that can be configured to implement custom hardware functionality, being able also to change almost instantly its behaviour by recompiling a new circuitry configuration. Recent advances in technology have permitted to construct FPGAs with considerable large amounts of processing power and memory storage, and as so they have been applied in several domains (Telecommunications, Robotics, Pattern recognition tasks, Infrastructure monitoring, etc.) [3–5]. In particular FPGAs seem quite suitable for Neural Network implementations as they can be programmed to operate in a parallel way [6–8].

Within the area of supervised pattern recognition, the efficient implementation of neurocomputational models into hardware can be done in principle following two different strategies: Modifying and adapting the traditional Back-Propagation algorithm or developing new algorithms that are better suited to the hardware constraints. In this work these two different possibilities are explored, first by doing a hardware optimization of the standard Back-Propagation algorithm [9], and secondly through the implementation of an alternative algorithm (C-Mantec) based on an incremental constructive architecture [10]. The overall idea of the work is to do a comparative analysis of the two approaches in order to identify the pros and cons for each case, and with this information take a decision depending on the specific application and the resources available. The Back-Propagation algorithm (BP) is the standard learning procedure for training multilayer neural networks architectures [11] [12] but one of the main problems associated to its implementation is the lack of a clear methodology for determining the network topology before training starts. On the other hand, C-Mantec [13] is a novel neural network constructive algorithm that utilizes competition between the neurons and a modified perceptron learning rule (thermal perceptron [14]) to build single hidden layer compact architectures with good prediction capabilities for the supervised classification problems. The organization of the present work is as follows: Section 2 includes the hardware implementation details and description about the Back-propagation and C-Mantec algorithms. Results from several comparison features are presented in Section 3, to finally present the discussion of the results and the conclusions obtained.

2   Algorithms Description and Implementation Details

We describe in this section the main functioning aspects of both algorithms, including also specific details of the FPGA implementation. An important issue and a big difference in relationship to standard PC implementations regards the number representation used in the FPGA. While for standard PCs floating point number representation is the standard choice, this type is not usually the most efficient for FPGAs and a fixed point number representation is preferred [15].

2.1   The Back-Propagation Algorithm

The Back-Propagation algorithm is a supervised learning method for training multilayer artificial neural networks based on the gradient descent strategy.

The network architecture for the implementation of the algorithm has to be decided previously and there is no standard methodology for this step, being the trial-and-error method one of the most used strategies. In the most general case the neural architecture comprises an input layer with a number of inputs determined by the problem at hand, several hidden layers, and one or many output neurons depending whether a binary or multi-output problem is analysed (for simplicity the first case is considered in this work).

The objective of the BP supervised learning algorithm is to minimize the difference between given outputs (targets) for a set of input data and the output of the network. This error depends on the values of the synaptic weights, and so these should be adjusted in order to minimize the error. The error function computed for the case of a single output neuron can be defined as:

$$E = \frac{1}{2} \sum_{i=1}^{M} (z_i - y_i)^2, \tag{1}$$

where the sum is over all training patterns, and z_i and y_i refers to target and network outputs for a given pattern i.

If we consider the neurons belonging to a hidden or output layer, the activation of these units, denoted by y_i, can be written as:

$$y_i = g \left(\sum_{j=1}^{L} w_{ij} \cdot s_j \right) = g(h) , \tag{2}$$

where w_{ij} are the synaptic weights between neuron i in the current layer and the neurons of the previous layer with activation s_j. In the previous equation, we have introduced h as the synaptic potential of a neuron. g is a sigmoid activation function given by:

$$g(x) = \frac{1}{1 - e^{-\beta x}} \tag{3}$$

By using the method of *gradient descent*, the BP algorithm attempts to minimize the error Eq. 1 in an iterative process by updating the synaptic weights upon the presentation of a given pattern. The synaptic weights between two last layers of neurons are updated as:

$$\Delta w_{ij}(k) = -\eta \frac{\partial E}{\partial w_{ij}(k)} = \eta [z_i(k) - y_i(k)] g_i'(h_i) s_j(k), \tag{4}$$

where η is the learning rate that has to be set in advance (a parameter of the algorithm), g' is the derivative of the sigmoid function and h is the synaptic potential previously defined, while the rest of the weights are modified according to similar equations by the introduction of a set of values called the "deltas" (δ), that propagate the error form the last layer into the inner ones.

The BP neural networks have been trained under a training / validation / testing strategy to avoid overfitting effects, caused mainly by excessive training

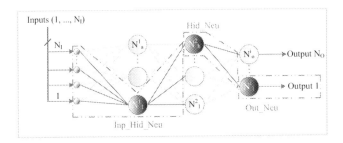

Fig. 1. Scheme of the FPGA architecture design for the implementation of the BP algorithm

iterations in the learning process, and thus a validation set is used to check the evolution of the mean square error between target and output values.

Regarding the FPGA implementation of the BP algorithm three main aspects have been carefully analyzed for increasing the efficiency of resource utilization: a) The introduction of a new input-hidden neurons block, b) A new scheme for computing the sigmoid transfer function, and c) A strategy of time division for using only a single multiplier block for each neuron.

Fig. 1 shows a scheme of the neural architecture, where three types of blocks are used for the implementation of the different parts of the neural architecture. The proposed implementation do not consider the input layer of neurons separately as this is included together with the first hidden layer neurons in a module named input-hidden neurons ("inp-hid"). The definition of this new type of module is possible because the input layer neurons do not process the information as they simple act as input to the network.

Another important FPGA design aspect regarding the hardware implementation of a neural network algorithm is the way of computing the activation function of the neurons, usually a sigmoid-type function. An scheme based on a lookup table approach plus linear interpolation scheme permits to obtain an efficient representation in terms of the resources needed together with low absolute and relative errors. In the previous work [16] a complete study about size of the table, employed resources and precision results has been presented. As conclusion the more efficient dimension of the table for a sigmoid is 2 bits for the decimal part, 3 bits for the integer part and one more bit foe the sign of the function. This parameters produce a table of $2^6 = 64$ inputs with 16 bits of word length as size of each input. The total resources necessary to implement this table are 32 bits or 1 block memory. Plus, Fig. 2 shows the approximation obtained for the sigmoid function (top graph) and the errors committed in its approximation (bottom graph).

Furthermore, a third important aspect considered during the FPGA implementation regards a time division scheme for performing the multiplications involved in the algorithm [16]. The multiplier blocks can be implemented both as a combination of logic cells or using specific DSP blocks. We have selected the first choice in this work for a fair comparison with the C-Mantec algorithm

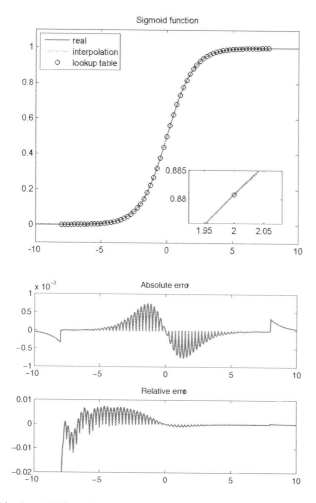

Fig. 2. FPGA sigmoid function approximation based on a lookup table plus linear interpolation scheme (top graph). Absolute (middle graph) and relative errors (bottom graph) committed in the approximation of the function.

implementation, as this was done without DSPs. Further the implementation using DSP is specific to each board and thus the present choice gives more generality to the results. The strategy consists in using a single multiplier for each neuron (built using logic blocks [17]) and then through using a time division multiplexing scheme compute all the multiplications related to the neuron.

2.2   The Constructive Neural Network Algorithm C-Mantec

C-Mantec as a constructive neural network algorithm generates the network topology in an on-line manner during the learning phase, avoiding the complex problem of selecting an adequate neural architecture [13]. The novelty of

C-Mantec in comparison to previous proposed constructive algorithms is that the neurons in the single hidden layer compete for learning the incoming data, and this process permits the creation of very compact neural architectures. The binary activation state (S_j) of each of the neurons in the hidden layer depends on N input signals, ψ_i, and on the actual value of the N synaptic weights (ω_{ji}) and bias (b_j) as follows:

$$S_j = \begin{cases} 1 & \text{if } h_j \geq 0 \\ 0 & \text{otherwise} \end{cases} \tag{5}$$

where h is the synaptic potential of the neuron defined as:

$$h_j = \sum_{i=1}^{N} \omega_{ji}\, \psi_i - b_j \tag{6}$$

The weight updating in the C-Mantec algorithm at the single neuron level is done using the thermal perceptron rule [14], in which the modification of the synaptic weights, $\Delta\omega_i$, is done on-line (after the presentation of a single input pattern) according to the following equation:

$$\Delta\omega_{ji} = (t - S_j)\, \psi_i\, T_{fac}, \tag{7}$$

where t is the target value (desired output of the whole network for the presented input), and ψ represents the value of input unit i connected to the hidden neuron S_j by synaptic weight ω_{ji}. The difference to the standard perceptron learning rule is that the thermal perceptron incorporates the T_{fac} factor. This factor, whose value is computed as shown in Eq. 8, depends on the value of the synaptic potential and on an artificially introduced temperature (T):

$$T_{fac} = \frac{T}{T_0} e^{-\frac{|h|}{T}}, \tag{8}$$

The computation of the T_{fac} factor involves the FPGA implementation of the exponential function, task that was done using the same approach applied for the computation of the sigmoid function needed for the BP algorithm. Section 3 includes Table 2 that shows a comparison between the approximation results obtained for both functions (the sigmoid and exponential functions).

Following with the description of the C-Mantec algorithm, the value of the temperature T decreases as the learning process advances according to Eq. 9, similarly to a simulated annealing process.

$$T = T_0 \cdot \left(1 - \frac{I}{I_{max}}\right), \tag{9}$$

where I is a cycle counter that defines an iteration of the algorithm on one learning cycle, and I_{max} is the maximum number of iterations allowed. One learning cycle of the algorithm is the process that starts when a chosen pattern is presented to the network and finishes after checking that all neurons respond correctly to the input or when the synaptic weights of the neuron chosen to

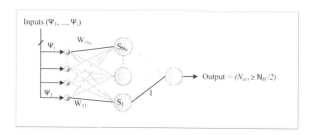

Fig. 3. Example of network architecture constructed by the C-Mantec algorithm

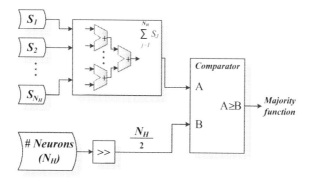

Fig. 4. Hardware implementation of the majority function that corresponds to the output neuron activation function of a network trained by the C-Mantec algorithm

learn the actual pattern (whether an existing or a new neuron) modifies its synaptic weights. The C-Mantec algorithm has three parameters to be set at the time of starting the learning procedure, and several experiments have shown the robustness of the algorithm that operates fairly well in a wide range of parameter values.

The output of a C-Mantec network consists in a single output that computes the majority function (see Eq.10) of the neuron activation of the hidden layer units, like in a voting process. The network output is active (1) if more than half of the N_H hidden neurons are active:

$$\text{Output} = \begin{cases} 1 & \text{if } \sum_j^{N_H} S_j \geq \frac{N_H}{2} \\ 0 & \text{otherwise} \end{cases} \tag{10}$$

Fig. 3 shows a network architecture of the type built by the C-Mantec algorithm. The network contains a single hidden layer of threshold neurons (S_j) with output values $\{0, 1\}$.

The FPGA implementation of the majority function is shown in Fig. 4. On the left part of the figure the activation value of all N_H hidden layer neurons S_i are shown, followed by the computation of the sum of their activation. In the module indicated by "comparator" the obtained value is compared with the value of $\frac{N_H}{2}$ and the whole network output is computed following Eq. 10. The whole process can be executed in less than one clock cycle of the FPGA because all operations involved are implemented with logic cells that introduce only minor delays.

3 Results

We present in this section results from the implementation of both algorithms (BP and C-Mantec) in a Xilinx Virtex-5 board. Table 1 shows some characteristics of the Virtex-5 XC5VLX110T FPGA, indicating its main logic resources. VHDL [17,18] (VHSIC Hardware Description Language) language was used for programming the FPGA, under the "Xilinx ISE Design Suite 12.4" environment using the "ISim M.81d" simulator.

Table 1. Main specifications of the Xilinx Virtex-5 XC5VLX110T FPGA board

Device	Slice Registers	Slice LUTs	Bonded IOBs	Block RAM
Virtex-5 XC5VLX110T	69,120	69,120	34	148

Table 2 shows the Maximum and Root Mean Square errors for different values of the integer N_a and decimal parts N_b obtained for the implementation of the exponential and sigmoidal functions used in the C-Mantec and Back-Propagation algorithms respectively through a lookup table plus linear interpolation scheme. As it can be appreciated from the the table both errors are quite low for both functions for almost all values of N_a and N_b being lower for the Sigmoidal function.

Table 2. Maximum error (Max) and Root Mean Square Error (RMSE) for different values of N_a and N_b in the implementation of the Exponential and Sigmoidal functions

N_a	N_b	Exponential		Sigmoidal	
		Max	RMSE	Max	RMSE
3	3	$1.833 \cdot 10^{-3}$	$3.249 \cdot 10^{-4}$	$3.353 \cdot 10^{-4}$	$9.477 \cdot 10^{-5}$
3	4	$4.704 \cdot 10^{-4}$	$7.632 \cdot 10^{-5}$	$1.982 \cdot 10^{-4}$	$4.428 \cdot 10^{-5}$
4	4	$4.704 \cdot 10^{-4}$	$5.501 \cdot 10^{-5}$	$6.091 \cdot 10^{-5}$	$1.394 \cdot 10^{-5}$
4	5	$1.151 \cdot 10^{-4}$	$1.358 \cdot 10^{-5}$	$2.669 \cdot 10^{-5}$	$8.783 \cdot 10^{-6}$
4	6	$2.530 \cdot 10^{-5}$	$6.493 \cdot 10^{-6}$	$1.755 \cdot 10^{-5}$	$8.362 \cdot 10^{-6}$

One of the most interesting results of this work is the comparison done for the maximum number of neurons that can be implemented with the board used, and also the resources associated with the implementation of a single neuron in

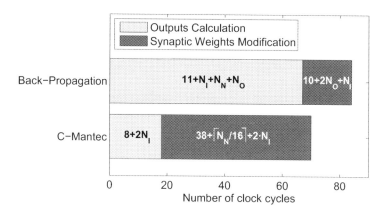

Fig. 5. Execution cycles needed to learn and compute the output for a single input pattern for the case of a 5-50-1 neural network architecture

relationship to both algorithms. Table 3 shows the values obtained for BP and C-Mantec for different values for the integer N_1 and decimal part N_2 of the fixed point representations tested.

Table 3. Number of LUTs and maximum number of neurons that can be implemented in a Virtex-5 board for the two algorithms and as a function of different fixed-point representations

N_1	N_2	LUTs/Neuron		# Neurons	
		BP	C-M	BP	C-M
8	8	787	689	82	94
8	12	967	757	67	85
8	16	1124	943	57	68
12	12	1057	826	61	78
12	16	1223	1033	53	62
16	16	1382	1299	47	50

We have also analysed the number of FPGA clock cycles involved in the computations related to training a network with one input pattern. The total number of cycles is divided in two parts related to the number cycles related to compute the output of the network for a given input and for the modification of the synaptic weights (cf. Eqs. 4 and 7). The results displayed in the table correspond to two neural networks with 50 neurons in the single hidden layer of the architecture.

Using a set of benchmark functions from the UCI repository, we have computed the generalization ability and computation time (ms) for both algorithms. Table 4 shows these results together with the number of neurons used in each

Table 4. Generalization ability (%) obtained and number of neurons used in the architectures for the implementation of seven classic benchmark problems

Function	C-Mantec			Back-Propagation		
	Gen.	# Neu.	time (ms)	Gen.	# Neu.	time (ms)
Diabetes	76.6	5	97	79.3	5	227
Cancer	96.9	2	52	95.7	5	210
Heart	82.6	3	71	78.2	5	104
Ionosphere	87.4	2	56	87.5	5	210
Heart-c	82.5	2	55	80.1	5	190
Card	85.2	3	72	83.1	5	195
Sonar	75.0	1	43	75.2	5	223
Average	83.7	3	63	82.7	5	194

case, noting that C-Mantec sets this number automatically while a constant size architecture comprising 5 neurons was used for Back-Propagation.

4 Conclusion

We have presented and analysed the implementation in a FPGA board of two neural network learning algorithms: Back-Propagation and C-Mantec. The algorithms operates from different principles as BP is a gradient based algorithm minimizing an error function for pre-determined architecture that has to be defined in advance, while C-Mantec is an error correcting method that constructs the network architecture automatically as it learns the input patterns. In terms of the FPGA implementation, both methods require the implementation of continues functions (the sigmoid and the exponential functions), process that is very simple for standard computers (PCs) but much more complex for hardware devices using a fixed point representation like FPGAs. An analysis of the resources needed to implement both functions efficiently indicate that similar error levels are obtained for both cases when using a lookup table plus linear interpolation scheme, with slightly lower error values for the case of the sigmoid function used in the BP algorithm. Nevertheless, it is worth noting that as C-Mantec is an error correcting algorithm the precision needed for the arithmetic representation is lower that for BP that might require high precision levels as it involves the accurate computation of the derivatives of the activation functions for its correct operation.

Another very important issue regarding the comparison of both algorithms is the amount of hardware resources needed for the implementation of single neurons in both algorithms, and in this aspect the advantage is on the C-Mantec side as a lower number of LUTs is required, permitting for a given board the construction of larger neural network architectures, that in our case resulted in approximately a 18.7% increase in the maximum number of neurons that can be included (cf. Table 3). Further, we have also estimated the average computation time needed for training both algorithms using a set of benchmark functions, finding that C-Mantec operates faster than BP, needing in average a third of the computational time (cf. Table 4).

As an overall conclusion the present work shows a comparison regarding the possibilities of the application of neurocomputational algorithms using FPGA boards. The comparison of both algorithms is a little bit in favour of C-Mantec as first it does not need the a priori specification of the neural architecture to be used, and second as it is less demanding in terms of hardware resource utilization.

Acknowledgments. The authors acknowledge support from Junta de Andalucía through grants P10-TIC-5770 and P08-TIC-04026, and from CICYT (Spain) through grant TIN2010-16556 (all including FEDER funds).

References

1. Haykin, S.: Neural networks: a comprehensive foundation. Prentice Hall (1994)
2. Kilts, S.: Advanced FPGA Design: Architecture, Implementation, and Optimization. Wiley-IEEE Press (2007)
3. Monmasson, E., Idkhajine, L., Cirstea, M., Bahri, I., Tisan, A., Naouar, M.W.: Fpgas in industrial control applications. IEEE Transactions on Industrial Informatics **7**(2), 224–243 (2011)
4. Bacon, D., Rabbah, R., Shukla, S.: Fpga programming for the masses. Queue **11**, 40–52 (2013)
5. Conmy, P., Bate, I.: Component-based safety analysis of fpgas. IEEE Transactions on Industrial Informatics **6**(2), 195–205 (2010)
6. Zhu, J., Sutton, P.: Fpga implementations of neural networks - a survey of a decade of progress. In: Cheung, P.Y.K., Constantinides, G.A. (eds.) FPL 2003. LNCS, vol. 2778, pp. 1062–1066. Springer, Heidelberg (2003)
7. Gomperts, A., Ukil, A., Zurfluh, F.: Development and implementation of parameterized fpga-based general purpose neural networks for online applications. IEEE Trans. Industrial Informatics **7**(1), 78–89 (2011)
8. Le, Q., Jeon, J.: Neural-network-based low-speed-damping controller for stepper motor with an fpga. IEEE Transactions on Industrial Applications **57**, 3167–3180 (2010)
9. Ortega-Zamorano, F., Jerez, J., Urda, D., Luque-Baena, R., Franco., L.: Efficient implementation of the backpropagation algorithm in fpgas and microcontrollers. IEEE Transactions on Neural Networks and Learning Systems (2015) (in press)
10. Ortega-Zamorano, F., Jerez, J., Franco, L.: Fpga implementation of the c-mantec neural network constructive algorithm. IEEE Transactions on Industrial Informatics **10**(2), 1154–1161 (2014)
11. Werbos, P.J.: Beyond Regression: New Tools for Prediction and Analysis in the Behavioral Sciences. PhD thesis, Harvard University (1974)
12. Rumelhart, D., Hinton, G., Williams, R.: Learning representations by back-propagating errors. Nature **323**(6088), 533–536 (1986)
13. Subirats, J.L., Franco, L., Jerez, J.M.: C-mantec: A novel constructive neural network algorithm incorporating competition between neurons. Neural Netw. **26**, 130–140 (2012)
14. Frean, M.: The upstart algorithm: a method for constructing and training feedforward neural networks. Neural Computation **2**(2), 198–209 (1990)

15. Savich, A., Moussa, M., Areibi, S.: The impact of arithmetic representation on implementing mlp-bp on fpgas: A study. IEEE Transactions on Neural Networks **18**(1), 240–252 (2007)
16. Ortega-Zamorano, F., Jerez, J., Juarez, G., Perez, J., Franco, L.: High precision fpga implementation of neural network activation functions. In: 2014 IEEE Symposium on Intelligent Embedded Systems (IES), pp. 55–60, December 2014
17. Chu, P.P.: FPGA Prototyping by VHDL Examples: Xilinx Spartan-3 Version. John Wiley & Sons (2008)
18. Ashenden, P.: The Designer's Guide to VHDL (Systems on Silicon), vol. 3, 3rd edn. Morgan Kaufmann Publishers Inc., San Francisco (2008)

Expert Systems

Logic Programming and Artificial Neural Networks in Breast Cancer Detection

José Neves[1(✉)], Tiago Guimarães[2], Sabino Gomes[2], Henrique Vicente[3], Mariana Santos[4], João Neves[5], José Machado[1], and Paulo Novais[1]

[1] Algoritmi, Universidade Do Minho, Braga, Portugal
{jneves,jmac,pjon}@di.uminho.pt
[2] Departamento de Informática, Universidade Do Minho, Braga, Portugal
{tiago.sguimaraes,sabinogomes.antonio}@gmail.com
[3] Departamento de Química, Centro de Química de Évora, Escola de Ciências E Tecnologia,
Universidade de Évora, Évora, Portugal
hvicente@uevora.pt
[4] Escola de Ciências Da Saúde, Universidade Do Minho, Braga, Portugal
marianamltsantos@gmail.com
[5] DRS. NICOLAS&ASP, Dubai, United Arab Emirates
joaocpneves@gmail.com

Abstract. About 90% of breast cancers do not cause or are capable of producing death if detected at an early stage and treated properly. Indeed, it is still not known a specific cause for the illness. It may be not only a beginning, but also a set of associations that will determine the onset of the disease. Undeniably, there are some factors that seem to be associated with the boosted risk of the malady. Pondering the present study, different breast cancer risk assessment models where considered. It is our intention to develop a hybrid decision support system under a formal framework based on Logic Programming for knowledge representation and reasoning, complemented with an approach to computing centered on Artificial Neural Networks, to evaluate the risk of developing breast cancer and the respective Degree-of-Confidence that one has on such a happening.

Keywords: Breast cancer · Tyrer-cuzick model · Knowledge representation and reasoning · Logic programing · Artificial Neural Networks

1 Introduction

Breast cancer is the most frequent malignancy in female [1], affecting one million women worldwide and 4500 in Portugal, every year. In Portugal, even though the occurrence rate is high, i.e., 75/100000 in 2005 [2], the mortality rate was reduced, exposing the role of screening at an early stage of the illness [3]. This type of tumor has a large impact in our society, not only because of its frequency and severity, but also due to its social and domestic standing.

© Springer International Publishing Switzerland 2015
I. Rojas et al. (Eds.): IWANN 2015, Part II, LNCS 9095, pp. 211–224, 2015.
DOI: 10.1007/978-3-319-19222-2_18

Mammography is the most common screening tool since it uses low dose X-rays to create an image of the breast to find breast cancer. Screening mammograms are used to look for breast disease in women who are asymptomatic. The American Cancer Society recommends that all women aged over 40 should undergo screening mammography once in a year.

Masses and microcalcification are an important early signs of breast cancer. However, it is often difficult to distinguish abnormalities from normal breast tissues because of their subtle appearance and ambiguous margins. Younger women, who tend to have denser breasts, can make mammograms harder to interpret and can provoke false positive results. According Kolb et al. [4], depending on the density of the breasts radiologists may miss up to 30% of breast cancers. Moreover, mass lesions vary in appearance from patient to patient and similar attributes are shared by some benign and malignant masses, as it is the mammography classification of mass lesions is a difficult task. Indeed, even experienced radiologists have difficulty to interpret screening mammograms in large numbers. Hence, computer-aided diagnostics is a promising tool for radiologists to produce an accurate and faster diagnosis result for breast cancer patients. Some general computer-aided diagnostics systems have been presented based on Kohonen's self-organizing map [5], neuro-fuzzy approach [6, 7], support vector machines [8], Artificial Neural Networks (ANNs) [9] ANNs combined with techniques for reducing the dimension of initial database like association rules [10], sequential forward selection, sequential backward selection, and principal component analysis [11], or clustering [12]. Other authors have developed computer-aided diagnostics systems combining different methodologies like genetic algorithms and ANNs [13] or swarm intelligence and wavelet neural networks [14, 15].

The computer-aided diagnostics systems presented above are related with the problem of distinguishing between abnormalities and normal breast tissues. However, the development of an all-inclusive risk assessment models should include other characteristics of patient like genetic risk factors (about 5% to 10% of breast cancer cases can to be hereditary) or hormonal factors (women who have had more menstrual cycles have a slightly higher risk of breast cancer, due to a longer lifetime exposure to the hormones estrogen and progesterone) [16, 17, 18].

Breast cancer is typically asymptomatic until the development of clinical complications. Unfortunately, these complications appear at a relatively late stage of the progression of the disease. Thus, it is difficult to make an early diagnosis of the disease, since it needs to consider different conditions with intricate relations among them, where the available data may be incomplete, contradictory and even unknown, i.e., where the most common limitations are related with the poor quality of the available information. Those drawbacks are mainly due to their reliance on known risk factors like oral contraceptive pill use, ethnic group, breast density, and higher than second-degree relatives with breast cancer, which are yet to be studied.

In order to overcome the problems related with incomplete, contradictory and/or unknown information, the present work reports the founding of a computational framework that uses knowledge representation and reasoning techniques to set the structure of the information and the associate inference mechanisms. We will centre on a Logic Programming (LP) approach to knowledge representation and reasoning [19, 20], and look at a soft computing approach to data processing based on ANNs [21].

2   Knowledge Representations and Reasoning

Many approaches to knowledge representation and reasoning have been proposed using the Logic Programming (LP) archetype, namely in the area of Model Theory [22, 23], and Proof Theory [19, 20]. In this work it is followed the proof theoretical approach in terms of an extension to the LP language to knowledge representations and reasoning. An Extended Logic Program is a finite set of clauses in the form:

$\{$

$\quad p \leftarrow p_1, \cdots, p_n, not\ q_1, \cdots, not\ q_m$

$\quad ?\ (p_1, \cdots, p_n, not\ q_1, \cdots, not\ q_m)\ (n, m \geq 0)$

$\quad exception_{p_1}$

$\quad \cdots$

$\quad exception_{p_j}\ (j \leq m, n)$

$\}\ ::\ scoring_{value}$

where "?" is a domain atom denoting falsity, the p_i, q_j, and p are classical ground literals, i.e., either positive atoms or atoms preceded by the classical negation sign $\neg$ [19]. Under this formalism, every program is associated with a set of abducibles [22, 23], given here in the form of exceptions to the extensions of the predicates that make the program. The term $scoring_{value}$ stands for the relative weight of the extension of a specific *predicate* with respect to the extensions of the peers ones that make the overall program.

In order to evaluate the knowledge that stems from a logic program an evaluation of the *Quality-of-Information* (*QoI*) was set in dynamic environments aiming at decision-making purposes [24, 25]. The objective is to build a quantification process of *QoI* and an assessment of the argument values of a given predicate with relation to their domains (here understood as *Degree-of-Confidence* (*DoC*), which stands for one's belief that its unknown values fits into the arguments ranges, taking into account their domains).

The *QoI* with respect to the extension of a predicate i will be given by a truth-value in the interval [0, 1], i.e., if the information is *known* (*positive*) or *false* (*negative*) the *QoI* for the extension of *predicate*$_i$ is 1. For situations where the information is unknown, the *QoI* is given by:

$$QoI_{predicate_i} = \lim_{N \to \infty} \frac{1}{N} = 0 \quad (N \gg 0) \tag{1}$$

where N denotes the cardinality of the set of terms or clauses of the extension of *predicate*$_i$ that stand for the incompleteness under consideration. For situations where the extension of *predicate*$_i$ is unknown but can be taken from a set of values, the *QoI* is given by:

$$QoI_{predicate_i} = {}^1/_{Card} \qquad (2)$$

where $Card$ denotes the cardinality of the $abducible$ set for i, if the $abducible$ set is disjoint. If the $abducible$ set is not disjoint, the QoI is given by:

$$QoI_{predicate_i} = \frac{1}{C_1^{Card} + \cdots + C_{Card}^{Card}} \qquad (3)$$

where C_{Card}^{Card} is a card-combination subset, with $Card$ elements. The next element of the model to be considered is the relative importance that a predicate assigns to each of its attributes under observation, i.e., w_i^k, which stands for the relevance of attribute k in the extension of $predicate_i$. It is also assumed that the weights of all the attribute predicates are normalized, i.e.:

$$\sum_{1 \le k \le n} w_i^k = 1, \forall_i \qquad (4)$$

where $\forall$ denotes the universal quantifier. It is now possible to define a predicate's scoring function $V_i(x)$ so that, for a value $x = (x_1, \cdots, x_n)$, defined in terms of the attributes of $predicate_i$, one may have:

$$V_i(x) = \sum_{1 \le k \le n} w_i^k * QoI_i(x)/n \qquad (5)$$

allowing one to set:

$$predicate_i(x_1, \cdots, x_n) :: V_i(x) \qquad (6)$$

that denotes the inclusive quality of $predicate_i$ with respect to all the predicates that make the program. It is now possible to set a logic program (here understood as the predicates' extensions that make the program) scoring function, in the form:

$$LP_{Scoring\ Function} = \sum_{i=1}^{n} V_i(x) * p_i \qquad (7)$$

where p_i stands for the relevance of the $predicate_i$ in relation to the other predicates whose extensions denote the logic program. It is also assumed that the weights of all the predicates' extensions are normalized, i.e.:

$$\sum_{i=1}^{n} p_i = 1, \forall_i \qquad (8)$$

where $\forall$ denotes the universal quantifier.

It is now possible to engender the universe of discourse, according to the information given in the logic programs that endorse the information about the problem under consideration, according to productions of the type:

$$predicate_i - \bigcup_{1 \le j \le m} clause_j(x_1, \cdots, x_n) :: QoI_i :: DoC_i \qquad (9)$$

where U and m stand, respectively, for *set union* and the *cardinality* of the extension of *predicate_i*. DoC_i denotes one's confidence on the attribute's values of a particular term of the extension of *predicate_i*, whose evaluation is given in [26]. In order to advance with a broad-spectrum, let us suppose that the *Universe of Discourse* is described by the extension of the predicates:

$$f_1(\cdots), f_2(\cdots), \cdots, f_n(\cdots) \ where \ (n \geq 0) \tag{10}$$

Assuming that a clause denotes a happening, a clause has as argument all the attributes that make the event. The argument values may be of the type unknown or members of a set, or may be in the scope of a given interval, or may qualify a particular observation.

3   A Case Study

In order to exemplify the applicability of our method, we will look at the relational database model, since it provides a basic framework that fits into our prospects, and is understood as the genesis of the *LP* approach to Knowledge Representation and Reasoning [19].

As a case study, consider the scenario where a relational database is given in terms of the extensions of the relations (or tables) depicted in Fig. 1, which stands for a situation where one has to manage information about breast cancer. Under this scenario some incomplete and/or default data is also available. For instance, in the *Hormonal Factors* table, the age of menarche in case 2 is unknown, while in the last case the age of menopause ranges in the interval [1, 2].

In Fig. 1 one may find four large groups of variables, i.e., *Personal Information*, *Hormonal Factors*, *Family History* and *Personal Breast Disease*. The *Personal Information* can be affected by the *Age* (whose value is either 0 (zero) if the age is below 50 (fifty) or 1 (one) if it is above 50 (fifty); and by the *Body Mass Index* (whose value is 0 (zero) if it is below 25 (twenty five), 1 (one) if it is above 25 (twenty five) but below 35 (thirty five), and 2 (two) if it is above 35 (thirty five).

The *Hormonal Factors* are affected by the age when menarche happened (0 (zero) if it occurred below 12 (twelve) or 1 (one) if it is above); by the age when an individual had her first born child (0 (zero) if it is nulliparous or with an age below 30 (thirty), or 1 (one) if it is above); and by the age when menopause happened (0 (zero) if the individual is not yet in the menopause, 1 (one) if it happened before the age of 50 (fifty), and 2 (two) if the age is above 50 (fifty)).

The *Family History* and *Personal Breast Disease* columns can be affected by five and three parameters, respectively. Each one can be either 0 (zero) or 1 (one), where 0 (zero) means the absence and 1 (one) the presence of the disease on the first and second degree relatives, an early onset of cancer in any relative, bilateral breast cancer and ovarian cancer for the first column and breast biopsies, atypical hyperplasia and Lobular Carcinoma In Situ (LCIS) for the second [16, 17].

Personal Information		
#	Age	Body Mass Index
1	37	24
2	58	28
...	...	...
217	55	32

Hormonal Factor			
#	Menarche	First Live Birth	Menopause
1	0	1	0
2	$\perp$	1	1
...	...	...	...
217	1	0	[1,2]

Breast Cancer Diagnosis					
#	Age	Body Mass Index	Hormonal Factors	Family History	Personal Breast Disease
1	0	0	1	1	0
2	1	1	$\perp$	3	2
...	...	...	...	...	...
217	1	2	[2,3]	$\perp$	2

Personal Breast Disease			
#	Breast Biopsies	Atypical Hyperplasia	LCIS
1	0	0	0
2	0	1	1
...	...	...	...
217	1	1	0

Family History					
#	First Degree Relatives	Second Degree Relatives	Early onset of Cancer	Bilateral Breast Cancer	Ovarian Cancer
1	0	1	0	0	0
2	0	1	0	1	1
...	...	...	...	...	...
217	1	$\perp$	1	1	0

Fig. 1. An extension of the relational database model

Now, applying the rewritten algorithm presented in [26], to all the tables that make the Extension of the Relational Database model for Liver Diseases Diagnosis (Fig. 1), excluding of such a process the *Breast Cancer Diagnosis* one, and looking to the DoC_s values obtained in this manner, it is possible to set the arguments of the predicate referred to below, that also denotes the objective function with respect to the problem under analyze.

$$breast\_cancer: Age, B_{ody}M_{ass}I_{ndex}, H_{ormonal}F_{actors},$$

$$F_{amily}H_{istory}, P_{ersonal}B_{reast}D_{isease} \rightarrow \{0,1\}$$

where 0 (zero) and 1 (one) denote, respectively, the truth values *false* and *true*. Indeed, the arguments of this predicate where set by a process of sensibility analysis, where the arguments chosen where those that present the higher DoC_s values, i.e., the ones that have a greater influence on the output of the objective function referred to above. Their terms also make the training and test sets of the Artificial Neural Network (ANN) given in Fig. 2.

Now, let us consider a patient that presents the symptoms $Age = 1$, $BMI = 2$, $HF = [3, 4]$, $FH = \perp$, $PBD = 1$, to which it is applied the rewritten algorithm presented in [23]. One may get:

Begin,

The predicate's extensions that make the Universe-of-Discourse for the patient under observation are set $\leftarrow$

{

   $\neg breast\_cancer(Age, BMI, HF, FH, PBD)$

                 $\leftarrow not\ breast\_cancer(Age, BMI, HF, FH, PBD)$

$$breast\_cancer \left(\underbrace{1, \quad 2, \quad [3,4], \quad \perp, \quad 1}_{attribute's\ values} \right) :: 1 :: DoC$$

$$\underbrace{[0,1][0,2][0,4]\ [0,5][0,3]}_{attribute's\ domains}$$

} :: 1

The attribute's values ranges are rewritten $\leftarrow$

{

   $\neg breast\_cancer(Age, BMI, HF, FH, PBD)$

                 $\leftarrow not\ breast\_cancer(Age, BMI, HF, FH, PBD)$

$$breast\_cancer \left(\underbrace{[1,1], [2,2], [3,4], [0,5], [1,1]}_{attribute's\ values\ ranges} \right) :: 1 :: DoC$$

$$\underbrace{[0,1]\ [0,2]\ [0,4]\ [0,5]\ [0,3]}_{attribute's\ domains}$$

} :: 1

The attribute's boundaries are set to the interval [0,1] $\leftarrow$

{

   $\neg breast\_cancer(Age, BMI, HF, FH, PBD)$

                 $\leftarrow not\ breast\_cancer(Age, BMI, HF, FH, PBD)$

$$breast\_cancer \left(\underbrace{[1,1], [1,1], [0.75,1], [0,1], [0.33,0.33]}_{attribute's\ values\ ranges\ once\ normalized} \right) :: 1 :: DoC$$

$$\underbrace{[0,1]\ [0,1]\ \ [0,1]\ \ \ [0,1]\ \ \ \ [0,1]}_{attribute's\ domains\ once\ normalized}$$

} :: 1

The DoC's values are evaluated ←

{

    ¬*breast_cancer*(*Age*, *BMI*, *HF*, *FH*, *PBD*)

$$\leftarrow not\ breast\_cancer(Age, BMI, HF, FH, PBD)$$

$$breast\_cancer \left(\underbrace{1,\quad 1,\quad 0.97,\quad 0,\quad\quad 1}_{attribute's\ confidence\ values} \right) :: 1 :: 0.79$$

$$\underbrace{[1,1][1,1][0.75,1][0,1][0.33,0.33]}_{attribute's\ values\ ranges\ once\ normalized}$$

$$\underbrace{[0,1][0,1]\quad[0,1]\quad[0,1]\quad\quad[0,1]}_{attribute's\ domains\ once\ normalized}$$

} :: 1

End.

where its argument values, i.e., (1, 1, 0.97, 0, 1) make the input to the Artificial Neural Network (ANN) given in Fig. 2. The output of ANN stands for the patient diagnosis.

4   Artificial Neural Networks

It was set a soft computing approach to model the universe of discourse of any patient suffering from breast cancer, based on Artificial Neural Networks (ANNs), which are used to structure data and capture complex relationships between inputs and outputs [27, 28, 29]. ANNs simulate the structure of the human brain, being populated by multiple layers of neurons, with a valuable set of activation functions. As an example, let us consider the case listed above, where one may have a situation in which the diagnosis of breast cancer is needed. In Fig. 2 it is shown how the normalized values of the interval boundaries and their DoC_s and QoI_s values work as inputs to the ANN. The output depicts a breast cancer diagnostic, plus the confidence that one has on such a happening.

In this study were considered 217 patients with an age average of 58.2 years, ranging from 28 to 87 years old. Breast cancer was diagnosed in 41 patients, i.e., 18.9% of the analysed population. The data came from a main health care center in the north of Portugal. The dataset holds information about risk factors considered critical in the prediction of breast cancer. Thirteen variables were selected allowing one to have a multivariable dataset with 217 records (Fig. 1). Table 1 shows a brief description of each variable and the data type, i.e., numeric or nominal.

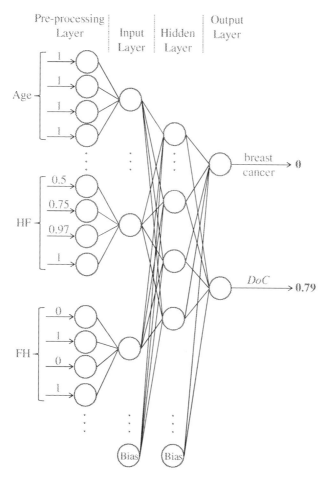

Fig. 2. The Artificial Neural Network topology

To ensure statistical significance of the attained results, 25 (twenty five) experiments were applied in all tests. In each simulation, the available data was randomly divided into two mutually exclusive partitions, i.e., the training set with 70% of the available data and, the test set with the remaining 30% of the cases. The back propagation algorithm was used in the learning process of the ANN. As the output function in the pre-processing layer it was used the identity one. In the other layers we used the sigmoid function.

A common tool to evaluate the results presented by the classification models is the coincidence matrix, a matrix of size $L \times L$, where L denotes the number of possible classes. This matrix is created by matching the predicted and target values. L was set to 2 (two) in the present case. Table 2 present the coincidence matrix (the values denote the average of the 25 experiments). Table 2 shows that the model accuracy was 96.1% for the training set (146 correctly classified in 152) and 95.4% for test set (62 correctly classified in 65).

Table 1. Variables characterization

Variable	Description	Data type
Age	Patient's age	Numeric
Body Mass Index	Patient's body mass index	Numeric
Menarche	Age when menarche happened	Nominal
First Live Birth	Age when patient had her first born child	Nominal
Menopause	Age when menarche happened	Nominal
Breast Biopsies	Had breast biopsies	Nominal
Atypical Hyperplasia	Has atypical hyperplasia	Nominal
LCIS	Has lobular carcinoma in situ	Nominal
First Degree Relatives	Presence of disease on first degree relatives	Nominal
Second Degree Relatives	Presence of disease on second degree relatives	Nominal
Early onset of Cancer	Early onset of cancer in any relative	Nominal
Bilateral Breast Cancer	Presence of disease in any relative	Nominal
Ovarian Cancer	Presence of disease in any relative	Nominal

Based on coincidence matrix it is possible to compute sensitivity, specificity, Positive Predictive Value (PPV) and Negative Predictive Value (NPV) of the classifier. Briefly, sensitivity and specificity are statistical measures of the performance of a binary classifier, while sensitivity measures the proportion of true positives that are correctly identified as such. Specificity measures the proportion of true negatives that are correctly identified. Moreover, it is necessary to know the probability of the classifier that give the correct diagnosis. Thus, it is also calculated both PPV and NPV, while PPV stands for the proportion of cases with positive results which are correctly diagnosed, NPV is the proportion of cases with negative results which are successfully labeled. The corresponding sensitivity, specificity, PPV and NPV values are displayed in Table 3 for training and test sets.

Table 2. The coincidence matrix for the ANN model

Target	Predictive			
	Training set		Test set	
	True (1)	False (0)	True (1)	False (0)
True (1)	28	1	12	0
False (0)	5	118	3	50

Table 3. Sensitivity, specificity, positive predictive value (PPV) and negative predictive value (NPV) for for the ANN model

	Sensitivity (%)	Specificity (%)	PPV (%)	NPV (%)
Training set	96.6	95.9	84.8	99.2
Test set	100.0	94.3	80.0	100.0

A perusal of Table 3 shows that the sensitivity ranges from 96.6% to 100%, while the specificity ranges from 94.3% to 95.9%. The sensitivity of the proposed model is higher than the reported in literature for medical imaging diagnosis, ranging from 70.21% to 94.17% [13, 14, 15]. PPV ranges from 80.0% to 84.8%, while NPV ranges from 99.2% to 100%. For comparison, the reported PPV for medical imaging diagnosis ranges between 64.03% and 87.09%, while the corresponding NPV ranges between 90.96% and 94.71% [13]. Thus, it is our claim that the proposed model is able to predict breast cancer predisposition properly. The inclusion of other patient's characteristics, like genetic risk and hormonal factors may be responsible for the good performance exhibited by the presented model.

5   Conclusions and Future Work

This risk assessment system is able to give an adequate response to the need for a good method of breast cancer prediction. To go around the problem, more effectively, much more variables must be studied and considered, thus fulfilling important gaps in the existent risk assessment methods.

Being an area filled with incomplete and unknown data it may be tackled by Artificial Intelligence based methodologies and techniques for problem solving. This work presents the founding of a computational framework that uses powerful knowledge representation and reasoning techniques to set the structure of the information and the associate inference mechanisms.

The knowledge representation and reasoning techniques presented above are very versatile and capable of covering almost every possible instance, namely by considering incomplete, contradictory, and even unknown data, a marker that is not present in existing systems. Indeed, this method brings a new approach that can revolutionize prediction tools in all its variants, making it more complete than the existing methodologies and tools available. The new paradigm of knowledge representation and reasoning enables the use of the normalized values of the interval boundaries and their *DoC* values, as inputs to the ANN. The output translates a diagnosis of liver disease and the confidence that one has on such a happening.

Our main contribution relies on the fact that at the end, the extensions of the predicates that make the universe of discourse are given in terms of *DoCs* that stand for one's confidence that the predicates arguments values fit into their respective domains. It also encapsulates, in itself, a new vision of Multi-value Logics, once a proof of a theorem in a conventional way, is evaluated to the interval [0, 1]. Indeed, some interesting results have been obtained, namely in the fields of Coronary Risk Evaluation [30], Hyperactivity Disorder [31] and Length of Hospital Stay [32] among others. Furthermore, this approach potentiates the use of diverse computational paradigms, in particular the *Logic Programming* one. Future work may recommend that the same problem must be approached using others computational frameworks like Genetic Programming [20], Case Based Reasoning [33] or Particle Swarm [34], just to name a few.

Acknowledgments. This work has been supported by FCT – Fundação para a Ciência e Tecnologia within the Project Scope UID/CEC/00319/2013.

References

1. McPherson, K., Steel, C.M., Dixon, J.M.: ABC of Breast Diseases: Breast Cancer Epidemiology, Risk Factors, and Genetics. British Medical Journal **321**, 624–628 (2000)
2. National Oncological Registry 2001 (in Portuguese). Instituto Português de Oncologia de Francisco Gentil Edition, Lisbon (2003)
3. Gøtzsche, P.C., Jørgensen, K.J.: Screening for breast cancer with mammography. Cochrane Database of Systematic Reviews, Issue 6, Art. N°. CD001877 (2013)
4. Kolb, T.M., Lichy, J., Newhouse, J.H.: Comparison of the performance of screening mammography, physical examination, and breast US and evaluation of factors that influence them: an analysis of 27,825 patient evaluations. Radiology **225**, 165–175 (2002)
5. Markey, M.K., Lo, J.Y., Tourassi, G.D., Floyd Jr., C.E.: Self-organizing map for cluster analysis of a breast cancer database. Artificial Intelligence in Medicine **27**, 113–127 (2003)
6. Keles, A., Keles, A., Yavuz, U.: Expert system based on neuro-fuzzy rules for diagnosis breast cancer. Expert Systems with Applications **38**, 5719–5726 (2011)
7. Nieto, J., Torres, A.: Midpoint for fuzzy sets and their application in medicine. Artificial Intelligence in Medicine **27**, 321–355 (2003)
8. Addeh, J., Ebrahimzadeh, A.: Breast cancer recognition using a novel hydride intelligent method. Journal of Medical Signals and Sensors **2**, 95–102 (2012)
9. Ubeyli, E.D.: Implementing automated diagnostic systems for breast cancer detection. Expert Systems with Applications **33**, 1054–1062 (2007)
10. Karabatak, M., Ince, M.C.: An expert system for detection of breast cancer based on association rules and neural network. Expert Systems with Applications **36**, 3465–3469 (2009)
11. Uzer, M.S., Inan, O., Yilmaz, N.: A hybrid breast cancer detection system via neural network and feature selection based on SBS, SFS and PCA. Neural Computing and Applications **23**, 719–728 (2013)
12. Kalteh, A.A., Zarbakhsh, P., Jirabadi, M., Addeh, J.: A research about breast cancer detection using different neural networks and K-MICA algorithm. Journal of Cancer Research and Therapeutics **9**, 456–466 (2013)
13. Belciug, S., Gorunescu, F.: A hybrid neural network/genetic algorithm applied to breast cancer detection and recurrence. Expert Systems **30**, 243–254 (2013)
14. Dheeba, J., Selvi, S.T.: An Improved Decision Support System for Detection of Lesions in Mammograms Using Differential Evolution Optimized Wavelet Neural Network. Journal of Medical Systems **36**, 3223–3232 (2012)
15. Dheeba, J., Singh, N.A., Selvi, S.T.: Computer-aided detection of breast cancer on mammograms: A swarm intelligence optimized wavelet neural network approach. Journal of Biomedical Informatics **49**, 45–52 (2014)
16. Powell, M., Jamshidian, F., Cheyne, K., Nititham, J., Prebil, L.A., Ereman, R.: Assessing Breast Cancer Risk Models in Marin County, a Population With High Rates of Delayed Childbirth. Clinical Breast Cancer **14**, 212–220 (2014)
17. Amir, E., Freedman, O.C., Seruga, B., Evans, G.G.: Assessing Women at High Risk of Breast Cancer: A Review of Risk Assessment Models. Journal of the National Cancer Institute **102**, 680–691 (2010)
18. Jacobi, C.E., de Bock, G.H., Siegerink, B., van Asperen, C.J.: Differences and similarities in breast cancer risk assessment models in clinical practice: which model to choose? Breast Cancer Research and Treatment **115**, 381–390 (2009)

19. Neves, J.: A logic interpreter to handle time and negation in logic databases. In: Muller, R.L., Pottmyer, J.J. (eds.) ACM 1984 Proceedings of the 1984 annual conference of the ACM on The Fifth Generation Challenge, pp. 50–54. Association for Computing Machinery, New York (1984)

20. Neves, J., Machado, J., Analide, C., Abelha, A., Brito, L.: The halt condition in genetic programming. In: Neves, J., Santos, M.F., Machado, J.M. (eds.) EPIA 2007. LNCS (LNAI), vol. 4874, pp. 160–169. Springer, Heidelberg (2007)

21. Cortez, P., Rocha, M., Neves, J.: Evolving Time Series Forecasting ARMA Models. Journal of Heuristics **10**, 415–429 (2004)

22. Kakas, A., Kowalski, R., Toni, F.: The role of abduction in logic programming. In: Gabbay, D., Hogger, C., Robinson, I. (eds.) Handbook of Logic in Artificial Intelligence and Logic Programming, vol. 5, pp. 235–324. Oxford University Press, Oxford (1998)

23. Pereira, L.M., Anh, H.T.: Evolution prospection. In: Nakamatsu, K., Phillips-Wren, G., Jain, L.C., Howlett, R.J. (eds.) New Advances in Intelligent Decision Technologies. SCI, vol. 199, pp. 51–63. Springer, Heidelberg (2009)

24. Lucas, P.: Quality checking of medical guidelines through logical abduction. In: Coenen, F., Preece, A., Mackintosh, A. (eds.) Proceedings of AI-2003 (Research and Developments in Intelligent Systems XX), pp. 309–321. Springer, London (2003)

25. Machado, J., Abelha, A., Novais, P., Neves, J., Neves, J.: Quality of Service in healthcare units. International Journal of Computer Aided Engineering and Technology **2**, 436–449 (2010)

26. Cardoso, L., Marins, F., Magalhães, R., Marins, N., Oliveira, T., Vicente, H., Abelha, A., Machado, J., Neves, J.: Abstract Computation in Schizophrenia Detection through Artificial Neural Network based Systems. The Scientific World Journal **2015**, 1–10 (2015). Article ID 467178

27. Caldeira, A.T., Arteiro, J., Roseiro, J., Neves, J., Vicente, H.: An Artificial Intelligence Approach to Bacillus amyloliquefaciens CCMI 1051 Cultures: Application to the Production of Antifungal Compounds. Bioresource Technology **102**, 1496–1502 (2011)

28. Vicente, H., Dias, S., Fernandes, A., Abelha, A., Machado, J., Neves, J.: Prediction of the Quality of Public Water Supply using Artificial Neural Networks. Journal of Water Supply: Research and Technology – AQUA **61**, 446–459 (2012)

29. Salvador, C., Martins, M.R., Vicente, H., Neves, J., Arteiro, J.M., Caldeira, A.T.: Modelling Molecular and Inorganic Data of Amanita ponderosa Mushrooms using Artificial Neural Networks. Agroforestry Systems **87**, 295–302 (2013)

30. Rodrigues, B., Gomes, S., Vicente, H., Abelha, A., Novais, P., Machado, J., Neves, J.: Systematic coronary risk evaluation through artificial neural networks based systems. In: Goto, T. (ed.) Proceedings of the 27th International Conference on Computer Applications in Industry and Engineering – CAINE 2014, pp. 21–26. ISCA, Winona (2014)

31. Pereira, S., Gomes, S., Vicente, H., Ribeiro, J., Abelha, A., Novais, P., Machado, J., Neves, J.: An artificial neuronal network approach to diagnosis of attention deficit hyperactivity disorder. In: Proceedings of 2014 IEEE International Conference on Imaging Systems and Techniques – IST 2014, pp. 410–415. Institute of Electrical and Electronics Engineers, Inc., New Jersey (2014)

32. Abelha, V., Vicente, H., Machado, J., Neves, J.: An Assessment on the Length of Hospital Stay through Artificial Neural Networks. In: Papadopoulos, G. (ed.) Proceedings of the 9th International Conference on Knowledge, Information and Creativity Support Systems – KICSS 2014, pp. 219–230. Cyprus Library, Nicosia (2014)
33. Carneiro, D., Novais, P., Andrade, F., Zeleznikow, J., Neves, J.: Using Case-Based Reasoning and Principled Negotiation to provide decision support for dispute resolution. Knowledge and Information Systems 36, 789–826 (2013)
34. Mendes, R., Kennedy, J., Neves, J.: The Fully Informed Particle Swarm: Simpler, Maybe Better. IEEE Transactions on Evolutionary Computation 8, 204–210 (2004)

An ANFIS-Based Fault Classification Approach in Double-Circuit Transmission Line Using Current Samples

Mohammad Amin Jarrahi, Haidar Samet[✉], Hossein Raayatpisheh,
Ahmad Jafari, and Mohsen Rakhshan

School of Electrical and Computer Engineering, Shiraz University, Shiraz, Iran
samet@shirazu.ac.ir

Abstract. Transmission line protective relaying is an essential feature of a reliable power system operation. Fast detecting, isolating, locating and repairing of the different faults are critical in maintaining a reliable power system operation. On the other hand, classification of the different fault types plays very significant role in digital distance protection of the transmission line. Accurate and fast fault classification can prevent from more damages in the power system. In this paper, an approach is presented to classify the fault in a double-circuit transmission line based on the adaptive Neuro- Fuzzy Inference System (ANFIS) using three phase current samples of only one terminal. This method is independent of effects of variation of fault inception angle, fault location, fault resistance and load angle. MATLAB/Simulink is used to produce fault signals. The proposed method is tested by simulating different scenarios on a given transmission line model. The simulation results denote that the proposed approach for fault identification is able to classify all the faults on the parallel transmission line within half cycle after the inception of fault.

Keywords: ANFIS · Sugeno fuzzy system · Fault classification · Double-circuit transmission lines

1   Introduction

The type of fault and the faulty phase/phases identification in electrical power systems is known as fault classification. Accurate and fast fault classification is one of main issues for protection engineers and because of its importance various methods have been proposed. Effectiveness of these methods in different working conditions is also making them an essential subject for detailed study. When faults happen in the power system, they usually causes significant variations in the system quantities such as current, voltage, power, power factor and impedance. The most common and also the one used in this paper is the current and so over-current protection is generally used.

In a three phase system, different types of faults are classified as: line-to-line fault, single line-to-ground fault, double line fault, double line to-ground fault, three phase fault and three phase to ground fault. When the type of fault is recognized, the probable corrective action can rapidly be arranged to resolve the problem.

© Springer International Publishing Switzerland 2015
I. Rojas et al. (Eds.): IWANN 2015, Part II, LNCS 9095, pp. 225–236, 2015.
DOI: 10.1007/978-3-319-19222-2_19

Double circuit transmission lines have been widely used in modern power systems to improve the power transfer, reliability and security for the transmission of electrical energy. The different probable structures of parallel lines combined with the influence of mutual coupling make their protection a challenging problem [1].

Many studies have been concentrated on the fault classification problem in transmission lines, especially on double-circuit transmission lines. Recently, some techniques have been proposed for this problem, such as; wavelet transform based techniques [1-3], neural network based techniques [4-5] and fuzzy logic techniques [6-7]. These techniques are also combined because of the developments and conditions of power systems and they have their own advantages and disadvantages [8-12].

The wavelet based techniques have some drawbacks due to their complexity and high computational tasks and the techniques based on neural network include some difficulties because of accurate training data. Fuzzy logic techniques introduce an easier way for fault classification but it cannot detect the faulty phase/phases in certain circumstances.

Considering the strengths and weaknesses of neural network and fuzzy logic techniques, in this paper, a fault classification approach based on the Adaptive Neuro-Fuzzy Inference System (ANFIS) is proposed.

The inputs to ANFIS are post-fault current signals obtainable at the relay location in one terminal which the sampling frequency is 1 KHz. The outputs of ANFIS are set of numbers that classify type of faults. The ANFIS were trained and tested using numerous sets of field data. The field data are took from widespread simulation studies that have been performed using MATLAB/Simulink. Various fault scenarios (fault types, fault location, fault inception angle, load angle and fault resistance) are considered in this paper.

Simulation results denote that all ten fault types (AG, BG, CG, ABG, ACG, BCG, ABC/ABCG, AB, AC and BC) can be correctly classified after half cycle from the inception of fault. It is shown that the proposed algorithm implements a high speed faulty phase selection scheme which operates correctly in variety of conditions.

In comparison with other approaches, the proposed method in this paper shows a better and reliable way for classifying the faults on double circuit transmission lines. Delay times of other techniques are relatively more than the proposed approach, for example in [1] delay time of detection and classification of faults is one cycle. Accuracy of the proposed method is also an advantage for protective relaying.

2   Fault Classification Methodology

The proposed method has been developed on the basis of simulation studies carried out on the power system model shown in Figure 1 using MATLAB/ SIMULINK. The transmission line data are shown in Table 1 [4].

Fig. 1. Single line diagram of the simulated power system

The system under study is consist of 220 kV double circuit transmission line 100 km in length, connected to sources at each end. Short circuit capacity at the two sides is 1.25 GVA and Xs /Rs is 10. The transmission line is simulated using distributed parameter line model.

Table 1. Double circuit transmission line's parameters

Positive sequence resistance R1, Ω/km	0.01809
Zero sequence resistance R0, Ω/km	0.2188
Zero sequence mutual resistance R0m, Ω/km	0.20052
Positive sequence inductance L1, H/km	0.00092974
Zero sequence inductance L0, H/km	0.0032829
Zero sequence mutual inductance L0m,H/km	0.0020802
Positive sequence capacitance C1, F/km	1.2571e-008
Zero sequence capacitance C0, F/km	7.8555e-009
Zero sequence mutual capacitance C0m,F/km	-2.0444e-009

The three phase current data after fault inception obtained from MATLAB/ SIMULINK is used for classification of fault which the sampling frequency is 1 KHz and the data window of sampling interval is half cycle after the fault.

The proposed method of fault classification is as follows:

The characteristic features of different types of faults are defined in terms of Δs, which are calculated as described below. First, the amount of r_{11}, r_{12} and r_{13} for the circuit1 are calculated as follows using post-fault current samples:

$$r_{11} = \frac{\max\{abs(I_{a1})\}}{\max\{abs(I_{b1})\}} \quad , \quad r_{12} = \frac{\max\{abs(I_{b1})\}}{\max\{abs(I_{c1})\}} \quad , \quad r_{13} = \frac{\max\{abs(I_{c1})\}}{\max\{abs(I_{a1})\}} \tag{1}$$

As the above definitions, the amount of r_{21}, r_{22} and r_{23} for circuit 2 are defined. I_{a1}, I_{b1} and I_{c1} are post-fault currents of bus 1 for circuit 1 and I_{a2}, I_{b2} and I_{c2} are post-fault currents of bus1 for circuit 2.

After that, the normalized amounts of r_{11}, r_{12} and r_{13} are defined as follows:

$$r_{12n} = \frac{r_{12}}{\max\{r_{11},r_{12},r_{13}\}} \quad , \quad r_{11n} = \frac{r_{11}}{\max\{r_{11},r_{12},r_{13}\}} \quad , \quad r_{13n} = \frac{r_{13}}{\max\{r_{11},r_{12},r_{13}\}} \tag{2}$$

The amounts of r_{21n}, r_{22n} and r_{23n} are also defined as above. At the end the differences of the normalized values are calculated:

$$\Delta_{11} = r_{11n} - r_{12n} \quad , \quad \Delta_{12} = r_{12n} - r_{13n} \quad , \quad \Delta_{13} = r_{13n} - r_{11n} \tag{3}$$

As other features for circuit 2, values of Δ_{21}, Δ_{22} and Δ_{23} are calculated like above equations.

The characteristic features of different types of fault are calculated in terms Δs for circuit 1 and circuit 2 and ANFIS fuzzy rules for fault classification are established on the basis of amounts of Δs.

The values Δ_{11}, Δ_{12} and Δ_{13} and also Δ_{21}, Δ_{22} and Δ_{23} for line-to-line fault, single line-to-ground fault, double line fault, double line to-ground fault and three phase fault at variable operating situation and different types of faults on double-circuit transmission line are calculated. For example, in Tables2 to 5, the values of Δ_{11}, Δ_{12} and Δ_{13} for a-g, a-b-g, a-b and a-b-c for circuit 1 are listed.

The required parameters for the fuzzy logic of ANFIS for the grounded faults are defined as follows:

- For a-g fault: $\Delta_{11} = high_g, \Delta_{12} = medium_g, \Delta_{13} = low_g$
- For b-g fault: $\Delta_{11} = low_g, \Delta_{12} = high_g, \Delta_{13} = medium_g$
- For c-g fault: $\Delta_{11} = medium_g, \Delta_{12} = low_g, \Delta_{13} = high_g$
- For a-b-g fault: $\Delta_{11} = low_g, \Delta_{12} = high_g, \Delta_{13} = low_g$
- For b-c-g fault: $\Delta_{11} = low_g, \Delta_{12} = low_g, \Delta_{13} = high_g$
- For c-a-g fault: $\Delta_{11} = high_g, \Delta_{12} = low_g, \Delta_{13} = low_g$

d= fault location in p.u. of line length from bus 1, FIA = fault inception angle, Rf=fault point resistance and δ= load angle.

In the above expressions $high_g$ is a value between 0.2 and 1.0, $medium_g$ is a value between 0.02 and 0.3 and low_g is between -1.0 and -0.005. The suffix "g" is used to characterize a grounded fault. The expressions for circuit 2's grounded faults are similar to the circuit 1.

The required parameters for the fuzzy logic of ANFIS for phase faults are as defined follows:

- For a-b fault: $\Delta_{11} = low_{ph}, \Delta_{12} = high_{ph}, \Delta_{13} = low_{ph}$
- For b-c fault: $\Delta_{11} = low_{ph}, \Delta_{12} = low_{ph}, \Delta_{13} = high_{ph}$
- For c-a fault: $\Delta_{11} = high_{ph}, \Delta_{12} = low_{ph}, \Delta_{13} = low_{ph}$
- For a-b-c fault: $\Delta_{11} = medium_{ph}, \Delta_{12} = medium_{ph}, \Delta_{13} = low_{ph}$
- Or: $\Delta_{11} = low_{ph}, \Delta_{12} = medium_{ph}, \Delta_{13} = medium_{ph}$
- Or: $\Delta_{11} = low_{ph}, \Delta_{12} = medium_{ph}, \Delta_{13} = low_{ph}$
- Or: $\Delta_{11} = medium_{ph}, \Delta_{12} = low_{ph}, \Delta_{13} = medium_{ph}$
- Or: $\Delta_{11} = low_{ph}, \Delta_{12} = low_{ph}, \Delta_{13} = medium_{ph}$
- Or: $\Delta_{11} = medium_{ph}, \Delta_{12} = low_{ph}, \Delta_{13} = low_{ph}$

In the above expressions $high_{ph}$ is a value between 0.5 and 1.0, $medium_{ph}$ is a value between 0.01 and 0.6 and low_{ph} is a value between -1.0 and -0.005. The suffix "ph" is used to symbolize a phase fault.

Table 2. Values of Δ_{11}, Δ_{12} and Δ_{13} for a-g fault in circuit 1 with different variable operating situation

Fault condition: d, Rf, FIA, δ	Δ_{11}	Δ_{12}	Δ_{13}
$0.15, 0\,\Omega, 0^0, 10^0$	0.944	0.053	-0.998
$0.15, 0\,\Omega, 0^0, 30^0$	0.899	0.084	-0.983
$0.15, 200\,\Omega, 0^0, 10^0$	0.684	0.208	-0.892
$0.15, 200\,\Omega, 0^0, 30^0$	0.441	0.219	-0.660
$0.15, 0\,\Omega, 90^0, 10^0$	0.909	0.082	-0.992
$0.15, 0\,\Omega, 90^0, 30^0$	0.871	0.094	-0.966
$0.15, 200\,\Omega, 90^0, 10^0$	0.682	0.206	-0.887
$0.15, 200\,\Omega, 90^0, 30^0$	0.438	0.229	-0.667
$0.85, 0\,\Omega, 0^0, 10^0$	0.924	0.054	-0.978
$0.85, 0\,\Omega, 0^0, 30^0$	0.719	0.123	-0.914
$0.85, 200\,\Omega, 0^0, 10^0$	0.599	0.190	-0.790
$0.85, 200\,\Omega, 0^0, 30^0$	0.327	0.242	-0.570
$0.85, 0\,\Omega, 90^0, 10^0$	0.814	0.127	-0.942
$0.85, 0\,\Omega, 90^0, 30^0$	0.661	0.174	-0.835
$0.85, 200\,\Omega, 90^0, 10^0$	0.615	0.181	-0.797
$0.85, 200\,\Omega, 90^0, 30^0$	0.315	0.214	-0.555

Table 3. Values of Δ_{11}, Δ_{12} and Δ_{13} for a-b-g fault in circuit 1 with different variable operating situation

Fault condition: d, Rf, FIA, δ	Δ_{11}	Δ_{12}	Δ_{13}
$0.15, 0\,\Omega, 0^0, 10^0$	-0.953	0.999	-0.046
$0.15, 0\,\Omega, 0^0, 30^0$	-0.875	0.993	-0.117
$0.15, 200\,\Omega, 0^0, 10^0$	-0.650	0.966	-0.316
$0.15, 200\,\Omega, 0^0, 30^0$	-0.343	0.841	-0.498
$0.15, 0\,\Omega, 90^0, 10^0$	-0.964	0.998	-0.035
$0.15, 0\,\Omega, 90^0, 30^0$	-0.899	0.991	-0.091
$0.15, 200\,\Omega, 90^0, 10^0$	-0.661	0.968	-0.307
$0.15, 200\,\Omega, 90^0, 30^0$	-0.339	0.838	-0.499
$0.85, 0\,\Omega, 0^0, 10^0$	-0.831	0.988	-0.157
$0.85, 0\,\Omega, 0^0, 30^0$	-0.520	0.924	-0.404
$0.85, 200\,\Omega, 0^0, 10^0$	-0.229	0.808	-0.579
$0.85, 200\,\Omega, 0^0, 30^0$	-0.131	0.713	-0.583
$0.85, 0\,\Omega, 90^0, 10^0$	-0.892	0.988	-0.096
$0.85, 0\,\Omega, 90^0, 30^0$	-0.600	0.875	-0.275
$0.85, 200\,\Omega, 90^0, 10^0$	-0.221	0.816	-0.595
$0.85, 200\,\Omega, 90^0, 30^0$	-0.580	0.720	-0.140

Table 4. Values of Δ_{11}, Δ_{12} and Δ_{13} for a-b fault in circuit 1 with different variable operating situation

Fault condition: d, Rf, FIA, δ	Δ_{11}	Δ_{12}	Δ_{13}
0.15, 0 Ω, 0^0 , 10^0	-0.971	0.999	-0.029
0.15, 0 Ω, 0^0 , 30^0	-0.913	0.994	-0.081
0.15, 200 Ω, 0^0 , 10^0	-0.737	0.949	-0.211
0.15, 200 Ω, 0^0 , 30^0	-0.577	0.723	-0.286
0.15, 0 Ω, 90^0 , 10^0	-0.964	0.999	-0.035
0.15, 0 Ω, 90^0 , 30^0	-0.884	0.989	-0.106
0.15, 200 Ω, 90^0 , 10^0	-0.738	0.949	-0.211
0.15, 200 Ω, 90^0 , 30^0	-0.569	0.733	-0.284
0.85, 0 Ω, 0^0, 10^0	-0.888	0.990	-0.102
0.85, 0 Ω, 0^0 , 30^0	-0.612	0.913	-0.319
0.85, 200 Ω, 0^0 , 10^0	-0.551	0.883	-0.331
0.85, 200 Ω, 0^0 , 30^0	-0.313	0.847	-0.534
0.85, 0 Ω, 90^0 , 10^0	-0.856	0.985	-0.129
0.85, 0 Ω, 90^0 , 30^0	-0.313	0.847	-0.534
0.85, 200 Ω, 90^0 , 10^0	-0.537	0.887	-0.350
0.85, 200 Ω, 90^0 , 30^0	-0.312	0.720	-0.408

Table 5. Values of Δ_{11}, Δ_{12} and Δ_{13} for a-b-c fault in circuit 1 with different variable operating situation

Fault condition: d, Rf, FIA, δ	Δ_{11}	Δ_{12}	Δ_{13}
0.15, 0 Ω, 0^0 , 10^0	0.501	0.029	-0.530
0.15, 0 Ω, 0^0 , 30^0	0.435	0.087	-0.523
0.15, 200 Ω, 0^0 , 10^0	-0.662	0.571	0.091
0.15, 200 Ω, 0^0 , 30^0	-0.518	0.428	0.090
0.15, 0 Ω, 90^0 , 10^0	-0.085	-0.322	0.408
0.15, 0 Ω, 90^0 , 30^0	-0.076	-0.350	0.426
0.15, 200 Ω, 90^0 , 10^0	-0.609	0.516	0.093
0.15, 200 Ω, 90^0 , 30^0	-0.523	0.421	0.102
0.85, 0 Ω, 0^0, 10^0	0.581	0.019	-0.600
0.85, 0 Ω, 0^0 , 30^0	0.271	0.259	-0.530
0.85, 200 Ω, 0^0 , 10^0	-0.516	0.476	-0.040
0.85, 200 Ω, 0^0 , 30^0	-0.350	0.335	-0.016
0.85, 0 Ω, 90^0 , 10^0	-0.059	-0.344	0.403
0.85, 0 Ω, 90^0 , 30^0	-0.026	-0.433	0.459
0.85, 200 Ω, 90^0 , 10^0	-0.509	0.472	-0.038
0.85, 200 Ω, 90^0 , 30^0	-0.328	0.331	-0.002

From the mentioned discussions, it is clear that it should be a difference between phase and grounded faults in the formation of fuzzy logic uses in ANFIS algorithm for fault classification and it should be detected by a technique. The technique is discussed below.

To define the participation of ground in the fault, the value of $\sigma_1 = \max\{I_{a1} + I_{b1} + I_{c1}\}$ for circuit 1 is considered. Similarly this value for circuit 2 is defined as $\sigma_2 = \max\{I_{a2} + I_{b2} + I_{c2}\}$.

It is observed that the values of σ are greater than 100 for grounded faults and less than 1 for ungrounded faults. Table6 confirms that the involvement of ground in a fault can be easily detected on the basis of the value of σ.

Table 6. Values of σ_1 in amps for different types of fault corresponding to $\delta = 15^o$

Fault condition: d, Rf, FIA,	Type of fault			
	a-g	a-b	a-b-g	a-b-c
$0.1, 0\,\Omega, 0^0$	2.37×10^4	0.057	1.35×10^4	0.100
$0.1, 200\,\Omega, 0^0$	1.42×10^3	0.010	1.45×10^3	0.010
$0.1, 0\,\Omega, 90^0$	1.61×10^4	0.080	1.36×10^4	0.080
$0.1, 200\,\Omega, 90^0$	1.53×10^3	0.006	1.52×10^3	0.010
$0.5, 0\,\Omega, 0^0$	5.50×10^3	0.040	2.96×10^3	0.030
$0.5, 200\,\Omega, 0^0$	7.91×10^2	0.008	8.24×10^2	0.010
$0.5, 0\,\Omega, 90^0$	3.60×10^3	0.010	3.24×10^3	0.020
$0.5, 200\,\Omega, 0^0$	8.29×10^2	0.006	8.05×10^2	0.008
$0.9, 0\,\Omega, 0^0$	2.42×10^3	0.005	1.53×10^3	0.008
$0.9, 200\,\Omega, 0^0$	2.06×10^2	0.004	2.09×10^2	0.006
$0.9, 0\,\Omega, 90^0$	1.91×10^3	0.007	1.74×10^3	0.010
$0.9, 200\,\Omega, 90^0$	2.94×10^2	0.005	2.27×10^2	0.006

3   ANFIS Basic Concepts

In this section ANFIS is described. As a simple data learning procedure, ANFIS uses a fuzzy inference system model to convert a given input into a target output. This transformation includes membership functions, fuzzy logic operators and if-then rules. ANFIS is a Sugeno model from a development of fuzzy inference system (FIS). Sugeno-type ANFIS accomplish first-order polynomial to the output system so as to substitute zero-order in Sugeno FIS model. There are five principal processing steps in ANFIS procedure including input fuzzification, application of fuzzy operators, application method, output aggregation and defuzzification. ANFIS included the advantage of both neural network and fuzzy system, which not only have good learning proficiency, but can be construed simply also [13].

Basic ANFIS architecture based on Sugeno fuzzy model is shown in Figure 2. For simplicity, a fuzzy system with two inputs and one output is considered, also five steps that are verified in the ANFIS structure as shown in Figure 2.

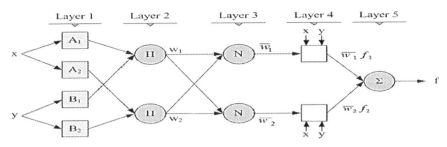

Fig. 2. Structure of ANFIS network

Each ANFIS layer has specific functions for calculating input and output parameter that described below:

Layer1: Each node in this layer is adaptive and the output is the rate of membership function of its input (input fuzzification).

Layer2: This layer involves fuzzy operators that uses the product operator (AND) to fuzzify the inputs.

Layer3: Each node in this layer is adaptive. The output of each node in this layer is the result of fuzzy system for that special rule.

Layer4: The function of this layer is an sum operation.

Layer5: This layer calculates the last result.

To design an ANFIS, the designer should perform the following steps:

1. Design of Sugeno FIS under the terms of problem.
2. Optimize FIS with the real data.
3. Form the training data and test according to input and output.
4. Train FIS with training data using ANFIS algorithm.
5. Test the training system with test data [14].

Now, to classify the faults in double-circuit transmission line, according to step 1 that mentioned above, we need to model the initial fuzzy system. Because the characteristic features of faults are set based on the value of Δ, the structure of the fuzzy system is defined based on it.

After detection of phase fault from grounded fault the structure of fuzzy system is designed as following:

Primary fuzzy rules for grounded faults of circuit 1:

- If Δ_{11} is $high_g$, Δ_{12} is $medium_g$ and Δ_{13} is low_g, it is an a-g fault;
- If Δ_{11} is low_g, Δ_{12} is $high_g$ and Δ_{13} is $medium_g$, it is a b-g fault;
- If Δ_{11} is $medium_g$, Δ_{12} is low_g and Δ_{13} is $high_g$, it is a c-g fault;
- If Δ_{11} is low_g, Δ_{12} is $high_g$ and Δ_{13} is low_g, it is an a-b-g fault;
- If Δ_{11} is low_g, Δ_{12} is low_g and Δ_{13} is $high_g$, it is a b-c-g fault;
- If Δ_{11} is $high_g$, Δ_{12} is low_g and Δ_{13} is low_g, it is a a-c-g fault;
    Primary fuzzy rules for phase faults of circuit 1:
- If Δ_{11} is low_{ph}, Δ_{12} is $high_{ph}$ and Δ_{13} is low_{ph}, it is an a-b fault;
- If Δ_{11} is low_{ph}, Δ_{12} is low_{ph} and Δ_{13} is $high_{ph}$, it is a b-c fault;

- If Δ_{11} is $high_{ph}$, Δ_{12} is low_{ph} and Δ_{13} is low_{ph}, it is a c-a fault;
- If Δ_{11} is $medium_{ph}$, Δ_{12} is $medium_{ph}$ and Δ_{13} is low_{ph}, it is an a-b-c fault;
- If Δ_{11} is low_{ph}, Δ_{12} is $medium_{ph}$ and Δ_{13} is $medium_{ph}$, it is an a-b-c fault;
- If Δ_{11} is $medium_{ph}$, Δ_{12} is low_{ph} and Δ_{13} is $medium_{ph}$, it is an a-b-c fault;
- If Δ_{11} is $medium_{ph}$, Δ_{12} is low_{ph} and Δ_{13} is low_{ph}, it is an a-b-c fault;
- If Δ_{11} is low_{ph}, Δ_{12} is $medium_{ph}$ and Δ_{13} is low_{ph}, it is an a-b-c fault;
- If Δ_{11} is low_{ph}, Δ_{12} is low_{ph} and Δ_{13} is $medium_{ph}$, it is an a-b-c fault;

Fuzzy rules for circuit 2 are defined as above.

The triangular membership function, shown in Figure 3, has been used to symbol-ize the different fuzzy variables in the previous parts of the fuzzy rules. The triangular membership function can be defined with reference to the points A, B and C, referred to as triplets [6]. As shown in Figure 3, the points A and C have membership value of 0.00 while the point B has a membership value of 1.00.

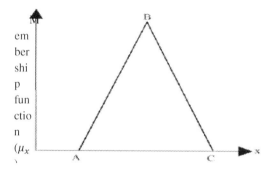

Fig. 3. The triangular fuzzy membership function

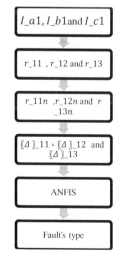

Fig. 4. Proposed method block diagram for circuit 1

The summary of proposed method is as following: first the post-fault line's current sampled half cycle after fault inception with 1kHz frequency, then the amount of r is calculated according to equation 1, after that r_n is calculated according to equation 2 and at last by calculating the amount of Δ according to equation 3, these values are applied as input to the ANFIS and the output is the type of fault. The proposed method block diagram for circuit 1 is shown in figure 4.

Next steps of ANFIS algorithm is studied in next section.

4   Simulation and Discussion of the Results

To show the capability of the proposed method for fault classification, a 220 kV, 3 phase, 100 km double-circuit transmission line is simulated. Different conditions of fault such as fault resistance (Rf), fault inception angle (FIA), fault distance (d) and load angle (δ) are considered. As the transmission line model with sources on both sides is a commonly recognized model for progress of line relaying algorithms, it has been considered for the development of the proposed method.

Intended for using ANFIS for classifying the faults of double-circuit transmission line, as mentioned in the previous section, it is necessary to design two similar Sugeno fuzzy systems that each system has three inputs with triangular membership function. The primary parameters of membership functions for circuit 1 are shown in table 7. In this table, A_{lg}, B_{lg} and C_{lg} are the membership function's parameters of low_g, A_{mg}, B_{mg} and C_{mg} are the membership function's parameters of $medium_g$ and at last A_{hg}, B_{hg} and C_{hg} are the membership function's parameters of $high_g$.

Table 7. Primary parameers of membership functions

values	A_{lg}	B_{lg}	C_{lg}	A_{mg}	B_{mg}	C_{mg}	A_{hg}	B_{hg}	C_{hg}
$\Delta_{11}, \Delta_{12}, \Delta_{13}$	-1	-1	0.1	0.02	0.15	0.3	0.2	1	1
	A_{lph}	B_{lph}	C_{lph}	A_{mph}	B_{mph}	C_{mph}	A_{hph}	B_{hph}	C_{hph}
$\Delta_{11}, \Delta_{12}, \Delta_{13}$	-1	-0.5	0	0.01	0.3	0.6	0.5	0.75	1

For circuit 2, the parameters defined totally similar like table7. The parameters of phase faults are also defined in this table.

For the simulation, 1800 data are recorded which 70% of them considered for training of fuzzy system using neural network algorithms. After training, 15% of the remaining data used for evaluation and another 15% applied to system for testing.

The membership function's value of the table7 has changed after applying the ANFIS. These parameters are shown in table8 for grounded and phase faults.

Table 8. Membership function's value of the fuzzy variable of the faults after applying ANFIS

values	A_{lg}	B_{lg}	C_{lg}	A_{mg}	B_{mg}	C_{mg}	A_{hg}	B_{hg}	C_{hg}
$\Delta_{11}, \Delta_{12}, \Delta_{13}$	-1	-1	0.13	-0.01	0.22	0.35	0.16	1	1
	A_{lph}	B_{lph}	C_{lph}	A_{mph}	B_{mph}	C_{mph}	A_{hph}	B_{hph}	C_{hph}
$\Delta_{11}, \Delta_{12}, \Delta_{13}$	-1	-0.48	0	0.01	0.33	0.64	0.47	0.7	1

ANFIS system is designed in such a way that if the output is in the range given in Table9, the type of fault is attained. The output's values of the ANFIS for different conditions are shown in table10. Because of brevity, only one condition for each type of fault is shown in this table. These values indicate that the proposed method is capable to detect the fault type, accurately.

For example, 5.120 in the output of ground fault means a-g fault for circuit 1. These amounts for circuit 2 are similar to table10.

The validity of the proposed approach has also been tested on simulated results in case of a 400 kV, 3 phase, 300 km double-circuit transmission line. Simulation studies extended to the second system demonstrates same results as concluded in the first case.

Also, due to the adjustment of this method, changes in the ratio of Xs/Rs of the source's impedance from 10 to 40 do not cause disturbance in finding type of fault and the proposed method indicates the output accurately.

Table 9. Output ranges for classifying grounded and phase faults

Grounded faults	a-g	b-g	c-g	a-b-g	b-c-g	c-a-g
Output range	[5-7]	[9-11]	[14-16]	[19-21]	[23-25]	[28-30]
Phase faults	a-b		b-c		c-a	a-b-c
Output range	[34-36]		[39-41]		[44-46]	[49-51]

Table 10. Simulation results in case of grounded and phase faults for circuit 1

Fault type	Fault conditions: d, Rf, FIA, δ	ANFIS inputs: $\Delta_{11}, \Delta_{12}, \Delta_{13}$	ANFIS output
a-g	$0.1, 5\,\Omega, 30^0, 15^0$	0.889,0.106,-0.995	5.120
b-g	$0.2, 200\,\Omega, 20^0, 30^0$	-0.662,0.405,0.257	10.007
c-g	$0.5, 100\,\Omega, 0^0, 30^0$	0.262,-0.689,0.427	15.000
a-b-g	$0.1, 20\,\Omega, 45^0, 20^0$	-0.868,0.996,-0.127	20.000
b-c-g	$0.2, 75\,\Omega, 30^0, 15^0$	-0.238,-.0746,0.984	24.960
c-a-g	$0.1, 0.1\,\Omega, 90^0, 30^0$	0.997,-0.056,-0.941	29.920
a-b	$0.1, 50\,\Omega, 45^0, 30^0$	-0.816,0.972,-0.156	35.029
b-c	$0.2, 0.1\,\Omega, 30^0, 10^0$	-0.049,0.972,-0.156	40.020
c-a	$0.15, 0.1\,\Omega, 90^0, 10^0$	0.999,-0.025,-0.974	44.980
a-b-c	$0.1, 30\,\Omega, 10^0, 10^0$	-0.407,0.490,-0.083	49.885

5   Conclusion

In this paper fault classification approach in double circuit transmission line is proposed based on ANFIS. The suggested scheme detects all types of faults by using three-phase current samples of only one terminal. Also, a separate technique is used to determine the participation of ground in fault. Simulation studies carried out considering wide variations in fault location, fault inception angle, fault resistance and load angle for different types of fault have proved the validity of the proposed approach, and also The simulation results denote that the proposed approach for fault identification is able to classify all the faults on the parallel transmission line within half cycle after the inception of fault.

References

1. Osman, A.H., Malik, O.P.: Protection of Parallel Transmission Lines Using Wavelet Transform. IEEE Trans. On Power Delivery, **19**(1) (2004)
2. Jana, S., De, A.: Transmission Line Fault Detection and Classification using Wavelet Analysis. Annual IEEE India Conference (INDICON) (2013)
3. A Valsan S.P., Swarup K.S: Wavelet transform based digital protection for transmission line. Electrical Power and Energy Systems (2009)
4. Jain, A., Thoke, A.S., Koley, E., Patel, R.N.: Fault classification and fault distance location of double circui transmission lines for phase to phase faults using only one terminal data. In: Third International Conference on Power Systems, Kharagpur, India (2009)
5. Martins, L.S., Martins, J.F., Pires, V.F., Alegria, C.M.: A neural network fault classification for parallel double-circuit distribution lines. International Journal of Electrical Power and Energy Systems **27**(3), 225–231 (2005)
6. Das, B., Reddy, J.V.: Fuzzy-logic-based fault classification scheme for digital distance protection. IEEE Trans. Power Deliv. **20**(2), 609–616 (2005)
7. Ferrero, A., Sangiovanni, S., Zapitelli, E.: A fuzzy set approach to fault type identification in digital relayin. IEEE Trans. Power Deliv. **10**(1), 169–175 (1995)
8. Omar, A.S.: Youssef: Combined Fuzzy-Logic Wavelet-Based Fault Classification Technique for Power System Relaying. IEEE Trans. On Power Delivery, **19**(2) (2004)
9. Vasilic, S., Kezunovic, M.: Fuzzy ART Neural Network Algorithm for Classifying the Power System Faults. IEEE Trans. On Power Delivery, **20**(2) (2005)
10. Bhowmika, P.S., Purkaitb, P., Bhattacharyac, K.: A novel wavelet transform aided neural network based transmission line fault analysis method. Int. J. Electr. Power Energy Syst. **31**(5), 213–222 (2009)
11. Reddy, M.J., Mohanta, D.K.: Adaptive-neuro-fuzzy inference system approach for transmission line fault classification and location incorporating effects of power swings. IET Gener. Transm. Dis. **2**(2), 235–244 (2008)
12. Zhang, J., He, Z.Y., Lin, S., Zhang, Y.B., Qian, Q.Q.: An ANFIS-based fault classification approach in power distribution system. Electrical Power and Energy Systems. **49**, 243–252 (2013)
13. Ghani, R.A., Othman, M.F., Hashim, A.M.: ANFIS Approach for Locating Precise Fault Points in Distribution System. International Journal of Emerging Technology and Advanced Engineering **2**(6), 2250–2459 (2012)
14. Rakhshan, M., Shabaninia, F., Shasadeghi, M.: ANFIS Approach for Tracking Control of MEMS Triaxial Gyroscope. MSEEE Journal, **1** (2014)

Evolutionary Hybrid Configuration Applied to a Polymerization Process Modelling

Silvia Curteanu[1], Elena-Niculina Dragoi[1(✉)], and Vlad Dafinescu[2]

[1] Faculty of Chemical Engineering and Environmental Protection,
"Gheorghe Asachi" Technical University of Iasi, 73,
Prof. Dr. Doc. D. Mangeron Blvd. 700050, Iasi, Romania
elenan.dragoi@gmail.com
[2] Clinical Emergency Hospital "Prof. Dr. N. Oblu", Str. Ateneului 700309, Iasi, Romania

Abstract. A modelling procedure based on hybrid configuration composed of artificial neural networks, differential evolution and clonal selection algorithms is developed and applied in this work. The neural network represents the model of the system, while the differential evolution and clonal selection algorithms perform a simultaneous topological and parametric optimization of the model. The results indicated that the combination of the two optimizers produces better results compared with each of them working separately. As case study, styrene polymerization, a complex process which is difficult to model when taking into consideration all the internal interactions, was chosen. Neural networks, designed in an optimal form, proved to be adequate tools for modelling this system.

Keywords: Neural network · Differential evolution · Clonal selection · Styrene polymerization

1 Introduction

In the case of real world problems, the correct modelling approach, leading to good solutions and acceptable errors, is an important aspect that can have an influential impact on the chemical system. Although specialists tried to identify the chemical and physical laws governing these systems, due to the complexity involved, the high number of parameters required and their inter-dependency, the mathematical models developed are either simplified variants which do not take into account different interactions (so significant errors might occur) or are based on constants and variables empirically determined. In addition, there are situations in which the mathematical models are impossible to determine or are so complex that they cannot be efficiently used in monitoring and control procedures where speed is of essence.

In this context, the application of artificial neural networks (ANNs) as alternative modelling tools can help researchers to solve such problems, which are otherwise difficult to tackle. The property of universal approximation of ANNs is the main characteristic that makes them suitable for modelling complex system with a high

© Springer International Publishing Switzerland 2015
I. Rojas et al. (Eds.): IWANN 2015, Part II, LNCS 9095, pp. 237–249, 2015.
DOI: 10.1007/978-3-319-19222-2_20

degree of nonlinearity [1]. In addition, the ANNs can estimate the behaviour of a system even when dealing with incomplete information and they are easy to use and develop [2;3]. When modelling real-world systems, the main advantages of the ANNs are related to the facts that: i) they can be constructed using only experimental data; ii) there is no requirement for extensive knowledge of the process; iii) once trained, an ANN with a good generalization capability is able to predict outputs for new input data; iv) ANNs can model simultaneously multiple input-multiple output relations [4].

These properties make ANNs a sought after solution, the multitude of paper encountered in literature suggesting that researchers found numerous problems that can be effectively solved based on this approach.

Although ANNs are widely spread, they have a series of problems related to: i) limited theory in assisting the topology design; ii) finding an acceptable solution to a problem is not guaranteed; iii) the rationalization of the solutions provided is difficult [5]. In addition, the wrong application of the training algorithms can lead to overfitting and lack of robustness, factors which negatively influence the network's performance. In order to reduce the occurrence of these problems and to reduce their effects, one of the approaches that can be successfully applied is represented by neuro-evolution or evolutionary artificial neural networks (eANNs). eANNs represent a special class of ANNs in which evolution is a form of adaptation that can be used for optimizing architecture and/or internal ANN parameters [6;7]. Evolution could be generally applied at three different levels: i) connection weights; ii) architecture and iii) learning rules [8]. The most encountered approaches in literature involve connection weight evolution, as this corresponds to the training step.

From the multitude of evolutionary algorithms developed over time, differential evolution (DE) distinguishes as a high performance, robust, simple and straightforward to implement algorithm [9]. In addition, it has a low computational complexity and only a few control parameters (F – mutation probability, Cr –crossover probability, Np – number of individuals in the population, G - number of generations) [10]. It was applied with great success for solving different types of problems from different areas. The problems tackled are both benchmarks [11-17] and real-life [18-21]. Some examples for chemical engineering applications include: oxidation processes [22-24], energy, fuels and petrol derivatives [25-29], fermentation [30-32].

In this work, along with DE, another optimization algorithm was used for evolving ANNs. It is represented by Clonal Selection (CS), an algorithm inspired from the response of the vertebrate immune system when attacked by antigens. CS is a subclass of the Artificial Immune System (AIS) algorithms, its characteristics being represented by diversity, optimization and exploration [33]. The types of problems solved with CS include: function optimization, pattern recognition, design problems, scheduling and classification [34]. The domains of these problems are varied, in the area of chemical engineering only a few applications being encountered.

For the modelling of the considered process, the feed forward multilayer perceptron neural networks were selected. In order to obtain acceptable errors, the ANNs where then optimized based on an optimization procedure combining DE and CS. The role of these algorithms is to evolve the architecture (number of hidden layers and neurons in the hidden layers) and the internal parameters (weights, biases and activation functions), so that the model predictions are closest to the experimental data. The idea of hybridization of the two optimizers is related to the possibility of combining their advantages and, sometimes, to the fact that the benefit obtained could be greater than the simple summation of the individual advantages.

The paper is organized as follows. Section two presents the method employed for data collection and the pre-processing procedures applied for preparing the data for modelling. Section 3 details the inner workings of the modelling and optimization procedures based on the combination of DE, CS and ANN. The simulation results are presented in Section 4. The last section concludes the paper.

2   Database

The case study approached here for applying the optimization procedures based on DE, CS and ANN was a complex polymerization process – free radical polymerization of styrene performed through batch suspension technique. The phenomenological model for this process was proposed in [35], where a series of kinetics specific to free radical polymerization are considered: initiation step (chemical and thermal), propagation, termination by recombination and chain transfer to monomer. Although this is a complete model, it was observed that the diffusion phenomena associated with the radical polymerization (gel and glass effects) are not modelled with precision and, in this way, significant errors affected the final results. Consequently, the determination of an alternative model is a recommended solution.

With the existing phenomenological model (phM), a series of simulations were performed. The data from these simulations was gathered in a complete database (3552 samples) which was further used for neural network modelling purposes (75% for training and 25% for testing).

The main parameters varied in this study are represented by temperature and initial concentration of initiator. Since the dynamic of the process is dependent on the values of these parameters, different combinations were considered. The entire interval of variation is represented by 60-100°C for temperature and 10-50 mol/l for the initiator. Taking into consideration that a process can be best modelled when the data available are evenly distributed in the entire search space, a variation of 5°C for temperature and 5 mol/l from initiator were taken into account.

For each temperature-initiator combination, different measures can be taken, in this study a variable frequency is considered. Since the reaction time is relatively slow (more than 30 hours) related to the default frequency value (30 s), the number of data gathered can be considerable. This aspect is favourable for the modelling procedure, but having too much data can slow down the model determination due to the amount of computational resources required. Consequently, a data processing technique is introduced to reduce the number of available information, keeping only the relevant data. The main idea is to change the frequency of data gathering based on the speed of the reaction. For example, when a sudden increase of the values of a reaction parameter occurs, then a high frequency of the data sampling is chosen.

In order to have all the parameters within the same scale [-0.9,0.9], for each data-set, a normalization procedure is applied. This represents one of the most common tools for improving results of modelling applications when inputs are scattered on different scales [36;37]. The formula used for normalization is described by Eq. 1:

$$p_{norm} = min_t + (max_t - min_t) \frac{p - min}{max - min} \qquad (1)$$

where p_{norm} is the normalized value of p parameter, $min_t = -0.9$ and $max_t = 0.9$ are the limits of the target interval, and min, max describe the interval in which p takes values.

For the styrene polymerization process considered here, the model input variables were chosen as: initiator concentration, I_0, temperature, T and reaction time, t. The other two variables, monomer conversion, x, and numerical average molecular weight, Mn represent the outputs of the models. The modelling techniques aim to provide predictions about the main properties (molecular mass) and reaction charac-teristics (conversion) as a function of the working conditions.

3   Differential Evolution and Clonal Selection Based Optimization Procedure

The methodology proposed and applied in this work is based on a combination of ANNs with two powerful optimization algorithms represented by DE and CS. In the following subsections the characteristics of each algorithm, with advantages and dis-advantages, as well as their combination, will be detailed.

3.1   Differential Evolution

DE is an optimization algorithm belonging to the Evolutionary Algorithms (EA) class. As almost every EA, DE has a series of steps represented by: initialization, mutation, crossover and selection. These steps are repeated until a stop criteria is reached. While some of this steps are a very attractive subject when considering per-formance improvement, the initialization and crossover are somewhat less studied.

Initialization represents the first step and plays a crucial role, affecting the search and influencing the final solution [38]. The result of this step is represented by an initial population solution, its generation depending on the knowledge available about the search space. In the absence of the knowledge, the uniform random number generation is usually the only choice [39]. In order to improve the likelihood of gen-erating good initial solutions, researcher proposed different strategies including oppo-sition and quasi-opposition [40].

Mutation represents the second step of the algorithm and, in the DE case, is per-formed by adding a scaled differential term to a base vector. Its role is to introduce new genetic material into the population, the perturbation factor (F) representing the control parameter specific to this operation. Due to the differential term, this step is also called differentiation and, depending on how the individuals participate at the mutation phase, on the number of differential terms and other aspects, there are a lot of mutation strategies.

After the mutation population was generated, in the crossover step, the mutation and target vectors were mixed in order to create the trial individuals, the diversity of the population being increased. Distinctively from the genetic algorithms, who usually generate two individuals, in the case of DE only a single trial vector is created. Also, although in literature various crossover types specific to EAs are encountered, the majority of DE variants employ only binomial or exponential crossover.

In the last step of the algorithm, a one to one survivor selection criterion is applied to select the individuals for the next generation. The competition is performed between the individuals from the trial population and the corresponding ones from the initial population. The survivor is usually selected based on its suitability to the environment, this assessment being performed using a fitness function.

These steps are repeated until a stop criteria is reached. Usually, the algorithm stops when the current generation reaches a predefined value, but some researches combined different stop criteria in order to create a more feasible algorithm which stops when a balance between exploration and exploitation is obtained.

3.2   Clonal Selection

The CS algorithm employs the principles of clonal expansion and affinity maturation and evolves a set of antibodies [41]. The steps of this algorithm are similar to the ones specific to a EAs, each individual (antibody) being a candidate solution belonging to the search space. The algorithm starts with an initialization procedure, followed by selection, cloning, and hypermutation.

The initialization step of CS does not differ from other types of initialization used for the other population based algorithms. The population of antibodies is usually generated using a uniform probability distribution, the entire search space being covered uniformly [42]. In order to assess the performance of each antibody (also known as affinity), an affinity function is used.

Based on their affinity value, the fittest antibodies will be independently cloned. In the case of CS, there are multiple approaches for setting the number of clones specific to each antibody, the most used ones being represented by: i) static cloning operator (fixed number of clones for each individual) and ii) proportional cloning operator, where the number of clones is proportional with their affinity [43]. This step is also known as proliferation.

Each clone will undergo a hypermutation procedure, which explores the search space and introduces innovation into the pool of potential solutions [43]. Because in the CS algorithm a crossover operator does not exist, the probability with which hypermutation is applied is higher than in the GA case [44].

After the affinity of the hypermutated clones is computed, the best individuals are added to the population, while the ones with the lowest performance are removed.

3.3   DE-CS-ANN

Our proposed modelling procedure (noted as DE-CS-ANN) combines the features of DE and CS in order to identify the optimum ANN for the approached process. The parameters of the ANN selected for evolution are structural and parametric as the automatic approach is applied in both topology selection and training steps. The Simultaneous structural and parametric ANN optimization performed leads to a relatively large search space. This high dimensionality was one of the factors for choosing not one, but two optimization algorithms, which behave differently and which together can more efficiently explore the search space.

The variants of DE and CS used for optimizing the ANN that models the polymerization process were proposed in our previous works [45;46], the novelty of the actual approach consisting in the hybridization of the two algorithms and its application for a difficult to solve real-world chemical engineering problem. The DE based algorithm combined with ANN is called SADE-NN-2 [45] and, compared with the classical DE proposed by Storn and Price [47], has a series of improvements consisting in: i) introduction of a simple-self adaptive procedure which applies the same principle of evolution to the algorithm control parameters as to other parameters of the individuals; ii) introduction of a mutation principle which arranges the individuals selected for mutation based on their fitness value; and iii) use of backpropagation algorithm as a local search procedure. This variant was successfully applied to solve a monitoring and prediction problem specific to a freeze-drying process [45].

The CS-based algorithm, also combined with ANN, is called CS-ANN [46] and, compared with the CLONALG [48] version it is based on, has a series of improvements including: i) introduction of a hybrid hypermutation operator combining three hypermutation principles (Gaussian, non-uniform and pair wise interchange hypermutation); and ii) a new mechanism for removing low affinity antibodies based on mixing age and affinity criteria.

The implementation of the two optimization procedures differed in terms of types of individuals in the population (chromosomes for DE and antibodies for CS) and after the manner in which the fitness/affinity was computed. In order to create a link between the two, these aspects were solved by implementing a core individual with properties belonging to both, antibody and chromosome, a single population being evolved by the two optimization algorithms. The individuals were represented by vectors containing real values extracted from a direct encoding of ANNs which contained information like: number of hidden layers, number of neurons in each hidden layer, weights, biases, activation functions and parameters of activation functions (whenever the case).

In addition, the fitness and the affinity functions (which will be further referred as fitness function as both indicate the fitness of the individual to the environment represented by the search landscape) were modified to be based on the mean squared error (MSE) in the training phase.

In order to perform the optimization of ANN, DE and CS were applied alternatively, as it can be seen in Figure 1. Initially, the population is randomly generated using the limits imposed to the network architecture. After that, in order to identify the odd and the non-odd iteration, an index (i) is introduced, and, based on its value, one of the two optimization procedure is selected.

In this case, the only information shared between the two components (DE and respectively CS) is represented by the population, the characteristics of the optimizers and their inner workings remaining unchanged.

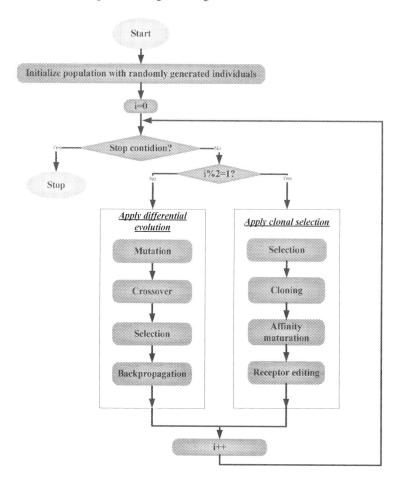

Fig. 1. Simplified schema of the optimization procedure

4   Results and Discussion

In order to assess the efficiency of the proposed hybrid method, a series of simulations was performed using DE and CS as separate optimization tool and then applying their hybrid configuration. Since the search space is quite large (as it includes number of hidden layer, number of neurons in the hidden layers, weights, biases, activation functions and parameters of the activation functions), a set of limits to the neural network topology were imposed based on practical an theoretical considerations. Therefore, the number of hidden layer was set to maximum 2, while the neurons in the first hidden layer was maximum 30 and the neurons in the second hidden layer was maximum 15. In this manner the length of the individuals is 1 (no of hidden layer) +1 (neurons in the first hidden layer) + 1 (neurons in the second hidden layer) + inputs * 30 (weights between the input and the first hidden layer) + 30 * 15 (weights between the first and the second hidden layers) + 15 * outputs (weights between the second hidden layer and the output layer) + 3 * (45 + outputs) (biases, activation functions and parameters of the activation functions). In total results 583 + 30 * inputs + 18 * outputs (709 for our application).

Because the CS variant implemented has a variable number of individuals in the population (as it adds and removes individuals based on internal rules), the entire procedure starts with a relatively low number of individuals (50). In what concerns the stop criterion, the algorithm stops when a low MSE in the training phase is obtained (10^{-6}) or the number of iterations reaches a predefined value, which was set to 1000 (based on the authors experience and practical considerations).

With these settings, the five best results obtained with the three algorithms are listed in Table 1. In this table, Fitness is invers proportional with the MSE in the training phase, MSE (for both training and testing) is computed using the normalized dataset and Topology represents the architecture of the ANN noted as: *inputs : number of neurons in the first hidden layer : number of neurons in the second hidden layer: outputs*. Although the algorithms were set to determine ANNs with one or two hidden layers, in all the cases, the models determined have only one hidden layer.

Table 1. Results obtained with the three methods employed for modelling the polymerization process

	Fitness	MSE training	MSE testing	Topology	Function Evaluation
CS	171.1167	0.005844	0.026465	3:05:2	1990990
	111.6214	0.008959	0.040368	3:17:2	183273
	86.1003	0.011614	0.051691	3:03:2	266287
	78.64243	0.012716	0.056687	3:03:2	1854103
	72.67783	0.013759	0.060911	3:03:2	1813471
CS average	104.0317	0.010578	0.047225	-	1221624.8
DE	293.3398	0.003409	0.013361	3:14:2	324012
	286.038	0.003496	0.013432	3:16:2	324012
	282.515	0.00354	0.014898	3:18:2	324012
	279.4237	0.003579	0.013839	3:09:2	324012
	275.4731	0.00363	0.014616	3:08:2	324012
DE average	283.3579	0.003531	0.014029	-	324012
DE+CS	301.7697	0.003314	0.012077	3:16:2	314014
	299.637	0.003337	0.012558	3:13:2	419873
	299.1274	0.003343	0.012958	3:17:2	366911
	294.4507	0.003396	0.013337	3:13:2	297262
	291.0381	0.003436	0.012724	3:11:2	437329
DE + CS average	297.2046	0.003365	0.012731	-	367077.8

As it can be observed from Table 1, the CS based algorithm had a tendency to over-simplify the network architecture, this being one of the main factors that lead to such poor performance compared with the DE based approach. In this case, although the number of function evaluations for CS is quite high compared with DE and DE+CS, CS was not able to efficiently model the process. The big difference in number evaluations from a simulation to the other, in case of CS algorithm, is explained by the inner structure of the algorithm, where individuals are added and removed based on the characteristics of the current population.

The best results were obtained with the combination of DE and CS, the alternate application of these two algorithms improving the performance for both the training and testing datasets. The best suited solution was chosen as the one having the highest fitness value - (15:16:2). Considering a significant set of values from the testing dataset, a comparison between expected (phM) and predictions provided for the three algorithms is presented in Fig. 2 (for x) and Fig 3 (for Mn).

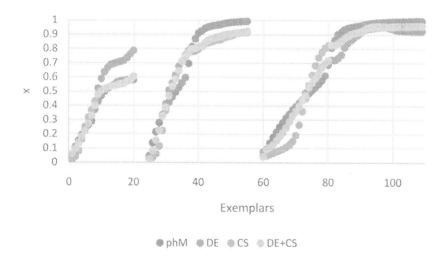

Fig. 2. Comparison between expected and predicted data, for the x output, in case of a set of testing points

In Figs. 2 - 3, the testing data belongs to three I_0 - T combinations: 368°C – 25mol/l (first 20 exemplars), 358°C – 50 mol/l (exemplars 20 through 60) and 348°C – 40 mol/l (exemplars 60 through 110). From the two figures, it is evident that DE-CS-ANN has superior performance, but for monomer conversion, x, the results are better than for Mn. This aspect is also indicated by the average relative errors: 8.12 % for x, and 10.06 % for Mn (for the whole testing set of data). An explanation of these results could be related to the complex dynamic of Mn.

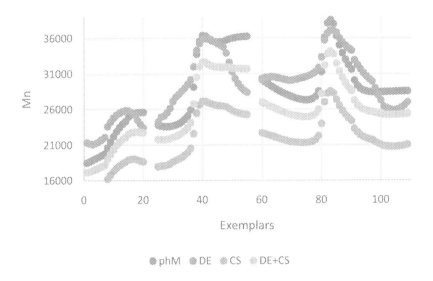

Fig. 3. Comparison between expected and predicted data, for the *Mn* output, in case of a set of testing points

When comparing the number of function evaluations for the best solutions obtained with DE and DE+CS, it is observed that the former has fewer function evaluations, although it has a slightly higher performance. This points out that DE+CS has the potential of being a good optimization procedure, with higher performance than its individual components, but a detail analysis must be performed in order to determine what are the elements that influence its behaviour.

5   Conclusions

In this work, a complex polymerization process was modelled using three different techniques: DE with ANN (called SADE-NN-2), CS with ANN (CS-NN) and a new combination of DE with CS and ANN (DE-CS-ANN). In all cases, the ANN acts as a model for the considered process, while DE and/or CS simultaneously optimize the topology and internal parameters of the model.

The idea of combining the two optimizers, DE and CS, with the goal of obtaining an optimal ANN model, seems to lead to an improvement of the algorithm performance. This hybrid configuration proved to be efficient and reliable when employed to real-world complex processes. It was developed in a general form, which allows it to be successfully applied to other systems.

Acknowledgment. This work was supported by the "Partnership in priority areas – PN-II" program, financed by ANCS, CNDI - UEFISCDI, project PN-II-PT-PCCA-2011-3.2-0732, No. 23/2012.

References

1. Subudhi, B., Jena, D.: A differential evolution based neural network approach to nonlinear system identification. Appl. Soft Comput. **11**(1), 861–871 (2011)
2. Kisi, O.: River suspended sediment concentration modeling using a neural differential evolution approach. Journal of Hydrology **389**(1–2), 227–235 (2010)
3. Noor, R.A.M., Ahmad, Z., Don, M.M., Uzir, M.H.: Modelling and control of different types of polymerization processes using neural networks technique: A review. Can. J. Chem. Eng. **88**(6), 1065–1084 (2010)
4. Lahiri, S.K., Ghanta, K.C.: Artificial neural network model with the parameter tuning assisted by a differential evolution technique: The study of the hold up of the slurry flow in a pipeline. Chemical Industry and Chemical Engineering Quarterly **15**(2), 103–117 (2009)
5. Yardimci, A.: Soft computing in medicine. Appl. Soft Comput. **9**(3), 1029–1043 (2009)
6. Xin, Y.: Evolving artificial neural networks. Proceedings of the IEEE **87**(9), 1423–1447 (1999)
7. Montana, D., VanWyk, E., Brinn, M., Montana, J., Milligan, S.: Evolution of internal dynamics for neural network nodes. Evolutionary Intelligence **1**(4), 233–251 (2009)
8. Islam, M., Yao, X.: Evolving artificial neural network ensembles. In: Fulcher, J., Jain, L. (eds.) Computational Intelligence: A Compendium, 115th edn, pp. 851–880. Springer, Heidelberg (2008)
9. Das, S., Suganthan, P.N.: Differential Evolution A Survey of the State-of-the-Art. IEEE Trans. Evol. Comput. **15**(1), 4–31 (2011)
10. Bedri Ozer, A.: CIDE: Chaotically Initialized Differential Evolution. Expert Syst. Appl. **37**(6), 4632–4641 (2010)
11. Islam, S.M., Das, S., Ghosh, S., Roy, S., Suganthan, P.N.: An adaptive differential evolution algorithm with novel mutation and crossover strategies for global numerical optimization. IEEE Transactions on Systems, Man, and Cybernetics, Part B: Cybernetics **42**(2), 482–500 (2012)
12. Xue, F., Sanderson, A.C., Bonissone, P.P., Graves, R.J.: Fuzzy logic controlled multiobjective differential evolution. In: IEEE, pp. 720–725 (2005)
13. Nobakhti, A., Wang, H.: A simple self-adaptive Differential Evolution algorithm with application on the ALSTOM gasifier. Appl. Soft Comput. **8**(1), 350–370 (2008)
14. Guo, J., Zhou, J., Zou, Q., Liu, Y., Song, L.: A novel multi-objective shuffled complex differential evolution algorithm with application to hydrological model parameter optimization. Water Resour. Manage. **27**(8), 2923–2946 (2013)
15. Wang, Y., Cai, Z., Zhang, Q.: Enhancing the search ability of differential evolution through orthogonal crossover. Inf. Sci. **185**(1), 153–177 (2012)
16. Li, H., Zhang, Q.: Multiobjective optimization problems with complicated Pareto sets, MOEA/D and NSGA-II. IEEE transations on Evolutionary Computation **13**(2), 284–302 (2009)
17. Zamuda, A., Brest, J.: Population reduction differential evolution with multiple mutation strategies in real world industry challenges. In: Rutkowski, L., Korytkowski, M., Scherer, R., Tadeusiewicz, R., Zadeh, L.A., Zurada, J.M. (eds.) EC 2012 and SIDE 2012. LNCS, vol. 7269, pp. 154–161. Springer, Heidelberg (2012)
18. Dong, M.G., Wang, N.: A novel hybrid differential evolution approach to scheduling of large-scale zero-wait batch processes with setup times. Computers & Chemical Engineering **45**, 72–83 (2012)

19. Lai, J.C.Y., Leung, F.H.F., Ling, S.H., Nguyen, H.T.: Hypoglycaemia detection using fuzzy inference system with multi-objective double wavelet mutation Differential Evolution. Appl. Soft Comput. 13(5), 2803–2811 (2013)
20. Maleki, R., Keikha, V., Rezaei, H.: Using Differential Evolution Algorithm and Rough Set Theory to Reduce the Features of Cataract Disease in a Medical Diagnosis System. Trans. Electrical Electronic Circuits Syst. 3(1) (2013)
21. Lei, B., Tan, E.L., Chen, S., Ni, D., Wang, T., Lei, H.: Reversible watermarking scheme for medical image based on differential evolution. Expert Syst. Appl. 41(7), 3178–3188 (2014)
22. Gujarathi, A.M., Babu, B.V.: Improved Multiobjective Differential Evolution (MODE) Approach for Purified Terephthalic Acid (PTA) Oxidation Process. Mater. Manuf. Processes 24(3), 303–319 (2009)
23. Hu, C., Yan, X.: An Immune Self-adaptive Differential Evolution Algorithm with Application to Estimate Kinetic Parameters for Homogeneous Mercury Oxidation. Chin. J. Chem. Eng. 17(2), 232–240 (2009)
24. Wu, Y., Lu, J., Sun, Y.: An Improved Differential Evolution for Optimization of Chemical Process. Chin. J. Chem. Eng. 16(2), 228–234 (2008)
25. Huang, S.R., Wu, C.C., Lin, C.Y., Chen, H.T.: Parameter optimization of the biohydrogen real time power generating system using differential evolution algorithm. Int. J. Hydrogen Energy 35(13), 6629–6633 (2010)
26. Khademi, M.H., Rahimpour, M.R., Jahanmiri, A.: Differential evolution (DE) strategy for optimization of hydrogen production, cyclohexane dehydrogenation and methanol synthesis in a hydrogen-permselective membrane thermally coupled reactor. Int. J. Hydrogen Energy 35(5), 1936–1950 (2010)
27. Iranshahi, D., Pourazadi, E., Paymooni, K., Rahimpour, M.R.: Utilizing DE optimization approach to boost hydrogen and octane number in a novel radial-flow assisted membrane naphtha reactor. Chem. Eng. Sci. 68(1), 236–249 (2012)
28. Vakili, R., Setoodeh, P., Pourazadi, E., Iranshahi, D., Rahimpour, M.R.: Utilizing differential evolution (DE) technique to optimize operating conditions of an integrated thermally coupled direct DME synthesis reactor. Chem. Eng. J. 168(1), 321–332 (2011)
29. Vakili, R., Eslamloueyan, R.: Optimal design of an industrial scale dual-type reactor for direct dimethyl ether (DME) production from syngas. Chemical Engineering and Processing: Process Intensification 62, 78–88 (2012)
30. Yuzgec, U.: Performance comparison of differential evolution techniques on optimization of feeding profile for an industrial scale baker's yeast fermentation process. ISA Transactions 49(1), 167–176 (2010)
31. Da Ros, S., Colusso, G., Weschenfelder, T.A., de Marsillac Terra, L., de Castilhos, F., Corazza, M.L., Schwaab, M.: A comparison among stochastic optimization algorithms for parameter estimation of biochemical kinetic models. Appl. Soft Comput. 13(5), 2205–2214 (2013)
32. Mendes, R., Rocha, I., Pinto, J., Ferreira, E., Rocha, M.: Differential evolution for the offline and online optimization of fed-batch fermentation processes. In: Chakraborty, U. (ed.) Advances in Differential Evolution, 143rd edn, pp. 299–317. Springer, Heidelberg (2008)
33. Abdul Hamid, M.B., Abdul Rahman, T.K.: Short Term Load Forecasting Using an Artificial Neural Network Trained by Artificial Immune System Learning Algorithm, pp. 408–413 (2010)
34. Haktanirlar Ulutas, B., Kulturel-Konak, S.: A review of clonal selection algorithm and its applications. Artif. Intell. Rev. 36(2), 117–138 (2011)

35. Curteanu, S.: Modeling and simulation of free radical polymerization of styrene under semibatch reactor conditions. Central European Journal of Chemistry 1(1), 69–90 (2003)
36. Priddy, K., Keller, P.: Artificial Neural Networks: An introduction. SPIE Press, Washington (2005)
37. Snyman, J.: Practical Mathematical Optimization. An introduction to basic optimization theory and classical and new gradien-based algorithms. Springer, New York (2005)
38. Ali, M., Pant, M., Abraham, A.: Unconventional initialization methods for differential evolution. Appl. Math. Comput. 219(9), 4474–4494 (2013)
39. Rahnamayan, S., Tizhoosh, H.: Differential evolution via exploiting opposite populations. In: Tizhoosh, H., Ventresca, M. (eds.) Oppositional Concepts in Computational Intelligence, 155th edn, pp. 143–160. Springer, Heidelberg (2008)
40. de Melo, V.V., Botazzo Delbem, A.C.: Investigating Smart Sampling as a population initialization method for Differential Evolution in continuous problems. Inf. Sci., 193, 36–53
41. Yap, D., Koh, S.P., Tiong, S.K., Prajindra, S.K.: A hybrid artificial immune systems for multimodal function optimization and its application in engineering problem. Artif Intell Rev 38(4), 291–301 (2012)
42. Swain, R.K., Barisal, A.K., Hota, P.K., Chakrabarti, R.: Short-term hydrothermal scheduling using clonal selection algorithm. Int. J. Electric Power Energ. Syst. 33(3), 647–656 (2011)
43. Cutello, V., Nicosia, G., Pavone, M.: Exploring the capability of immune algorithms: A characterization of hypermutation operators, pp. 263–276. Springer-Verlag, Berlin (2004)
44. Liu, R., Zhang, X., Yang, N., Lei, Q., Jiao, L.: Immunodomaince based Clonal Selection Clustering Algorithm. Appl. Soft Comput. 12(1), 302–312 (2012)
45. Dragoi, E.N., Curteanu, S., Fissore, D.: Freeze-drying modeling and monitoring using a new neuro-evolutive technique. Chem. Eng. Sci. 72, 195–204 (2012)
46. Dragoi, E.N., Suditu, G.D., Curteanu, S.: Modeling methodology based on artificial immune system algorithm and neural networks applied to removal of heavy metals from residual waters. Environmental Engineering and Management Journal 11(11), 1907–1914 (2012)
47. Storn, R., Price, K.: Differential Evolution – A Simple and Efficient Heuristic for Global Optimization over Continuous Spaces. J. Global Optim. 11(4), 341–359 (1997)
48. de Castro, L.N., Von Zuben, F.J.: Learning and optimization using the clonal selection principle. IEEE Transactions on Evolutionary Computation 6(3), 239–251 (2002)

Multi-layer Perceptrons for Voxel-Based Classification of Point Clouds from Natural Environments

Victoria Plaza[1], Jose Antonio Gomez-Ruiz[2(✉)], Anthony Mandow[1],
and Alfonso J. Garcia-Cerezo[1]

[1] Departamento de Ingeniería de Sistemas y Automática,
Universidad de Málaga, Andalucía Tech, 29071 Málaga, Spain
{victoriaplaza,amandow,ajgarcia}@uma.es
[2] Departamento de Lenguajes y Ciencias de la Computación,
Universidad de Málaga, Andalucía Tech, 29071 Málaga, Spain
janto@uma.es

Abstract. This paper addresses classification of 3D point cloud data from natural environments based on voxels. The proposed model uses multi-layer perceptrons to classify voxels based on a statistic geometric analysis of the spatial distribution of inner points. Geometric features such as tubular structures or flat surfaces are identified regardless of their orientation, which is useful for unstructured or natural environments. Furthermore, the combination of voxels and neural networks pursues faster computation than alternative strategies. The model has been successfully tested with 3D laser scans from natural environments.

Keywords: Multi-layer perceptron · 3D classification · Mobile robot · Voxel map

1 Introduction

Knowledge of geometric features in three-dimensional (3D) scenes is useful for object recognition in challenging applications in natural and unstructured environments, such as robotics for search and rescue (SAR) and planetary exploration [1, 2]. In these applications, scenes are usually obtained through laser scanners [3] and stereo vision [4] as large and complex point clouds. Object recognition from point clouds usually involves three main steps: segmentation, feature extraction, and classification.

Three main approaches have been adopted for object recognition in point clouds. The first approach avoids point-based computations by reducing the scene to a 2D representation which can be processed with standard artificial vision algorithms. The objects may be classified based on local and global statistics features of each object from a range image [5] or from a 2D deviation map when classification is based on texture analysis [6]. The classification can be also performed with image with depth data (RGB-D) by fusing results from separate 2D and 3D segmentation and feature extraction processes [7].

© Springer International Publishing Switzerland 2015
I. Rojas et al. (Eds.): IWANN 2015, Part II, LNCS 9095, pp. 250–261, 2015.
DOI: 10.1007/978-3-319-19222-2_21

The second approach consists on point-based techniques, where each 3D point is analyzed based on the features from its local neighborhood. In this case, segmentation becomes a costly step in 3D object recognition. Two major point-based segmentation techniques have been proposed: *i*) graph based algorithms such as Normalized Cut (NCut) [8] and Felzenszwalb-Huttenlocher (FH) [9, 10]; and *ii*) merging points with similar geometric features within *k*-Nearest Neighbors (KNN). Following the second option, segmentation may be based on the normal similarity with respect the KNN points [11, 12]. Then urban objects can be classified depending on the height [11] or geometric features from the bounding box of each object [12]. In contrast with the object classification, a geometrical classification may be performed at each point. Points may be classified as planar or non planar through a Support Vector Machine (SVM) [11] or in tubular structures, planes surfaces, and scatter through a classifier based on a Gaussian mixture model (GMM) [13]. The main drawback of this approach is the memory and execution time required to process each point in the point cloud. Therefore these methods may be incompatible with real-time applications such as SAR.

Another group of techniques for 3D classification is group-based techniques, which aim to solve the time-consumption problem of the point-based techniques by reducing the scene to a voxel model. Features are extracted at each neighborhood which means each voxel instead of the neighborhood at each point as the previous approach. After the classification for each voxel, a clustering process generates objects which can be also classified. Ground, short structures and tall structures may be classified from a voxel map through height thresholds, also vegetation may be detected with the multiecho in the laser rangefinder [14]. However, this method is restricted to urban scenarios where no slopes on the ground are assumed. Otherwise, voxels may be classified as linear, scatter, horizontal surfaces or vertical surfaces [15, 16], obtaining a Geometric-Featured Voxel map (GFV) where traversable ground can be detected [15]. Each class is defined by an equation with eigenvalues and eigenvectors from the covariance matrix of the inner 3D points in the voxel, the class with the largest value is applied to the voxel. These classes are not enough for natural environments where surfaces may have other slope than horizontal or vertical. In [17] the voxels are still with constant size but dynamically positioned, not a 3D grid as the previous techniques, which involves a larger computational cost.

This paper proposes a multi-layer perceptron-based model for 3D classification which can be applied to natural environments. A point cloud is modelled as a voxel map where each voxel is classified as a tubular structure, a flat surface of arbitrary inclination, or a scatter shape based on geometric features. This voxel-based solution reduces data processing in comparison with point-based techniques. Furthermore, geometric features are identified regardless of their orientation, which extends applicability to unstructured environments.

This paper is organized as follows. Next section describes the data structures needed for generating the voxel map. Section 3 proposes the classifier based in neural networks. Section 4 shows several experiments performed and, finally, conclusions are presented.

2   Data Structures for the Voxel Map

This section describes the data structures that will be employed in the proposed classification model. The flowchart of this model is depicted in Fig 1. The first step is building data structures that will be used for feature extraction and voxel classification by neural networks.

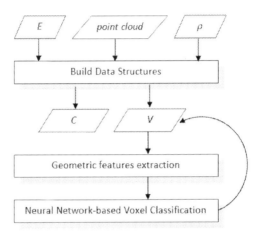

Fig. 1. Flowchart for the proposed 3D classification model

The process to build data structures produces C and V based on an input point cloud in Cartesian coordinates. The V data structure is based on coarse binary cubes (CBC) [18]. This structure represents those voxels in a 3D spatial grid with constant size that have a representative number of points. C represents information of all the points from the original point cloud and it keeps redundant information with respect V, however will be useful in the implementation of optional applications as visualization of results. Two additional parameters are required: E which is the edge length of the voxels and ρ which is a density threshold, the minimum number of points within a voxel to be considered.

The input *point cloud* is defined as a list of n points where each point p_i is defined as a vector with the Cartesian location (x_i, y_i, z_i), as describes

$$point\ cloud = \begin{bmatrix} x_1 & y_1 & z_1 \\ \vdots & \vdots & \vdots \\ x_n & y_n & z_n \end{bmatrix} \tag{1}$$

where its Z axis pointing upwards. In addition the vectors p_{min} and p_{max} are defined as the minimum and maximum Cartesian coordinates of the minimum bounding box for the point cloud respectively. This way, p_{min} is defined with the vector $(x_{min}, y_{min}, z_{min})$ where $x_{min} = minimum(x_i), \forall x_i \in point\ cloud$. In the same way, y_{min} and z_{min} are computed in the different axes Y and Z respectively and p_{max} with the function $maximum()$.

Furthermore, a vector $Idmax$ with the length of voxels in each axis will be computed as

$$Idmax = [Idmax_x \quad Idmax_y \quad Idmax_z] = round\left(\frac{p_{max}-p_{min}}{E}\right) + 1 \tag{2}$$

where $round()$ returns the nearest integer for each element from the input vector or matrix.

This is the distance in number of voxels between the first and last voxel in each axis (not all the voxels in the middle have to exist because may not have inner points). For each point of point cloud, the index of its corresponding voxel in each axis is computed. Using the expression

$$Id = \begin{bmatrix} Idx_1 & Idy_1 & Idz_1 \\ \vdots & \vdots & \vdots \\ Idx_n & Idy_n & Idz_n \end{bmatrix} = round\left(\frac{point\ cloud-(U \cdot p_{min})}{E}\right) + 1 \tag{3}$$

three values are obtained for each point forming the matrix Id, where U is a column vector of n ones.

With these computed values, for each point can be obtained an index value I_i in I of its corresponding voxel using the expression

$$I = \begin{bmatrix} I_1 \\ \vdots \\ I_n \end{bmatrix} = Id \times \begin{bmatrix} 1 \\ Idmax_x \\ Idmax_x \cdot Idmax_y \end{bmatrix} \tag{4}$$

Several points can have the same I_i value, meaning that they belong to the same voxel.

In this way, an indexed Cartesian point can be defined by coordinates (x, y, z, I), and they can be sorted by the voxel index I getting the data structure C in its initial state:

$$C = \begin{bmatrix} x_j & y_j & z_j & I_1 \\ \vdots & \vdots & \vdots & \vdots \\ x_k & y_k & z_k & I_n \end{bmatrix} \forall j, k \in \{1, ..., n\}, j \neq k \tag{5}$$

The voxels data structure, V, is created based on the initial state of C storing only those voxel with enough density. The voxels with less points than the predefined density threshold, ρ, are not taking into account. For example, a voxel with one or very few points it may not be enough representative in the scene, in addition the extraction of geometric features and classification may return noisy results.

Each voxel is defined in the voxels data structure V as a vector with the following elements:

- Voxel index, I_i.
- Indices in C of the first and last occurrence of a point inside the current voxel.
- Cartesian coordinates of the voxel center, ccv_i, computed with the expression in (6) based on the voxel index I_i.
- Three features for the subsequent classification input (see Section 3.1).

- The subsequent classification output which is the resulting class: scatter, tubular or surface (see Section 3.3).

$$ccv_{i,x} = \text{remainder} \left(\frac{I_i}{Idmax_x} \right)$$

$$ccv_{i,y} = \text{remainder} \left(\frac{(I_i - Idx)/Idmax_x}{Idmax_y} \right) \tag{6}$$

$$ccv_{i,z} = \frac{I_i - Idx - Idy \cdot Idmax_x}{Idmax_x \cdot Idmax_y}$$

3 Neural Networks for a Voxel-Based Classification

3.1 Geometric Features Extraction

Starting from the voxel data structure, geometric features used are inspired by the tensor voting approach [19] but instead of using the distribution of surface orientation, the distribution of the 3D points are used directly [13]. The classification algorithm relies on geometric features to classify each voxel as: planar surface, tubular structure, or scatter regions. The local spatial distribution of the points within a voxel is obtained by decomposition in the principal components from the covariance matrix of the points' Cartesian coordinates. The positive symmetric covariance matrix for each voxel is determined by a set of the N inner 3D points $\{X_i\} = \{(x_i, y_i, z_i)^T\}$ with $\bar{X} = \frac{1}{N} \sum_{i=1}^{N} X_i$, and it is defined in the expression (7):

$$\frac{1}{N} \sum_{i=1}^{N} (X_i - \bar{X}) \cdot (X_i - \bar{X})^T \tag{7}$$

The matrix is decomposed into its main components and its eigenvalues are sorted in ascending order, $\lambda_0 \geq \lambda_1 \geq \lambda_2$. Eigenvectors $\vec{e_0}, \vec{e_1}, \vec{e_2}$ correspond to eigenvalues $\lambda_0, \lambda_1, \lambda_2$ respectively. For the scatter voxels, it is satisfied that $\lambda_0 \simeq \lambda_1 \simeq \lambda_2$ and no dominant direction can be found. For the tubular structure case, the main direction will be tangent at the curve, with $\lambda_0 \gg \lambda_1 \simeq \lambda_2$. Finally, for the solid surfaces case, the main direction is aligned with the normal surface, that is $\lambda_0 \simeq \lambda_1 \gg \lambda_2$. A linear combination of the eigenvalues is used, see expression (8), to represent the three saliency features named *scatter-ness*, *tubular-ness* and *surface-ness*.

$$\begin{bmatrix} scatter - ness \\ tubular - ness \\ surface - ness \end{bmatrix} = \begin{bmatrix} \lambda_0 \\ \lambda_0 - \lambda_1 \\ \lambda_1 - \lambda_2 \end{bmatrix} \tag{8}$$

In practice, the decision of which class represents a voxel may be performed manually with thresholds. If $\lambda_0 - \lambda_1$ is higher than a threshold θ_l indicates *tubular-ness*, and $\lambda_1 - \lambda_2$ higher than a threshold θ_2 represents *surface-ness*; otherwise *scatter-ness*. Nevertheless these thresholds can vary depending on the scene, sensor characteristics, or point cloud density. An approach to make this process automatically is training a classifier that maximizes the probability of correct classification on a training data set. In [13], a classifier based on a GMM was trained using the expectation maximization algorithm [20]. However this classifier involves a higher computational cost than real time applications requires.

3.2   Approximating Bayes' Decision Rule by Neural Networks

Given a voxel containing a 3D points cloud, represented by the three saliency features shown in expression (8), a decision rule is required to classify it as scatter, tubular or surface. That is, classifying an observation $x \in R^n$ as belonging to one of a set of populations is desired, such that if x belongs to the i-th population, x occurs according to the density function $p(x/C_i)$. When maximizing the correct classification probability is desired, the minimum probability of error decision rule (Bayes' rule [21]) is defined by the function

$$\phi_i(x) = \begin{cases} 1 & \text{if } p(C_i / x) \geq p(C_k / x) \ \forall k \\ 0 & \text{in otherwise} \end{cases} \tag{9}$$

where $p(C_i/x)$ is the a posteriori density function and ϕ_i is the probability of classifying the pattern x in class C_i. The a posteriori density function $p(C_i/x)$ is unknown, and can be estimated in order to compute the probability of success.

It is theoretically proved [22] that a three-layer neural network (Multi-Layer Perceptron, MLP), with at least $2n$ hidden units (n is the dimension of input patterns), one output unit, and using the usual logistic function $\sigma(x)=1/(1+exp(-x))$ for both output and hidden layer, approximates the a posteriori probability in the two-category classification problem with arbitrary accuracy, and that it tends to the a posteriori probability as back-propagation learning proceeds ideally. Thus,

$$F(x, w_i) \cong p(C_i / x) \tag{10}$$

where $F(x, w_i)$ is the network output for an input pattern x and w_i are the synaptic weight matrices for class C_i.

Hence, the approximate Bayes' decision rule is given by

$$\tilde{\phi}_i(x) = \begin{cases} 1 & \text{if } F(x, w_i) \geq F(x, w_k) \ \forall k \\ 0 & \text{in otherwise} \end{cases} \tag{11}$$

That is, x will be classified in class C_i if $\tilde{\phi}_i(x) = 1$.

3.3   The MLP-Based Model

Three different MLPs are needed to implement the decision rule in expression (11), i.e., one for every voxel class: scatter (class C_1), tubular (class C_2) or surface (class C_3). Before using the MLPs to classify voxels, a labelling process and the MLPs training process are required. The labelling process was performed by a human supervisor using a graphical interface which allows labelling every voxel, stored in the data structure explained in section 2, in one of the three classes (scatter, tubular or surface). Several data sets manually classified or labelled, usually called training data, are used to train all the MLPs. When training a MLP corresponding to one class C_i, all voxels labelled as C_i are considered with value 1 in the supervisory variable (expected networks output) and the rest voxels, labeled as a different class, with value 0. The labeling of data sets and the training process is performed off-line and only once.

The main common characteristics of the MLPs employed are shown in Table 1. Input layers have three units, corresponding to the three saliency features shown in expression (8). The middle or hidden layers have the order of 100 units with logistic transfer functions. These numbers of units were determined using a cascade learning constructive process, adding neurons to the hidden layer one at time until there is no further improvement in network performance. The output layers have one logistic unit corresponding to the single dependent variable. The output units predict the a posteriori probability of classify the input patterns in the corresponding class by means of its numerical output (ranging from 0 to 1). Connection weights are changed using a Levenberg-Marquardt errors back-propagation algorithm [23] and the learning constant was set to 0.02.

Table 1. Common characteristics of MLPs

Network topology	Multilayer perceptron – Full connectivity
Learning algorithm	Levenberg-Marquardt
Learning rule	Generalized delta rule
Input data	The three saliency features of voxels
Output data	Probability of classification in the corresponding class

Weights initialization is crucial in the learning process with artificial neural networks. In order to obtain a realistic estimation of the probability of success in the classification probability, 30 weights initializations were carried out and the average and standard deviation of the executions are computed.

Giving the information to the MLP input layer requires a data preprocessing process. It is important to normalize the three saliency features ranges to lie within the central range of the hidden layer transfer function in the neural network.

A crucial aspect of carrying out learning and prediction analysis with a neural network system is to split the database into two independent sets: the training set (80% of the data set), which is used to train the neural network, and the test set (20% of the data set) to validate its predictive performance. During training, the input patterns of each training set are repetitively presented to the network which attempts to generate 1 at the output unit when the pattern must be classified is the corresponding class, and 0 when not.

The MLP were trained and the mean square errors between the classification variable (supervisory variable) and the dependent output variables decreased with an increasing number of epochs during training: first, it decreases rapidly and then continues to decrease slowly as the network makes its way to local minimum. With good generalization as the goal, the network ended up overfitting the training data since the training session was not stopped at the correct point.

The procedure used to avoid overfitting was the early stopping method of training [24], which leads to identifying the onset of overfitting through the use of the hold-out method, for which the training set is split into an estimation subset (80% of the training set), and a validation subset (20% of the training set). The estimation subset of examples was used to train each network of the system until 1000 epochs, but the training sessions were stopped periodically, weights matrices were saved to files, and the networks were tested on the validation subsets after each training period.

The early stopping points were found by plotting together the estimation learning curve, which decreased monolithically, and the validation learning curve, which decreased monolithically to a minimum, then started to increase as the training continued. The optimally trained MLPs were tested and, to evaluate the proposed models, a standard technique of stratified tenfold cross-validation was used [25] for each class (scatter, tubular and surface). This technique divides the input data set into ten sets of approximately equal size and equal distributions of input patterns. Each of the 10 random subsets of the data serves as a test set for the model trained with the remaining nine partitions. The overall prediction accuracy for the system is then assessed as an average of ten experiments.

Once the training process is over, which is performed off-line and only once, the sets of synaptic weights matrices w_1, w_2, w_3 are obtained for scatter, tubular and surface classes, respectively. Thus, the MLP-based model can be used to classify any unknown point cloud (any unused scene for training). That is, given a scene, for each voxel x, a classification into one of three classes described is performed, depending to the output of the MLP-based model that implement the expression (11).

4   Experiments and Results

This section presents the results of the proposed 3D classification model tested in several point clouds captured with a 3D laser rangefinder in natural environments.

The first step of our proposed method involves transforming the RAW data from the laser rangefinder to a point cloud such as list of Cartesian 3D points. This process is different depending on the 3D device used [26]. In this work, the low-cost laser rangefinder UnoLaser 30M135Y has been used [27]. It is a stop-and-go laser system [28] based on pitching a commercial 2D rangefinder, Hokuyo UTM-30LX. It has a maximum scanning range of 30 meters, a field of view 270° in horizontal and 131° in vertical. A spherical scan is obtained where each point in the cloud is defined by two angles ϕ and ψ around Z and X axes, respectively, and a distance r. The Cartesian coordinates $[x,y,z]^T$ are computed based on the spherical coordinates $[r, \phi, \psi]$ as describes the expression (12).

$$\begin{bmatrix} x \\ y \\ z \end{bmatrix} = \begin{bmatrix} cos(\phi)\,r \\ cos(\psi) \cdot sin(\phi) \cdot r \\ sin(\psi) \cdot cos(\phi) \cdot r \end{bmatrix} \tag{12}$$

Point clouds used in experiments corresponds to scenarios captured during a realistic emergency drill in natural environments in Marbella [29], Spain. Voxels of 0.5 meters edge ($E=0.5$) were used as a compromise between computational efficiency, memory management and scene reconstruction accuracy and voxels were processed only if contains at least 10 points ($\rho =10$).

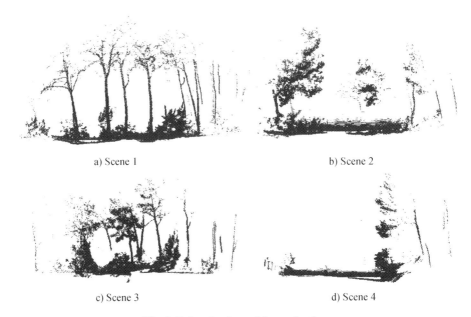

a) Scene 1   b) Scene 2

c) Scene 3   d) Scene 4

Fig. 2. Point clouds used for evaluation

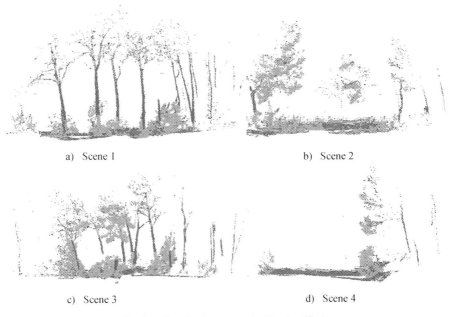

a) Scene 1   b) Scene 2

c) Scene 3   d) Scene 4

Fig. 3. Point clouds automatically classified

Fig. 2 shows point clouds obtained directly from the 3D laser rangefinder without any processing. Original scenarios appeared in this figure have a maximum size of 30 x 15 x 20 meters in width, height and depth respectively.

Table 2 shows the features for each scene used in simulation: quantity of points in the point cloud, quantity of dense voxels in the data structure obtained (see section 2), execution time, in seconds, for the generation of the data structures and execution time for the classification procedure. Simulations have been executed on a work station with an i7 processor, 64 bits with a clock frequency of 3.9 GHz and 8 GB in RAM.

When a voxel is classified as scatter, then all the points components of this voxel are considered as scatter too and they are represented in green. Same way, blue color is used for tubular and red for surface class.

For each point cloud in the Fig.2, the Fig.3 shows the same point cloud colored according to the classification obtained from the MLP-based model described in section 3.3. The automatic classification agree with the expected: scatter for the vegetation as bushes and tree crowns, tubular for trunks and branches and surface for the ground. In addition, the sum of the execution times, for the data structures generation and the scenes classification (see Table 2), allows carrying out the algorithm on real time on the site. Hence, a traversability map may be generated for the autonomous navigation system on a mobile robot.

Table 2. Scenes used in experiments

	# points	# voxels	Time to generate the voxel map	Time to perform the classification
Scene 1	293484	1625	0.2685	0.8978
Scene 2	225245	1494	0.2259	0.6792
Scene 3	333048	1898	0.3103	1.1594
Scene 4	263597	1407	0.2402	0.6976

5   Conclusions

This paper has reviewed classification techniques based on 3D data where the widest disadvantage is the large computational cost. Many of them compute a classification based on the neighborhood of each point in the cloud, compute the normal vector at each point or neighborhood or use a complex classifier, hence they involve a very high time-consuming algorithm. In other cases the problem is reduced by using a 3D grid to compute the neighborhood features, but these features can be not enough for unstructured environments. In this paper a model for 3D classification has been proposed which generates a voxel map where each voxel is classified as tubular structure, flat surface or scatter shape.

The classification task is done by a neuronal networks-based model which is a lower computational algorithm than other classifiers from the literature. The classification is based on geometric features, in particular the local spatial distribution around each point which is defined by the principal components from the covariance matrix of the point's positions. The model has been successfully tested on several point clouds in natural environments captured by a 3D laser rangefinder and achieves low execution time as required in SAR missions.

Acknowledgements. This work has been partially supported by the Spanish CICYT project DPI 2011-22443.

References

1. Cole, D., Newman, P.: Using laser range data for 3D SLAM in outdoor environments. Proceedings - IEEE International Conference on Robotics and Automation. 2006, 1556–1563 (2006)
2. Rusu, R., Sundaresan, A., Morisset, B., Hauser, K., Agrawal, M., Jean-Claude, L., Beetz, M.: Leaving flatland: Efficient real-time three-dimensional perception and motion planning. Journal of Field Robotics **26**(10), 841–862 (2009)
3. Mandow, A., Martínez, J., Reina, A., Morales, J.: Fast range-independent spherical subsampling of 3D laser scanner points and data reduction performance evaluation for scene registration. Pattern Recognition Letters **31**(11), 1239–1250 (2010)
4. Linhui, L., Mingheng, Z., Lie, G., Yibing, Z.: Stereo vision based obstacle avoidance path-planning for cross-country intelligent vehicle. In: 6th International Conference on Fuzzy Systems and Knowledge Discovery, FSKD 2009, vol. 5, pp. 463–467 (2009)
5. Zhu, X., Zhao, H., Liu, Y., Zhao, Y., Zha, H.: Segmentation and classification of range image from an intelligent vehicle in urban environment. In: 2010 IEEE/RSJ International Conference on Intelligent Robots and Systems (IROS), pp. 1457–1462 (2010)
6. Othmani, A., Piboule, A., Dalmau, O., Lomenie, N., Mokrani, S., Voon, L.F.C.L.Y.: Tree species classification based on 3D bark texture analysis. In: Klette, R., Rivera, M., Satoh, S. (eds.) PSIVT 2013. LNCS, vol. 8333, pp. 279–289. Springer, Heidelberg (2014)
7. Douillard, B., Brooks, A., Ramos, F.: A 3D laser and vision based classifier. In: 2009 5th International Conference on Intelligent Sensors, Sensor Networks and Information Processing (ISSNIP), pp. 295–300 (2009)
8. Yu, Y., Li, J., Guan, H., Wang, C., Yu, J.: Semiautomated extraction of street light poles from mobile LIDAR point-clouds. IEEE Transactions on Geoscience and Remote Sensing **53**(3), 1374–1386 (2015)
9. Felzenszwalb, P.F., Huttenlocher, D.P.: Efficient graph-based image segmentation. Int. J. Comput. Vision **59**(2), 167–181 (2004)
10. Sima, M.C., Nüchter, A.: An extension of the Felzenszwalb-Huttenlocher segmentation to 3D point clouds. In: Society of Photo-Optical Instrumentation Engineers (SPIE) Conference Series, vol. 8783, pp. 878302–878306 (2013)
11. Hao, W., Wang, Y.: Classification-based scene modeling for urban point clouds. Optical Engineering **53**(3), 033110 (2014)
12. Zhou, Y., Yu, Y., Lu, G., Du, S.: Super-segments based classification of 3D urban street scenes. International Journal of Advanced Robotic Systems **9**(248), 1–8 (2012)
13. Lalonde, J., Vandapel, N., Huber, D., Hebert, M.: Natural terrain classification using three-dimensional ladar data for ground robot mobility. Journal of Field Robotics **23**(10), 839–861 (2006)
14. Borcs, A., Jozsa, O., Benedek, C.: Object extraction in urban environments from large-scale dynamic point cloud datasets. In: Proceedings - International Workshop on Content-Based Multimedia Indexing, pp. 191–194 (2013)
15. Seo, B., Chung, M.: Traversable ground detection based on geometric-featured voxel map. In: FCV 2013 - Proceedings of the 19th Korea-Japan Joint Workshop on Frontiers of Computer Vision, pp. 31–35 (2013)

16. Choe, Y., Shim, I., Chung, M.: Geometric-featured voxel maps for 3D mapping in urban environments. In: 9th IEEE International Symposium on Safety, Security, and Rescue Robotics, SSRR 2011, pp. 110–115 (2011)
17. Habermann, D., Hata, A., Wolf, D., Osorio, F.: Artificial neural nets object recognition for 3D point clouds. In: 2013 Brazilian Conference on Intelligent Systems (BRACIS), pp. 101–106 (2013)
18. Reina, A., Martínez, J., Mandow, A., Morales, J., García-Cerezo, A.: Collapsible cubes: Removing overhangs from 3D point clouds to build local navigable elevation maps. In: IEEE Conference on Advanced Intelligence Mecatronics (AIM), pp. 1012–1017 (2014)
19. Medioni, G., Lee, M., Tang, C.: A computational framework for segmentation and grouping. Elsevier Science Inc., New York (2000)
20. Bilmes, J.: A gentle tutorial on the EM algorithm and its application to parameter estimation for Gaussian mixture and hidden Markov models. Tech. Rep. ICSI-TR-97-021. The International Computer Science Institute, University of Berkeley, Berkeley (1997)
21. Duda, R., Hart, P., Stork, D.: Pattern classification. Wiley Interscience (2000)
22. Funahashi, K.: Multilayer neural networks and Bayes decision theory. Neural Networks 11(2), 209–213 (1998)
23. Patterson, D.W.: Artificial neural networks, theory and applications. Prentice-Hall Series in Advanced Communications, Prentice Hall (1996)
24. Amari, S., Murata, N., Muller, K.R., Finke, M., Yang, H.: Statistical theory of overtraining - is cross-validation asymptotically effective? Adv Neural Inform Process Syst. 8, 176–182 (1996)
25. Janssen, P., Stoica, P., Söderström, T., Eykhoff, P.: Model structure selection for multivariable systems by cross-validation. International Journal of Control 47(6), 1737–1758 (1988)
26. Plaza-Leiva, V.: Local Semantic Map based on 3D data for navigability. Master's thesis, Univeristy of Málaga, ETSII (2013)
27. Morales, J., Martinez, J., Mandow, A., Pequeno-Boter, A., Garcia-Cerezo, A.: Design and development of a fast and precise low-cost 3D laser rangefinder. In: 2011 IEEE International Conference on Mechatronics (ICM), pp. 621–626 (2011)
28. Elhabiby, M., Teskey, W.J.: Stop-and-go 3D laser scanning and mobile mapping. B. Sc. Thesis (2010)
29. Cátedra de Seguridad Emergencias y Catástrofes y Catástrofes de la Universidad de Málaga: VIII Jornadas sobre Seguridad, Emergencias y Catástrofes (2014). http://www.umaemergencias.es/formacion-y-actividades/jornadas-seguridademergencias-y-catastrofes/viii-jornadas-mayo-2014/

An Improved RBF Neural Network
Approach to Nonlinear Curve Fitting

Michael M. Li[✉] and Brijesh Verma

CINS and School of Engineering and Technology,
Central Queensland University, Rockhampton, QLD 4702, Australia
m.li@cqu.edu.au

Abstract. This article presents a new framework for fitting measured scientific data to a simple empirical formula by introducing an additional linear neuron to the standard Gaussian kernel radial basis function (RBF) neural networks. The proposed method is first used to evaluate two benchmark datasets (Preschool boy and titanium heat) and then is applied to fit a set of stopping power data (MeV energetic carbon projectiles in elemental target materials C, Al, Si, Ti, Ni, Cu, Ag and Au) from high energy physics experiments. Without increasing computational complexity, the proposed approach significantly improves accuracy of fitting. Based on this type RBF neural network, a simple 6-parameter empirical formula is developed for various potential applications in curve fitting and nonlinear regression problems.

Keywords: RBF · Curve fitting · Neural networks · Stopping power

1   Introduction

The need for data analysis is nearly ubiquitous in numerous fields of science and engineering, ranging from atomic spectroscopy, physicochemical reaction dynamics, to atmospheric sampling survey. A very frequently encountered problem of data analysis is curve fitting, where a set of measured data is taken from experiments and generating a curve to fit this dataset is required so that underlying characteristics represented by the data can be determined accurately. Despite the noise of data (due to uncontrollable randomness error of measurements), the desired output curve as fitting result should be a continuous smooth curve. There are a few reasons for requiring a smooth fitted curve with the data set, rather than a curve by directly connected existed data points. Firstly, the smooth curve itself provides an intuitional visualization and therefore the functional dependency and trend can be clearly displayed. Secondly, analytical properties of curve such as the rate of change, the local minimum and maximum points, and the asymptotic limit can be conveniently calculated. Thirdly, the fitted smooth curve allows for an easy comparison of the different theoretical algorithms to the measured points.

The goal of curve fitting is to find a parametric function that is as close as possible in containing all the data points. Traditionally, curve fitting can be performed by using the least-square method in which the sum of the squares between the observed points and those fitted by the model are minimized by a numerical analysis method.

© Springer International Publishing Switzerland 2015
I. Rojas et al. (Eds.): IWANN 2015, Part II, LNCS 9095, pp. 262–275, 2015.
DOI: 10.1007/978-3-319-19222-2_22

Based on the conventional least-square method, the fitting of data to a proper curve in the presentence of an assumed function could be achieved through a simple balance of intuition and experience in quantitative data analysis. However, the resultant curve or model may be not necessarily optimal. Furthermore, since the determination of fitting parameters in the least-square procedure is made by an optimization process in which a numerical iteration algorithm (for example Levenberg-Marquartd) is usually used, the iteration procedure requires a good initial guess. The initial value of solution largely affects the convergence of iterative process. Without an appropriate initial value the iterative procedure may not converge towards a correct solution.

Mathematically, the problem of curve fitting can be regarded as a class of function approximation problem, where an unknown target function instead of an explicit functional form is expected to be constructed by a simpler known function (for example a polynomial, an exponential or an irrational) or a set of basis function (such as a set of radial basis function), with a given set of data points. In principle, any theory or approach for function approximation can also be applied to curve fitting. With the emerging of intelligent techniques and the progressing of statistical learning theory, the neural network approach is able to provide a satisfactory solution for certain types of function approximation problems. Both multilayer perceptron (MLP) and radial basis function (RBF) neural networks have been proved to be a universal function approximator [1][2], provided existence of sufficient neuron in the hidden layers. While MLP networks are good at fitting, there are two obvious disadvantages: (1) To achieve a high accuracy of fitting, it needs considerable number of neurons in the hidden layers and subsequently excessive parameters in terms of weight and bias appear in a model, which is not desired as a good empirical model; (2) Suffering over-fitting where generalization could be very poor. On the other hand, RBF networks possess the property of the best approximation which is not shared by MLP networks [3]. Selecting a limited number of centers by algorithms, RBF network offers an alternative solution with relatively fewer number of parameters. In addition, the Tikhonov regularization in RBF ensures it a unique and optimal solution.

The study of empirical fitting of stopping power data has been overwhelmingly driven by application requirements. There are two prominent application fields - Ion beam analysis techniques and Radiation therapy, requiring accurate knowledge of stopping powers for practical uses. In a quantitative ion beam analytical technique, it is the stopping power that determines the depth scale and hence resolves the accuracy of analysis [4]. In the radiation therapy, stopping power is essential data for dosimetry. Ion beam therapy has the advantage of depositing most of its energy per unit distance, compared to the traditional photon treatment. Particularly a carbon ion beam is able to apply a large dose to deeply seated tumors to kill more malignant cells and decrease toxicity sediment due to its superior depth dose distribution. Obviously, without an accurate dose delivery, any treatment using ion beams would be extremely risky. In addition, it has reported that the United States also took place using other species heavy ions like neon, silicon and argon to clinical treatment and radiobiology experiment [5].

In despite of many pioneering theoretical computations and experimental measurements, it is still impossible to provide high precision stopping power data for up to thousand combinations of elemental projectile/target in the required energy regions.

In past decades, many works have been done to build reliable and easy-use empirical models. These models were mainly based on classical least-square algorithms incorporated quantum theory [6-8]. Overall these studies have led up to discoveries of a few of parametric equations that are able to predict stopping power data with varying level of agreement to corresponding experimental values.

In this study, we propose using a standard RBF neural network with an additional linear neuron to fit stopping power curve versus the energy of projectile. Conceptually this method approximates the unknown stopping power function as a linear combination of a few simple functions represented by hybridizing neurons of several Gaussians, a linear term and a constant term. The presentence of additional linear term helps improve the global behavior of the function. Our investigation mainly focuses on the energetic carbon projectiles in different targets – C, Al, Si, Ti, Ni, Cu, Ag and Au.

The organization of this paper is as follows: the fitting problem is defined and the methodology is described in Section 2. Next, computer simulations and results are discussed in Section 3. Finally, Section 4 concludes the paper.

2   Methodology

2.1   Definition of Curve Fitting Problem

A curve to be fitted from data points is a regression function with parameters. The task of curve fitting is not only to find a right function but also to attempt the determination of the optimal values of parameters. Consider the problem of fitting the data set $\{(\mathbf{x}_i, y_i)\}^N_{i=1}$ with a regression function $f(\mathbf{x})$,

$$y_i = f(\mathbf{x}_i, \theta) + \varepsilon_i \tag{1}$$

where θ and ε_i are a parameter set and random errors respectively.

The optimal parameters of function $f(\mathbf{x}, \theta)$ can be obtained by minimizing the sum of weighted square residuals,

$$S(\theta) = \sum_{i=1}^{N} w_i [y_i - f(\mathbf{x}_i, \theta)]^2 \tag{2}$$

By making differentiation of $S(\theta)$ over θ, we obtain

$$\frac{\partial S(\theta)}{\partial \theta} = -2 \sum_{i=1}^{N} w_i [y_i - f(\mathbf{x}_i, \theta)] \frac{\partial f(\mathbf{x}_i, \theta)}{\partial \theta} \tag{3}$$

Setting the partial derivatives to 0, it produces a set of equations concerning the fitting parameters θ. These equations require solution by numerical algorithms due to their nonlinearity. Following this philosophy, various functions with empirical parameterization have been proposed [6-8] to fit stopping power data. Overall, the functions commonly used are power-like, logarithm, Weibull or their combinations. Generally, some satisfactory performances have reached when the above classical method was used. Nonetheless, it was observed that there were still some significant underestimates or overestimates of stopping power values near the peak.

2.2 RBF Neural Networks with Additional Linear Neuron

Mathematically, an RBF network is a linear combination of a set of weighted radial basis functions. It typically consists of three layers – an input layer, a hidden layer with activation functions and an output layer. The network represents a nonlinear mapping that transforms the input vector into the output scalar through a set of basis functions. When it is applied to the problem of regression, the unknown function $f(\mathbf{x})$ to be fitted can be directly approximated by an RBF as below,

$$f(\mathbf{x}) \approx \sum_{i=1}^{N} w_i \varphi(\|\mathbf{x} - \mathbf{c}_i\|) \tag{4}$$

where $\mathbf{x}$ is the input vector, w_i are weights, φ is a set of N basis functions with centers $\mathbf{c}_i$ $(i=1 \ldots N)$ and $\|.\|$ denotes the Euclidian distance.

A number of functions have been tested as basis functions. The Gaussian function is typically selected as the basis function in many applications [9,10], due to its outstanding analytical properties such as smoothness and infinite differentiability. In the case of Gaussian as the basis function, we have the following form RBF network,

$$y(\mathbf{x}) = \sum_{i=1}^{N} w_i \exp(-\frac{(\|\mathbf{x} - \mathbf{c}_i\|)^2}{2\sigma^2}) \tag{5}$$

where σ is the width, and y is the output of the network.

The Gaussian function is a type of localized function. It decays quickly from the locations nearby the centre. More generally, its influence decreases according to the Mahalanobis distance from the centre. This suggests that data points far from centers with a large Mahalanobis distance will fail to activate that basis function. Particularly, provided the centers sorted in ascending, basis functions at the margins of lower boundary of first center and upper boundary of last center will not be able to efficiently to deal with the behavior of function to be approximated since other Gaussians influence little in the margin regions. In addition, in the vicinities near centers, the Gaussian might overestimate a bit for the function. To overcome these problems, we proposed to add an extra linear term to the usual RBF network so that it not only can help efficiently to reproduce the global behavior of function but also plays a role of corrections.

With adding the extra linear term, the RBF network has the following form,

$$y(\mathbf{x}) = \sum_{i=1}^{N} w_i \varphi(\|\mathbf{x} - \mathbf{c}_i\|) + a\mathbf{x} + b \tag{6}$$

where a is the linear coefficient, and b is the constant term.

The diagram in Figure 1 shows the architecture of the customized RBF network, where the linear term can be regarded as a special neuron with the coefficient as its weight.

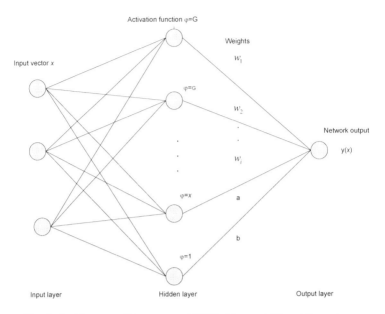

Fig. 1. Architecture of the proposed RBF with an additional linear term

The training of RBF network requires determinations of parameters for centers and weights. It usually is a two-stage process. In first the stage, the number of centers and their locations are determined and then follow a process to evaluate the weights. There are a few ways to determine the centers which include the random selection from subset of training data, clustering techniques and a supervised learning process. We only consider the relative simple one - the random selection from the training set for our applications. Following the determination of centers, a simple and straightforward equation for evaluating weights can be derived as below. By re-writing the equation (6) in the form of matrix,

$$y(\mathbf{x}) = \sum_{i=1}^{N} w_i \varphi(\| \mathbf{x} - \mathbf{c}_i \|) + a\mathbf{x} + b$$
$$= \mathbf{\Phi}^T(r)\mathbf{W} \tag{7}$$

where

$$r = \| \mathbf{x} - \mathbf{c}_i \|$$
$$\varphi_i(r) = \varphi(\| \mathbf{x} - \mathbf{c}_i \|)$$
$$\mathbf{\Phi}^T(r) = [\varphi_1(r), \varphi_2(r), \ldots \ldots \varphi_N(r), \mathbf{x}, 1]$$
$$\mathbf{W} = [w_1, w_2, \ldots \ldots w_N, a, b]^T$$

Considering each point in the training set $\{\mathbf{x}_j, \mathbf{y}_j\}_{j=1}^{M}$ for equation (7), we obtain

$$\sum_{i=1}^{N} w_i \varphi_i (r_j) + a\mathbf{x}_j + b = y_j \qquad j=1, 2,....M \qquad (8)$$

The set of linear equations of (8) can be re-written in terms of matrix form,

$$\mathbf{\Phi}^T W = Y \qquad (9)$$

where

$$\mathbf{\Phi} = \begin{bmatrix} \varphi_1(r_1) & \varphi_2(r_1) & .... & \varphi_N(r_1) & x_1 & 1 \\ \varphi_1(r_2) & \varphi_2(r_2) & .... & \varphi_N(r_2) & x_2 & 1 \\ . & & & & & \\ . & & & & & \\ . & & & & & \\ \varphi_1(r_M) & \varphi_2(r_M) & .... & \varphi_N(r_M) & x_M & 1 \end{bmatrix}$$

$$\varphi_i(r_j) = \exp(-\frac{\| \mathbf{x}_j - \mathbf{c}_i \|^2}{2\sigma^2})$$

$$Y = [y_1, y_2,........y_M]^T$$

The solution of the matrix equation (9) is

$$W = (\mathbf{\Phi}^T \mathbf{\Phi})^{-1} \mathbf{\Phi}^T Y$$
$$= \mathbf{\Phi}^+ Y \qquad (10)$$

where $\mathbf{\Phi}^+$ is the pseudoinverse of matrix $\mathbf{\Phi}$; it is defined as

$$\mathbf{\Phi}^+ = (\mathbf{\Phi}^T \mathbf{\Phi})^{-1} \mathbf{\Phi}^T \qquad (11)$$

The computation of the pseudoinverse $\mathbf{\Phi}^+$ can be conveniently performed by using the algorithm known as singular-value decomposition (SVD).

2.3   Numerical Examples

Before the proposed method is applied to a batch of stopping power data from experimental measurements in high energy physics, two numerical examples are used to illustrate the linear term effects in the developed approach. As the first example, preschool boy dataset (the ratio of weight to height varying with age) [11] is demonstrated. This dataset is often used as a benchmark data in nonlinear regression statistics. The plot below clearly shows RBF with linear term fits well while the RBF without a linear term fails the fitting.

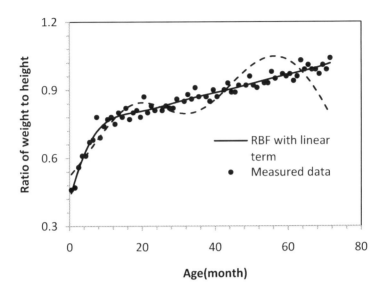

Fig. 2. Testing preschool boy benchmark dataset using the developed RBF fitting

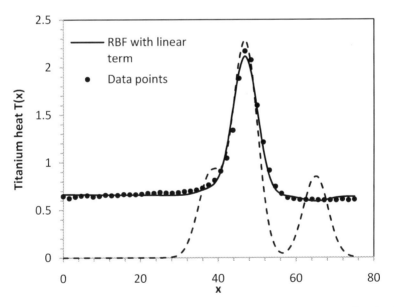

Fig. 3. Comparison of RBF fitting with and without a linear term for titanium data (The coefficient of determination R^2=0.9834)

The second illustrative example is another typical nonlinear regression problem with the titanium heat data [12]. The data have been known to be difficult to fit using classical numerical techniques because of its noise and a sharp peak, and therefore

have often been cited as a benchmark dataset to test new algorithms in B-Spline inter-
polations. Using a 3-centered RBF network with a linear term, the data set is fitted
reasonably well as shown in Figure 3. However, without including the linear term, the
RBF produces a big oscillation. It is interesting that we noted a similar situation occurs when the Support Vector Regression (SVR) method was used to demonstrate the
regression of using Gaussian RBF kernel, where the result of SVR regression displays
a poor fitting with a dramatic fluctuation of the fitted curve outside the peak area [13].
It appears to us that the developed method exhibits a formidable robustness in the
situation of difficult data.

3   Experimental Results and Discussion

Using the proposed method, the stopping power curves versus incident energy in the
range $0.01<E<1.0$ (MeV/u) carbon projectiles in target materials C, Al, Si, Ti, Ni, Cu,
Ag and Au are fitted. Carbon is chosen as the projectile because its stopping data have
extensively practical interest in ion beam analysis technique and heavy ion radiation
therapy. The data considered are mainly from Paul's collections and Zhang's measurements [7,14]. The experimental measurement data in Paul's collection from different laboratories worldwide have spanned over decades. To make the collected data
more objective and consistent, it requires a filtering principle to exclude some measured data with greater uncertainties.

We adopted the procedure of random subset selection to determine the center locations of Gaussians. In the curve fitting problems, it should be favorable if there is a
minimal amount of number centers. Our simulations have indicated that a set of 4
centers is minimal for our problem and a different set of centers selected randomly
has little impact to the fitting outcomes but gives the different weight values. The
center locations along the energy axis for the following fittings are set to {0.01548,
0.06025, 0.1967, 0.6025}, and parameter σ is set to 1. After this, parameters of
weights are computed by the equations (10) and (11).

The results of fitting parameters w_1, w_2, w_3, w_4, a, and b along with the normalized
root mean squared error $\sqrt{\chi^2}$ are listed in Table 1. In all fittings, the values of R^2
(the coefficient of determination) are greater than 0.99, indicating the data points
close to the fitted curve. The quality of fitting can also be inspected by looking at the
residuals plots, which provide a certain insight into the fitted results. As an example,
we look at the normalized residuals plot (Figure 4) for the projectile carbon in carbon
target. The plot clearly shows that the data points deviate the fitted curve randomly,
presenting a good random behavior of the residual distribution with a small error.

The obtained parameters can be used as a set of coefficients thus a simple empirical formula of stopping power (S) upon the projectile energy (E) is available for any
applications

$$S(E) = \sum_{i=1}^{4} w_i e^{-(E-E_i)^2/2} + aE + b \qquad (12)$$

where E_i ($i=1,4$) are constants, and values of coefficients w_i, a and b can be found
from the tabulation for various targets.

Table 1. Fitting parameters for carbon projectiles in targets C, Al, ... and Au

Target	w_1	w_2	w_3	w_4	a	b	$\sqrt{\chi^2}$
C	149.29	-345.48	402.38	-213.79	27.96	36.04	0.02575
Al	47.65	-131.58	164.55	-99.01	12.95	26.64	0.03760
Si	44.11	-104.53	125.17	-68.73	9.95	14.75	0.01966
Ti	417.85	-2548.1	2800.7	-650.27	25.43	6.14	0.02334
Ni	-29.40	69.72	-77.67	51.73	-6.45	-12.60	0.04090
Cu	-90.86	223.29	-236.85	89.92	-10.34	-3.98	0.05232
Ag	8.73	-2.56	-9.51	17.61	-1.63	-8.73	0.04522
Au	11.83	-13.31	9.18	2.54	1.03	-4.07	0.03209

Figures 5(a)-5(e) are plots of fitted stopping power curves along with the original measured data. Due to limited pages, only partial fitting results are presented in this paper. From these figures, it can be seen that for most cases the agreement is excellent; the fitted curves exhibit the general shape of data points and typical features including peaks. The peak of curves reflects the underlying micro-scale collisions of projectile particles against the atoms of target matter where the shift of the peak position depends on the atomic number of the projectile. This is consistent with the computations from the sophisticated quantum physics model. On average, the deviations of fitted curves from the corresponding experimental data sets are within 3.5%. However, we noticed that systematic error of measured data, resulting from uncertainties in measurements etc, ranges between 2.5% and 10%. Only a few isolated points with large discrepancies are observed.

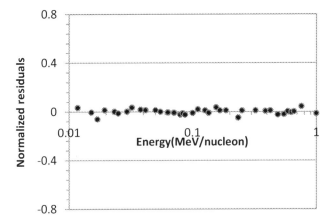

Fig. 4. Normalized residuals ploted for C projectiles in carbon as a function of ion energy per nucleon

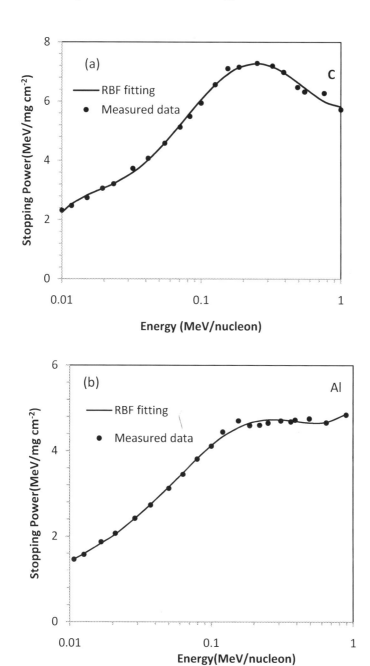

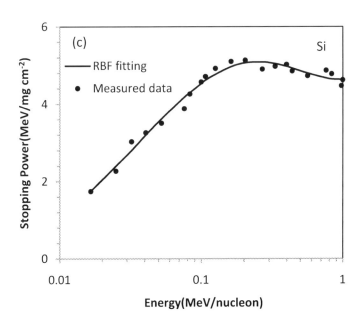

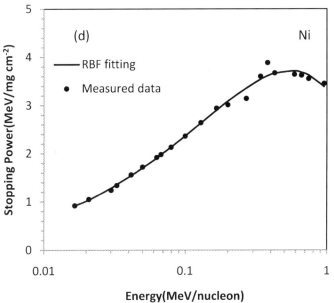

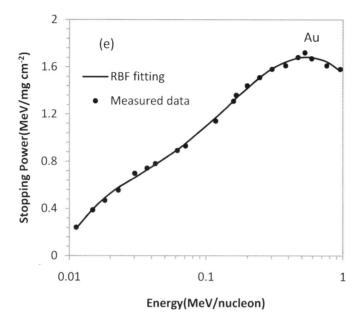

Fig. 5(a)-(e). Stopping power for carbon projectiles in the target C, Al, Si, Ni and Au, Comparison of RBF fittings with the measured data

Finally, to test the universality of the method in the stopping power fitting and compare our RBF fitting with those by using MLP and Paul's empirical formula, Figure 6 presents the fitting results of another type projectile Li in carbon target, where measured data points are well fitted by the proposed RBF method and a three-layers MLP [15] while the result of Paul's empirical fitting critically underestimates the stopping power around the peak area with quite a large error. Paul's work provides a framework aiming at reproducing stopping power universally for nearly all ion-target combination. In their fittings, the majority of the combinations of projectile-target has a reasonable accuracy but considerable amount of the projectile-target combinations such as Li in carbon didn't give a satisfactory accuracy. This appears to users of the empirical stopping power formula that in a global model the issue of accuracy and universality still needs a certain degree of compromise.

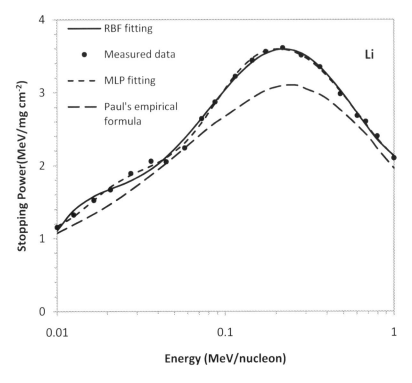

Fig. 6. Stopping power for Li in carbon versus energy per nucleon. The measured data are compared to RBF method, MLP fitting and Pauls' empirical prediction.

4   Conclusions

This work explores the feasibility using RBF neural networks to fit stopping power data so that an accurate and relative simple empirical stopping power formula with 6 parameters for carbon projectiles in elemental targets C, Al, Si, Ti, Ni, Cu, Ag and Au has been developed. With one additional linear term, the proposed method significantly improves accuracy of fitting without compromising computational complexity. The results show that a 4-centered RBF network with an additional linear term can sufficiently fit the stopping power versus projectile energy with high accuracy. An extension of this work that makes a further fitting or interpolation versus projectile atomic number (Z_1) is undertaking. It appears to us that the approach as a generic method has a great potential to extend its applications for any different combinations of projectile and matter, as well as other type data in physics. It is also possible from fitted results using the RBF method to guide future stopping power measurements for physicists so that the measurement expenditure could be saved.

References

1. Hornik, K., Stinchcomb, M., White, H.: Multilayer feedforward networks are universal approximators. Neural Networks **2**, 359–366 (1989)
2. Park, J., Sandberg, I.W.: Universal approximation using radial basis function. Neural Computation **3**, 246–257 (1991)
3. Poggio, T., Girosi, F.: Networks for approximation and learning. Proc. IEEE **78**, 1481–1497 (1990)
4. Bird, J.R., Williams, J.S.: Ion beams for materials analysis. Academic Press, New York (1989)
5. Schulz-Ertner, D., Tsujii, H.: Particle radiation therapy using proton and heavier ion beams. J. Clin. Oncol. **25**, 953–964 (2007)
6. Ziegler, J.F., Biersack, J.P., Ziegler, M.D.: SRIM - The Stopping and Range of Ions in Matter. SRIM Co., Chester (2008)
7. Paul, H., Schinner, A.: Empirical stopping power tables for ions from 3_{Li} to 18_{Ar} and from 0.001 to 1000 MeV/nucleon in solids and gases. Atomic Data and Nuclear Data Tables **85**, 377–452 (2003)
8. Konac, G., Klatt, Ch., Kalbitzer, S.: Universal fit formula for electronic stopping power of all ions in carbon and silicon. Nuclear Instruments and Methods in Physics Research B **146**, 106–113 (1998)
9. Haykin, S.: Neural Networks: A Comprehensive Foundation, 2nd edn. Prentice Hall (1998)
10. Li, M.M., Verma, B., Fan, X., Tickle, K.: RBF neural networks for solving the inverse problem of backscattering spectra. Neural Computing and Applications **17**, 391–399 (2008)
11. Gallant, A.R.: Nonlinear statistical models, pp. 142–146. John Wiley, Canada (1987)
12. Dierckx, P.: Curve and surface fitting with splines. Monograph on Numerical Analysis. Clarendon Press, London (1993)
13. Gunn, S.R.: Support vector machine for classification and regression. Technical Report. University of Southhampton, UK (1998)
14. Zhang, Y., Possnert, G., Whitlow, H.J.: Measurement of the mean energy-loss of swift heavy ions in carbon with high precision. Nuclear Instruments and Methods in Physics Research B **183**, 34–47 (2001)
15. Li, M., Guo, W., Verma, B., Lee, H.: A neural networks-based fitting to high energy stopping power data for heavy ion in solid matter. In: Proceedings of WCCI 2012 IEEE, Brisbane, Australia, pp. 832–837 (2012)

QSVM: A Support Vector Machine for Rule Extraction

Guido Bologna[1][(✉)] and Yoichi Hayashi[2]

[1] University of Applied Sciences of Western Switzerland,
Rue de la Prairie 4, 1202 Geneva, Switzerland
Guido.Bologna@hesge.ch
[2] Department of Computer Science, Meiji University,
Tama-ku, Kawasaki, Kanagawa 214-8571, Japan
hayashiy@cs.meiji.ac.jp

Abstract. Rule extraction from neural networks represents a difficult research problem, which is NP-hard. In this work we show how a special Multi Layer Perceptron architecture denoted as DIMLP can be used to extract rules from ensembles of DIMLPs and Quantized Support Vector Machines (QSVMs). The key idea for rule extraction is that the locations of discriminative hyperplanes are known, precisely. Based on ten repetitions of stratified 10-fold cross validation trials and with the use of default learning parameters we generated symbolic rules from five datasets. The obtained results compared favorably with respect to another state of the art technique applied to Support Vector Machines.

Keywords: Rule extraction · Ensembles · SVM

1 Introduction

In various domain applications, explaining answers provided by neural network models is crucial. For instance a physician cannot trust any model without explanation. A natural way to elucidate the knowledge embedded within neural network connections is to extract symbolic rules. However, producing rules from Multi Layer Perceptrons (MLPs) is a NP-hard problem [12]. Since the earliest work of Gallant on rule extraction from neural networks [11], many techniques have been introduced. In the nineties Andrews et al. introduced a taxonomy to characterize rule extraction techniques [1]. Later, Duch et al. published a survey article on this topic [10]. Finally, Diederich et al. published a book on techniques to extract symbolic rules from Support Vector Machines (SVMs) [9] and Barakat and Bradley reviewed a number of rule extraction techniques applied to SVMs [2].

More than twenty years ago Hansen and Salamon demonstrated that combining several neural networks in an ensemble can improve the predictive accuracy with respect to a single model [13]. Nevertheless, only a few authors started to extract rules from neural network ensembles. Bologna proposed the Discretized

© Springer International Publishing Switzerland 2015
I. Rojas et al. (Eds.): IWANN 2015, Part II, LNCS 9095, pp. 276–289, 2015.
DOI: 10.1007/978-3-319-19222-2_23

Interpretable Multi Layer Perceptron (DIMLP) to produce unordered symbolic rules from both single networks and ensembles [5]. With the DIMLP architecture rule extraction is performed by determining the precise location of axis-parallel discriminative hyperplanes. Zhou et al. introduced the REFNE algorithm (Rule Extraction from Neural Network Ensemble) [19], which utilizes the trained ensembles to generate instances and then extracts symbolic rules from those instances. Attributes are discretized during rule extraction and it also uses particular fidelity evaluation mechanisms. Moreover, rules have been limited to only three antecedents. More recently Hara and Hayashi proposed the two-MLP ensembles by using the "Recursive-Rule eXtraction" (Re-RX) algorithm [17] for data with mixed attributes [14]. Re-RX utilizes C4.5 decision trees and back-propagation to train the MLPs recursively. Here, the rule antecedents for discrete attributes are disjointed from those for continuous attributes. Subsequently Hayashi at al. presented the "three-MLP Ensemble" by the Re-RX algorithm [15].

In this work we introduce the Quantized Support Vector Machine (QSVM), which is a novel DIMLP network trained by an SVM learning algorithm. This special architecture makes it possible to apply the rule extraction algorithm introduced in [5]. Based on stratified 10-fold cross validation trials, rules are generated and compared from both QSVMs and DIMLP ensembles on five datasets. Finally, we compare the accuracy and complexity of rules extracted from QSVMs with those generated in another work [16]. In the following sections we present the DIMLP model for which QSVM represents a particular case, the experiments, followed by the conclusion.

2   The DIMLP Model

A symbolic rule is defined as: "if tests on antecedents are true then conclusion"; where "tests on antecedents" are in the form $x_i \leq v_i$ or $x_i \geq v_i$; with x_i as an attribute value and v_i as a real value. Generally, extracted rulesets are ordered or unordered. Ordered rules correspond to "if ... then ... else if ... " logical arrangement. Because of the presence of "else", a rule implicitly negates the antecedents of the previous rule. Thus, with the exception of the first rule all the other rules depend also on a number of hidden antecedents corresponding to the negation of the previous antecedents. Unordered rules correspond to "if ... then ... " structure. Contrary to ordered rules, a sample can activate more than one rule. Generally, unordered rulesets present more rules and antecedents, since all rule antecedents are explicitly displayed.

The extraction of symbolic rules is based on a special MLP architecture. We first introduced the *Interpretable Multi Layer Perceptron* (IMLP) [3]. Later we presented the DIMLP model [5] and finally we demonstrated how to generate symbolic rules from DIMLP ensembles. The main novelty of this work consists in making it possible to transform a single DIMLP into a particular SVM network. Thus, the training is achieved by a typical SVM learning algorithm [18] and rule extraction is carried out by the same method used for DIMLPs. This new model is denoted as *Quantized Support Vector Machine* (QSVM).

2.1   IMLP Networks

An IMLP differs from an MLP in the connectivity between the input layer and
the first hidden layer. Specifically, any hidden neuron receives only a connection
from an input neuron and the bias neuron, as shown in Figure 1.

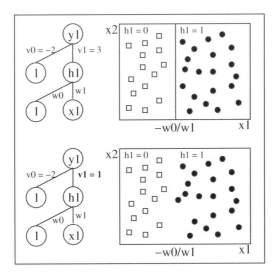

Fig. 1. Example of simple IMLP networks. Depending on weight values v_0 and v_1, at
the top a discriminative hyperplane is created, but not at the bottom.

In addition, the activation function in the hidden layer is a threshold function.
The output h_k of the kth neuron of the first hidden layer is:

$$h_k = \begin{cases} 1 \text{ if } \sum_l w_{kl}.x_l > 0 \\ 0 \text{ otherwise} \end{cases} \tag{1}$$

In the other layers the activation function is a sigmoid $\sigma(x)$:

$$\sigma(x) = \frac{1}{1 + \exp(-x)} \tag{2}$$

Note that the threshold activation function makes it possible to precisely locate
possible discriminative hyperplanes. Specifically, in Figure 1 assuming two different classes, the first is being selected when $y_1 > 0.5$ and the second with $y_1 \leq 0.5$.
Hence, a possible hyperplane split could be located in $-w_0/w_1$. However, this
hyperplane will result discriminative only when $v_1 > |v_0|$.

Another example is shown in figure 2. Weights v_0, v_1 and v_2 linearly separate
white squares and black circles, since the possible values of h_1 and h_2 define a
logical OR classification problem.

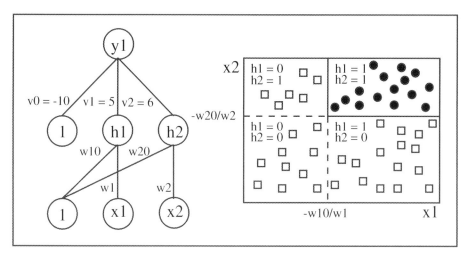

Fig. 2. An IMLP network creating two discriminative hyperplanes

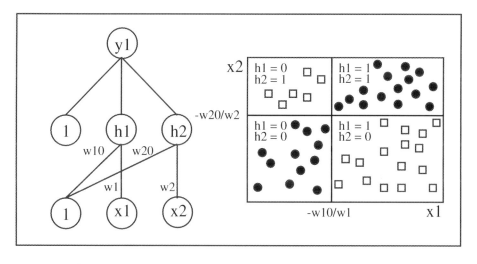

Fig. 3. An IMLP network unable to correctly classify a problem of two classes

Figure 3 depicts an IMLP that will not be able to learn the classification problem represented in the right part of the figure. The reason is that weights between h_1, h_2 and y_1 cannot create two discriminative hyperplanes, since the possible values of h_1 and h_2 defines a logical XOR problem.

2.2 DIMLP Networks and Ensembles

DIMLP networks represent a generalization of IMLPs. Essentially, they usually have two hidden layers, with the first one related to a stair activation function, instead of a threshold function. The stair function $S(x)$ is given as

$$\begin{cases} S(x) = \sigma(R_{min}) & \text{If } x < R_{min} \\ S(x) = \sigma(R_{min} + \left[d \cdot \frac{x - R_{min}}{R_{max} - R_{min}}\right] \cdot (\frac{R_{max} - R_{min}}{d})) & \text{If } R_{min} \leq x \leq R_{max} \quad (3) \\ S(x) = \sigma(R_{max}) & \text{If } x > R_{max} \end{cases}$$

where σ is the sigmoid function, d is the number of stairs and R_{min} and R_{max} forms a range of $d - 1$ stairs. Finally, "[]" denotes the integer part notation.

Each neuron of the first hidden layer virtually creates a number of virtual parallel hyperplanes that is equal to the number of stairs of its stair activation function. As a consequence, the rule extraction algorithm corresponds to a covering algorithm for which the goal is to determine whether a virtual hyperplane is virtual or effective. A distinctive feature of this rule extraction technique is that fidelity, which is the degree of matching between network classifications and rules' classifications is equal to 100%, with respect to the training set. Here we describe the general idea behind the rule extraction algorithm, since the details are provided in [6].

According to the 100% fidelity criterion, a decision tree is generated from the list of discriminative hyperplanes; then symbolic rules are extracted by simply following each path of the tree. At this point rules generally present too many antecedents; hence, a pruning process is carried out. Moreover, a "greedy" algorithm is performed in order to maximize the number of covered samples by modifying the thresholds of the antecedents. Overall, the computational complexity of the rule extraction algorithm is polynomial with respect to the number of stairs in the stair function, the number of inputs and the number of training samples.

Ensembles of DIMLP networks can be trained by bagging [7] or arcing [8]. Bagging and arcing are based on resampling techniques. On one hand, assuming a training set of size p, bagging selects for each classifier included in an ensemble p samples drawn with replacement from the original training set. Hence, for each DIMLP network many of the generated samples may be repeated while others may be left out.

On the other hand, arcing defines a probability with each example of the original training set. For each classifier the examples contained in the training set are selected according to these probabilities. Before learning all the training samples have the same probability to belong to a new training set $(= 1/p)$. Then, after the first classifier has been trained the probability of sample selection in a new training set is increased for all unlearned samples and decreased for the others.

Rule extraction from ensembles can still be performed, since an ensemble of DIMLP networks can be viewed as a single DIMLP network with one more hidden layer. For instance, in Figure 4 a "Transparent Box" corresponds to a single DIMLP that can be translated into symbolic rules. The linear combination is again a transparent box with one more layer of weights.

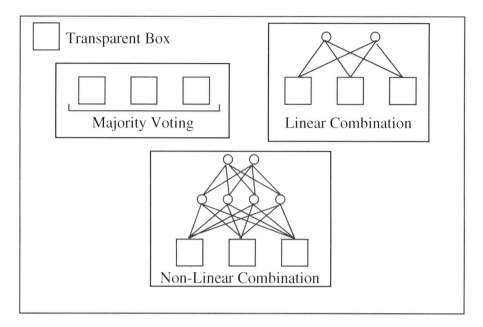

Fig. 4. Ensembling DIMLP networks by majority voting, linear combination and non-linear combination

2.3 QSVM

The classification decision function of an SVM model is given by

$$C(x) = sign(\sum_i \alpha_i y_i K(x_i, x) + b);$$
(4)

α_i and b being real values, $y_i \in \{-1, 1\}$ corresponding to the target values of the support vectors, and $K(x_i, x)$ representing a kernel function with x_i as the vector components of the support vectors. The following kernels are used:

- dot;
- polynomial;
- Gaussian.

Specifically, for the dot and polynomial cases we have:

$$K(x_i, x) = (x_i \cdot x)^d;$$
(5)

with $d = 1$ for the dot kernel and $d = 3$ for the polynomial kernel. The Gaussain kernel is:

$$K(x_i, x) = \exp(-\gamma ||x_i - x||^2);$$
(6)

with $\gamma > 0$, a parameter.

We define a Quantified Support Vector Machine (QSVM) as a DIMLP network with two hidden layers. The activation function of the neurons in the second hidden layer is related to the SVM kernel. For instance, with a dot kernel the corresponding activation function is the identity, while with a Gaussian kernel the activation function is Gaussian. The number of neurons in this layer is equal to the number of support vectors, with the incoming weight connections corresponding to the components of the support vectors.

Figure 5 presents a QSVM with a Gaussian activation function in the second hidden layer. The activation function of the output neuron (only for classification

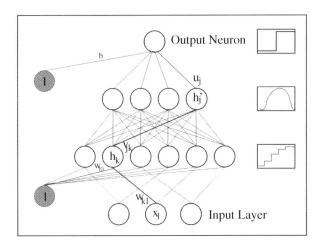

Fig. 5. A QSVM network

problems of two classes) is a threshold function. Finally, neurons in the first hidden layer use a stair function.

Weights between the second hidden layer and the output neuron denoted as u_j correspond to α_j coefficients in equation 4. Moreover, a weight between the first and second hidden layers denoted as v_{jk} corresponds to the jth component of the kth support vector. Finally, the role of neurons of the first hidden layer is to perform a normalization of the input attributes. Thus, during the training phase weights between the input layer and the first hidden layer remain unchanged. For clarity, let us assume that we have the same number of input neurons and hidden neurons in the first hidden layer. These weights are defined as:

- $w_{kl} = 1/\sigma_l$, with σ_l as the standard deviation of input variable l;
- $w_{l0} = -\mu_l/\sigma_l$, with μ_l as the average on the training set of the lth attribute.

According to these weight values, when the stair function is a threshold function, neurons of the first hidden layer are activated with values 0 or 1, depending on whether the input attributes are greater than their corresponding average values μ_l. With an arbitrary number of stairs the activations of the first hidden layer will correspond to discrete values between 0 and 1. During training, weights above the first hidden layer are modified according to the SVM training algorithm [18]. After training, since QSVM is also a DIMLP network, rules can be extracted by performing the DIMLP rule extraction algorithm.

3   Experiments

In the experiments we used five datasets representing classification problems of two classes. Table 1 illustrates their main characteristics in terms of number of samples, number of input features, type of features, and source[1]. All the results are based on ten repetitions of stratified 10-fold cross-validation trials with normalized attributes and default learning parameters.

Table 1. Datasets used in the experiments

Dataset Name	Number of Samples	Number of Inputs	Attribute Types	Ref.
Istambul Stock Exchange	536	8	real	UCI
Australian Credit Approval	690	14	bool, discr, real	UCI
Breast Cancer	683	9	discrete	UCI
Coronary Heart Disease	884	16	bool, real	[4]
Vertebral Column	310	6	real	UCI

3.1   Models and Learning Parameters

The following models were trained on the five datasets:

– DIMLP ensembles trained by bagging (DIMLP-B);
– DIMLP ensembles trained by arcing (DIMLP-A);
– QSVM with dot kernel (QSVM-L);
– QSVM with polynomial kernel of third degree (QSVM-P3);
– QSVM with Gaussian kernel (QSVM-G).

Learning parameters were assigned to default values for both DIMLP ensembles and QSVMs. For DIMLP ensembles the learning parameters are:

– the learning parameter η ($\eta = 0.1$);
– the momentum μ ($\mu = 0.6$);
– the *Flat Spot Elimination* ($FSE = 0.01$);
– the number of stairs q in the stair function ($q = 50$).

---

[1] In this column UCI designates the Machine Learning Repository at the University of California, Irvine (www.archive.ics.uci.edu/ml/).

Table 2. Average accuracy on the training set

Dataset Name	DIMLP-B	DIMLP-A	QSVM-L	QSVM-P3	QSVM-G
Istambul Stock Exchange	79.2 0.1	77.0 1.6	79.0 0.2	79.3 0.2	80.2 0.1
Australian Credit Approval	88.8 0.2	99.9 0.1	85.5 0.2	86.7 0.1	86.2 0.0
Breast Cancer	98.0 0.1	100.0 0.0	95.2 0.0	92.3 0.0	97.2 0.0
Coronary Heart Disease	95.9 0.1	100.0 0.0	87.1 0.1	88.3 0.1	87.8 0.1
Vertebral Column	88.2 0.2	95.2 0.7	86.0 0.1	86.8 0.3	85.6 0.2
AVERAGE	90.0	94.4	86.6	86.7	87.4

Table 3. Average predictive accuracy (on the testing sets)

Dataset Name	DIMLP-B	DIMLP-A	QSVM-L	QSVM-P3	QSVM-G
Istambul Stock Exchange	77.6 0.6	75.3 1.8	77.7 1.0	76.4 0.6	77.1 1.0
Australian Credit Approval	86.7 0.4	86.2 0.4	85.6 0.2	86.1 0.4	85.8 0.2
Breast Cancer	97.1 0.2	96.6 0.3	95.1 0.2	92.1 0.2	97.2 0.1
Coronary Heart Disease	93.1 0.4	94.6 0.6	86.3 0.3	87.5 0.2	86.9 0.2
Vertebral Column	85.5 0.8	84.2 1.1	84.9 0.4	85.0 0.9	84.4 0.7
AVERAGE	88.0	87.4	85.9	85.4	86.3

The default number of neurons in the first hidden layer is equal to the number of input neurons and the number of neurons in the second hidden layer is empirically defined in order to obtain a number of weight connections that is less than the number of training samples. Finally, the default number of DIMLPs in an ensemble is equal to 25, since it has been observed many times that for bagging and arcing the most substantial improvement in accuracy is achieved with the first 25 networks.

In QSVMs the default learning parameters are those defined in the *libSVM* library[2]. Here our goal was not to optimize the predictive accuracy of the models, but just to use default configurations and to determine how would vary accuracy and complexity of the models.

3.2   Results

For all the models table 2 shows the average accuracy on the training set, while table 3 illustrates the average predictive accuracy.

---

[2] This software is available at http://www.csie.ntu.edu.tw/~cjlin/libsvm/

Table 4. Average predictive accuracy of the extracted rules

Dataset Name	DIMLP-B	DIMLP-A	QSVM-L	QSVM-P3	QSVM-G
Istambul Stock Exchange	77.2 0.8	75.1 1.9	77.6 1.0	76.8 0.6	77.1 0.6
Australian Credit Approval	86.5 0.5	84.9 0.7	85.6 0.2	85.7 0.6	85.6 0.3
Breast Cancer	96.5 0.3	96.2 0.3	95.1 0.4	92.0 0.5	96.7 0.4
Coronary Heart Disease	91.6 0.5	92.3 0.6	85.4 0.4	87.0 0.5	86.5 0.5
Vertebral Column	84.0 0.6	82.7 1.1	83.9 0.8	84.0 0.7	83.5 1.2
AVERAGE	87.2	86.2	85.5	85.1	85.9

Table 5. Average fidelity of the extracted rules

Dataset Name	DIMLP-B	DIMLP-A	QSVM-L	QSVM-P3	QSVM-G
Istambul Stock Exchange	96.4 0.9	96.7 0.7	94.4 0.8	96.1 0.6	95.9 0.9
Australian Credit Approval	97.9 0.5	96 0.5	100.0 0.0	98.1 0.6	99.3 0.2
Breast Cancer	98.8 0.4	98.9 0.3	98.9 0.4	98.1 0.3	98.7 0.4
Coronary Heart Disease	97.1 0.4	96.2 0.5	97.5 0.5	97.3 0.6	97.8 0.4
Vertebral Column	96.1 0.9	93.3 1.8	96.2 0.7	95.9 1.2	95.6 0.7
AVERAGE	97.3	96.2	97.4	97.1	97.5

Table 4 shows the average predictive accuracy of the extracted rules. Note that the classification decision was determined by the neural network model when a testing sample was not covered by any rule. Moreover, in case of conflicting rules (i.e. rules of two different classes), the selected class is again the one determined by the model. Table 5 shows the average fidelity of the extracted rules.

Table 6 shows the average predictive accuracy when the models and the extracted rules agree. For instance, for DIMLP-B on the "Vertebral Column" problem and with respect to the fidelity rate of 96.1% the average predictive accuracy is equal to 86.2 %.

Table 7 shows the complexity of extracted rules in terms of average number of rules and number of antecedents per rule. Note that the "AVERAGE" row represents the average product of the number of rules times the number of antecedents.

3.3 An Example of Extracted Ruleset

We illustrate in Figure 6 rules extracted from the Breast Cancer dataset from a QSVM-G network. In this problem we have 9 attributes (x_1 to x_9 with discrete values from 1 to 10. "Class1" is "Benign" and "Class2" is "Malignant".

Table 6. Average predictive accuracy when rules and models agree

Dataset Name	DIMLP-B	DIMLP-A	QSVM-L	QSVM-P3	QSVM-G
Istambul Stock Exchange	78.4 0.6	76.1 1.8	78.7 1.0	77.7 0.4	78.3 0.9
Australian Credit Approval	87.4 0.4	87.1 0.3	85.6 0.2	86.6 0.5	86.0 0.2
Breast Cancer	97.4 0.2	96.9 0.3	95.6 0.3	92.9 0.3	97.5 0.2
Coronary Heart Disease	93.6 0.4	95.2 0.3	86.8 0.3	88.3 0.3	87.6 0.3
Vertebral Column	86.2 0.7	85.8 1.0	85.8 0.6	86.0 0.9	85.5 0.9
AVERAGE	88.6	88.2	86.5	86.3	87.0

Table 7. Average complexity of the extracted rules (number of rules and number of antecedents per rule)

Dataset Name	DIMLP-B	DIMLP-A	QSVM-L	QSVM-P3	QSVM-G
Istambul Stock Exchange	21.4 2.8	23.4 2.9	30.0 3.0	26.6 3.1	26.1 3.1
Australian Credit Approval	22.7 3.7	82.7 5.1	2 1	20.5 3.7	8.3 2.6
Breast Cancer	12.5 2.7	25.2 3.6	13.2 3.0	22.7 3.5	11.6 2.9
Coronary Heart Disease	44.8 4.0	71.6 4.6	41.9 3.8	42.0 3.7	34.7 3.6
Vertebral Column	16.7 2.7	27.8 3.2	17.1 2.8	18.3 2.8	17.9 2.8
AVERAGE	80.4	199.7	67.7	88.9	62.2

Figure 7 shows the accuracy of the rules on the training set. The first column of numbers indicates the number of covered training examples, then is showed the number of correct answers, the number of wrong answers and the accuracy.

Finally, we illustrate the predictive accuracy of the ruleset in Figure 8.

3.4   Discussion

The extracted rulesets were a bit less accurate than the models themselves, on average. A possible reason is that hyper-rectangles try to approximate the complex manifolds related to the classification problems, but their precision is limited with respect to the better power of expression of the corresponding models. In two out of five classification problems the highest predictive accuracy was provided by SVMs (Instanbul Stock Exchange and Breast Cancer).

Globally, the fidelity of the extracted rules was always above 94%. On four out of five datasets the most complex rulesets in terms of number of extracted rules and number of antecedents per rule were obtained by DIMLP ensembles trained by arcing (DIMLP-A). Among the rulesets related to QSVMs, the Gaussian

Rule 1: $(x1 < 9.99462)(x3 < 3.98945)(x6 < 6.92955)(x7 < 6.95644)(x8 < 9.94272)$ Class1
Rule 2: $(x1 < 6.99821)(x2 < 5.96902)(x5 < 4.96611)(x6 < 4.93843)(x8 < 9.94272)$ Class1
Rule 3: $(x3 > 2.97063)(x6 > 4.93843)$ Class2
Rule 4: $(x1 > 4.96292)(x2 > 3.9883)$ Class2
Rule 5: $(x2 > 4.97866)$ Class2
Rule 6: $(x1 > 2.98416)(x6 > 6.92955)$ Class2
Rule 7: $(x1 > 5.98057)(x2 > 2.99794)(x3 > 3.98945)$ Class2
Rule 8: $(x3 > 2.97063)(x5 > 4.96611)$ Class2
Rule 9: $(x8 > 8.964)$ Class2
Rule 10: $(x2 < 2.99794)(x3 > 3.98945)(x5 < 4.96611)$ Class1
Rule 11: $(x1 < 2.98416)(x3 < 4.94833)(x6 > 9.9531)$ Class1
Rule 12: $(x1 > 9.99462)(x2 < 2.99794)(x6 < 7.96198)$ Class1

Fig. 6. An example of ruleset extracted from the breast cancer dataset. "Class1" indicates Benign, while "Class2" is Malignant.

Rule 1:	370	365	5	0.986486	Class = 1
Rule 2:	365	361	4	0.989041	Class = 1
Rule 3:	170	163	7	0.958824	Class = 2
Rule 4:	164	158	6	0.963415	Class = 2
Rule 5:	156	153	3	0.980769	Class = 2
Rule 6:	150	145	5	0.966667	Class = 2
Rule 7:	127	124	3	0.976378	Class = 2
Rule 8:	122	116	6	0.950820	Class = 2
Rule 9:	65	65	0	1	Class = 2
Rule 10:	7	6	1	0.857143	Class = 1
Rule 11:	1	1	0	1	Class = 1
Rule 12:	1	0	1	0	Class = 1

Accuracy of rules on training set = 0.971619

Fig. 7. Accuracy of each rule on a training set of the breast cancer problem (see text)

kernel provided the less complex rules in four out five datasets. Finally, the predictive accuracy of the extracted rules when networks and rules agreed was always better than that obtained by the models (cf. tables 3 and 6).

Based on stratified 10-fold cross-validation trials, Nunez et al. reported an average predictive accuracy of the extracted rulesets equal to 96.5% for the Breast Cancer dataset [16]. Fidelity was equal to 98.4% and the average number of rules was equal to 9. These results are very close to ours with respect to SVM-G (96.7%; 98.7%; and 11.6). On the Australian Credit Approval dataset the same authors generated rules with average predictive accuracy equal to 86.3%, average fidelity equal to 93.2% and 21.6 extracted rules, on average [16]. Our corresponding values with QSVM-P3 are very similar: 86.5%; 97.9%; and 22.7. Note however that in the technique introduced by Nunez et al. the number of antecedents per rule is always equal to the number of dataset attributes. Hence, these numbers are equal to 9 in the Breast Cancer problem and 14 in the

Rule 1: 54 54 0   1   Class = 1
Rule 2: 56 56 0   1   Class = 1
Rule 3: 19 19 0   1   Class = 2
Rule 4: 20 19 1   0.950000 Class = 2
Rule 5: 19 19 0   1   Class = 2
Rule 6: 13 13 0   1   Class = 2
Rule 7: 17 16 1   0.941176 Class = 2
Rule 8: 14 14 0   1   Class = 2
Rule 9: 10 10 0   1   Class = 2
Rule 10: 0 0 0   Class = 1
Rule 11: 1 0 1   0   Class = 1
Rule 12: 0 0 0   Class = 1

Accuracy on testing set = 0.988095
Fidelity(83/84) = 0.988095
Accuracy when rules and network agree (82/83) = 0.987952
Number of default rule activations (network classification) = 0

Fig. 8. Accuracy of each rule on a testing set of the breast cancer problem (see text)

Australian Credit dataset. The corresponding values obtained for QSVMs are 2.9 and 3.7, respectively. Thus, our extracted rules are more comprehensible, on average, since their complexity is lower.

4   Conclusion

In this work we applied ensembles of multi layer perceptrons and support vector machines to five classification problems. Unordered symbolic rules of high fidelity were extracted and compared to those obtained with another rule extraction technique. The obtained results were very similar on two datasets for the average predictive accuracy. However, we obtained a significant lower number of antecedents per rule. This is very encouraging, since all the results were obtained with default learning parameters, which is easy to accomplish. Although the DIMLP model is a particular MLP architecture it can also be used to generate rules from ensembles of decision trees. For instance, it will be interesting in a future work to characterize the complexity of boosted shallow trees.

References

1. Andrews, R., Diederich, J., Tickle, A.B.: Survey and critique of techniques for extracting rules from trained artificial neural networks. Knowledge-Based Systems **8**(6), 373–389 (1995)
2. Barakat, N., Bradley, A.P.: Rule extraction from support vector machines: a review. Neurocomputing **74**(1), 178–190 (2010)

3. Bologna, G.: Rule extraction from the IMLP neural network: a comparative study. In: Proceedings of the Workshop of Rule Extraction from Trained Artificial Neural Networks (after the Neural Information Processing Conference) (1996)
4. Bologna, G., Rida, A., Pellegrini, C.: Intelligent assistance for coronary heart disease diagnosis: a comparison study. In: Keravnou, E., Garbay, C., Baud, R., Wyatt, J. (eds.) AIME 1997. LNCS, vol. 1211, pp. 199–210. Springer, Heidelberg (1997)
5. Bologna, G.: A study on rule extraction from several combined neural networks. International Journal of Neural Systems **11**(3), 247–255 (2001)
6. Bologna, G.: Is it worth generating rules from neural network ensembles? J. of Applied Logic **2**, 325–348 (2004)
7. Breiman, L.: Bagging predictors. Machine Learning **26**, 123–40 (1996)
8. Breiman, L.: Bias, variance, and arcing classifiers. California: Technical Report, Statistics Department, University of California (1996)
9. Diederich, J. (ed.): Rule extraction from support vector machines, vol. 80. Springer Science and Business Media (2008)
10. Duch, W., Rafal, A., Grabczewski, K.: A new methodology of extraction, optimization and application of crisp and fuzzy logical rules. IEEE Trans. Neural Networks **12**(2), 277–306 (2001)
11. Gallant, S.I.: Connectionist expert systems. Commun. ACM **31**(2), 152–169 (1988)
12. Golea, M.: On the complexity of rule extraction from neural networks and network querying. In: Proceedings of the Rule Extraction From Trained Artificial Neural Networks Workshop, Society For the Study of Artificial Intelligence and Simulation of Behavior Workshop Series (AISB 1996), University of Sussex, Brighton, UK, pp. 51–59, April 1996
13. Hansen, L.K., Salamon, P.: Neural network ensembles. IEEE Trans. Pattern Anal. **12**(10), 993–1001 (1990)
14. Hara, A., Hayashi, Y.: Ensemble neural network rule extraction using Re-RX algorithm. In: Proc. of WCCI (IJCNN) 2012, Brisbane, Australia, June 10–15, pp. 604–609 (2012)
15. Hayashi, Y., Sato, R., Mitra, S.: A new approach to three ensemble neural network rule extraction using recursive-rule eXtraction algorithm. In: Proc. of International Joint Conference on Neural Networks (IJCNN), Dallas, pp. 835–841 (2013)
16. Nunez, H., Angulo, C., Catala, A.: Rule-based learning systems for support vector machines. Neural Processing Letters **24**(1), 1–18 (2006)
17. Setiono, R., Baesens, B., Mues, C.: Recursive neural network rule extraction for data with mixed attributes. IEEE Trans. Neural Netw. **19**(2), 299–307 (2008)
18. Vapnik, V.N.: The Nature of Statistical Learning Theory. Springer (1995). ISBN 0-387-98780-0
19. Zhou, Z.-H., Yuan, J., Shi-Fu, C.: Extracting symbolic rules from trained neural network ensembles. AI Communications **16**, 3–15 (2003)

Multiwindow Fusion for Wearable Activity Recognition

Oresti Baños[1]([✉]), Juan-Manuel Galvez[2], Miguel Damas[2], Alberto Guillén[2],
Luis-Javier Herrera[2], Hector Pomares[2], Ignacio Rojas[2], Claudia Villalonga[2],
Choong Seon Hong[1], and Sungyoung Lee[1]

[1] Department of Computer Engineering, Kyung Hee University, Yongin-si, Korea
{oresti,cshong,sylee}@khu.ac.kr
[2] Department of Computer Architecture and Computer Technology,
University of Granada, Granada, Spain
{jonas,mdamas,aguillen,jherrera,hector,irojas,cvillalonga}@ugr.es

Abstract. The recognition of human activity has been extensively investigated in the last decades. Typically, wearable sensors are used to register body motion signals that are analyzed by following a set of signal processing and machine learning steps to recognize the activity performed by the user. One of the most important steps refers to the signal segmentation, which is mainly performed through windowing approaches. In fact, it has been proved that the choice of window size directly conditions the performance of the recognition system. Thus, instead of limiting to a specific window configuration, this work proposes the use of multiple recognition systems operating on multiple window sizes. The suggested model employs a weighted decision fusion mechanism to fairly leverage the potential yielded by each recognition system based on the target activity set. This novel technique is benchmarked on a well-known activity recognition dataset. The obtained results show a significant improvement in terms of performance with respect to common systems operating on a single window size.

Keywords: Activity recognition · Segmentation · Windowing · Wearable sensors · Ensemble methods · Data fusion

1 Introduction

The identification of human behavior based on body-worn sensors, also known as wearable activity recognition, has attracted very much attention during the last years. Wearable activity recognition systems have been proven of particular interest, for example, to promote healthier lifestyles [1,2,26], detect anomalous behaviors [20,23] or track on conditions [16]. A set of steps combining signal processing and machine learning techniques are normally used in the activity recognition process. Concretely, one or various sensors are typically placed on limbs and trunk to register and translate human body motion into digital signals

© Springer International Publishing Switzerland 2015
I. Rojas et al. (Eds.): IWANN 2015, Part II, LNCS 9095, pp. 290–297, 2015.
DOI: 10.1007/978-3-319-19222-2_24

representing the magnitude measured, normally acceleration. The registered signals are sometimes filtered when these are found to be disturbed by electronic noise or other type of artifacts [17]. To capture the dynamics of the movement, the signals are subsequently partitioned in segments or data windows [8]. Then a feature extraction process is performed on each data window to provide a handler representation of the signals for the pattern recognition stage. Diverse heuristics [18], time-frequency domain [19,22] and other sophisticated mathematical and statistical functions [4] are commonly used to that end. In some cases, a feature selector is used to reduce redundancy among features as well as to minimize dimensionality [21]. The resulting feature vector is provided as input of a classifier, which ultimately yields the recognized activity or class to one of the considered for the particular application. All these steps, commonly referred as to activity recognition chain, are extensively reviewed in [10].

Although all stages of the activity recognition process are undoubtedly important, a recent work [8] showed the particular impact of the segmentation phase on the accuracy of the recognition models. Amongst other findings, this work showed the existing relation among activity categories and involved body parts with the window size utilized during the segmentation process. As a result, specific design figures are proposed, which in principle allow developers to set a certain window size value to optimize the recognition system capabilities. Nevertheless, these values are very specific to each application domain; thus, no particular window size may exist for systems intended to recognize multiple diverse activities. In that vein, this paper proposes the use of fusion mechanisms to benefit from the utilization of several window sizes instead of restricting to a single one. Fusion strategies have been already used in previous activity recognition systems for diverse purposes, such as dealing with sensor displacement [6,9], anomalies [3] and power management [27]. This work presents an innovative multiwindow fusion technique that weights and combines the decisions provided by multiple activity recognizers configured to operate on different windows sizes of the same input data. The rest of the paper is organized as follows. Section 2 describes the multiwindow fusion method. Section 3 presents and discusses the results obtained after benchmarking the proposed method on a well-known activity recognition dataset. Final conclusions are summarized in Section 4.

2   Multiwindow Fusion

As stated before, the recognition of activities of diverse characteristics potentially require the use of various levels of segmentation. Therefore, the model proposed here consists in the combination of multiple activity recognition chains, every one utilizing a different window size configuration. Each of these chains builds on the same input signals, and for the sake of simplicity, all are considered to use similar feature extraction and classification procedures. For practical reasons, the selected window sizes should be divisors of the largest one among considered, which is defined according to the particular needs posed by the target activity set and system recognition period. The key challenge of this approach consists in the intelligent aggregation of the decisions, i.e., recognized activities, delivered by each chain. To that end, a two-step fusion process is here suggested. First, the decisions provided by each individual activity recognizer are locally weighted and aggregated to yield a sole recognized activity per chain. The activities identified for each chain are then combined in a second stage to eventually deliver a unique recognized activity. The complete structure of the proposed model is depicted in Figure 1, while its mathematical foundation is described in the following.

Let us consider a problem with N classes or activities, $n = 1, ..., N$. Given a set of raw, u, or preprocessed, p, sensor data, these are segmented by using Q different window sizes, $\{W_1, ..., W_{Q-1}, W_Q\}$, with $W_Q | W_{Q-1} | ... | W_1$ divisors of W_Q, and W_Q formally representing the system recognition period. This leads to the creation of Q independent recognition chains, in which every data window of size W_Q, i.e., s^{W_Q}, is split into W_Q/W_k segments of size W_k, i.e., $\{s_1^{W_k}, ..., s_i^{W_k}, ..., s_{W_Q/W_k}^{W_k}\}$, for all $k = 1, ..., Q$ and $i = 1, ..., W_Q/W_k$. Each segment $s_i^{W_k}$ is transformed into features, $f(s_i^{W_k})$, which are input to each respective classifier, yielding a recognized activity or class, $c_i^{W_k}$.

At this point the multiwindow fusion technique is employed. First, the decisions of each individual classifier are weighted and averaged across all segments and for all classes:

$$WD_n^{W_k} = \sum_{i=1}^{W_Q/W_k} \lambda_n^{W_k} \quad \forall\, c_i^{W_k} = n \qquad (1)$$

where the weight factor $\lambda_n^{W_k}$ represents the capabilities of the classifier k, that operates on data windows of size W_k, for the recognition of the activity or class n. This factor is different for each class and window size, and can be calculated from a prior evaluation of the performance of each respective activity recognition chain, similarly as it is proposed in [5]. Now, the class c^{W_k} predicted for each classifier after fusion is the class n for which $WD_n^{W_k}$ is maximized:

$$c^{W_k} = \underset{n}{\operatorname{argmax}} \left(WD_n^{W_k} \right) \qquad (2)$$

This process is repeated in a second level by weighting and averaging the decisions obtained in the previous fusion step for each respective window size:

$$WD_n = \sum_{k=1}^{Q} \lambda_n^{W_k} \quad \forall\, c^{W_k} = n \qquad (3)$$

The eventual recognized class is defined as the one obtaining the highest weighted sum:

$$c = \underset{n}{\operatorname{argmax}} \left(WD_n \right) \qquad (4)$$

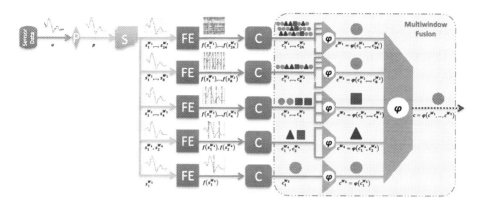

Fig. 1. Multiwindow fusion schema (example for Q=5 different window sizes). The raw *sensor data*, u, is *preprocessed*, p, and *segmented* into data windows, $s_i^{W_k}$, of size W_k, thus defining Q independent recognition chains. For each chain k and window i, a set of *features* are *extracted*, $f(s_i^{W_k})$, which are input to each respective classifier, yielding a recognized activity or class, $c_i^{W_k}$. The recognized classes are weighted and fused in a first stage to predict the most likely activity for each chain, $c_i^{W_k}$. The classes predicted after fusion for each chain are again weighted and fused to deliver the eventual recognized activity.

3   Results and Discussion

3.1   Experimental Setup

One of the most complete available activity recognition datasets [7] is used for evaluation. This dataset comprises motion data, namely, acceleration, rate of turn and magnetic field orientation, recorded for 17 volunteers while performing 33 fitness activities. A set of nine inertial sensors attached to different parts of their bodies was used for the motion recording. From all measured magnitudes only the acceleration data is here considered since this demonstrates as the most prevalent sensor modality in previous activity recognition contributions. The potential of this dataset stems from the number of considered activities, diversity of body parts involved, as well as the variety in intensity and dynamicity of the actions. Moreover, all the recordings were collected in an out-of-lab environment with no constraints on the way the activities must be executed.

The activity recognition models devised for evaluation are described next. No preprocessing of the data is applied to avoid the removal of relevant information. This is normal practice when the activities are of a diverse nature. Five window sizes used in previous works are considered for study, respectively, $W_1 = 0.25$, $W_2 = 0.75$, $W_3 = 1.5$, $W_4 = 3$ and $W_5 = 6$, all in seconds. Mean and standard deviation are used for the feature extraction, given their discrimination potential and ease of interpretation in the acceleration domain [13,14]. Four well-known machine learning techniques widely utilized in previous activity recognition problems are considered for classification, namely, C4.5 decision trees

(DT, [12]), k-nearest neighbors (KNN, [11]), naive Bayes (NB, [25]) and nearest centroid classifier (NCC, [15]). The k-value for the KNN model is particularly set to three as it has been shown to provide good results in related works. The $\lambda_n^{W_k}$ weights used in the fusion process correspond to the *F-score* [24] values obtained in [8] from the analysis of similar recognition systems operating on the window sizes and activities considered in this experiment. The evaluation of the multiwindow fusion models is performed through a ten-fold random-partitioning cross validation process applied across all subjects and activities. The process is repeated 100 times for each method to ensure statistical robustness.

3.2   Multiwindow Fusion Evaluation

The results obtained for the multiwindow fusion process after assessment of all possible combinations of the selected window sizes are presented in Table 1. No fusion is explicitly performed for the single-window-based recognition models; thus, the results presented for this case refer to the performance obtained at the classification level, i.e., before fusion.

In broad strokes, the use of multiple window sizes certainly improves the recognition capabilities of the considered systems. This result is observed for all classification paradigms. For example, an enhancement of more than 7% is attained when using the combination $W_1W_2W_4$ and DT with respect to the best results obtained by using a single window size, here for W_3. More modest improvements, around 2%, are achieved for NB, NCC, and KNN in similar conditions. The differences are more striking when compared with the worst performing single-window-based recognition models, with improvements of up to 30%.

Another fact to be noted corresponds to the number of windows required for improving the performance of the recognition system. Best results are not necessarily obtained for the combination that involves the highest number of windows. Conversely, in some cases such as for NCC and KNN, the combination of simply two windows turns to be enough to neatly improve the recognition capabilities of the system. This demonstrates the potential of the fusion mechanism even for small sets of decision makers.

As it may be apparent, the use of multiple windows translates into a higher computation complexity, therefore might not be justified under some circumstances or not be recommended when the improvement is negligible. However, in some cases it is observed that the use of multiple windows can actually reduce the recognition time, a key characteristic in applications that require a fast response (e.g., fall detector). This is the case, for example, of the combination W_1W_4 in NB, which enhances the accuracy with respect to the best single-window-based recognition system, W_5, thus permitting to reduce the recognition period from 6s to 3s. The importance of this effect is also observed for the case of W_1W_2 in KNN, which improves the performance of W_4 while reducing the recognition time from 3s to 0.75s.

Finally, it is worth noting that the combination of two or more window sizes generally translates into a recognition performance greater or equal to the one

Table 1. Multiwindow fusion performance ($F - score$) for all possible combinations of considered window sizes ($W_1 = 0.25s$, $W_2 = 0.75s$, $W_3 = 1.5s$, $W_4 = 3s$ and $W_5 = 6s$) and diverse classification paradigms (DT, NB, NCC, KNN)

Combined window sizes	DT	NB	NCC	KNN
W_1	0.835	0.702	0.596	0.976
W_2	0.879	0.868	0.807	0.979
W_3	0.895	0.900	0.864	0.981
W_4	0.886	0.908	0.873	0.984
W_5	0.869	0.910	0.870	0.942
W_1W_2	0.878	0.855	0.760	0.991
W_1W_3	0.915	0.905	0.856	**0.996**
W_1W_4	0.920	0.922	0.870	0.976
W_1W_5	0.915	0.917	0.867	0.967
W_2W_3	0.877	0.905	0.858	0.991
W_2W_4	0.910	0.925	0.878	0.981
W_2W_5	0.917	0.918	0.866	0.958
W_3W_4	0.876	0.925	**0.881**	0.976
W_3W_5	0.893	0.922	0.876	0.954
W_4W_5	0.861	0.923	0.878	0.952
$W_1W_2W_3$	0.954	0.893	0.832	0.995
$W_1W_2W_4$	**0.968**	0.916	0.855	0.990
$W_1W_2W_5$	0.960	0.911	0.855	0.968
$W_1W_3W_4$	0.956	0.927	0.880	0.989
$W_1W_3W_5$	0.956	0.922	0.876	0.967
$W_1W_4W_5$	0.936	0.926	0.878	0.966
$W_2W_3W_4$	0.945	**0.928**	0.880	0.989
$W_2W_3W_5$	0.944	0.923	0.876	0.966
$W_2W_4W_5$	0.928	0.924	0.873	0.966
$W_3W_4W_5$	0.925	0.926	0.878	0.959
$W_1W_2W_3W_4$	0.967	0.926	0.879	0.989
$W_1W_2W_3W_5$	0.961	0.923	0.873	0.967
$W_1W_2W_4W_5$	0.955	0.926	0.874	0.967
$W_1W_3W_4W_5$	0.953	0.927	0.878	0.964
$W_2W_3W_4W_5$	0.945	0.924	0.878	0.962
$W_1W_2W_3W_4W_5$	0.960	0.924	0.878	0.968

of best characteristics among considered. This fact is of special importance since it proves the stability and consistency of the proposed fusion mechanism.

4   Conclusions

The choice of window size used in typical activity recognition applications is highly coupled to the particular characteristics of the activities to be recognized. Previous works proved that a single window size value can be considered in recognition systems devised for a very specific domain including a few similar activities. However, no clear value can be determined for problems involving several activities of a more diverse nature. To overcome this limitation, the simultaneous use of multiple window sizes is here suggested. Concretely, this work proposes a novel multiwindow fusion technique that weights and combines the decisions provided by multiple activity recognizers configured to operate on different windows sizes of the same input data. The proposed approach is shown to significantly outperform classical single-window-based recognition models. Moreover, the performed evaluation also shows that using several windows sizes not necessarily translates into best results, but that considering a few ones might be enough for obtaining a highly accurate recognition system.

Acknowledgments. This work was partially supported by the Industrial Core Technology Development Program, funded by the Korean Ministry of Trade, Industry and Energy, under grant number #10049079. This work was also funded by the Junta de Andalucia Project P12-TIC-2082.

References

1. Alshurafa, N., Xu, W., Liu, J.J., Huang, M.-C., Mortazavi, B., Roberts, C.K., Sarrafzadeh, M.: Designing a robust activity recognition framework for health and exergaming using wearable sensors. IEEE Journal of Biomedical and Health Informatics **18**(5), 1636–1646 (2014)
2. Banos, O., Bilal-Amin, M., Ali-Khan, W., Afzel, M., Ali, T., Kang, B.-H., Lee, S.: Mining minds: an innovative framework for personalized health and wellness support. In: Int. Conf. on Pervasive Computing Technologies for Healthcare (2015)
3. Banos, O., Damas, M., Guillen, A., Herrera, L.-J., Pomares, H., Rojas, I., Villalonga, C.: Multi-sensor fusion based on asymmetric decision weighting for robust activity recognition. Neural Processing Letters, 1–22 (2014)
4. Banos, O., Damas, M., Pomares, H., Prieto, A., Rojas, I.: Daily living activity recognition based on statistical feature quality group selection. Expert Systems with Applications **39**(9), 8013–8021 (2012)
5. Banos, O., Damas, M., Pomares, H., Rojas, F., Delgado-Marquez, B., Valenzuela, O.: Human activity recognition based on a sensor weighting hierarchical classifier. Soft Computing **17**, 333–343 (2013)
6. Banos, O., Damas, M., Pomares, H., Rojas, I.: On the use of sensor fusion to reduce the impact of rotational and additive noise in human activity recognition. Sensors **12**(6), 8039–8054 (2012)
7. Banos, O., Damas, M., Pomares, H., Rojas, I., Toth, M.A., Amft, O.: A benchmark dataset to evaluate sensor displacement in activity recognition. In: Proceedings of the ACM Conference on Ubiquitous Computing, pp. 1026–1035 (2012)
8. Banos, O., Galvez, J.-M., Damas, M., Pomares, H., Rojas, I.: Window size impact in human activity recognition. Sensors **14**(4), 6474–6499 (2014)
9. Banos, O., Toth, M.A., Damas, M., Pomares, H., Rojas, I.: Dealing with the effects of sensor displacement in wearable activity recognition. Sensors **14**(6), 9995–10023 (2014)
10. Bulling, A., Blanke, U., Schiele, B.: A tutorial on human activity recognition using body-worn inertial sensors. ACM Comput. Surv. **46**(3), 33:1–33:33 (2014)

11. Cover, T., Hart, P.: Nearest neighbor pattern classification. IEEE Transactions on Information Theory **13**(1), 21–27 (1967)
12. Duda, R.O., Hart, P.E., Stork, D.G.: Pattern Classification, 2nd edn. Wiley-Interscience (2000)
13. Figo, D., Diniz, P.C., Ferreira, D.R., Cardoso, J.M.P.: Preprocessing techniques for context recognition from accelerometer data. Personal and Ubiquitous Computing **14**(7), 645–662 (2010)
14. Kwapisz, J.R., Weiss, G.M., Moore, S.A.: Activity recognition using cell phone accelerometers. Conference on Knowledge Discovery and Data Mining **12**(2), 74–82 (2011)
15. Lam, W., Keung, C.-K., Ling, C.X.: Learning good prototypes for classification using filtering and abstraction of instances. Pattern Recognition **35**(7), 1491–1506 (2002)
16. Laudanski, A., Brouwer, B., Li, Q.: Activity classification in persons with stroke based on frequency features. Medical Engineering & Physics **37**(2), 180–186 (2015)
17. Mannini, A., Intille, S.S., Rosenberger, M., Sabatini, A.M., Haskell, W.: Activity recognition using a single accelerometer placed at the wrist or ankle. Medicine and Science in Sports and Exercise **45**(11), 2193–2203 (2013)
18. Mathie, M.J., Coster, A.C.F., Lovell, N.H., Celler, B.G.: Accelerometry: providing an integrated, practical method for long-term, ambulatory monitoring of human movement. Physiological Measurement **25**(2), 1–20 (2004)
19. Maurer, U., Smailagic, A., Siewiorek, D.P., Deisher, M.: Activity recognition and monitoring using multiple sensors on different body positions. In: International Workshop on Wearable and Implantable Body Sensor Networks, pp. 113–116 (2006)
20. Mazilu, S., Blanke, U., Hardegger, M., Tröster, G., Gazit, E., Hausdorff, J.M.: Gaitassist: a daily-life support and training system for parkinson's disease patients with freezing of gait. In: Proceedings of the SIGCHI Conference on Human Factors in Computing Systems, pp. 2531–2540 (2014)
21. Pirttikangas, S., Fujinami, K., Nakajima, T.: Feature selection and activity recognition from wearable sensors. In: Youn, H.Y., Kim, M., Morikawa, H. (eds.) UCS 2006. LNCS, vol. 4239, pp. 516–527. Springer, Heidelberg (2006)
22. Ravi, N., Mysore, P., Littman, M.L.: Activity recognition from accelerometer data. In: Proceedings of the Conference on Innovative Applications of Artificial Intelligence, pp. 1541–1546 (2005)
23. Sama, A., Perez-Lopez, C., Romagosa, J., Rodriguez-Martin, D., Catala, A., Cabestany, J., Perez-Martinez, D.A., Rodriguez-Molinero, A.: Dyskinesia and motor state detection in parkinson's disease patients with a single movement sensor. In: Annual International Conference of the IEEE Engineering in Medicine and Biology Society, pp. 1194–1197 (2012)
24. Sokolova, M., Lapalme, G.: A systematic analysis of performance measures for classification tasks. Information Processing & Management **45**(4), 427–437 (2009)
25. Theodoridis, S., Koutroumbas, K.: Pattern Recognition, 4th edn. Academic Press (2008)
26. Weiss, G.M., Lockhart, J.W., Pulickal, T.T., McHugh, P.T., Ronan, I.H., Timko, J.L.: Actitracker: a smartphone-based activity recognition system for improving health and well-being. SIGKDD Exploration Newsletter (2014)
27. Zappi, P., Roggen, D., Farella, E., Tröster, G., Benini, L.: Network-level power-performance trade-off in wearable activity recognition: A dynamic sensor selection approach. ACM Trans. Embed. Comput. Syst. **11**(3), 68:1–68:30 (2012)

Ontological Sensor Selection
for Wearable Activity Recognition

Claudia Villalonga[1]([✉]), Oresti Baños[2], Hector Pomares[1], and Ignacio Rojas[1]

[1] Research Center for Information and Communications Technologies
of the University of Granada, C/Periodista Rafael Gomez Montero 2,
Granada, Spain
cvillalonga@correo.ugr.es, {hector,irojas}@ugr.es
[2] Department of Computer Engineering, Kyung Hee University, Seoul, Korea
oresti@oslab.khu.ac.kr

Abstract. Wearable activity recognition has attracted very much attention in the recent years. Although many contributions have been provided so far, most solutions are developed to operate on predefined settings and fixed sensor setups. Real-world activity recognition applications and users demand more flexible sensor configurations, which may deal with potential adverse situations such as defective or missing sensors. A novel method to intelligently select the best replacement for an anomalous or nonrecoverable sensor is presented in this work. The proposed method builds on an ontology defined to neatly describe wearable sensors and their main properties, such as measured magnitude, location and internal characteristics. SPARQL queries are used to retrieve the ontological sensor descriptions for the selection of the best sensor replacement. The on-body location proximity of the sensors is considered during the sensor search process to determine the most adequate alternative.

Keywords: Ontologies · Activity recognition · Wearable sensors · Sensor selection · Sensor placement · Human anatomy

1 Introduction

The recognition of human activity by means of wearable systems has lately attracted much attention. These systems consist of mobile and portable devices normally worn on diverse body parts to quantify physical activity patterns. Although many solutions have been provided to this respect [4,7,8,10], most of them are designed to operate in closed environments, where the sensors are predefined, well-known and steady. However, these conditions cannot be guaranteed in practical situations, where sensors may be subject to diverse types of anomalies, such as failures [3] or deployment changes [5,6]. Hence, the support of anomalous sensor replacement is seen to be a key requirement to be met by realistic activity recognition systems in order to ensure a fully functional operation. To enable sensor replacement functionalities in an activity recognition system, mechanisms to

© Springer International Publishing Switzerland 2015
I. Rojas et al. (Eds.): IWANN 2015, Part II, LNCS 9095, pp. 298–306, 2015.
DOI: 10.1007/978-3-319-19222-2_25

abstract the selection of the most adequate sensors are needed. To that end, a comprehensive and interoperable description of the available sensors is required, so that the best ones could be selected to replace the anomalous ones. For an accurate sensor selection during runtime use of the recognition systems, practical definitions such as sensor location or availability are required. Consequently, there is a clear need of models that may integrate these heterogeneous sensor descriptions, as well as techniques that may enable the discovery and selection of the most adequate sensors.

This work proposes the use of ontologies to comprehensively describe the wearable sensors available to the user. This fairly reflects an innovative utilization of ontologies in the activity recognition domain, which goes beyond their typical use to detect activities in a knowledge-based recognition approach. Moreover, an iterative search mechanism is also defined to determine the best replacement for a given sensor based on the analysis of the on-body location of the existing ones and their proximity to the anomalous sensor.

2   Related Work

Ontologies have been primarily used in human behavior recognition to detect activities in a knowledge-based oriented fashion. Concretely, ontologies are utilized to describe the activities, while reasoning and inference methods are considered for the recognition process. This is the case of [9], in which ontologies are used to both represent and reason activities based on the analysis of the user interaction with smart objects in pervasive environments. Similarly, in [11] the authors use ontologies to detect office activities based on the analysis of the outputs of binary sensors. Another example is reported in [1], which presents an ontology-based smart home system that discovers and monitors activities of the daily living. Binary or very simple sensors are considered in related works to detect primitives or atomic activities, which are described in an ontological model and used for ontological reasoning to detect high level activities. However, they do no exploit the potential of data-driven approaches in activity recognition, where the sensor data is analyzed using machine learning techniques to detect patterns matching known activities. Therefore, and in order to move one step forward, knowledge-driven approaches have been combined with data-driven approaches to recognize activities. For example, a hybrid model using machine learning techniques applied to body motion data and reasoning based on the ontological representation of the activities is defined in [2]. Ontological reasoning is also used in [12] to recognize complex activities based on simple actions, which are detected via supervised learning algorithms building on data from wearable sensors.

3   SS4RWWAR Ontology

The Sensor Selection for Real-World Wearable Activity Recognition Ontology (SS4RWWAR Ontology), firstly introduced in [13], is an extensible and evolvable ontology that describes heterogeneous wearable sensors in order to enable the selection of replacements for anomalous sensors in activity recognition systems. The SS4RWWAR Ontology comprises an upper ontology defining the basic common concepts for the description of wearable sensors and several plugable domain ontologies which inherit from the concepts in the upper ontology and define them in more detail.

The SS4RWWAR Upper Ontology (see Fig. 1) specifies the sensor description through the *WearableSensor* class. This class links to the *Magnitude* class which lists the magnitudes measured by the sensor -acceleration or rate of turn in inertial measurement units- via the *measures* property; the location where the sensor is placed (*Location*) via the *placedOn* property; the sensor internal characteristics (*Characteristic*) via the *hasInternalCharacteristic* property; and a human readable description (*rdfs:Literal*) via the *hasReadableDescription* property.

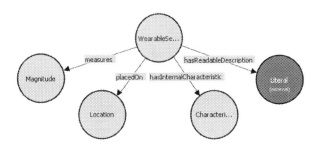

Fig. 1. SS4RWWAR Upper Ontology representing the description of wearable sensors

The SS4RWWAR Human Body Ontology describes the human body parts and uses them as sensor locations, by defining the *HumanBodyPart* class as a subclass of the *Location* class. Fig. 2 shows how the division of the body is done through the subclasses of *HumanBodyPart* -*Head*, *Trunk*, *UpperLimb* and *LowerLimb*- and their subclasses representing the subdivisions of each body part. The different body parts are linked to others through the *connectedTo* property and its eight subproperties defining the connections according to the standard human directional terminology: superior or inferior, anterior or posterior, medial or lateral, proximal or distal. Since the head has located the trunk below, the *Head* is defined as *inferiorlyConnectedTo some Trunk*; and inversely, the *Trunk* is defined as *superiorlyConnectedTo some Head*. The connections between the *Trunk* and the *UpperLimb* are also established via the *inferiorlyConnectedTo* and the *superiorlyConnectedTo* properties. The *Trunk* is also defined as *laterallyConnectedTo some UpperLimb* and the *UpperLimb* as *mediallyConnectedTo some Trunk* since the upper limbs are in a lateral position from the trunk. Similarly, the subdivisions of the main body parts are also linked via the subproperties of *connectedTo*. The face is the anterior part of the head and the scalp the posterior part of it; thus, the *Face* is defined as *posteriorlyConnectedTo some Scalp* and the *Scalp* as *anteriorlyConnectedTo some Face*. The subdivisions of the trunk, *Thorax*, *Back* and *Abdomen*, are also linked through the *anteriorlyConnectedTo*, the *posteriorlyConnectedTo*, the *inferiorlyConnectedTo* and the *superiorlyConnectedTo* properties. Last, to link the subdivisions of the upper and lower limbs the *distallyConnectedTo* and the *proximallyConnectedTo* properties are used. For example, the elbow is connected to the arm in the direction of the main body mass and to the forearm in the opposite direction. Thus, the *Elbow* is defined as *distallyConnectedTo some Forearm* and *proximallyConnectedTo some Arm*.

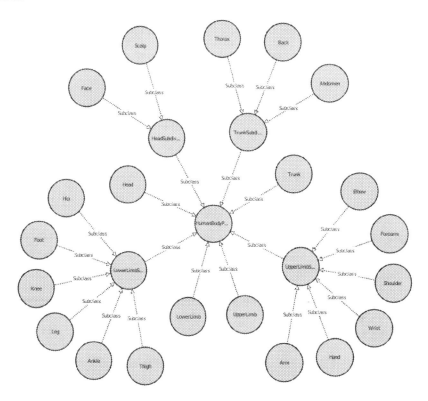

Fig. 2. SS4RWWAR Human Body Ontology defining the human body parts where the wearable sensor can be located

4 SS4RWWAR Ontology-Based Sensor Selection Method

Sensor selection to enable the replacement of anomalous sensors in the activity recognition system can be supported by the usage of the previously presented SS4RWWAR Ontology. The sensor selection method is based on the inference features provided by the SS4RWWAR Ontology and requires that all the available wearable sensors are described using this ontology. Posing the adequate SPARQL queries on the wearable sensor descriptions will allow finding the best sensors which could replace the ones suffering from anomalies.

The novel selection method for the replacement of anomalous sensors proposed in this work is based on an iterative query process triggered once a sensor is detected to have failed. The first option in order to replace an anomalous sensor would be trying to find another sensor located on the same body part. Under this condition, the measurements of the two sensors would be very similar and the activity recognition system would continue working with a similar performance after replacing the anomalous sensor with the other one in the same body part. Hence, the first step of the method is posing to the SS4RWWAR Ontology a SPARQL query to select a replacement sensor located on the same body part. This query, shown in Listing 1.1, is generic and applies to any sensor located on any body part. This abstraction is possible since the query does not require to include where the anomalous sensor is located. The sensor location is inferred from the ontology; thus, only the identifier of the anomalous sensor is needed. In fact, the `<sensor-id>` string in the query must be replaced with the actual identifier of the anomalous sensor as indicated in the SS4RWWAR Ontology. In case no sensor is found on the body part where the anomalous sensor is located, no results would be returned. In this scenario, it would be logical trying to replace the sensor with another one located on any of the adjacent body parts. If two parts are connected, one could expect that their movements are similar and the activity recognition system could still work with the sensor on the adjacent body part. Therefore, the second step of the method is posing to the SS4RWWAR Ontology a SPARQL query to select a replacement sensor located on an adjacent part. This generic query is shown in Listing 1.2, where the `<sensor-id>` string must be replaced with the identifier of the anomalous sensor. In case no sensor is found on the adjacent parts, one could think of trying to find the sensor placed on the closest body location. The closest sensor should be the one that could provide the most similar measurements and therefore be the best to replace the anomalous one. Thus, the third step of the method is posing to the SS4RWWAR Ontology a SPARQL query to select a replacement sensor located on a part directly connected to the adjacent part. This generic query is shown in Listing 1.3, where the `<sensor-id>` string must be replaced with the identifier of the anomalous sensor. This query process would be repeated increasing the number of hops on the body anatomy until a sensor is found. The SS4RWWAR Ontology would exploit the power of inference to find the replacement sensor placed in the closer body part.

```
SELECT ?replacementsensor
WHERE { <sensor-id> ss4rwwar:placedOn ?part.
    ?replacementsensor ss4rwwar:placedOn ?part.
    FILTER (<sensor-id> != ?replacementsensor) }
```

Listing 1.1. SPARQL query for the selection of a replacement sensor located on the same body part where the anomalous sensor is worn.

```
SELECT ?replacementsensor
WHERE { <sensor-id> ss4rwwar:placedOn ?part.
    ?part rdf:type ?parttype.
    ?parttype rdfs:subClassOf ?restriction.
    ?restriction rdf:type owl:Restriction.
    ?restriction owl:onProperty ?connected.
    ?connected rdfs:subPropertyOf body:connectedTo.
    ?restriction owl:someValuesFrom ?conparttype.
    ?conpart rdf:type ?conparttype.
    ?bodypart body:hasPart ?part.
    ?bodypart body:hasPart ?conpart.
    ?replacementsensor ss4rwwar:placedOn ?conpart }
```

Listing 1.2. SPARQL query for the selection of a replacement sensor located on a body part adjacent to where the anomalous sensor is worn.

```
SELECT ?replacementsensor
WHERE { <sensor-id> ss4rwwar:placedOn ?part.
    ?part rdf:type ?parttype.
    ?parttype rdfs:subClassOf ?restriction1.
    ?restriction1 rdf:type owl:Restriction.
    ?restriction1 owl:onProperty ?connected1.
    ?connected1 rdfs:subPropertyOf body:connectedTo.
    ?restriction1 owl:someValuesFrom ?conparttype1.
    ?conpart1 rdf:type ?conparttype1.
    ?conparttype1 rdfs:subClassOf ?restriction2.
    ?restriction2 rdf:type owl:Restriction.
    ?restriction2 owl:onProperty ?connected2.
    ?connected2 rdfs:subPropertyOf body:connectedTo.
    ?restriction2 owl:someValuesFrom ?conparttype2.
    ?conpart2 rdf:type ?conparttype2.
    ?bodypart body:hasPart ?part.
    ?bodypart body:hasPart ?conpart1.
    ?bodypart body:hasPart ?conpart2.
    ?replacementsensor ss4rwwar:placedOn ?conpart2.
    FILTER (<sensor-id> != ?replacementsensor) }
```

Listing 1.3. SPARQL query for the selection of a replacement sensor located on a body part directly connected to a part adjacent to where the anomalous sensor is worn.

In order to provide a clearer view of the functioning of the proposed sensor selection method, let us imagine a sensor setup where the user is wearing six acceleration sensors: S1 and S2 on the left wrist, S3 on the thorax, S4 on the abdomen, S5 on the right thigh, and S6 on the right leg. This scenario and therefore the wearable sensors are described using the SS4RWWAR Ontology. The sensor descriptions are actually instances of the *WearableSensor* class, concretely *S1, S2, S3, S4, S5* and *S6* in this example. The links between the sensors and their locations are established through the *placedOn* property. Therefore, the *S1* and *S2* instances are asserted to have the property *placedOn* with value *User-LeftWrist* which is an instance of the *Wrist* class. The *S3* instance is asserted to have the property *placedOn* with value *UserThorax* which is an instance of the *Thorax* class. The *S4* instance is asserted to have the property *placedOn* with value *UserAbdomen* which is an instance of the *Abdomen* class. The *S5* instance is asserted to have the property *placedOn* with value *UserRightThigh* which is an instance of the *Thigh* class. Last, the *S6* instance is asserted to have the property *placedOn* with value *UserRightLeg* which is an instance of the *Leg* class. Let us suppose that S1 is detected to suffer from some anomaly and the proposed sensor selection method is triggered. Therefore, the ontology is retrieved with the SPARQL query presented in Listing 1.1 where the string <sensor-id> is replaced with the string S1 which is the actual identifier of the anomalous sensor. This query returns S2 as replacement for S1 since the two sensors are located on the same body part, concretely on the left wrist. This result is produced because the ontology describes that both the *S1* and the *S2* instances have the same value *UserLeftWrist* for the property *placedOn*. Let us now suppose that it is S3 which suffers from an anomalous behavior. In this case the SPARQL query presented in Listing 1.1 (replacing <sensor-id> with S3) does not produce any results since there is no other sensor on the thorax. Therefore, the query presented in Listing 1.2 (replacing <sensor-id> with S3) is posed to the ontology. S4 is returned as a result since this sensor is located on the abdomen which is a body part adjacent to the thorax where S3 is worn. This query result can be explained thanks to the structure of the ontology. The property *placedOn* has value *UserThorax* for the *S3* instance. The *UserThorax* is an instance of the *Thorax* class, and any member of this class is *inferiorlyConnectedTo* a member of the *Abdomen* class. Since the *S4* instance asserts the property *placedOn* to have the value *UserAbdomen*, which is an instance of the *Abdomen* class; then, the *UserThorax* is *inferiorlyConnectedTo* the *UserAbdomen*. Therefore, it can be concluded that the *S3* and the *S4* instances have for the property *placedOn* values that are instances of adjacent body parts, concretely *UserThorax* and *UserAbdomen*. Finally, let us suppose that S5 behaves anomalously. In this case, the first and the second SPARQL queries presented in Listing 1.1 and in Listing 1.2 (replacing <sensor-id> with S5) do not produce any results since there is no sensor on the thigh neither on the hip or the knee which are the two parts adjacent to the thigh. Therefore, the query presented in Listing 1.3 (replacing <sensor-id> with S5) is posed to the ontology. S6 is returned as a result since this sensor is located on the leg which is a body part directly connected to the knee, that is an adjacent part to the thigh. Like in the other examples, the explanation of this result is based on the logical of the ontology definition. For the *S5* instance, the *placedOn* property has asserted the value *UserThigh* which is an instance the *Thigh* class. Any member of *Thigh* class is *distallyConnectedTo* a member of the *Knee* class, and any member of this class is *distallyConnectedTo* a member of the *Leg* class. Moreover, the *S6* instance asserts for the property *placedOn* the value *UserLeg*, which is an instance of the *Leg* class. Therefore, the *UserThigh* is *distallyConnectedTo* the *UserKnee* which is *distallyConnectedTo* the *UserLeg*. So, it can be concluded that the *S5* and the *S6* instances have for the property *placedOn* values that are instances of close by body parts, concretely *UserThigh* and *UserLeg*.

5   Conclusions

Wearable activity recognition systems operating in realistic conditions are subject to malfunctioning due to changes suffered by body-worn sensor devices. Hence, mechanisms to support the selection of adequate alternative sensors are required. This work has presented an innovative use of ontologies in the activity recognition domain, which goes beyond their typical use to detect activities in a knowledge-based recognition fashion. Concretely, ontologies are used here to enhance the machine learning activity recognition used in data-driven approaches. The novel sensor selection method provides an iterative query mechanism to discover the best replacement sensor for an anomalous one by analyzing the on-body location of the available sensors and their proximity to the anomalous or nonrecoverable one. The intelligent selection method builds on an ontology defined to comprehensively and unequivocally describe wearable sensors and their main properties, including their exact placement on the human body. Future work includes the extension of the ontology towards the magnitude and sensor characteristics domains, as well as the application of ontological reasoning techniques to improve the sensor selection method and to reduce the increasing complexity of the queries.

Acknowledgments. This work was supported by the Junta de Andalucia Project P12-TIC-2082 and the grant "Movilidad Internacional de Jóvenes Investigadores de Programas de Doctorado Universidad de Granada y CEI BioTic".

References

1. Bae, I.-H.: An ontology-based approach to adl recognition in smart homes. Future Generation Computer Systems **33**, 32–41 (2014)
2. BakhshandehAbkenar, A., Loke, S.-W.: Myactivity: cloud-hosted continuous activity recognition using ontology-based stream reasoning. In: IEEE International Conference on Mobile Cloud Computing, Services, and Engineering (2014)
3. Banos, O., Damas, M., Guillen, A., Herrera, L.-J., Pomares, H., Rojas, I., Villalonga, C.: Multi-sensor fusion based on asymmetric decision weighting for robust activity recognition. Neural Processing Letters, 1–22 (2014)
4. Banos, O., Damas, M., Pomares, H., Prieto, A., Rojas, I.: Daily living activity recognition based on statistical feature quality group selection. Expert Systems with Applications **39**(9), 8013–8021 (2012)
5. Banos, O., Damas, M., Pomares, H., Rojas, I.: On the use of sensor fusion to reduce the impact of rotational and additive noise in human activity recognition. Sensors **12**(6), 8039–8054 (2012)
6. Banos, O., Toth, M.-A., Damas, M., Pomares, H., Rojas, I.: Dealing with the effects of sensor displacement in wearable activity recognition. Sensors **14**(6), 9995–10023 (2014)
7. Bao, L., Intille, S.S.: Activity recognition from user-annotated acceleration data. In: Ferscha, A., Mattern, F. (eds.) PERVASIVE 2004. LNCS, vol. 3001, pp. 1–17. Springer, Heidelberg (2004)
8. Ermes, M., Parkka, J., Mantyjarvi, J., Korhonen, I.: Detection of daily activities and sports with wearable sensors in controlled and uncontrolled conditions. IEEE Trans. on Inform. Tech. Biomed. **12**(1), 20–26 (2008)
9. Liming, C., Nugent, C., Okeyo, G.: An ontology-based hybrid approach to activity modeling for smart homes. IEEE T. Human-Machine Systems **44**(1), 92–105 (2014)
10. Mannini, A., Intille, S.S., Rosenberger, M., Sabatini, A.M., Haskell, W.: Activity recognition using a single accelerometer placed at the wrist or ankle. Medicine and Science in Sports and Exercise **45**(11), 2193–2203 (2013)

11. Nguyen, T., Raspitzu, A., Aiello, M.: Ontology-based office activity recognition with applications for energy savings. Journal of Ambient Intelligence and Humanized Computing **5**(5), 667–681 (2014)
12. Riboni, D., Bettini, C.: Owl 2 modeling and reasoning with complex human activities. Pervasive Mobile Computing **7**(3), 379–395 (2011)
13. Villalonga, C., Banos, O., Pomares, H., Rojas, I.: An ontology for dynamic sensor selection in wearable activity recognition. In: Ortuño, F., Rojas, I. (eds.) IWBBIO 2015, Part II. LNCS, vol. 9044, pp. 141–152. Springer, Heidelberg (2015)

Short-Term Spanish Aggregated Solar Energy Forecast

Nicolas Perez-Mora[(✉)], Vincent Canals, and Víctor Martínez-Moll

University of Balearic Islands, Ctra. Valldemossa Km 7.5, 07122 Palma of Majorca, Spain
nicolasperezmora@gmail.com

Abstract. This work presents and compare six short-term forecasting methods for hourly aggregated solar generation. The methods forecast one day ahead hourly values of Spanish solar generation. Three of the models are based on MLP network and the other three are based on NARX. The two different types of NN use to forecast the same NWP data, comprising solar radiation, solar irradiation and the cloudiness index weighted with the installed solar power for the whole country. In addition of the NWP data the models are fed with the aggregated solar energy generation in hourly step given by the System Operator.

The results of the two types of NN are compared and discussed in the conclusions as much as the error variability along the day hours. The results obtained by the six methods are evaluated, concluding that the most accurate result is the one given by the developed NARX irradiance forecast method; achieving the lowest one day-ahead Mean Average Daily Error of 16.64%.

Keywords: Forecasting · Solar radiation · Solar forecasting · Energy market · Neural networks · Time series

1   Introduction

The installation of power plants based on renewable energy sources have increased exponentially in the recent years [1]. The fuel-based energy generation price increase along with the subsidies promoting the construction of clean energies made the installation of renewables a profitable business for many particulars and companies [2]. The expansion of these technologies and the policies adopted by the Spanish national government to integrate them into the electrical power system has brought a fussy market situation.

The integration of the electric energy generated by these power plants into electric power system is priority. According with the current legislation, the power generated by green sources is fed into the system in preferential order within the Spanish energy generation mix. This energy mix is result of a counter clockwise energy auction held one day-ahead.

In order to always ensure the renewable energy acceptance into the system the energy bids [2] for such technologies are placed at the legal minimum, 0€/MWh and its retribution would be an addition of the final auction price and the generation subsidy agreed with the government. This fact may lead into situations where the energy demand is fewer than the energy generation by technologies offering their energy at the minimum   price, being the final auction price 0€/MWh.

© Springer International Publishing Switzerland 2015
I. Rojas et al. (Eds.): IWANN 2015, Part II, LNCS 9095, pp. 307–319, 2015.
DOI: 10.1007/978-3-319-19222-2_26

To understand the variability of the energy market is necessary understand the variability of the green energy technologies since they have direct implications on the economic operation of power system [3,4]. This variability causes the necessity of having accurate power forecast for wind and solar power [5,6] being those essential for an effective integration in the system.

A major drawback of solar energy lies in the non-continuous generation, for both technologies harvesting electricity out from the solar resource, PV power [7] and electric thermo solar plants. Solar energy depends critically on the variability of irradiance [8], typically cloud cover will cause many rapid changes in the irradiance during the day [7] which will bring along generation fluctuations. A proper irradiance forecast will result into a more accurate solar generation forecast since the two variables are directly and strongly related.

This paper presents six forecast methods based on ANN, three of them based on Nonlinear AutoRegressive models with eXogenous (NARX) and other three based on Multilayer Perceptron (MLP) neural networks. In order to compare the both neural models the same three explanatory variables would be used in the methods. After analyzing the relation between variables and aggregated energy generation the explanatory variables to use are cloudiness, radiation and irradiation since these variables seem suitable for aggregated energy generation forecasting.

This paper is organized as follows: Next section provides an overview of the problem to approach. Section 3 describes the obtained data to use. In Section 4 the methodology used to approach the problem is explained. Section 5 presents the result obtained from the methodology and the conclusions from those would be given in Section 6.

The generators place their power generation sale offers (bids) to the market operator, for the energy demand and the cheapest combination of energies will be granted as generators. The retribution for all the accepted generators will be the same and equal to the last and highest accepted bid in the auction.

2   Problem Description

The aim of this work is forecast the Spanish solar generation in hourly step for one day ahead. A natural approach to energy generation forecast over a wide geographical area is to calculate individual forecast and sum the results up [9]. Hence, is expectable that individual errors to be relatively uncorrelated and, thus, to partially cancel out when their forecasts are added. This, however, is an extremely difficult approach due the large amount of installations, the variety of technologies and the lack of historical generation information. To achieve results aggregated solar generation data is obtained and the irradiation associated to the generation is calculated.

Accurate solar power forecasts can be obtained using computational models such as Artificial Neural Networks (ANN) [10] or autoregressive models such as ARIMA [11]. Both forecast models will be improved by using a sufficient data series which may be supported by an explanatory variable. As mentioned before the relation between irradiation and solar generation is direct for a certain power plant, therefore a precise irradiation value is presumed to be an outstanding explanatory variable. In the same way, cloudiness index has a negative relation with solar energy generation, hence, is presumed to be a useful explanatory variable. Solar generation forecast based on cloudiness is deemed to be the most successful method for a long term solar forecasting [12].

The importance on forecasting energy generation lies in the markedly different generation between stations changing from a 540 MW of power peck in winter to a 5600MW of power peck in summer and achieving a maximum of 12% in the total renewable power injected into the grid. It is also important to understand that the solar electricity generation in Spain includes two technologies able to produce electricity:

- The energy from photovoltaic technology which generation is influenced only by the solar irradiation on the panel surface on a certain time and is not supported with storage. The installed power of this technology is 4,16GW [13].
- The thermo solar power plants are able to generate electricity through a process dependent on solar irradiance and temperature; additionally these power plants count with storage system. The installed power of this technology is 2,3GW [13].

Hence, to provide a reliable explanatory data in Spain is required to develop a Numerical Weather Prediction (NWP). The NWP would obtain the cloudiness index for a certain location. This information would help modeling solar radiation with consideration of geographical information. The model is based on a clear-sky radiation calculation and satellite cloudiness index in different locations in the country. The calculation of the irradiation for a given location is done subtracting the fraction blocked by the clouds to the clear-sky radiation [14,15]. The calculation of the average irradiation in Spain would be the weighted irradiation with the installed power in the area under study [16].

2.1   Clear-Sky Radiation

The first step for obtaining a reliable data consists on calculating the value of the extraterrestrial hourly radiation (G0) [17]. This calculation depends upon the location of the plant, the time of the year and the slope of the solar collectors. The radiation (G0) is calculated according to (1):

$$G_0 = G_{SC}\left(1 + 0.033\cos\frac{360n}{365}\right)\cos\theta_z \tag{1}$$

Where GSC is a solar constant (1367 W/m^2), n is the number of day within a year, θ_z is the zenith angle calculated in (2):

$$\cos\theta_z = \cos(\emptyset - \alpha)\cos\delta\cos\omega + \sin(\emptyset - \alpha)\sin\delta \tag{2}$$

The latitude of the location is denoted by $\emptyset$, a is the slope of the collecting surface, δ is the declination or angular position of the sun calculated in (3) and ω is the hour angle or the angle of displacement of the sun calculated in (4).

$$\delta = 23.45\sin\left(360\frac{284+n}{365}\right) \tag{3}$$

$$\omega = (h_s - 12)\cdot 15 \tag{4}$$

The solar time is denoted by h_s. The difference between the solar time and the standard time is calculated through the formula (5) and the result is given in minutes:

$$Solar\ time - standard\ time = 4(L_{st} - L_{loc}) + E + DLS \tag{5}$$

Where L_{st} and L_{loc} are the longitudes for the standard meridian and the location, DLS references the possibility of having Day Light Savings and E is a value calculated through (6):

$$E = 229.2(0.000075 + 0.001868 \cdot cos(B) - 0.032077 \cdot sin(B) -$$

$$0.014615\ cos(2B) - 0.04089 \cdot sin(2B) \tag{6}$$

And finally, B is calculated according (7):

$$B = (n-1)\frac{360}{365} \tag{7}$$

2.2   Forecasted Irradiation

Solar intensity generally decreases with increasing values of sky cover [18]. Therefore forecasted irradiation (I_f) for a given location depends upon the extraterrestrial hourly radiation (G_0) and the symmetrical of the forecasted cloudiness measured in 0 to 1 range (N_f). Both values are calculated in an hourly step and calculated according to (8):

$$I_f = G_0 \cdot (1 - N_f) \tag{8}$$

A similar model was presented in [19], where global irradiance is obtained by the transmissivity of the clouds measurements.

2.3   Weighted Irradiation

The value of the weighted irradiation in Spain (I_E) is calculated taking into account the forecasted irradiations in the measured places (I_{fn}) and its installed powers (P_n), sum of both PV and thermo electric. Such value of irradiation is calculated according to (9):

$$I_E = \frac{1}{P_S}\sum_{n=1}^{50} I_{fn} \cdot P_n \tag{9}$$

P_n is the installed sun power for a given area, P_s is the total power accumulated in Spain and is calculated using (10):

$$P_S = \sum_{n=1}^{50} P_n \tag{10}$$

2.4   Weighted Cloudiness

The value of the weighted cloudiness in Spain (N_E) is calculated taking into account the forecasted cloudiness index in the measured places (N_{fn}) and its installed powers in those places (Pn), sum of both PV and thermo electric. Such value of cloudiness is calculated according to (11):

$$N_E = \frac{1}{P_S} \sum_{n=1}^{50} N_{fn} \cdot P_n \qquad (11)$$

Where P_n is the installed sun power for a given area, P_s is the total power accumulated in Spain and is calculated using (10). For practical proposes the opposite of cloudiness would be used (1-N) as variable.

3   Data Acquisition

The forecasting of solar energy generation is based on the historical data obtained from the Spanish grid operator, *Red Eléctrica de España* (REE). This data is available from 24/07/2013 and provides the Spanish national solar production in 10min steps.

The amount of installed power plants and the aggregated power per province is given by the Ministry of Industry. Detailed information of the installations, such as, location, technology, power and whether or not is connected to grid or in progress is available in the website [13].

Cloudiness forecast is given by a web service in hourly step. The NWP is obtained using an algorithm that analyses satellite images from the vast historical archives to predict the climatic data [20]. There is data availability for the current day and three days ahead. The cloudiness database has data from 2014, 8 September to 2015, 23 February. In order to have a homogeneous database in the area of Spain, one station per province is used; those stations are located, mainly, in the province's main city, therefore 50 stations are selected for Spain. Stations are marked as dots in Fig.1.

Fig. 1. Weather Stations Localizations

The fact of having one station per province and using this data for the whole province may lead into errors since the province value is used in the whole province area and it is not taking into consideration the exact location of the plant.

For the radiance calculation a value of slope (a) is estimated, this value remains the same for all the power plants and fix for the whole year. The slope value is the optimum value for the region of Spain and it is established in the average latitude value for Spain, 40°. The optimum yearly slope value for an installation could be taken as the same as the latitude where it is located [21].

The acquisition and processing of the information and calculations are made by an automatic tool designed in Matlab®.

As result, three variables are obtained to allow the forecast, cloudiness, radiation and irradiation. After taking as valid the previous conditions, the national calculated irradiation using the cloudiness, and the solar national generation in hourly step show a strong relation (Fig. 2a). It is important to bear in mind that the units and factors do not match for the variables, as the radiation is measured in [W/m^2] and its maximum value is 1367 W/m^2, in case of power is measured in [MW] and its maximum value is 6460MW. This relation, measured with the Pearson factor reaches 0.86 and a R^2 = 0.741.

In cloudiness index case, the relation with the aggregated energy generation is not so obvious when it comes to indexes or graphs (Fig.2b). Pearson factor gives a result of 0.09 and R2 = 0.008. Nevertheless the relation between cloudiness and solar generation is easily understandable from a physics point of view, the existence of clouds blocks the solar energy generation.

For the irradiation, the relation with the aggregated energy generation (Fig.2c) reaches the highest value. Pearson factor gives a result of 0.90 and R2 = 0.816.

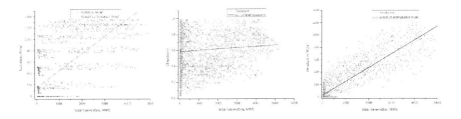

Fig. 2. Variables relation with Solar Generation: (a) Radiation Vs Solar Generation, (b) Cloudiness Vs Solar Generation, (c) Irradiation Vs Solar Generation

4   Methodology

The direct relation between the calculated irradiation and the generation is reason enough to think that this likeness would be useful and sufficient to calculate the aggregated energy generation (Fig. 3). Nevertheless the variability and differences between the different installed power plants, technologies and performances precludes an analytical reliable model; such calculation would require detailed information of the installed plants comprising the Spanish aggregated solar power. Ergo, a forecasting method has to be developed in order to obtain the amount of power which is generated hourly in Spain.

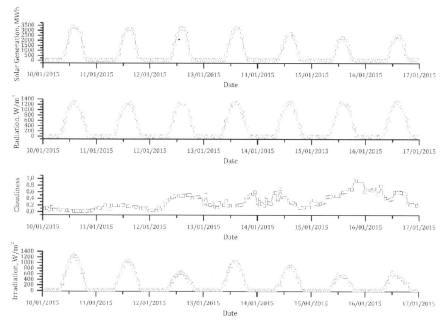

Fig. 3. Generation and proposed explanatory variables

An approach to the problem is given by using neural networks, these are found to outperform the regressions models when it comes to high resolutions [11].

Since the late 1990s ANN have been used as application in the field of solar forecasting using climatological variables as inputs to an ANN to predict generation values [22]. Although the design of ANN models using specific sets of design constraints relies so much on the expertise with similar applications and is subjected to trial and error processes [23]. Furthermore, there is a substantial difficulty in training nets which may extent to a large amount of required iterations before the net would converge [11].

The ANN may use solar radiation, ambient temperature and cloudiness as its inputs whereas power is given as output [24]. Past measurements of aggregated power and NWP forecast for cloudiness, radiation and irradiation would be used as inputs to an autoregressive model with exogenous input (ARX) building therefore a NARX recurrent neural networks. NARX model relates the current value of a time series to current and past values of the exogenous series influencing, therefore the series of interest.

Forecasts has been carried out using cloudiness and is concluded than the use of this data improves the accuracy of power production forecasts [25]. Hence, weighted radiation, weighted cloudiness and weighted irradiation would be used as support to forecast the aggregated generation. Consequently three NARX methods are proposed; the first one (NARX1) uses as exogenous input the theoretical radiation, the second method (NARX2) uses the value of cloudiness as input and the third method (NARX3) uses the calculated irradiation as input. The methods are otherwise, equal.

Other common way to approach the approximation which allow to find the relationship between the inputs and the output data is the MLP network [10]. In a MLP neurons are grouped in layers and only forward connections exist creating a structure able to learn and to model a phenomenon. In order to forecast a fixed number of past values are set as inputs, the output is the forecast of the future value in the time series [26]. For the MLP model another three methods are proposed; the first one (MLP1) uses as input the theoretical radiation, the second method (MLP2) uses as input the cloudiness index and the third method (MLP3) uses as input the calculated irradiation. In principle the methods are equal.

5   Results

Many neural networks configurations have been tried out to forecast the aggregated solar power generation. The best performing set of configurations and explanatory variables results obtained, are shown in this section. These methods and configurations reach the lowest error values and fit better into the sought purpose of forecasting energy in order to obtain valuable information to use in the energy market. The data is split in training, 70%; validation to avoid over fitting, 15%; and result comparison and testing, 15% sets. The training method used in both cases is Levenberg-Marquardt back-propagation algorithm.

The best performing tried methods and neural network configurations are shown in Table 1; where "L", stands for Linear; "S", stands for Sigmoid; "ST", stands for Sigmoid Tan-gent and "I" stands for Input layer.

Table 1. Neural Networks Configurations & Models

NN Method	Number of layers	Neurons in each layer	Activation function	Explanatory Variable
MLP1	3	1-5-1	I-S-L	Radiation
MLP2	3	1-5-1	I-S-L	Cloudiness
MLP3	3	1-5-1	I-S-L	Irradiation
NARX1	3	48-24-1	I-ST-L	Radiation
NARX2	3	48-24-1	I-ST-L	Cloudiness
NARX3	3	48-24-1	I-ST-L	Irradiation

The forecast results obtained with the different methods are shown in hourly steps in Fig. 4 where they are compared with the real aggregated solar generation. The forecasts are clustered using as criteria the explanatory variable used, either radiation in Fig. 4a, cloudiness index in Fig. 4b or irradiation in Fig. 4c.

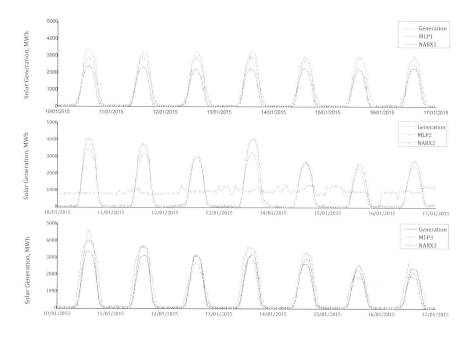

Fig. 4. Different forecast methods results comparison

In order to measure the accuracy of the predictions, the error between the fore-casted values and real data are analyzed here. In this work, Mean Absolute Error (MAE), Mean Absolute Percentage Error (MAPE) and Mean Daily Absolute Percentage Error (MADPE), are used as the errors to validate the prediction methods. These calculations are described in (12) (13) (14).

$$MAE = \frac{1}{n}\sum_{t=1}^{n}\left|S_{(t)} - F_{(t)}\right| \qquad [\text{MWh}] \qquad (12)$$

$$MAPE = \frac{100}{n}\sum_{t=1}^{n}\left|\frac{S_{(t)}-F_{(t)}}{S_{(t)}}\right| \qquad [\%] \qquad (13)$$

$$MADPE = \frac{100}{n}\sum_{d=1}^{n}\frac{\sum_{t=1}^{24}\left|S_{(t)}-F_{(t)}\right|}{\sum_{t=1}^{24}\left|S_{(t)}\right|} \qquad [\%] \qquad (14)$$

MAE calculation is useful showing the deviation of the forecast in MWh and gives a perception of the accuracy of the method in terms of energy, this value would be useful while selling energy in the market. MAPE shows the same error in percentage in order to relativize this error along the daily generation, the problem with this calcu-lation comes in low values of energy, for example, during the nights when even low energy error brings and undue error. Hence a proposed calculation is here presented with MADPE, where the daily MAE is divided by the daily generation and thus mak-ing relative the daily amount of generation with the daily forecasted error. Table 2 shows the three error calculations for one of the representative weeks evaluated.

Table 2. Comparation of the Forecasted Errors

		10/1/15	11/1/15	12/1/15	13/1/15	14/1/15	15/1/15	16/1/15	**Mean**
MLP 1	MAE	187	148	163	157	157	306	217	**190.61**
	MAPE	64%	45%	107%	85%	125%	146%	84%	**93.73%**
	MADPE	19%	16%	19%	18%	21%	53%	32%	**25.51%**
MLP 2	MAE	1088	1024	1080	1114	939	742	926	**987.66**
	MAPE	965%	797%	1446%	1530%	1871%	1439%	885%	**1276.29%**
	MADPE	111%	113%	129%	126%	128%	128%	137%	**124.52%**
MLP 3	MAE	486	372	272	338	332	167	270	**319.50**
	MAPE	118%	85%	134%	115%	167%	105%	116%	**119.97%**
	MADPE	50%	41%	33%	38%	45%	29%	40%	**39.32%**
NAR X1	MAE	347	276	292	292	160	70	85	**217.20**
	MAPE	53%	35%	76%	38%	50%	29%	33%	**44.89%**
	MADPE	36%	30%	35%	33%	22%	12%	12%	**25.71%**
NAR X2	MAE	173	149	109	330	50	140	116	**152.37**
	MAPE	43%	27%	77%	106%	97%	45%	37%	**61.62%**
	MADPE	18%	16%	13%	37%	7%	24%	17%	**18.91%**
NAR X3	MAE	178	151	72	142	108	92	179	**131.70**
	MAPE	26%	30%	45%	48%	60%	39%	71%	**45.53%**
	MADPE	18%	17%	9%	16%	15%	16%	26%	**16.64%**

An hourly error study has been carried out. In this study the same hours in the day are clustered and analysed as a whole. This study will allow determine which time frame has the biggest propensity to errors. The results are shown in Fig. 5.

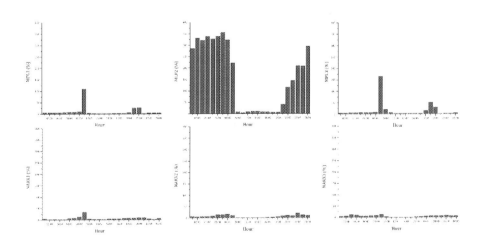

Fig. 5. Forecasted Hourly Errors

6   Conclusions

In this work a MLP and NARX models for solar energy generation for one day-ahead forecast have been developed. Both of the proposed models accepts as inputs either the hourly radiation, cloudiness index or irradiation. The output is always the solar generation in Spain. After several simulations for the six proposed methods the best configuration was found and results were given.

When analysing those results three errors figures are taken into consideration, MAE, MAPE and MADPE; each one of them show a different feature of error, being MADPE the one which provides the most neat result.

Looking to those error figures, the results are very clear, being the NARX fed with solar irradiation and solar generation historical data (NARX3) the one which always outperform the other methods in all the different figures of error. In terms of MADPE it reaches quite a low error of 16.64%. If the MAE is taking into consideration, the value in energy is equal to 132MW which would be equivalent to 0.42% of the Spanish total daily mean demand.

In contrast the method with higher error is the MLP fed with cloudiness index (MLP2), the inexistence of temporal relation between cloudiness and generation penalizes this method which does not incorporate generation information at the inputs while forecasting. The forecasts yields from this method do not have either the shape or magnitude of the aim data, and is, therefore, useless at all.

If neural network type is studied the most successful forecaster is the NARX. Both variants outperform any of the variants on MLP. This information may lead into the conclusion that solar generation information is quite precise while forecasting and that external weather output is not sufficient to give an accurate result.

In order to understand the source of error an hourly study was carried out. This study shows that the lowest amount of error in percentage (MADPE) is place in the central hours of the day, coinciding with the highest amount of generation. Even though the MAE in those hours is higher, as is the generation, the accuracy of all the methods increase as generation follows a predictable shape. In the other hand, the highest amount of error in MADPE is placed during the night hours and especially a spike of error takes place on the rise of solar energy generation, around 8-9AM.

A similar research in solar energy forecast with NN is found, [2] in this case a real PV plant generation was forecasted, the results achieved, in best case of MADPE was 23.26%. However the case is not the same. The study presented in this work forecasts a whole country generation with heterogeneous plants and not a single generation point, the result obtained outperforms [2] obtaining a value of 16.64%.

Acknowledgements. This work is included inside the Solar Heat Integration Network (SHINE) project which is supported by the European Union, as an initial research training network (ITN) in the framework of the Marie-Curie program, FP7. Project reference 317085, at Sampol Ingeniería y Obras S.A.

References

1. Monteiro, C., Santos, T., Fernandez-Jimenez, L.A., Ramirez-Rosado, I.J., Terreros-Olarte, M.S.: Short-term power forecasting model for photovoltaic plants based on historical similarity. Energies **6**, 2624–2643 (2013)
2. Monteiro, C., Fernandez-Jimenez, L.A., Ramirez-Rosado, I.J., Muñoz-Jimenez, A., Lara-Santillan, P.M.: Short-term forecasting models for photovoltaic plants: Analytical versus soft-computing techniques. Math. Probl. Eng. **2013** (2013)
3. Parsons, B., Milligan, M., Zavadil, B., Brooks, D., Kirby, B., Dragoon, K., Caldwell, J.: Grid impacts of wind power: A summary of recent studies in the United States. Wind Energy **7**, 87–108 (2004)
4. Ortega-Vazquez, M.A., Kirschen, D.S.: Assessing the impact of wind power generation on operating costs. Smart Grid, IEEE Trans. **1**(3), 295–301 (2010). Angarita-Márquez, J.L., Hernandez-Aramburo, C.A., Usaola-Garcia, J.: Analysis of a wind farm's revenue in the British and Spanish markets. Energy Policy **35**, 5051–5059 (2007)
5. Lange, M., Focken, U.: Physical Approach to Short-Term Wind Power Prediction, pp. 1–208. Springer, Berlin Heidelberg (2006)
6. Yang, D., Jirutitijaroen, P., Walsh, W.M.: Hourly solar irradiance time series forecasting using cloud cover index. Sol. Energy **86**(12), 3531–3543 (2012)
7. Chen, C., Duan, S., Cai, T., Liu, B.: Online 24-h solar power forecasting based on weather type classification using artificial neural network. Sol. Energy **85**(11), 2856–2870 (2011)
8. Gala, Y., Fernández, A., Dorronsoro, J.: Machine learning prediction of global photovoltaic energy in spain. In: International Conference on Renewable Energies and Power Quality (ICREPQ 2014), no. 12
9. Mellit, A., Pavan, A.M.: A 24-h forecast of solar irradiance using artificial neural network: Application for performance prediction of a grid-connected PV plant at Trieste, Italy. Sol. Energy **84**(5), 807–821 (2010)
10. Reikard, G.: Predicting solar radiation at high resolutions: A comparison of time series forecasts. Sol. Energy **83**(3), 342–349 (2009)
11. Huang, R., Huang, T., Gadh, R., Li, N.: Solar generation prediction using the ARMA model in a laboratory-level micro-grid. In: 2012 IEEE 3rd International Conference on Smart Grid Communications, SmartGridComm 2012, pp. 528–533 (2012)
12. Ministerio Industria Energía y Turismo, Gobierno de España. http://www.minetur.gob.es/. Enero (2015). https://oficinavirtual.mityc.es/ripre/informes/informeinstalaciones.aspx
13. Davies, J.A., McKay, D.C.: Evaluation of selected models for estimating solar radiation on horizontal surfaces. Solar Energy **43**, 153–168 (1989)
14. Biga, A.J., Rosa, R.: Estimating solar irradiation sums from sunshine and cloudiness observations. Solar Energy **25**, 265–272 (1980)
15. Lorenz, E., Hurka, J., Heinemann, D., Beyer, H.G.: Irradiance forecasting for the power prediction of grid-connected photovoltaic systems. IEEE J Sel Top Appl Earth Obs Remote Sens **2**(1), 2–10 (2009)
16. Duffie, J.A., Beckman, W.A.: Solar Engineering of Thermal Processes: Fourth Edition (2013)
17. Sharma, N., Sharma, P., Irwin, D., Shenoy, P.: Predicting solar generation from weather forecasts using machine learning. In: 2011 IEEE International Conference on Smart Grid Communications, SmartGridComm 2011, pp. 528–533 (2011)
18. Bacher, P., Madsen, H., Nielsen, H.A.: Online short-term solar power forecasting. Sol. Energy **83**(10), 1772–1783 (2009)

19. Zhang, Y., Wistar, S., Piedra-Fernandez, J.A., Li, J., Steinberg, M.A., Wang, J.Z.: Locating visual storm signatures from satellite images", in. IEEE International Conference on Big Data (Big Data) **2014**, 711–720 (2014)

20. Benghanem, M.: Optimization of tilt angle for solar panel: Case study for Madinah, Saudi Arabia. Appl. Energy **88**(4), 1427–1433 (2011)

21. Inman, R.H., Pedro, H.T.C., Coimbra, C.F.M.: Solar Forecasting Methods for Renewable Energy Integration. Prog. Energy Combust. Sci. **39**(6), 535–576 (2013)

22. Yao, X.: Evolving artificial neural networks. Proc. IEEE **87**, 1423–1447 (1999)

23. Sulaiman, S., Rahman, T.A., Musirin, I.: Partial Evolutionary ANN for Output Prediction of a Grid-Connected Photovoltaic System. Int. J. Comput. Electr. Eng. **1**(1), 40–45 (2009)

24. Da Silva Fonseca, J.G., Oozeki, T., Takashima, T., Koshimizu, G., Uchida, Y., Ogimoto, K.: Use of support vector regression and numerically predicted cloudiness to forecast power output of a photovoltaic power plant in Kitakyushu, Japan. Prog. Photovoltaics Res. Appl. vol. 20, no. July 2011, pp. 874–882, 2012

25. Paoli, C., Voyant, C., Muselli, M., Nivet, M.-L.: Forecasting of preprocessed daily solar radiation time series using neural networks. Sol. Energy **84**(12), 2146–2160 (2010)

Intelligent Presentation Skills Trainer Analyses Body Movement

Anh-Tuan Nguyen[✉], Wei Chen, and Matthias Rauterberg

Department of Industrial Design, Eindhoven University of Technology,
Postbus 513, 5600MB Eindhoven, The Netherlands
{a.nguyen,w.chen,g.w.m.rauterberg}@tue.nl

Abstract. Public speaking is a non-trivial task since it is affected by how nonverbal behaviors are expressed. Practicing to deliver the appropriate expressions is difficult while they are mostly given subconsciously. This paper presents our empirical study on the nonverbal behaviors of presenters. Such information was used as the ground truth to develop an intelligent tutoring system. The system can capture bodily characteristics of presenters via a depth camera, interpret this information in order to assess the quality of the presentation, and then give feedbacks to users. Feedbacks are delivered immediately through a virtual conference room, in which the reactions of the simulated avatars can be controlled based on the performance of presenters.

Keywords: Body motion analysis · Depth vision · Nonverbal behavior · Social signal processing

1 Introduction

Public speaking is the art of persuasion. It has the tremendous impact on the success of everyone [1, p.102]. Unfortunately, delivering an oral presentation is not as simple as computer data transmission. Instead, the audience simultaneously perceives the messages via various non-spoken channels, which are known as nonverbal behaviors. On one hand, the content of a presentation must be *clear, vivid and appropriate* [2]. On the other hand, the significant component of a presentation lies upon nonverbal cues, which *has the power to change the meaning assigned to the spoken words* [1, p.241].

Nonverbal behaviors of public speakers are expressed via several channels such as voice, gesture and facial expression. They have been proven to have greater influence than verbal cues. For example, a research by [3] showed that, nonverbal messages are twelve to thirteen times more powerful than verbal ones. Similarly, according to [4], the audience receives more than half of information from body language. The same result was found during the study of [1], in which most people unconsciously more believe in nonverbal than verbal communication.

Practicing to express the effective nonverbal behaviors is difficult due to the fact that, they are mostly expressed subconsciously. Thus, in order to achieve the

© Springer International Publishing Switzerland 2015
I. Rojas et al. (Eds.): IWANN 2015, Part II, LNCS 9095, pp. 320–332, 2015.
DOI: 10.1007/978-3-319-19222-2_27

positive learning results, learners must be provided with the appropriate feedbacks from skilled experts, which in most cases might be expensive to achieve. In parallel, the role of nonverbal behaviors in computing is becoming increasingly recognized by the development of the emerging fields, such as social signal processing [5] and affective computing [6]. Therefore, computers have been equipped with the abilities to decode the complexity of humans non-spoken channels.

In the literature, there are several approaches toward the automatic recognition of nonverbal cues from presenters, such as [7–10]. These approaches analyze some vocal and visual channels of presenters, thus can provide them with the information about their performance. For example, the system in [10] was built solely on vocal cues, by analyzing the physical characteristics of voice such as pitch or tempo. It was similar to the approach of [9], which was originated from a vocal emotion detection module. The authors applied the support vector machine [11] to analyze one presentation based on a set of 6 qualities and achieved the accuracy of 81%. The approach introduced in [7] might be the simplest. By relying on the importance of pitch variance in oral presentations, the system measured the changes in vocal pitch, and then give visual feedbacks to promote pitch variation.

On the other hand, there are three systems that include visual cues in the analysis. In [8], the authors added face position and orientation as the approximation of eye contact, together with utterance, pitch, filled pauses and speaking rate. In contrast, [12] introduced the method that only based on visual information. Similar to [8], face orientation was used as an indication for eye contact. The authors tracked the trajectories of global body movement and head position. This information helped their system to rank the performance of the whole presentation using the RankBoost algorithm [13], achieved promising results. However, they did not consider the complexity of body parts. To the best of our knowledge, the system that was presented in [14] is the only one that included the configurations of single body parts.

The common drawback of most existing systems is that, they were not implemented based on empirical research of nonverbal behaviors. Moreover, although most of them provided mechanisms to deliver feedbacks to the presenters (except [9]), the forms of feedbacks are rather simple. They are text/images [8], sound [10] or lightning [7]. These methods can only provide users with solely assessment information, without concerning the entertaining aspect of the system, which might be valuable for educational purposes.

This paper presents our progress in developing a tutoring system for public speaking, which assesses presentations based solely on the visual behaviors of presenters. Firstly, an empirical study was performed in order to investigate on the nonverbal cues that impact a presentation, serving as the ground truth. Next, a Microsoft Kinect was implemented for capturing skeletal representations of the presenters' body as input data for the analysis. The recognition process can detect if the behaviors appeared in real-time. Multi-class support vector machine was used to classify the quality of presentations into a four-degree scale with the recognition rate of 73.9% on a training/test database that includes

76 presentations. For the feedback, the system allows presenters to review their presentation, together with the analysis results. In parallel, we developed a simulated conference room as the real-time feedback mechanism.

In the next section, we will explain our empirical results from the recorded presentations. The current development status will be introduced afterward. The last section is for conclusions and future works.

2 Nonverbal Behaviour of Presenters

In order to gather the ground truth as the guidance for our system, an observation was performed. We collected data from a training class about public speaking skills for postgraduate students. Learners were asked to give short presentations (about one minute) in front of the audience, which includes about ten other learners and one or two coaches. The content of the presentations was freely chosen by the presenters. In fact, all presenters chose to talk about their own research, in the ways that it can be understood by all of the audience that might came from the different fields. After each presentation, the audience gave feedbacks and suggestions on how the presentation should be improved, in terms of nonverbal expressions. We set up a regular camera to record the presentations. In parallel, a Microsoft Kinect was used to capture the whole-body movement for our further signal processing, as well as behavioral studies (Figure 1). Data from Kinect was stored as the *.ONI files using the OpenNI SDK (http://www.openni.org/). Finally, after removed the unsatisfied videos (e.g. presenters moved out of the camera range), 39 presentations of 11 presenters (four females, seven males) were collected.

Regular videos were used for behavioral analysis. This task was done through the collaboration with an expert in public speaking. The role of the expert was to review the recorded videos, and then specifying the nonverbal cues that affected the performance of the speakers, together with the durations that they appeared. Thus, for each video, a set of behaviors was created. We collected the nonverbal cues and then annotated their appearance using the commercial software Noldus Observer XT [15]. Behaviors were categorized into either *State event* if their duration is necessary to be studied, or *Point event* otherwise. The software provided us with the statistical analysis on the appearance of these behaviors, including the number of presentations that contain the behaviors, the rate that they appeared (point events) and the percentage of time that they accounted for (Table 1).

The observed behaviors can be separated based on the nonverbal channels that they were generated: (1) Posture (the static configuration of body), (2) Voice (concerning the paralinguistic characteristics), (3) Eye contact, (4) Facial Expression, (5) Globe body movement, (6) Hand gesture. This method of categorization is similar to the literature of public speaking skills [2]. Due to the limited amount of space, we could not describe all of the observed behaviors in

Table 1. The list of observed nonverbal cues

# Behaviors	Event Type (S/P)	No.	Rate of occurrences (times/minute)			Percentage during observation of the occurrences (%)		
			M	SD	Range	M	SD	Range
Postural behaviors								
1 (-) Shoulders too tight	S	19				60.94	23.80	12.67 - 98.50
2 (-) Legs closed	S	12				73.02	36.44	5.15 - 100
3 (-) Legs too stretch	S	3				61.42	11.33	19.18 - 100
4 (-) Weight in one foot	S	20				65.42	28.69	5.20 - 100
5 (-) Chin too high	S	14				64.94	23.80	12.67 - 98.50
6 (-) Hands in pockets	S	3				11.85	4.89	12.76 - 96.20
7 (+) Lean forward	S	19				32.50	28.66	3.70 - 82.78
8 (-) Lean backward	S	17				62.80	28.07	12.73 - 96.20
Vocal behaviors								
9 (-) Speak too fast	S	19				45.88	36.55	7.32 - 100
10 (-) Start too fast	P	18						
11 (-) Energy decreases at the end	P	23	2.88	1.77	0.53 - 6.31			
12 (+) Vocal emphasis	P	33	5.51	4.51	0.59 - 17.50			
13 (+) Suitable pause	P	33	4.63	3.16	0.53 - 12.5			
14 (-) Unsuitable pause	P	20	1.73	1.14	0.53 - 5.19			
15 (-) Monotone	S	20				92.49	13.08	56.29 - 100
16 (-) Fillers	P	34	5.17	4.22	1.44 - 19.03			
17 (-) Stuttering	P	12	1.72	0.83	0.53 - 3.42			
Behaviors of eye contact								
18 (+) Make eye contact	S	39				93.81	8.24	75.00 - 100
19 (-) Contact avoidance	S	28				9.98	8.47	1.12 - 25.00
19.1 (-) Look up to ceiling	S	14				4.23	2.95	1.12 - 9.61
19.2 (-) Look down to floor	S	19				7.67	4.67	2.84 - 14.17
19.3 (-) Look at hands	S	11				10.24	3.15	4.40 - 13.15
Behaviors related to facial expression								
20 (+) Facial mimicry	S	30				39.31	25.97	4.50 - 91.81
21 (+) Smile	S	22				13.62	11.54	3.54 - 41.08
22 (-) Flat face	S	8				80.61	24.16	40.41 - 100
Behaviors related to whole body movement								
23 (-) Too much movement	P	11				42.21	25.87	4.68 - 89.32
24 (-) Too little movement	P	23				50.62	29.21	10.05 - 100
25 (-) Step backward	P	31	1.83	1.27	0.36 - 4.36			
26 (+) Step forward	P	34	2.06	1.04	0.59 - 4.61			
Behaviors related to hand gesture								
Amount of hand gesture								
27 Hand gesture occur	P	38	16.83	7.15	0.93 - 28.42			
28 (-) Too little gestures	S	20				69.55	34.64	17.21 - 100
29 (-) Too much gestures	S	10				61.49	31.82	27.34 - 96.10
Quality of hand gesture								
30 (-) Bounded gestures	P	30	6.75	5.33	1.00 - 19.77			
31 (+) Relaxed gestures	P	29	7.41	4.95	1.15 - 15.79			
32 (-) Casual gestures	P	10	5.16	3.14	1.56 - 10.28			
33 (-) Uncompleted gestures	P	27	3.23	2.78	0.93 - 10.27			
34 (+) Gestural emphasis	P	20	4.43	4.05	0.36 - 11.99			
35 (-) Repeated gestures	P	31	6.57	2.49	1.09 - 12.31			

detail in this section. Only the behaviors that support our current development will be further explained in the next section. On the other hand, although we aimed to observe all of the available nonverbal cues, the contributions of each

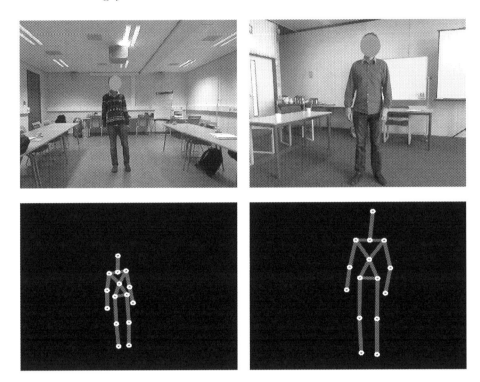

Fig. 1. Two samples from the database, with the color images (top row) and the skeletal representations of presenters' bodies, which were extracted and stored using Microsoft Kinect and the OpenNI SDK (bottom row)

individual to the success of a presentation are unequal. From our observation, as well as advices from the expert, the following aspects are the most important:

- *Eye Contact*: Similar to social interaction, maintaining good eye contact is the first thing the presenters must keep in mind. It initiates and strengthens the connection between them and the audience (#18, 19 in Table 1). It might have the first and foremost influence to the performance of a presentation, as well as regular communications [16].
- *Amount of energy*: This aspect concerns the dynamic characteristics of a presentation, thus can reflect the internal state of the presenters. It has impact in most behaviors that we have found (except posture as the static channel). For example, the amount of whole body movement (#23, 24), the amount of hand gesture (#28, 29), vocal behaviors (partly via tempo, emphases) and most features of hand gesture.
- *Variety*: The presentations with strong variations significantly increase the attention of the audience. Lacking variation results in monotone (#15), flat face (#22), and hand gesture repeated (#35). In fact, variety can be separated as one single measurement to analyze a presentation. It takes the role

as rhythm in music. Even a beautiful piece of music, without changes in rhythm will steadily lose attention from the audience.

3   Automatic Feedback System

In order to support presenters with an effective solution that can help them self-practice even at home, we aimed to implement the system with the following functions: (1)Automatic analyzes presenters performance; (2)Provides immediate feedback during the presentation; (3)Provides overall analysis about the whole presentation; (4)Lets users review their performance together with the analyzed results, thus allows them to keep track of their practicing progress. In order to achieve these purposes, we set up a Microsoft Kinect to extract body's skeletal representation as input for the analysis task. In parallel, a regular camera or webcam is positioned to simulate the audiences point of view. The automatic analysis, as well as recording is processed in real-time using a regular PC. The result is visualized on the PC or an external screen/projector (Figure 2). Users also have the chance to review their presentations, together with in-depth analysis about nonverbal cues in the end.

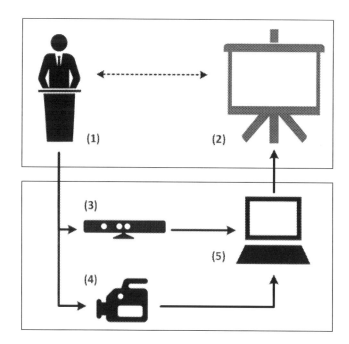

Fig. 2. Setup of the system

3.1   Recognition of Nonverbal Cues

In order to implement a system that mostly takes visual nonverbal channels into account, together with conclusions from our observation, we currently focused on the four aspects: (1)Eye contact; (2)Posture; (3)Gesture; (4)Whole body movement. For each aspect, several related behaviors were recognized, using solely data captured from the Kinect device. They are listed as follows, readers can refer to Table 1 to link to the behavior associated to each number:

– Eye Contact: #18, 19.
– Posture: #2, 3, 4, 7, 8.
– Gesture: #28, 29.
– Movement: #23, 24, 25, 26.

In the next subsections, we will explain the algorithms that were employed to detect these behaviors.

Eye Contact. Eye contact has significant impact on the creation and maintenance the connection between a presenter and the audience. Unfortunately, one common mistake was found in our observation is *eye contact avoidance*, which appeared in 28 over 39 presentations (Table 1, #19). Our observation found that, losing eye contact with the audience is the quickest way to ruin a presentation. Thus, we aimed to recognize the moments that the visual attention of the presenters is shifted away from the audience.

Visual focus of attention has been widely employed for human-computer interaction and usability research [17]. However, these applications require several constrain, regarding the minimum resolution of input data, or users are required to wear specific devices. Furthermore, in public speaking, eye contact does not really means knowing exactly the eye movements. Instead, it more concerns whether the audience can perceive that they are being addressed. In many applications, instead of localizing gaze exactly, head/face orientation was used as the effective approximation for subjects focus target [18,19]. The experiment in [20] also proved that, head orientation was the reliable indication of the visual focus of attention in 89% of the time.

Our system relies on faces 3D orientation, which is provided as part of the Kinect SDK as the measurement for visual attention. In our observation, when not making eye contact, presenters mostly looked up to ceiling or down to floor. When these behaviors appear, the pitch angle of their faces drops to lower/raise to higher than some specific ranges. These ranges can be set manually based on the observation. In our implementation, we set the range of face's pitch angle that according to having good facial orientation is within $[-15^0; 10^0]$. Being outside this range is assumed as *eye contact avoidance* appears.

Amount of Global Movement and Hand Gesture. Amount of movements and hand gesture can visually indicate the amount of energy that presenters use, also can reveal some hints about their internal states. Our observation results suggested that, this amount should be kept at the appropriate intensity to avoid negative impressions for the audience (Table 1, #23, 24, 28, 29). The next issue is, *how much movements/gesture is suitable?* In order to answer this question, we extracted from the annotation data the durations that movements and hand gesture were annotated as too much/little. Next, the EyesWeb XMI [21] was used on the Kinect-recorded database to extract the skeletal movement in these durations. We used central position of upper body to measure the amount of whole body movement, and total distance of two hands for the amount of hand gesture. Finally, the average results were computed (Table 2).

Table 2. Average velocity of body movements and hand gestures when being too much or little (in meter/s)

Behavior	N	Mean	Sd
Too much movements	42	0.34	0.08
Too little movements	67	0.11	0.03
Too much hand gesture	36	1.87	0.32
Too little hand gesture	53	0.17	0.09

We computed the average velocity of upper body and two hands in a moving time window of 5 seconds. Next, the mean values in Table 2 were used as hard thresholds to detect whether presenters' whole body and hands travelled too much/little.

Direction of Global Movement. Direction is one important characteristic of movement (Table 1, #25, 26). Moving small steps forward brings presenters closer to the audience, thus expresses the willing to make connection. In contrast, small steps backward can be seen in awkward presenters, when they subconsciously retreat from the stage.

We aimed to detect these behaviors via analyzing the trajectories of the skeletal joint *Spine*, being projected to the ground floor. At each frame, we calculated the direction of displacement. This direction was compared with the orientation of the upper body that was determined via shoulders positions. The comparison output the decision whether the presenter shifted backward/forward in this frame. We accumulated the number of these frames in a moving window of 1 second. All of frames inside a moving window is regarded as positive if such window contains more than 80% positive frames.

Body Posture. Posture is the static configuration of the presenters' body. We focused on the two postural aspects, including foot position and the focus of body weight. Foot position relates to the behaviors #2 and #3 in Table 1. It can be easily measured via Euclidean distance of *Foot_Left* (F_C) and *Foot_Right* (F_R). This number was normalized by the distance of *Shoulder_Left* (S_L) and *Shoulder_Right* (S_R) to eliminate the effect of body size (Equation 1).

$$Ratio_{FootDistance} = \frac{distance(F_L, F_R)}{distance(S_L, S_R)} \tag{1}$$

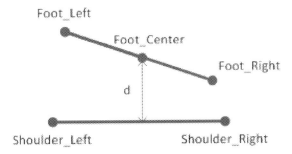

Fig. 3. Determine whether the presenter is leaning forward or backward via the distance d

Body weight relates to the behaviors #7, 8 in Table 1. It was measured via the the positions of shoulders, in comparison with foot, being projected to the ground floor (Figure 3). The distance d determines the degree of leaning, while the leaning backward or forward of body is determined via whether *Foot_Center* - *(F_C)* is in the left or right side of the vector created by *Shoulder_Left* and *Shoulder_Right*.

Another behavior relates to body weight is whether a presenter is standing in one leg/feet. This degree of leaning is determined via comparing the distance between *Hip_Center* - *(H_C)* with *Foot_Left* and *Foot_Right* (Equation 2).

$$Ratio_{Weight} = \frac{distance(H_C, F_L) - distance(H_C, F_R)}{distance(F_L, F_R)} \tag{2}$$

Again, we relied on our Kinect-recorded database to find the thresholds for the above postural metrics. The method similar to the one used in Section 3.1 was used, this time for every single frames that the behaviors appeared. The process produced the results as shown in Table 3.

Table 3. Average values for the postural metrics

Postural metric	N	Mean	Sd
Foot distance - closed (ratio)	3982	0.48	0.18
Foot distance - stretch (ratio)	1246	1.23	0.35
Weight forward (meters)	2108	0.31	0.12
Weight backward (meters)	2732	0.09	0.04
Weight on one feet (ratio)	4120	0.88	0.19

Table 4. Confusion matrix for the classification of the whole presentation

Ground Truth	Classified as			
	1	2	3	4
1 - Not good	**0.826**	0.043	0.087	0.044
2 - Good	0.087	**0.696**	0.174	0.043
3 - Very good	0.087	0.130	**0.652**	0.131
4 - Excellent	0.043	0.043	0.043	**0.783**
	Average sensitivity: **0.739**			

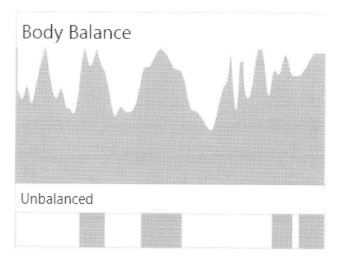

Fig. 4. Example of one behavior after being analyzed and displayed to presenters as off-line feedback. The upper graph visualizes the change of the metric used to measure body balance. The lower bar marks the durations that the behavior appeared.

Assessment of the Whole Presentation. After the separated behaviors can be recognized, we aimed to produce the final assessment for the whole presentation. Each presentation is assessed based on the four qualities, accordingly to the four nonverbal visual channels: (1) Posture; (2) Gesture; (3) Global Movement; (4) Eye contact. Additionally, one single quality is measured accordingly to the general performance. This is similar to what learners are given in most training classes about public speaking skills.

Firstly, in order to enrich the database, we asked for the permission to copy the DVDs of presentations from other students who attended the training class (each student is given one DVD that contains all of their presentations during the course in the last session). These video clips are not appear previously in our recorded database. Finally, we achieved extra 37 presentations (without Kinect recordings), in addition to our existing 39 videos that were used for the observation (which contain Kinect recordings). The expert was asked to assess the presentations based on a 4-level scale for each single aspect: (1) Posture; (2)

Fig. 5. The simulated conference room as the method to deliver immediate feedbacks. The avatars can perform different animations based on the performance of presenters.

Gesture; (3) Global Movement; (4) Eye contact, plus one score for the overall assessment. This data will be used for the later classification.

Based on the previous recognition process, a feature vector is created for each aspect, in which each component stands for the percentage that one single behavior appeared in the whole presentation. For example, feature vectors that represent hand gesture contain two dimensions for gesturing too much/little. Similarly, the number of dimensions of the feature vectors of movement, posture and eye contact are: 4, 5 and 1, respectively. Additionally, the whole presentations are represented via 12-component vectors, is the serialization of all features.

Amongst them, eye contact is represented via one single number and can be easily classified using hard thresholds. For the rest, feature vectors were used as the input for a multi-class Support Vector Machine system [11], using one-versus-all method and winner-take-all strategy. The algorithm was used to classify four classes, which represented the performance from [1 - Not good] to [4 - Excellent].

We applied 70-30% train/test split on the total of 76 presentations in order to evaluate the classification of the whole presentation, and then calculate the confusion matrix (Table 4). As the result, the system achieved the recognition rate of 73.9%.

The Two Methods of Giving Feedbacks. The system provides two ways of delivering feedbacks to the audience. The first one shows users their recorded presentation, the appearances of each behaviors (Figure 4) and results on the four nonverbal aspects, plus the overall result. In parallel, with purpose to give presenters the helpful feedbacks, also aim to provide them the experience as presenting for the real audience, we developed a virtual conference room as one method to deliver feedbacks. The environment was built using the Unity3D engine, simulates the classroom that we collected data for observation. Avatars are able to perform several animations that may bring either positive or negative feeling for presenters. These animation clips are sorted based on the increase of negative feeling: (1) Nodding; (2) Sitting still; (3) Sleeping; (4) Yawning.

In order to achieve real-time assessment, we applied a moving window to assess the overall quality for the most recent 5 seconds. This process outputted a single value from [1 - Not good] to [4 - Excellent]. The value then was used to manually adjust the distribution of the animation clips. Higher quality presentations resulted in higher proportion of positive animations and vice versa.

Acknowledgments. This work was supported in part by the Erasmus Mundus Joint Doctorate in Interactive and Cognitive Environments (ICE), which is funded by the Edu- cation, Audiovisual and Culture Executive Agency of the European Commission under EMJD ICE FPA 2010-0012.

References

1. Seiler, W.J., Beall, M.L.: Communication - Making connections. Allyn&Bacon (2004)
2. Rodman, G., Adler, R.B.: Style: delivery and language choices. In: The New Public Speaker, 1st edn. Wadsworth Publishing (1996)
3. Argyle, M., Alkema, F., Gilmour, R.: The communication of friendly and hostile attitudes by verbal and nonverbal signals. European Journal of Social Psychology **1**, 385–402 (1971)
4. D'Arcy, J.: Communicating with effective body language. In: Technically Speaking. Battelle Press, ch. 14 (1998)
5. Vinciarelli, A., Pantic, M., Bourlard, H.: Social signal processing: Survey of an emerging domain. Image and Vision Computing **27**(12), 1743–1759 (2009)
6. Picard, R.: Affective Computing, 1st edn. The MIT Press (2000)
7. Hincks, R., Edlund, J.: Promoting increased pitch variation in oral presentations with transient visual feedback. Language Learning & Technology **13**(3), 32–50 (2009)
8. Kurihara, K., Goto, M., Ogata, J.: Presentation sensei: a presentation training system using speech and image processing. In: Proceedings of the 9th International Conference on Multimodal interfaces, pp. 358–365 (2007)
9. Pfister, T., Robinson, P.: Real-Time Recognition of Affective States from Nonverbal Features of Speech and Its Application for Public Speaking Skill Analysis. IEEE Transactions on the Affective Computing, 1–14 (2011). http://ieeexplore.ieee.org/xpls/abs_all.jsp?arnumber=5740838

10. Silverstein, D.A., Tong, Z., Zhang, T.: System and method of providing evaluation feedback to a speaker while giving a real-time oral presentation. US Patent 7,050,978 (2003)
11. Duan, K.-B., Keerthi, S.S.: Which is the best multiclass SVM method? An empirical study. In: Oza, N.C., Polikar, R., Kittler, J., Roli, F. (eds.) MCS 2005. LNCS, vol. 3541, pp. 278–285. Springer, Heidelberg (2005). http://link.springer.com/chapter/10.1007/11494683_28
12. Gao, T., Wu, C., Aghajan, H.: User-centric speaker report: Ranking-based effectiveness evaluation and feedback. In: 2009 IEEE 12th International Conference on Computer Vision Workshops (ICCV Workshops), pp. 1004–1011. IEEE (2009)
13. Freund, Y., Iyer, R., Schapire, R., Singer, Y.: An efficient boosting algorithm for combining preferences. The Journal of Machine Learning Research 4, 933–969 (2003). http://dl.acm.org/citation.cfm?id=964285
14. Nguyen, A., Chen, W., Rauterberg, G.: Feedback system for presenters detects nonverbal expressions. In: SPIE Newsroom (2013). http://spie.org/x91885.xml?highlight=x2410&ArticleID=x91885
15. Zimmerman, P., Bolhuis, J.: The Observer XT: A tool for the integration and synchronization of multimodal signals. Behavior Research Methods 41(3), 731–735 (2009). http://link.springer.com/article/10.3758/BRM.41.3.731
16. Kleinke, C.L.: Gaze and eye contact: a research review. Psychological Bulletin 100(1), 78–100 (1986)
17. Jacob, R., Karn, K.: Eye tracking in human-computer interaction and usability research: Ready to deliver the promises. Work 2(3), 573–605 (2003). http://www.ee.uwa.edu.au/~roberto/research/projects2013/10.1.1.100.445.pdf
18. Sheikhi, S., Odobez, J.-M.: Recognizing the visual focus of attention for human robot interaction. In: Salah, A.A., Ruiz-del-Solar, J., Meriçli, Ç., Oudeyer, P.-Y. (eds.) HBU 2012. LNCS, vol. 7559, pp. 99–112. Springer, Heidelberg (2012). http://link.springer.com/chapter/10.1007/978-3-642-34014-7_9
19. Ba, S.O., Odobez, J.-M.: Recognizing visual focus of attention from head pose in natural meetings. IEEE transactions on systems, man, and cybernetics. Part B, Cybernetics : a publication of the IEEE Systems, Man, and Cybernetics Society 39(1), 16–33 (2009). http://www.ncbi.nlm.nih.gov/pubmed/19068430
20. Stiefelhagen, R.: Tracking focus of attention in meetings. In: Proceedings of the Fourth IEEE International Conference on Multimodal Interfaces, pp. 273–280. IEEE Comput. Soc (2002). http://ieeexplore.ieee.org/lpdocs/epic03/wrapper.htm?arnumber=1167006
21. Camurri, A., Hashimoto, S., Ricchetti, M., Ricci, A., Suzuki, K., Trocca, R., Volpe, G.: Eyesweb: Toward gesture and affect recognition in interactive dance and music systems. Computer Music Journal 24(1), 57–69 (2000)

Performing Variable Selection by Multiobjective Criterion: An Application to Mobile Payment

Alberto Guillén[1]([✉]), Luis-Javier Herrera[1], Francisco Liébana[2],
Oresti Baños[3], and Ignacio Rojas[1]

[1] Department of Computer Architecture and Technology,
University of Granada, Granada, Spain
aguillen@ugr.es
[2] Department of Marketing and Market Research,
University of Granada, Granada, Spain
[3] Ubiquitous Computing Lab (UCLab), Kyung Hee University, Seoul, Korea

Abstract. The rapid growth social networks have led many companies to use mobile payment systems as business sales tools. As these platforms have an increasing acceptance among the consumers, the main goal of this research is to analyze the individuals' use intention of these systems in a social network environment. The problem of variable selection arises in this context as key to understand user's behaviour. This paper compares several non-parametric criteria to perform variable selection and combines them in a multiobjective manner showing a good performance in the experiments carried out and validated by experts.

1 Introduction

Social commerce presents two advantages over existing forms of commerce [25]. First of all, it facilitates the interaction between Internet users, allowing direct interaction and exchange of opinions, purchase advice and experiences (participatory environment). Secondly, it makes web surfing and knowledge about a variety of products possible, both of which are limited in an offline context (unlimited access).Therefore, given the growth that Virtual Social Networks (VSNs) are experiencing in our society, it is necessary for both companies and users to analyze the intention to use the new tools implemented on social networks, to establish the level of general acceptance of these tools, as well as the factors that determine user behavior regarding payment systems that can generate profit for a company.

There are numerous behavioral decision-making theories and intention models in scientific literature, analyzing individuals' behavior when facing innovation, the majority of which are based on social psychology studies [20]. However, this problem can be reformulated so machine learning and data mining algorithms can be applied. The definition of the problem from these disciplines point of view is that given a set of input vectors $X = [x_1^1 x_1^2 x_1^d; x_2^1 x_2^2 x_2^d; x_m^1 x_m^2 x_m^d]$ and their corresponding output $Y = [y_m]$, it is necessary to find out which subset

© Springer International Publishing Switzerland 2015
I. Rojas et al. (Eds.): IWANN 2015, Part II, LNCS 9095, pp. 333–340, 2015.
DOI: 10.1007/978-3-319-19222-2_28

of variables are the most relevant to determine the output. Therefore, there are (2^d)-1 possible solutions only considering if the variable i is selected or not.

There are several ways to determine which subset of variables is the most adequate, but two classes can be defined: parametric versus non-parametric. Parametric criteria are the ones that design a model that adjusts the output Y, so the more accurate the model obtained using a particular subset of variables, the better this subset is. Examples of previous algorithms performing this task include models built simultaneously using a genetic algorithm[9]. The main problem with these approaches is that they are very expensive in terms of computation so very few solutions can be evaluated. Another problem that this approach presents is that the value obtained for each subset may change depending on the parameters chosen for the model, (ex. the kernel function, number of neurons, weights in the hidden layer, etc.) so the evaluation is not robust.

In order to overcome these drawbacks, several papers apply non-parametric values so the result is independent of the choices made beforehand. This paper proposes a new approach to tackle the variable selection problem by using multiobjective criteria that consider opposite non-parametric criteria as could be mutual information and delta test [10],[16], [8], [14], [11], [12].

The rest of the paper is organised as follows: Section 2 explines how the data were collected. Then, Section 3 introduces the procedures to rank variable selections. In Section 4, experiments are carried out and, afterwards in Section 5, conclusions are drawn.

2   Data Compilation

The research was carried out on the worlds best-known social network, Facebook, because of its widespread diffusion, the large number of users and the usefulness it confers. These users were exposed to an experimental scenario on a Facebook profile, in which they watched a video explaining the proposed new payment system. This system, called Zong, facilitates the purchase of both physical goods and virtual content using a mobile terminal. This very simple process accepts different payment methods for purchases via the Internet, social networks, television and even the point of sale itself). Payment can be charged to a users phone bill or to a credit card authorized during service activation.

The size of the final sample was 1840, using a quota sampling method based on the characteristics of users who participated in the Survey about Equipment and Use of Information and Communication Technologies in Households by the National Statistics Institute [17].

The variables considered in this research to define intention to use the new payment system were structured in five groups: behavioral variables, sociodemographic variables, user technological and social media comfort level and users previous experience with similar payment tools. First, social influence or social image and subjective norms , since in the context of this research there is a very close relationship between them due both to desired social value, and to the influence the social environment on the social network exerts on users of the new payment system [13]; secondly, ease of use based on individuals perception that using a certain system will be effortless or simply easy to use [2]; thirdly, perceived usefulness, defined as the potential users subjective probability that using a specific system will improve their job performance in an organizational context [3]; fourthly, attitude based on favorable or unfavorable feelings people express towards a given behavior [22]; fifthly, trust based on the psychological state that leads one to accept the vulnerability of those who trust and is based on favorable expectations about the intentions and behaviors of others [23]; sixthly, perceived risk considered in view of a potential unsafe use of the innovation [7]; seventhly, perceived quality based on users subjective comparison between the desired and the received quality of service [6]; and finally, perceived satisfaction, defined as the difference between the expectations and the feelings generated by the use experience [19].

Finally, the level of experience shows that a users positive experience with a similar tool in the past will have a decisive impact on their behavior [5].

Once the data was collected, in order to analyze it, preprocessing was required to obtain concrete values for each type of question. In order to do this, an average value was computed considering all the answers for a particular type of question. Afterwards, these values were normalized with a mean equal to zero and a standard deviation equal to one.

3   Criteria for Variable Selection

In order to determine the relevance of each variable measured, two criteria were selected: the Delta Test (DT), and Mutual Information (MI). Four different algorithmic approaches have been considered and compared: the minimum value of Delta Test; the maximum value of the estimation of the MI and a novel approach that combines the former two criteria in a multi-objective approach described later.

3.1   Mutual Information

Given a single-output multiple input function approximation or classification problem, with input variables $X = [x_1, x_2, \ldots, x_d]$ and output variable $Y = y$, the main goal of a modeling problem is to reduce the uncertainty on the dependent variable Y. According to the formulation of Shannon, and in the continuous case, the uncertainty on Y is given by its entropy defined as:

$$H(Y) = - \int \mu_Y(y) \log \mu_Y(y) dy, \tag{1}$$

considering that the marginal density function $\mu_Y(y)$ can be defined using the joint probability density function $\mu_{X,Y}$ of X and Y as $\mu_Y(y) = \int \mu_{X,Y}(x, y) dx$.

Given that we know X, the resulting uncertainty of Y conditioned to known X is given by the conditional entropy, defined by:

$$H(Y, X) = - \int \mu_X(x) \int \mu_Y(y|X = x) \log \mu_Y(y|X = x) dy dx. \tag{2}$$

The mutual information (also called cross-entropy) between X and Y can be defined as the amount of information that the group of variables X provide about Y, and can be expressed as $I(X,Y) = H(Y) - H(Y|X)$. In other words, the mutual information $I(X,Y)$ is the decrease of the uncertainty on Y once we know X. Due to the mutual information and entropy properties, the mutual information can also be defined as:

$$I(X,Y) = H(X) + H(Y) - H(X|Y), \tag{3}$$

leading to:

$$I(X,Y) = \int \mu_{X,Y}(x,y) \log \frac{\mu_{X,Y}(x,y)}{\mu_X(x)\mu_Y(y)} dxdy. \tag{4}$$

Thus, only the estimate of the joint PDF between X and Y is needed to estimate the mutual information between two groups of variables. Estimating the joint probability distribution can be performed using a number of techniques like histograms and kernel density estimators have been used for this purpose. Although there exists a variety of algorithms to calculate the mutual information between variables, this paper uses the approach presented in [26] which is based on the k-nearest neighbors.

3.2 Delta Test

The Delta Test (DT), introduced by Pi and Peterson for time series and proposed for variable selection in [4], is a technique to estimate the variance of the noise, or the mean squared error (MSE), that can be achieved without overfitting. Given N input-output pairs $(\mathbf{x}_i, y_i) \in \mathbb{R}R^d \times \mathbb{R}R$, the relationship between $\mathbf{x}_i$ and y_i can be expressed as

$$y_i = f(\mathbf{x}_i) + r_i, \quad i = 1, ..., N \tag{5}$$

where f is an unknown function and r is the noise. The DT estimates the variance of the noise r.

The DT is useful for evaluating the nonlinear correlation between two random variables, namely, input and output pairs. The DT can also be applied to input variable selection: the set of input variables that minimizes the DT is the one that is selected. Indeed, according to the DT, the selected set of input variables is the one that represents the relationship between input variables and the output variable in the most deterministic way.

This test based on a hypothesis coming from the continuity of the regression function. If two points $\mathbf{x}$ and $\mathbf{x}'$ are close in the input space, the continuity of the regression function implies that the outputs $f(\mathbf{x})$ and $f(\mathbf{x}')$ are also close enough in the output space. Alternatively, if the corresponding output values are not close in the output space, this is due to the influence of the noise.

3.3 MultiObjetive Criteria: MOmaxMI or MOminDT

This paper presents a novel approach that is able to combine the two criteria considered. Very little research has previously been done in this direction. In [21] two ways of computing MI are optimized simultaneously. However, the concept being computed is the same. Although the evaluation of a good concrete subset of variables should be similar for both DT and MI, the reality is that, for many solutions, these two values are opposites. The reason is because the MI measures the entropy and the DT measures the variance of the noise in the output so, if there is a high variance of the noise, there is high entropy. Therefore, it is interesting to optimize both values at the same time in order to obtain a compromise solution.

Multi-objective optimization is a well-known problem that is usually solved by using heuristics such as genetic algorithms. However, in this case, we can afford to compute the values for all the possible solutions and select the solutions from the complete Pareto front [1].

4   Experiments

In order to validate the experimental results a set of experts (with more than 10 years of experience) related to the different research topics were consulted.

The evaluations of the methods from the subsets of experts are shown in Table 1 using a Likert (1-7) scale. This table presents 4 criteria to perform the variable selection. The first one considers the criterion of maximizing the mutual information. The second one is one of the novel contributions of the paper as it tries to maximize the mutual information while also, selecting a solution that has a relatively small Delta test value. The third criterion is also a multi-objective approach that minimizes the Delta test value but it considers it as a solution that provides a high value for the MI. The forth one consists of the minimization of the Delta test. The two multiobjective solutions were taken from the Pareto front shown in Figure 1.

Table 1. Punctuation results from expert's opinion on the variables selected by the different criteria. Group A are marketing and market research experts and Group B corresponds to Financial and payment experts.

A	maxMI	MO-maxMI	MO-minDT	minDT
Expert 1	3	3	4	3
Expert 2	2	4	4	2
Expert 3	3	3	5	3
Mean(std)	2,6	3,3	2,6	2,6
B	maxMI	MO-maxMI	MO-minDT	minDT
Expert 1	3	7	3	3
Expert 2	2	6	4	2
Expert 3	4	7	5	2
Mean(std)	3	6,6	4	2,3
Global Mean (std)	2,8	4,95	3,3	2,45

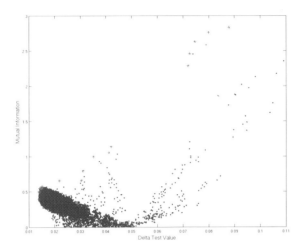

Fig. 1. Blue dots represent all posible solutions according to their DT and MI values. Red stars represent the Pareto front from the whole solution set.

As the table shows, the best total score is obtained by the second criterion. The other criteria are not highly rated in comparison with the first two although the second one is another MO approach.

Once all the results were collected, a second interview was carried out and the experts justified the scores given to each method. The reasons for the scores were: Regarding Max MI criterium (option 1): It only selects one variable (users experience with Internet). The experts argue that, although the variable is important, it is not the only one that should be taken into account. Regarding the MO approach that maximizes MI (option 2) and keeps a small value for the DT: It includes several variables such as the ones related to the users experience using Internet, mobile and social networks. Some experts comment that the method does not select variables that represent the behavior of the potential user as it does not consider the trajectory of the user with similar situations. On the other hand, the fact that it selects the elements related to the new payment platform it is highly scored by the experts in the marketing field. Regarding the MO approach that minimizes the DT (option 3) and tries to keep a high MI: It selects nearly all the variables presented as inputs so the total number of variables is very high and this is punished by the experts (in both the marketing and the financial fields) who could not make a decision based on so many factors. Regarding the minimization of the DT (option 4), it has the same problem as the previous method. The number of variables selected is too high and experts agree that it is not practical to make a decision.

From the machine learning point of view, the results are very interesting. In the first place, they confirm that the multi-objective way of thinking is more practical and provides better results than considering only one criterion. Specifically, in this case, maximizing the entropy but keeping in mind that the noise should be reduced, by considering a small value of DT, was a very successful approach.

5   Conclusions and Further Work

The developments that are taking place in the Information and Communication Technologies (ICT) sector in our society in recent years have had repercussions in terms of profitability, productivity, competitiveness and economic growth, at the company level. Consumers are increasingly using social networks to obtain recommendations and opinions from friends, relatives, experts and the entire social community. It is very important to be able to identify the key elements that drive the user behaviour. The paper has presented a fusion between two well-known criteria to perform variable selection and optimize them in a multi-objective way. The results obtained were validated by several experts showing that it is interesting to select the variables with high entropy, maintaining variables with small variance of the noise in the output.

Acknowledgments. This work has been supported by the GENIL-PYR-2014-12 project from the GENIL Program of the CEI BioTic, Granada, and the Junta de Andalucia Excellence Project P12-TIC-2082.

References

1. Coello, C.A.C., Lamont, G.B., Van Veldhuisen, D.A.: Evolutionary algorithms for solving multi-objective problems. Springer (2007)
2. Davis, F.D.: Perceived Usefulness, Perceived Ease of Use, and User Acceptance of Information Technology. MIS Quarterly **13**(3), 319–340 (1989)
3. Davis, F.D., Bagozzi, R.P., Warshaw, P.R.: User Acceptance of Computer Technology: A Comparison of Two Theoretical Models. Management Science **35**, 982–1003 (1989)
4. Eirola, E., Liitiinen, E., Lendasse, A., Corona, F., Verleysen, M.: Using the delta test for variable selection. In: ESANN, pp. 25–30 (2008)
5. Fishbein, M., Ajzen, I.: Belief, Attitude, Intention and Behavior: An Introduction to Theory and Research. Addison-Wesley, Reading (1975)
6. Gefen, D.: E-commerce: The role of familiarity and trust. The International Journal of Management Science **28**, 725–737 (2000)
7. Gerrard, P., Cunningham, J.B.: The diffusion of internet banking among Singapore consumers. International Journal of Bank Marketing **21**(1), 16–28 (2003)
8. Guillén, A., Del Moral, F.G., Herrera, L.J., Rubio, G., Rojas, I., Valenzuela, O., Pomares, H.: Using near-infrared spectroscopy in the classification of white and iberian pork with neural networks. Neural Computing and Applications **19**(3), 465–470 (2010)
9. Guillén, A., Pomares, H., González, J., Rojas, I., Valenzuela, O., Prieto, B.: Parallel multiobjective memetic rbfnns design and feature selection for function approximation problems. Neurocomputing **72**(16), 3541–3555 (2009)

10. Guillén, A., Sovilj, D., Lendasse, A., Mateo, F.: Minimising the delta test for variable selection in regression problems. International Journal of High Performance Systems Architecture **1**(4), 269–281 (2008)
11. Guillén, A., Sovilj, D., van Heeswijk, M., Herrera, L.J., Lendasse, A., Pomares, H., Rojas, I.: Evolutive approaches for variable selection using a non-parametric noise estimator. In: de Vega, F.F., Pérez, J.I.H., Lanchares, J. (eds.) Parallel Architectures and Bioinspired Algorithms. SCI, vol. 415, pp. 243–266. Springer, Heidelberg (2012)
12. Herrera, L.J., Fernandes, C.M., Mora, A.M., Migotina, D., Largo, R., Guillén, A., Rosa, A.C.: Combination of heterogeneous EEG feature extraction methods and stacked sequential learning for sleep stage classification. International journal of neural systems **23**(3) (2013)
13. Liébana-Cabanillas, F.: El papel de los sistemas de pago en los nuevos entornos electrónicos, Doctoral Thesis, Marketing and Market Research Department, University of Granada (2012)
14. Long, X.X., Li, H.D., Fan, W., Xu, Q.S., Liang, Y.Z.: A model population analysis method for variable selection based on mutual information. Chemometrics and Intelligent Laboratory Systems **121**, 75–81 (2012)
15. Mallat, N., Rossi, M., Tuunainen, V., Rni, A.: The impact of use context on mobile services acceptance: the case of mobile ticketing. Information & Management **46**(3), 190–195 (2009)
16. May, R.J., Maier, H.R., Dandy, G.C., Fernando, T.M.K.: Non-linear variable selection for artificial neural networks using partial mutual information. Environmental Modelling & Software **23**(10), 1312–1326 (2008)
17. National Statistics Institute, Survey about Equipment and Use of Information and Communication Technologies in Households (2012). www.ine.es
18. OCass, A., Fenech, T.: Web retailing adopction: exploring the nature of internet users web retailing behavior. Journal of Retailing and Consumer services **10**, 81–94 (2003)
19. Oliver, R.L.: A cognitive model of the antecedents and consequences of satisfaction decisions. Journal of Marketing Research **17**, 460–469 (1980)
20. Pavlou, P.A.: A theory of Planned Behavior Perspective to the Consumer Adoption of Electronic Commerce. MIS Quarterly **30**(1), 115–143 (2002)
21. Peng, H., Long, F., Ding, C.: Feature selection based on mutual information criteria of max-dependency, max-relevance, and min-redundancy. IEEE Transactions on Pattern Analysis and Machine Intelligence **27**(8), 1226–1238 (2005)
22. Premkumar, G., Rammurthy, K., Liu, H.: Internet Messaging: An Examination of the Impact of Attitudinal, Normative and Control Belief Systems. Information & Management **45**, 451–457 (2008)
23. Singh, J., Sirdeshmukh, D.: Agency and Trust Mechanisms in Consumer Satisfaction and Loyalty Judgments. Journal of the Academy of Marketing Science **28**(1), 150–167 (2000)
24. Smith, M., Brynjolfsson, E.: Consumer decision making at an internet shopbot: Brand still matters. The Journal of Industrial Economics **49**(4), 541–558 (2001)
25. Zhang, L.: Business model analysis for online social shopping companies. Case Company: Run To Shop Oy. Doctoral Thesis, Department of Business Technology/Logistics, Helsinki School of Economics (2009)
26. Herrera, L.J., Pomares, H., Rojas, I., Verleysen, M., Guilén, A.: Effective input variable selection for function approximation. In: Kollias, S.D., Stafylopatis, A., Duch, W., Oja, E. (eds.) ICANN 2006. LNCS, vol. 4131, pp. 41–50. Springer, Heidelberg (2006)

Advances in Computational Intelligence

Aggregation of Partial Rankings – An Approach Based on the Kemeny Ranking Problem

Gonzalo Nápoles[1,2(✉)], Zoumpoulia Dikopoulou[2], Elpiniki Papageorgiou[3],
Rafael Bello[1], and Koen Vanhoof[2]

[1] Universidad Central "Marta Abreu" de Las Villas, Santa Clara, Cuba
gnapoles@uclv.edu.cu
[2] Hasselt University, Diepenbeek, Belgium
[3] Technological Education Institute of Central Greece, Lamia, Greece

Abstract. Aggregating the preference of multiple experts is a very old problem which remains without an absolute solution. This assertion is supported by the Arrow's theorem: there is no aggregation method that simultaneously satisfies three fairness criteria (non-dictatorship, independence of irrelevant alternatives and Pareto efficiency). However, it is possible to find a solution having minimal distance to the consensus, although it involves a NP-hard problem even for only a few experts. This paper presents a model based on Ant Colony Optimization for facing this problem when input data are incomplete. It means that our model should build a complete ordering from partial rankings. Besides, we introduce a measure to determine the distance between items. It provides a more complete picture of the aggregated solution. In order to illustrate our contributions we use a real problem concerning Employer Branding issues in Belgium.

Keywords: Partial rankings · Aggregation · Ant colony optimization

1 Introduction

The aggregation of multiple preferences has been widely studied by economists under social choice theory. In recent years reasoning based on permutations has gained great attention due to the applications in solving decision-making problems. For example, a main aspect in some machine learning tasks is how to combine the output of multiple classifiers, in order to determine the most adequate class. Other relevant applications include: computational biology [1], multi-agent planning [2], voting in elections [3], information retrieval [4], among others interesting fields.

Formally the aggregation of preferences could be summarized as follows: given N orderings over M objects/items where each ranking denotes the preference of a single expert, then the goal is to build a consensus (aggregated) ranking taking into account all input rankings. Borda [5] and Condorcet [6] proposed diverse ways of aggregating the preferences of the multiple voters, and argued over which method is the right one. Unfortunately, Arrow [7] proved that there is no right approach, since there exists no aggregation method that concurrently satisfies three fairness criteria: non-dictatorship, Pareto efficiency and independence of irrelevant alternatives.

© Springer International Publishing Switzerland 2015
I. Rojas et al. (Eds.): IWANN 2015, Part II, LNCS 9095, pp. 343–355, 2015.
DOI: 10.1007/978-3-319-19222-2_29

Despite this negative result, it is possible to compute an aggregated ranking having minimal distance to the global consensus. This ranking/ordering is also known as the Kemeny ranking. In [8] the authors performed an extensive study including different methods (e.g. Branch and Bound, approximate algorithms) for computing the Kemeny ranking. It was also concluded that heuristic approaches are recommended in contexts having weak or no consensus. More recently, Aledo at al. [9] introduced an approach based on evolutionary computation, which clearly outperformed the remaining tested algorithms. However, this model is based on the Kendall distance and hence it cannot be directly applied to the aggregation of incomplete preferences.

Inspired on this work we introduce an approximate model for aggregating partial rankings which uses Ant Colony Optimization (ACO) as optimizer. The main reason behind this decision is moved by the strong ability of ACO for solving combinatorial problems [10] and also by its scheme for generating new states. In a few words, ACO exploits the heuristic information for improving the search, and this knowledge could be easily computed from input data. Moreover, we present a measure for computing the relative distance between two items/objects in the final ranking (which is entirely based on the induced deviation to the experts' consensus).

The rest of the paper is organized as follows. The next section makes an overview about the Kemeny ranking problem and describes an extension for aggregating partial rankings. It includes the design of the objective function to be optimized, and also the measure for computing the distance between two objects. Section 3 reviews the main ideas of ACO-based optimizers, which will be used next for generating the candidate ranking. In Section 4 we explain how to estimate the heuristic information from input data for two different scenarios. Section 5 presents a real study case about employees' preferences in Belgium when they look for an employer. In the last section the authors discuss conclusions and future research directions.

2   Extending the Kemeny Ranking Problem

Solving the Kemeny ranking problem is equivalent to compute the consensus ranking for a set of input rankings. The reader can find the formulation of this problem in [11] although it could be summarized as follows: given a set of N rankings $\pi_1, \pi_2, ..., \pi_N$ with M elements, the Kemeny ranking problem consists on finding the ranking π_* that satisfies the expression (1). Here π_* is the central permutation to be computed by the model, whereas $d_1(\pi, \pi_*)$ denotes the Kendall distance [12]. In brief, the Kendall-Tau distance between two rankings π_1 and π_2 is defined as the total number of items pairs over which they disagree. However, this function assumes that all objects are ranked where ties are not allowed, hence the Kendall-Tau distance is not a suitable alternative when we want to face the aggregation of partial rankings.

$$\pi_* = argmin_{\pi_i} \frac{1}{N} \sum_{i=1}^{N} d_1(\pi_i, \pi_*) \qquad (1)$$

In many important applications, rankings are only partial, in the sense that ties are allowed. This happens, for example, when only the top, say K, elements are ordered, while all the remaining elements are assumed to have rank $K + 1$ [13]. It is possible a second scenario where each expert X_i selects K_i items as relevant, leading to a ranking with $(M - K_i)$ elements are tied at the $K_i + 1$ position. To deal with such scenarios we need to extend the Kemeny ranking problem by replacing the Kendall-Tau distance by a measure capable to compare partial rankings. In this paper we adopt the Hausdorff distance [14] since it is based on the Kendall-Tau principle. Besides, this distance has been widely studied and shown to have especially nice mathematical and algorithmic properties, particularly with regard to rank aggregation [15].

$$d_2(\pi_i, \pi_*) = |\mathcal{D}(\pi_i, \pi_*)| + max \{|\mathcal{R}_1(\pi_i, \pi_*)|, |\mathcal{R}_2(\pi_i, \pi_*)|\} \tag{2}$$

In the equation (2) the set $\mathcal{D}(\pi_i, \pi_*)$ denotes all items pairs that appear in different order, $\mathcal{R}_1(\pi_i, \pi_*)$ represents the set of all objects pairs which are tied in π_i but not tied in the ordering π_*, whereas $\mathcal{R}_2(\pi_i, \pi_*)$ is the set of elements pairs which are tied in π_* but not tied in the ranking π_i. However, if we consider that the ranking π_* is complete then the expression (2) could be notably reduced as follows:

$$d_2(\pi_i, \pi_*) = |\mathcal{D}(\pi_i, \pi_*)| + \binom{M - K_i}{2} \tag{3}$$

In other words, $|\mathcal{R}_2(\pi_i, \pi_*)| = 0$ since π_* is a complete ordering (i.e. the candidate solution to be generated by the selected optimizer). Likewise, the reader could observe that $|\mathcal{R}_1(\pi_i, \pi_*)| = \binom{M - K_i}{2}$ where K_i is the number of relevant items, according to the ith respondent. Next equation shows the normalized objective function, which should be minimized during the search process. The closer to zero the evaluation, the closer the permutation π to the consensus. In the next Section we describe the main ideas of ACO-based algorithms, but first we present a new measure for computing the relative distance between two consecutive objects in a consensus ranking.

$$H(\pi) = \frac{1}{N} \sum_{i=1}^{N} \frac{|\mathcal{D}(\pi_i, \pi)| + \binom{M - K_i}{2}}{M(M - 1)/2} \tag{4}$$

2.1   Measuring the Distance Between Two Objects

As discussed, the goal of the Kemeny ranking problem is to find a complete ordering from a set of input rankings, having minimal distance to the consensus. In this scheme the inputs could be partial or complete. The aggregated ranking is a suitable tool when we want to face decision-making problems based on permutations, but sometimes this knowledge is not enough, and further analysis is often required. For instance, a central question when experts interpret a consensus ordering is: which is the relative distance between two consecutive objects? Next we introduce a strategy to solve this question, which is completely based on the objective function.

Let us consider a consensus ordering $\pi = \{\pi(1), \pi(2), ..., \pi(l), \pi(l + 1), ..., \pi(M)\}$ that minimizes (4) such as $x = \pi(l)$ and $y = \pi(l + 1)$. In other words, π is a solution for the Kemeny ranking problem where $x \prec y$. The equation (5) displays the *induced deviation* $\delta(y, x)_\pi$ which measures the distance between such items.

$$\delta(y,x)_\pi = \frac{|H(\pi) - H(\breve{\pi}_{y \prec x})|}{\sum_k |H(\pi) - H(\breve{\pi}_{\pi(k+1) \prec \pi(k)})|} \tag{5}$$

In this equation $\breve{\pi}_{y \prec x}$ represents a ranking obtained from π where objects x and y were exchanged, whereas $k = 1, \ldots, M$ indexes the elements. The reader could notice that $\delta(y,x)_\pi = 0$ if $\pi_{x \prec y}$ and $\breve{\pi}_{y \prec x}$ have the same heuristic value, and accordingly we can conclude that x and y are tied (i.e. they are at the same level). It should be stated that our model attempts to compute a complete ordering using a set of partial and/or complete rankings as input data. However, it is possible to obtain a solution implicitly having tied items. Even it is possible that $\sum |H(\pi) - H(\breve{\pi}_{\pi(k+1) \prec \pi(k)})| = 0, \forall k$, so we must assume that $\delta(y,x) = 0$ for all pairs of elements.

The central idea behind this measure could be summarized as follows: which is the induced deviation to the consensus if x and y are exchanged? In principle the ordering should have minimal distance to the general consensus, otherwise the results could be confusing. However, as was discussed before, finding this ranking involves a complex combinatorial problem for only a few experts and alternatives. That is why we prefer to adopt a heuristic approach based on Swarm Intelligence.

3   Ant Colony Optimization

A central component in the proposed model is the generation of feasible permutations using the objective function for guiding the search. In this paper we use ACO methods as optimizers, where each permutation comprises a possible solution.

The ACO metaheuristic is a search method for addressing combinatorial problems, which is inspired on a colony of agents (ants) [16]. Real ants in nature search for food in a random proximity to the nest. Once the ants found a source of food, they evaluate this source according to quality and quantity. In the path back to the nest, they deposit a chemical pheromone trail on the ground, in order to guide the rest of the colony to the food source [17]. Therefore, ACO is a fully constructive model where each ant builds a candidate solution of the problem by exploring a construction graph.

Each artificial ant moves from one state to another during the search process (here states are components of the solution). The preference of moving from one node to the other mainly depends on two values associated to each connection:

- The artificial information τ_{ij} is directly based on the pheromone trails, and it is iteratively updated by ants during the algorithm progress.
- The heuristic information η_{ij} denotes the preference of moving from one state to another. It should be specified that this knowledge is not modified during the algorithm execution, so it must be carefully estimated.

From the perspective of the Kemeny ranking problem, the equation (6) denotes the probability of accepting the jth state (i.e. elements to be ordered) at the ith position of the candidate ranking, $\mathcal{N}_i^k$ is the set of unvisited states for the kth ant, while α and β are two parameters which are used for controlling the strength of the pheromone trails and the heuristic information over the decision, respectively.

$$P_{ij}^k(t+1) = \frac{[\tau_{ij}(t)]^\alpha [\eta_{ij}]^\beta}{\sum_{r \in \mathcal{N}_i^k} [\tau_{ir}(t)]^\alpha [\eta_{ir}]^\beta}, j \in \mathcal{N}_i^k \tag{6}$$

After the construction process is complete, it is necessary to update all pheromone trails using the solutions found by agents. As a first stage, pheromone evaporation takes place uniformly reducing all pheromone trails. Subsequently, one or more solutions are used to increase the value of such paths included in selected solutions. It is a sensible issue in ACO-based algorithms. Essentially, most of ACO variants mainly differ in the strategy for updating the pheromone trail at each cycle.

3.1   Ant System

The Ant System (AS) was the first ACO algorithm [18]. In AS the pheromone trails is updated once all ants have completed their tours. As a first step all pheromone trails are uniformly evaporated using a constant factor $0 < \rho < 1$. After that, each ant k deposits a quantity of pheromone $\Delta\tau_{ij}$ on those connections that belong to its solution. It should be mentioned that the value $\Delta\tau_{ij}$ is calculated according to the quality of the solution found by the kth ant. The following equation shows both procedures, where ρ denotes the evaporation rate, whereas P is the number of agents.

$$\tau_{ij}(t+1) = (1-\rho)\,\tau_{ij}(t) + \sum_{k=1}^{P} \Delta\tau_{ij}^k \tag{7}$$

On arcs which are not regularly chosen by ants, the associated pheromone strength will decrease exponentially with the number of iterations, whereas arcs often chosen by agents will receive more pheromone and therefore they are more likely to be chosen in future cycles. However, deeper simulations reported in [18] proved that better results could be computed if only the global-best solution is used for updating the pheromone trails, instead of using all individuals belonging to the swarm.

3.2   Ant Colony System

The Ant Colony System (ACS) improves the AS method by exploiting the global-best solutions found by ants during the search stage [19]. As result, the algorithm enhances the exploitation features of ants when they build a solution, instead of exploring new areas of the solution space. This goal is achieved using three mechanism: a strong elitist strategy for updating pheromone trails, a rule for updating pheromone trails during the search phase, and a pseudo-random rule when selecting new states.

The following equation formalizes the strategy when updating the pheromone trails, where τ_{ij}^* denotes the pheromone quantity associated to the agent having better heuristic value. It means that the evaporation step takes place in all arcs, but the updating process only occurs in the tour traveled by the best individual.

$$\tau_{ij}(t+1) = (1-\rho)\,\tau_{ij}(t) + \rho\tau_{ij}^*(t) \tag{8}$$

In order to fully exploit the best knowledge discover by ants, ACS also introduces a pseudo-random proportional rule (see next equation). More specifically, if a random number $q \sim U(0,1)$ falls below q_0 then the agent will move to the state maximizing the product between pheromone trail and heuristic information, otherwise ACS will adopt the standard decision rule (6). The value q_0 is a parameter that should be fixed by the expert; when it is close to 1, exploitation is favored over exploration.

$$j = \underset{r \in \mathcal{N}_i^k}{\operatorname{argmax}} \left\{ [\tau_{ij}(t)]^\alpha [\eta_{ij}]^\beta \right\} \; if \; q \leq q_0 \tag{9}$$

Finally, in the ACS model ants use a further rule for updating the pheromone trails whey they are building the candidate solution (see next equation). This approach has the same effect of decreasing the probability of selecting the same path for all ants, as a way of introducing a balance between exploitation and exploration.

$$\tau_{ij}(t+1) = (1 - \rho) \, \tau_{ij}(t) + \rho \tau_{ij}(0) \tag{10}$$

The ACS algorithm regularly computes better solutions regarding the AS, since we know that ACO-based model performs better if artificial ants exploit the best solution found during the search process. In the next sub-section we revise another variant that adopt a similar principle attempting improving the performance.

3.3   MAX-MIN Ant System

The MAX-MIN Ant System (MMAS) was specifically developed to achieve stronger exploitation of solutions, avoiding stagnation states [20]. In a nutshell, we could define a stagnation state as the situation where ants construct the same solution over and over again and the exploration stops. This model has the following features.

Equally to the ACS, a strong elitist strategy regulates the agent which is allowed to update the pheromone trails. It could be the ant having better evaluation so far, or the agent with the best tour in the current iteration. Second, all pheromone trails are limited in the range $[\tau_{MIN}, \tau_{MAX}]$. If $\tau_{MIN} > 0$ for all solution components, then the probability of choosing a specific state will never be zero, which avoids stagnation configurations. As a final point, pheromone trails are initialized with τ_{MAX} to ensure further exploration of the search space at the beginning of the optimization phase.

4   Estimating the Heuristic Information

Another crucial component when solving combinatorial problems using an ACO-based algorithms is the estimation of the heuristic matrix. During the search process ants use this information to guide their movements (i.e. selection of a new state when they are building the candidate solution). If this matrix is appropriately estimated, then the ACO metaheuristic will lead to high-quality solutions, otherwise the algorithm will produce sub-optimal rankings. Next we explain two strategies to estimate this component from input data, assuming two partial rankings aggregation scenarios.

4.1   Aggregation of Multiple top-K Rankings

The first scenario takes place when each expert (hereinafter called respondent) selects the top-K objects (hereinafter called factors). It means that each input rankings will be partial in the sense that only the top-K factors are ordered, whereas the other $(M - K)$ factors are tied at the $K + 1$ position. The reader can notice that estimating the matrix for the first K factors across the M positions is equivalent of computing the total number of times that the jth factor *was observed* in the ith position. For the remaining $(M - K)$ places it cannot be directly used since these factors are tied, however, we could count how many times a factor *was not included* into the top-K. Next equation formalizes this reasoning, assuming N as the total number of respondents.

$$\eta_{ij} = \begin{cases} X_{ij}/N, & i \leq K \\ Y_j/(N(M - K)), & i > K \end{cases} \tag{11}$$

Example#1. Let us consider a ranking aggregation problem with 5 possible factors and 5 respondents, where each expert selected the top-3 factors. Next table shows this scenario. Observe that each row (respondent) comprises a partial ordering where factors are associated to specific positions. According to (11) the heuristic value of accepting the second factor at the first ranking position is $\eta_{12} = 2/5$.

Table 1. Example of a dataset when aggregating multiple top-K rankings

	F_1	F_2	F_3	F_4	F_5
R_1	1	2	$K + 1$	3	$K + 1$
R_2	2	1	$K + 1$	$K + 1$	3
R_3	1	3	$K + 1$	$K + 1$	2
R_4	1	3	$K + 1$	2	$K + 1$
R_5	2	1	3	$K + 1$	$K + 1$

Note that $\eta_{52} = 0$ since F_2 was always included into the top-3 sites and according to the equation (6) the probability P_{52} will be zero! However, this probability should not be zero because it is still possible building a solution having F_2 at the last position. That is why we replace all zero-values by $\eta_{MIN} = \min\{\eta_{ij}\}$ such as $\eta_{ij} \neq 0$, so we guarantee that all states could be visited by ants when they are building a solution.

4.2   Aggregation of Multiple Top-K_l Rankings

This scenario is more complex since each respondent is free of selecting K_l factors such as $1 \leq K_l \leq M$. Since the number of related factors is not fixed, we cannot simply count the number of times that a factor *was observed* at each site, we also must quantify the number of times that the factor *could be observed* at each position. It allows computing a more realistic heuristic matrix. Next equation summarizes this idea, where X_{ij} denotes the number of times the jth factor was observed at the ith position, Q_j is the set of input rankings where the jth factor was not included into the top-K_l positions, D_k represents the set of feasible ranking positions (i.e. they do not induce new tied elements) for the jth factor, while $E_{ki}(j)$ is a binary function. This function responds a simple question: could be the jth factor assigned to the ith ranking position?

$$\eta_{ij} = \frac{1}{N}\left(X_{ij} + \sum_{k \in Q_j} \frac{E_k(i)}{|D_k|}\right) \qquad (12)$$

Example#2. Let us consider a ranking aggregation problem with 5 possible factors and 5 respondents, where each expert R_l selected the top-K_l factors, as summarizes the next table. According to (12) the heuristic value of accepting F_2 at the first position is $\eta_{12} = 1/5(2 + 0/2 + 0/3) = 2/5$ because $Q_2 = \{R_1, R_4\}$, $D_1 = \{4,5\}$, $D_4 = \{3,4,5\}$ and $E_k(i) = 0, \forall i \in \{1,4\}$ since $1 \notin (D_1 \cup D_4)$. It means that F_2 cannot be assigned to the first position, without introducing a new tied element. It should be also mentioned that tied factors into the top-K_l relevant positions are not allowed.

Table 2. Example of a dataset when aggregating multiple top-K_l rankings

	F_1	F_2	F_3	F_4	F_5
R_1	1	$K_1 + 1$	2	$K_1 + 1$	3
R_2	3	1	$K_2 + 1$	2	$K_2 + 1$
R_3	2	1	4	5	3
R_4	2	$K_4 + 1$	$K_4 + 1$	$K_4 + 1$	1
R_5	5	3	1	2	4

Similarly to the above scenario, we must avoid zero-values in the heuristic matrix, although this situation is possible (i.e. the factor was never observed in a position and there is no chance to be observed without inducing a tied element). However, it is still possible to build a candidate solution with this feature having minimal distance to the general consensus, therefore it must be considered as well. In such cases the probability should not be zero but low (e.g. $\eta_{MIN} = \min\{\eta_{ij}\}$ such as $\eta_{ij} \neq 0$).

Notice that other scenarios are possible when aggregating partial rankings (e.g. tied elements in the top-K are allowed). In such cases our methodology is still useful since the modeling (i.e. generation of permutations) could be adopted, but some changes are required. In the following section, we explore the performance of our methodology in a real study case concerning Employer Branding issues in Belgium.

5   Numerical Simulations

Which are the most significant factors to be considered by employees when they look for an employer? The answer for this question embraces a valuable knowledge for any company since it provides the key for attracting the best employees, hence being more competitive. In this section we address this complex issue by solving a partial ranking aggregation problem. It includes two different scenarios:

a) Each respondent R_l selects the top-5 factors from $M = 17$ possible factors.

b) Each respondent R_l is free of selecting the most significant K_l factors (there is no limit regarding the number of relevant factors, but $1 \leq K_l \leq 17$).

In this study 14.585 respondents (aged between 18 and 65 years old) from Belgium where consulted[1]. As mentioned before, in the survey they evaluated 17 global factors elaborated by marketing experts, which are listed in the following table:

Table 3. Global factors evaluated by each respondent during the online survey

F_1	Financially sound
F_2	Offers quality training
F_3	Offers long-term job security
F_4	Offers international / global career
F_5	Future prospects / career opportunities
F_6	Strong management
F_7	Offers interesting jobs (job description)
F_8	Pleasant working environment
F_9	Competitive salary package
F_{10}	Good balance between life and work
F_{11}	Well located
F_{12}	Strong image / pursues strong values
F_{13}	Quality products / services offered
F_{14}	Deliberately handles the environment and society
F_{15}	Uses the latest technologies / innovative
F_{16}	Provides flexible working conditions
F_{17}	Encourages diversity (age, gender, ethnicity)

Besides, respondents are grouped according to the sector that they belong, resulting the following categories: automotive, business services, chemical and pharmaceutical industry, construction, education/government/care, energy services, human resources, informatics-consultancy, retail, travel/leisure/hospitality, industry and manufacturing, finance, food, transportation and logistics, and other.

For ACO-based methods we adopt the following parameters: $\alpha = 3$ and $\beta = 2$ since the knowledge learned by ants is often more confident, the evaporation constant

---

[1] Randstad was founded in 1960 by Frits Goldschmeding in the Netherlands. This company plays a pivotal role in the World of Work since it expanded its operations to 39 countries, representing more than 90 percent of the global HR services market. Actually, Randstad company is now the second largest HR services provider in the world. See http://www. randstad.com.

is set as $\rho = 0.6$, whereas the pheromone matrix is initialized with $\tau_{ij}(0) = 0.5$. In the case of the ACS method, we fix $q_0 = 0.7$; whereas the pheromone limits λ_{MIN} and λ_{MAX} in the MMAS algorithm are computed as suggested [19]. Finally, we use a swarm having 17 artificial ants (one ant per factor) and 150 generations.

5.1　First Scenario: Each Respondent R_l Selects the Top-K Factors

The first experiment consists on finding the best optimizer. With this purpose in mind we averaged the best heuristic value $H(\pi_*)$ over 10 independent trials for each sector, since our model is non-deterministic. Next figure shows the mean ranks achieved for each algorithm according to the Friedman test [21]. Using a significance level of 0.05, corresponding to the 95% confidence interval, the Friedman test suggests rejecting the null hypothesis H_0 (p-value = 0.0 < 0.05). As a result, we can conclude that there exist highly significant differences between at least two algorithms.

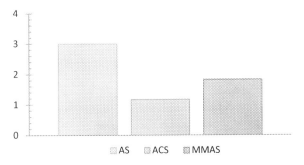

Fig. 1. Mean ranks achieved by the Friedman test for the first problem

However, we cannot ensure that ACS is the best optimizer (notice that ACS has the lowest mean rank). As second step it is necessary to proof that ACS is involved in the highly significant differences reported by the Friedman test. To do that, we compute the Wilcoxon signed rank test [22] with the purpose of identifying significant difference between pairs of algorithms. The Wilcoxon test attempts to answer a question: do two samples represent two different populations? It suggests rejecting the null hypothesis for all pairs (p-value < 0.05), assuming a significance level of 0.05. Hence, one can conclude that ACS is able of finding solutions closest to the consensus, outperforming the other algorithms when aggregating partial rankings.

Based on this result, next we select the best solution found by the ACS model for all respondents after 10 independent trials. Table 4 shows the averaged ranking π and also the distance between consecutive factors, according to (5). It can be noted that people in Belgium prefer interesting jobs, financially sound with competitive salary packages having pleasant working environment. However, the most desirable factor is *long-term security*. It is confirmed by the "distance" between the factor F_3 ranked as first and the factor F_7 ranked as second: $\delta(F_7, F_3)_\pi = 0.27$. It should be remarked that this measure has the greatest value between all pairs of consecutive factors.

Table 4. Best ranking for the first problem and distances between consecutive factors

Position	Factor	$\delta(y,x)_\pi$	Position	Factor	$\delta(y,x)_\pi$
$\pi(1)$	F_3	0.0000	$\pi(10)$	F_2	0.1048
$\pi(2)$	F_7	0.2717	$\pi(11)$	F_{13}	0.0886
$\pi(3)$	F_8	0.0638	$\pi(12)$	F_4	0.0026
$\pi(4)$	F_9	0.0419	$\pi(13)$	F_{14}	0.0190
$\pi(5)$	F_1	0.0870	$\pi(14)$	F_6	0.0159
$\pi(6)$	F_{10}	0.0962	$\pi(15)$	F_{12}	0.0026
$\pi(7)$	F_{11}	0.0602	$\pi(16)$	F_{17}	0.0012
$\pi(8)$	F_5	0.0090	$\pi(17)$	F_{15}	0.0289
$\pi(9)$	F_{16}	0.1059	-	-	-

5.2   Second Scenario: Each Respondent R_l Selects the Top-K_l Factors

In this scenario we assume that each respondent is free of selecting the most significant K_l factors such as $1 \leq K_l \leq M$. Here the same simulation design discussed in the
above section is assumed. The following figure summarizes the mean ranks achieved
for each algorithm according to the Friedman test [21]. Adopting a significance level
of 0.05, corresponding to the 95% confidence interval, the Friedman test suggests
rejecting the hypothesis H_0 (p-value = 0.0 < 0.05). These results show that the ACS
model computes the lowest mean rank, but we must analyze all pairs of optimizers.

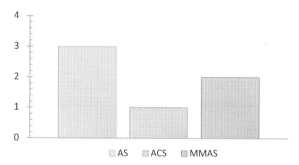

Fig. 2. Mean ranks achieved by the Friedman test for the second problem

In order to determine that the ACS algorithm is a responsible of the highly significant
differences reported by the Friedman test, next we compute the Wilcoxon signed rank
test [22] for all pairs of optimizers (i.e. AS-ACS, AS-MMAS, ACS-MMAS). Assuming
a significance level of 0.05, this test proposes rejecting the null hypothesis H_0 for all pairs
(p-value < 0.05). It confirms that the ACS method is also the best variant when facing
partial rankings aggregation problems with variable length.

Afterward we select the best solution found by the ACS model for all respondents
after 10 independent trials. Table 4 depicts the averaged ranking π and also the distance between consecutive factors. From these results we can surely conclude that the

most desirable factor is *long-term security*. In addition, this aggregated ranking has the same top-5 regarding the previous one, although the order is different.

Table 5. Best ranking for the second problem and distances between consecutive factors

Position	Factor	$\delta(y,x)_\pi$	Position	Factor	$\delta(y,x)_\pi$
$\pi(1)$	F_3	0.0000	$\pi(10)$	F_{16}	0.0128
$\pi(2)$	F_9	0.0852	$\pi(11)$	F_6	0.1857
$\pi(3)$	F_7	0.1419	$\pi(12)$	F_{13}	0.0349
$\pi(4)$	F_8	0.0171	$\pi(13)$	F_{14}	0.0262
$\pi(5)$	F_1	0.0474	$\pi(14)$	F_{12}	0.0615
$\pi(6)$	F_{10}	0.1410	$\pi(15)$	F_4	0.0092
$\pi(7)$	F_{11}	0.0750	$\pi(16)$	F_{17}	0.0072
$\pi(8)$	F_5	0.0009	$\pi(17)$	F_{15}	0.0031
$\pi(9)$	F_2	0.1502	-	-	-

In general terms we believe that both solutions are consistent, although the second scenario is more informed (i.e. there are less tied items) if we want to build a complete ranking. However, these solutions are not totally comparable because we use different input datasets, although we must expect some correspondence between them since the same respondent provided its best expertise for both scenarios. To overcome this issue we should formulate a "consistency" measure capable of computing the correspondence degree between both rankings (e.g. using the Hausdorff distance properties).

6   Conclusions

The aggregation of partial rankings could be faced using different approaches, although the absolute solution remains as an open problem. In this paper we presented a distance-based approach which directly extends the Kemeny ranking problem. In a few words, the objective is to find a permutation of factors minimizing the averaged distance to the general consensus. It involves a complex combinatorial problem for only a few experts and alternatives; that is why we prefer to use an approximate approach based on Swarm Intelligence. The proposal also includes other features such as:

- It replaces the Kendall distance by the Hausdorff distance, which allows to face partial rankings aggregation problems.
- It introduces a new measure for computing the relative distance between two consecutive items in the consensus (final) ranking.

From numerical results we concluded that the ACS method is the best variant when solving this kind of aggregation problems. It also showed that people in Belgium prefer stability (*long-term security*) instead of jobs financially sound. However, this outcome is not surprising and it could be a direct result of the economic crisis. As a future work we will be focused on extending the simulations by including other approaches such as the Genetic Algorithm discussed in [9]. While such experiments are absolutely required we are expecting promising results due to the ability of ACO-based methods for solving combinatorial problems, but this conjecture must be verified.

References

1. Jackson, B.N., et al.: Consensus genetic maps as median orders from inconsistent sources. IEEE/ACM Trans. on Computational Biology and Bioinformatics **5**, 161–171 (2008)
2. Ephrati, E., Rosenschein, J.: Multi-agent planning as a dynamic search for social consensus. In: Proceedings of the 13th International Joint Conference on Artificial Intelligence, IJCAI 1993, pp. 423–429. Morgan Kaufmann (1993)
3. Diaconis, P., et al.: A generalization of spectral analysis with application to ranked data. Annals of Statistics **5**, 949–979 (1989)
4. Chen, H., et al.: Global models of document structure using latent permutations. In: Proc. of Human Language Technologies: The 2009 Annual Conference of the North American Chapter of the Association for Computational Linguistics, NAACL, pp. 371–379 (2009)
5. Borda, J.: Memoire sur les elections au scrutin. In: Histoire de l'Academie Royal des Sciences (1781)
6. Condorcet, M.: Sur L'application de L'analyse à la Probabilité Des Décisions Rendues à la Pluralité Des Voix, L'Imprimerie Royale, Paris (1785)
7. Arrow, K.J.: Social Choice and Individual Values. Yale University Press (1963)
8. Ali, A., Meilǎ, M.: Experiments with Kemeny ranking: What works when? Mathematical Social Sciences **64**, 28–40 (2012)
9. Aledo, J., Gámez, J., Molina, D.: Tackling the rank aggregation problem with evolutionary algorithms. Applied Mathematics and Computation **222**, 632–644 (2013)
10. Puris, A., Bello, R., Herrera, F.: Analysis of the efficacy of a Two-Stage methodology for ant colony optimization: Case of study with TSP and QAP. Expert Systems with Applications **37**, 5443–5453 (2010)
11. Kemeny, J.L., Snell, J.G.: Mathematical Models in the Social Sciences, New York (1962)
12. Kendall, M.G.: A new measure of rank correlation. Biometrika **30**, 81–93 (1938)
13. Bansal, M., Fernández-Baca, D.: Computing distances between partial rankings. Information Processing Letters **109**, 238–241 (2009)
14. Hausdorff, F.: Grundzüge der Mengenlehre, Leipzig (1914)
15. Fagin, R., Kumar, R., Mahdian, M., Sivakumar, D., Vee, E.: Comparing partial rankings. Journal of Discrete Mathematics **20**, 628–648 (2006)
16. Dorigo, M., Caro, G.D., Gambardella, L.: Ant algorithms for discrete optimization. Artificial Life **5**, 137–172 (1999)
17. Dorigo, M., Bonabeau, E., Theraulaz, G.: Ant algorithms and stigmergy. Future Generation Computer Systems **16**, 851–871 (2000)
18. Dorigo, M., Colorni, A., Maniezzo, V.: The ant system: Optimization by a colony of cooperating agents. IEEE Transactions on Systems and Cybernetics **26**, 29–41 (1996)
19. Dorigo, M., Gambardella, L.: Ant colony system: A cooperative learning approach to the traveling salesman problem. IEEE Trans. on Evolutionary Computation **1**, 53–66 (1997)
20. Stützle, T., Hoos, H.H.: MAX–MIN ant system. Future Generation Computer System **16**, 889–914 (2000)
21. Friedman, M.: The use of ranks to avoid the assumption of normality implicit in the analysis of variance. Jour. of the American Statistical Association **32**, 674–701 (1937)
22. Wilcoxon, F.: Individual comparisons by ranking methods. Biometrics. **1**, 80–83 (1945)

Existence and Synthesis of Complex Hopfield Type Associative Memories

Garimella Rama Murthy[1(✉)] and Moncef Gabbouj[2]

[1] International Institute of Information Technology, Hyderabad, India
rammurthy@iiit.ac.in
[2] Tampere University of Technology, Tampere, Finland

Abstract. In this research paper, a complex valued generalization of associative memory synthesized by Hopfield is considered and it is proved that it is impossible to synthesize such a neural network with desired unitary stable states when the dimension of the network (number of neurons) is odd. The linear algebraic structure of such a neural network is discussed. Using Sylvester construction of Hadamard matrix of suitable dimension, an algorithm to synthesize such a complex Hopfield neural network is discussed. Also, it is discussed how to synthesize real / complex valued associative memories with desired energy landscape (i.e. desired stable states and desired energy values of associated quadratic energy function).

1 Introduction

Ever since the dawn of civilization, homosapiens innovated science and technology (through the habit of concentration). Researchers proposed mathematical models of various natural (physical / chemical / biological) systems. Specifically, in an effort to model the biological neural network, McCulloch and Pitts proposed a model of artificial neuron. Such a model of neuron and its variations / generalizations were utilized to build the comprehensive theory of Artificial Neural Networks (ANNs). In such a collaborative effort, some important conceptual innovations were the so called, Single Layer Perceptron (SLP) and Multi-Layer Perceptron (MLP). Following the principle of parsimony, Multi-Layer Perceptron constitutes a feed-forward neural network which is extensively utilized in applications. It continues to provide good results in various machine learning applications.

Motivated by the need to model the biological memory, Hopfield succeeded in innovating a feedback neural network based on McCulloch-Pitts neuron. It serves as a model of associative memory. Utilizing such a successful effort, researchers accelerated the attempts to arrive at more sophisticated Artificial Neural Networks (ANNs) which model associative memories. One such innovation was Bi-directional Associative Memory (BAM) [16].

Early efforts on ANNs effectively utilized real valued inputs / outputs and real valued synaptic weights. As a natural generalization, Complex Valued Neural Networks (CVNNs) were proposed in which the inputs / outputs, synaptic weights are allowed

© Springer International Publishing Switzerland 2015
I. Rojas et al. (Eds.): IWANN 2015, Part II, LNCS 9095, pp. 356–369, 2015.
DOI: 10.1007/978-3-319-19222-2_30

to be complex numbers [2], [6], [15]. It was demonstrated that CVNNs have certain advantages in terms of performance. The authors succeeded in proposing a natural complex valued generalization of Hopfield Neural Network (HNN) [3]. It is clearly evident that such a model of CVNN is different from earlier models.

In the case of real valued Hopfield network, researchers routinely assumed that synaptic weight matrix with desired / programmed stable states can be synthesized in all dimensions (i.e. networks with even or odd number of neurons). Questioning such an assumption, in this research paper, we investigate the question of existence and synthesis (constructing a synaptic weight matrix with programmed stable states which are mutually unitary to each other) of complex valued associative memory proposed in [3]. We also provide a constructive procedure to synthesize real /complex valued neural networks with desired energy landscape i.e. method of synthesizing synaptic weight matrix with certain stable states and energy values (of associated quadratic energy function).

This research paper is organized as follows. In Section 2, a clear and comprehensive description of real valued Hopfield Neural Network (HNN) is provided for completeness. Also, the complex valued HNN innovated by the authors [3] is discussed. In Section 3, existence of a complex valued Hopfield Neural Network innovated by the authors [3] is discussed. Also, adopting the approach utilized by Hopfield, one possible method of synthesis of such a CVNN [3] is discussed. In Section 4, using the spectral representation of symmetric / Hermitian matrix, synthesis of real / complex Hopfield type associative memories is discussed. The research paper concludes in Section 5.

2   Hopfield Neural Network: Review of Research Literature

Hopfield neural network is an Artificial Neural Network model. It is a nonlinear dynamical system represented by a weighted, undirected graph. The nodes of the graph represent artificial neurons and the edge weights correspond to synaptic weights. At each neuron / node, there is a threshold value. Thus, in summary, a Hopfield neural network can be represented by a synaptic weight matrix, M and a threshold vector, T. The order of the network corresponds to the number of neurons [1].

Every neuron is in one of the two possible states +1 or -1. Thus, the state space of Nth order Hopfield Neural Network (HNN) is the N-dimensional unit hypercube. Let the state of ith neuron at time 't' be denoted by $V_i(t) \in \{+1 \ or -1\}$. Thus, the state of the non-linear dynamical system is represented by the N x 1 vector, $\bar{V}(t)$. The state updation at the ith node is governed by the following equation:

$$V_i(t+1) = Sign\{\sum_{j=1}^{N} M_{ij} V_j(t) - T_i\} \qquad (2.1)$$

where Sign(.) isthe Signum function. In other words,$V_i(t+1)$ is +1 if the term in the brackets is non-negative, otherwise is -1. Depending on the set of nodes at which the state updation (2.1) is performed at any time t, the Hopfield Neural Network(HNN)operation is classified into the following modes.

- Serial Mode: The state updation in equation(2.1) is performed exactly at one of the nodes / neurons at time 't'.
- Fully Parallel Mode: The state updation in equation(2.1) is performed simultaneously at all the "N" nodes / neurons at time 't'.
  In the state space of Hopfield Neural Network, a non-linear dynamical system, there are certain distinguished states, called the "stable states".
- **Definition**: A state $\bar{V}(t)$ is called a stable state if and only if

$$\bar{V}(t) = Sign\{M\bar{V}(t) - T\} \tag{2.2}$$

Thus, once the Hopfield Neural Network (HNN) reaches the stable state, irrespective of the mode of operation of the network, the HNN will remain in that state forever. Thus, there is no further change of the state of HNN once a stable state is reached. The following convergence Theorem summarizes the dynamics of Hopfield Neural Network (HNN). It characterizes the operation of neural network as an associative memory.

Theorem 1. Let the pair N = (M,T) specify a Hopfield neural network. Then the following hold true:

[1] Hopfield : If N is operating in a serial mode and the elements of the diagonal of M are non-negative, the network will always converge to a stable state (i.e. There are no cyclesin the state space).

[13] Goles: If N is operating in the fully parallel mode, the network will always converge to a stable state or to a cycle of length 2 (i.e. the cycles in the state space are of length almost 2).

Proof. Refer the proof in [14].

Remark 1. It should be noted that in [4], it is shown that the synaptic weight matrix can be assumed to be an arbitrary symmetric matrix.

- **Novel Complex Hopfield Neural Network**

In research literature on artificial neural networks, there were attempts by many researchers to propose Complex Valued Neural Networks (CVNNs) in which the synaptic weights, inputs are necessarily complex numbers [2], [15]. In our research efforts on CVNNs, we proposed and studied one possible Complex Hopfield neural network first discussed in [3], [7-12]. The detailed descriptions of such an artificial neural network requires the following concepts.

- *Complex Hypercube*

Consider a vector of dimension "N" whose components assume values in the following set { 1+ j1 , 1 –j 1, -1 + j 1, -1 – j 1 }.Thus there are 4^N points as the corners of a set called the 'Complex Hypercube".

- *Complex Signum Function*

Consider a complex number "a + j b". The "Complex Signum Function" is defined as follows:

$$Csign(a + jb) = Sign(a) + jSign(b)$$

Now we briefly summarize the Complex Hopfield Neural Network first proposed in [3].

- The synaptic weight matrix of such an artificial neural network is a Hermitian matrix. The activation function in such a complex valued neural network is the complex signum function. The convergence Theorem associated with such a neural network is provided below.

Theorem 2. Let $N=(W, T)$ be a neural network, with W being a synaptic weight matrix which is a Hermiti an matrix with nonnegative diagonal elements and T being the threshold vector; then the following hold.

1) If N is operating in a serial mode and the consequent state computation of any arbitrary node(i) is as follows:

$$V_i(t + 1) = Csign(H_i(t)), \text{where} \qquad H_i(t) = \sum_{j=1}^{n} W_{i,j} V_j(t) - T_i,$$

Then the network will always converge to a stable state.

2) If N is operating in a fully parallel mode and the consequent state computation of any arbitrary node(i) is similar to that of the serial mode, then the network will always converge to a stable state or to a cycle of length 2.

Proof. Refer [3]

Remark 2: It has been shown in [5], the diagonal elements of W can be arbitrary real numbers and need not be necessarily non-negative.

3   Existence and Synthesis of a Complex Valued Hopfield Neural Network

- **Linear Algebraic Structure of a Structured Complex Hopfield Neural Network**

Hopfield synthesized a real valued symmetric synaptic weight matrix from the patterns to be stored in such a way that the network so obtained has these patterns as stable states. As discussed in Section 2, a novel complex valued Hopfield neural network was proposed in [3]. The synthesis of such an associative memory with specified desired stable states (lying on the complex unit hypercube) is now provided in the following discussion. Equivalently, we synthesize a complex valued Hermitian synaptic weight matrix from the patterns to be stored in such a way that the network so obtained has these patterns as stable states. The weight matrix is as follows:

$$W = \sum_{j=1}^{s} (X_j X_j^* - 2I) \tag{3.1}$$

Where S is the number of patterns to be stored (with $S < n$), I is the identity matrix and $X_1, X_2 \ldots X_S$ are the complex patterns(unitary to each other and also lying on the complex hypercube) to be stored.

Now, we would like to understand the linear algebraic structure of a complex Hopfield neural network based on Hermitian synaptic weight matrix, W. Some results related to W areas follows

- Trace(W) is zero. Thus, the sum of eigen values of W is zero.
- It is easy to see that:

$$W \, X_k \; = \; 2 \, (N - S) X_k \; with \; S < N$$

Now, we need the following definition in the succeeding discussion.

Definition: The local minimum vectors of the energy function associated with W are called **anti-stable states**. Thus, if u is an anti-stable state of matrix W, then it satisfies the condition

$$u \; = \; -CSign \, (W \, u)$$

Using the definitions, we have the following general result

Lemma 1: If a corner of unit hypercube is an eigenvector of W corresponding to positive / negative eigen value, then it is also a stable / anti-stable state.

Proof: Follows from the utilization of definitions of eigenvectors, stable / anti-stable states. Q.E.D.

Now we arrive at the spectral representation of W using the above results. Let $\{ X_1, X_2 \ldots X_S, \; Y_{S+1}, Y_{S+2} \ldots Y_N \}$ be unitary vectors that are corners of complex unit hypercube.

It is easy to see that
$$W \, Y_k \; = \; (-2) S Y_k \; for \; (S + 1) \leq k \leq N$$

Thus, the corner of hypercube is also an eigenvector corresponding to the negative eigenvalue '−2 S.

Hence, the spectral representation of the connection matrix of associative memory synthesized by Hopfield is given by:

$$W = \sum_{j=1}^{S} 2(N-S) \frac{X_j}{\sqrt{2N}} \frac{X_j^*}{\sqrt{2N}} + \sum_{j=S+1}^{N} (-2)S \frac{Y_j}{\sqrt{2N}} \frac{Y_j^*}{\sqrt{2N}}$$

$$= \sum_{j=1}^{S} \frac{(N-S)}{N} X_j X_j^* + \sum_{j=S+1}^{N} \frac{(-S)}{N} Y_j Y_j^*$$

Remark 3: We synthesized such a synaptic weight matrix to ensure that the diagonal elements of W are all zero (and necessarily non-negative) as required by this Proof of the convergence Theorem (i.e. Theorem 2).

In the above synthesis of associative memory, we implicitly assumed that the corners of complex hypercube that are also "unitary" exist in all dimensions. We questioned such an assumption as to the existence of unitary corners in all dimensions. The effort of the authors on this issue is summarized in the following. We now propose some interesting distance measures.

- **Some Interesting Distance Measures**

The following "distance" definitions are useful in the succeeding discussion. Consider $\{+1, -1\}$ vectors on the unit hypercube. Consider the mapping:

$$+1 \rightarrow +1$$
$$-1 \rightarrow 0$$

Definition: "Hamming—Like" distance between any two vectors on the unit hypercube is the Hamming distance between them (with the above mapping)

Remark: Consider any two vectors P, Q on the N-dimensional Euclidean space. By dividing each of the components of P, Q by the corresponding Euclidean norm, we arrive at vectors on the unit hyper sphere (i.e. after normalization by the corresponding Euclidean norm).

Using this approach, we now restrict consideration to vectors on the unit hyper sphere obtained from the corresponding vectors on the N-d Euclidean space.

Definition: Consider any two points X, Y on the unit hypersphere. Define

$$Z = Sign(X)$$

$$W = Sign(Y)$$

i.e. the components of vector Z are obtained as the sign of the corresponding components of X . It should be noted that the sign of a component which is "ZERO" is consistently defined as +1 or -1 (i.e. Sign(0) = +1 or -1).

The "induced Hamming like Distance" between { X, Y } i.e. $d_H(X, Y)$ is defined as the Hamming like distance between Z, W. Let the Euclidean distance between X, Y be denoted by $d_E(X, Y)$.

The following properties of the distance measures can be easily verified:

(I) $d_E(X, Y) = 0$ implies that $d_H(X, Y) = 0$. But $d_H(X, Y) = 0$ does not imply that $d_E(X, Y) = 0$.

(II) The distances $d_E(X, Y)$ and $d_H(X, Y)$ cannot be consistently ordered.

Since $d_H(X, Y) = 0$ does not imply that X = Y, it is only a pseudo-metric.

The above discussion is now generalized to determine the "generalized induced hamming distance" between two vectors on the N-dimensional Euclidean space.

We consider two points / vectors on the "bounded"(in the sense of Euclidean norm) N-dimensional Euclidean space. We quantize the components of the two vectors using the ceiling function

$$Y_1(j) = Ceiling[X_1(j)] \, for 1 \le j \le N \quad \text{and}$$

$$Y_2(j) = Ceiling[X_2(j)] \, for 1 \le j \le N$$

Note: In the above equations, we can use Lower Ceiling or Upper Ceiling function. Hence, in effect the components of vectors are rounded off / truncated to the nearest integer. It should be noted that after quantization, the components of the vectors will be positive or negative integers. Boundedness of vectors ensures that after quantization, all the components are below certain integer. Thus, the operation of quantization ensures that the vectors lie on the bounded lattice.

Definition: Consider any two bounded vectors X_1, X_2 lying on the N-dimensional Euclidean Space. Let

$$Y_1 = Ceiling (X_1)$$
$$Y_2 = Ceiling (X_2)$$

The "generalized induced Hamming Distance" between the bounded vectors X_1, X_2 is defined as the Hamming distance between the vectors Y_1, Y_2 .

Note: We can also define "induced Manhattan distance" between as the Manhattan distance between the vectors . In the above two definitions, a countable / finite set of vectors / points is extracted(through the process of appropriate quantization) from the uncountable set of vectors. Using Hamming / Manhattan distance on the countable / finite set, the "induced" Hamming / Manhattan distance is defined.

Remark 4: Using similar idea, distance between vectors on complex unit hypersphere can be defined and studied. Details are avoided for brevity.

First we provide the following well known definition of unitarity of vectors:

Definition: Two vectors X, Y lying on the complex unit hypercube are unitary if their inner product is zero i.e.
$$< X, Y > = 0$$

Note: We are interested in the existence of unitarity of vectors lying on the complex unit hypercube when the dimension is odd or even.

- **Conditions on Unitary Vectors on Complex Hypercube**

Thus, let us consider the complex unit hypercube. Let the two complex valued vectors lying on it be denoted by X, Y. They can be represented in the following manner:

$X = A + j B$, $Y = C + j D$, where A, B, C, D lie on the real valued unit hypercube.

- **Definition:** X and Y unitary / orthogonal when
$$X^* Y = 0, \text{where } X^* \text{ denotes the conjugate transpose of X.}$$

Now we find the conditions for unitarity / orthogonality of X, Y. We need the following definitions.
Let
d_1 is the number of places where A,C differ,
d_2 is the number of places where B,D differ,
d_3 is the number of places where A,D differ,
d_4 is the number of places where B,C differ.

Lemma 2: The vectors X and Y are unitary if and only if $d_1 = d_2 = d_3 = d_4 = \frac{N}{2}$.
Proof: It is easy to see that

$$X^*Y =(A^T - j B^T)(C + j D)= (A^T C + B^T D) + j (A^T D - B^T C)$$

Now, we have that:

$$X^*Y = [(N - 2d_1 + N - 2 d_2)] + j[(N - 2d_3) - (N - 2 d_4)]$$

Thus, we necessarily have that

$$X^*Y = [(2N - 2(d_1 + d_2)) + j \, 2(d_4 - d_3)]$$

Thus, for X, Y to be unitary / orthogonal to one another, we must have that $N = (d_1 + d_2)$ and $d_3 = d_4$.

Note: One sufficient condition for unitarity / orthogonality of X, Y is that the vectors A, B, C, D lying on the real valued unit hypercube are such that

$A^T C = 0, B^T D = 0, A^T D = 0, B$ i.e. Vector pairs(A, C),(B, D),(A, D),(B, C) are orthogonal. This can happen if

$$d_1 = d_2 = d_3 = d_4 = \frac{N}{2}$$

Now using the " identity swapping argument", it can be easily shown that the conditions

$$N = (d_1 + d_2) \text{ and } d_3 = d_4$$

Necessarily require that $d_1 = d_2 = d_3 = d_4 = \frac{N}{2}$

Hence this condition is necessary as well as sufficient.   **Q. E.D**

Suppose X, Y lie on the k-th order complex hypercube. Then, we have that the inner product of X, Y is expressed by the following formula.

$$X^*Y = 2K^2[(N-(d_1 + d_2)) + j(d_4 - d_3)]$$

Note: It should be noted that the above inner product is a complex number with even integer real and complex parts. It is a real number if and only if

$$d_3 = d_4 .$$

Corollary: If the dimension is odd, then there are no unitary vectors lying on any one of the countably many complex hypercubes.

- **On The Existence of Complex Valued Associative Memory (Synthesis in the Spirit of Hopfield)**

Thus, in view of the above Lemma 2, we have the following result.

Lemma 3: Hopfield type construction of complex valued associative memory exists only when the dimension of the hypercube is EVEN.

Proof: Follows from Lemma 2 and the synthesis procedure utilized for constructing the complex valued associative Memory.

- **Synthesis of Complex Hopfield Neural Network (Based on Procedure Used by Hopfield)**

One way of synthesizing unitary vectors X, Y lying on the complex hypercube is based on the following idea.

$$X = A + j A; Y = B + j B \text{ (or) } X = A- j A; Y = B - j B$$

with the condition that A, B are real valued orthogonal vectors lying on the unit hypercube. Thus, we are interested in providing an algorithm / procedure for determining real valued orthogonal vectors lying on the real unit hypercube. In effect we need to find "s" corners of hypercube that are mutually (any pair of them) orthogonal to one another. These "s" orthogonal corners of hypercube are utilized to determine desired stable states on the complex unit hypercube.

In literature, there is no algorithmic procedure to arrive at , say "s" orthogonal { +1 , -1} vectors. Thus, we need a constructive procedure to choose "s" corners of hypercube that are mutually (any two corners) orthogonal to one another. In solving this problem, some results on Hadamard matrices are very relevant.

Definition: A Hadamard matrix of order 'm', denoted by H_m , is an m x m matrix of +1's and -1's such that

$$H_m H_m^T = m I_m$$

Where I_m is the m x m identity matrix. This definition is equivalent to saying that any two rows of H_m are orthogonal.

In view of the above definition and Lemma 2, we have the following interesting result

Lemma 4: Hadamard matrices of odd order do not exist.

Proof: Follows directly from Lemma 2   Q.E.D.

Thus, we are interested in determining whether Hadamard matrices of any arbitrary EVEN ORDER exist. A partial answer to this question is well known. Specifically, it is known that Hadamard matrices of order 2^k exist for all. The so called Sylvester construction is provided below:

$$H_1 = [1]$$

$$H_2 = \begin{bmatrix} 1 & 1 \\ 1 & -1 \end{bmatrix}$$

$$H_{2^{n+1}} = \begin{bmatrix} H_{2^n} & H_{2^n} \\ H_{2^n} & -H_{2^n} \end{bmatrix}$$

Now we provide an algorithm to arrive at orthogonal corners lying on the real valued unit hypercube.

Step 1: Thus, to choose "s" pair wise (any two) orthogonal corners, compute a Hadamard matrix, H_m (using Sylvester construction) with $m = 2^k$ for some 'k' such that $m \geq s$.

Step 2: Pick any "s" rows of H_m that are pairwise orthogonal to one another. From those corners determine "s" unitary vectors lying on the unit complex hypercube.
Step 3: Synthesize Complex Valued Hopfield Associative memory as specified in equation (3.1).

4 Synthesis of Complex Hopfield Type Associative Memories

We first consider the synthesis of "generalized" real Hopfield associative memory. In view of the fact that the synaptic weight matrix can be an arbitrary symmetric matrix, we can generalize the Hopfield construction (of associative memory) in the following manner. It should be noted that the associative memory synthesized by Hopfield highly constrains the eigen values (and consequently the stable/anti-stable values). We do not have those constraints
In the following synthesis procedure.

Let $\{\frac{\mu_j}{N}\}_{j=11}^{S}$ be desired positive eigen values (with μ_j being the desired stable value) and let $\{X_j\}_{j=1}^{S}$ be the desired stable states. Then it is easy to see that the following symmetric matrix constitutes the desired synaptic weight matrix of Hopfield neural network (Using the spectral representation of symmetric matrix W):

$$W = \sum_{j=1}^{s} \mu_j X_j X_j^T$$

Once again, in this case N must be even. Also, it is clear that the synaptic weight matrix is positive definite.

Note

It should be noted that $X_j's$ are mutually at a Hamming distance of $\frac{N}{2}$ from each other. Thus the minimum distance between any two of them is $\frac{N}{2}$.

In the same spirit as above, we now synthesize a synaptic weight matrix, with desired stable/anti-stable values and the corresponding stable / anti-stable states. Let $\{X_j\}_{j=1}^{S}$ be desired orthogonal stable states and $\{Y_j\}_{j=1}^{L}$ be the desired orthogonal anti-stable states. Let the desired stable states be eigenvectors corresponding to positive eigen values and let the desired anti-stable states be eigenvectors corresponding to negative eigen values. The spectral representation of desired synaptic weight matrix is given by:

$$W = \sum_{j=1}^{s} \frac{\mu_j}{N} X_j X_j^T - \sum_{j=1}^{L} \frac{\beta_j}{N} Y_j Y_j^T$$

Where $\dfrac{\mu'_j s}{N}$ are desired positive eigen values and $\dfrac{-\beta_j' s}{N}$ are desired negative eigen values.

Hence the above construction provides a method of arriving at desired energy landscape (with orthogonal stable /anti-stable states and the corresponding positive / negative energy values).

Remark

It can be easily shown that Trace(W) is constant contribution (DC value) to the value of quadratic form $X^T W X$ at all corners, X of unit hypercube. Thus, the location of stable / anti-stable states is invariant under modification of Trace (W) value. Thus, from the standpoint of location /computation of stable / anti-stable states Trace (W) can be set to zero. Let $\widehat{W}$ be the matrix obtained from W by setting all the diagonal elements of W to zero. It can be easily seen that $\widehat{W}$ is anindefinite matrix.

In view of Lemma 2, when the dimension of the unit hypercube is odd, there are only the following two possibilities:

CASE A: Only one corner of the hypercube is an eigenvector in the spectral representation of W. Thus, the desired corneris astable state or anti-stable state.
CASE B: None of the corners of the hypercube is an eigenvector in the spectral representation of W.

- In case A, if the corresponding eigen value is positive, then the corner is a stable state, else, the corner is an anti-stable state.
- In case B, if the spectral representation of W is of the following form (Rank one matrix)
  $$W = \gamma f_i f_i^T$$ with all the other eigen values are zeroes, then
     F = Sign (f_i)
  is the global optimum stable state. The logical reasoning is fairly simple and avoided for brevity.

- **Synthesis of Novel Complex Hopfield Type Associative Memory**
As in the case of real valued Hopfield type associative memory, we now synthesize the synaptic weight matrix of complex Hopfield associative memory considered in this research paper. It is based on the spectral representation of Hermitian matrix i.e.

$$W = \sum_{j=1}^{s} \frac{\mu_j}{2N} X_j X_j^* - \sum_{j=1}^{L} \frac{\beta_j}{2N} Y_j Y_j^*$$

As in the real case such an associative memory exists only in the even dimension. Once again, as in the real case, when the dimension is odd, there are only two possible cases:

CASE A: Only one corner of the complex hypercube is an eigenvector in the spectral representation of W. Thus, the desired corneris astable state or anti-stable state.

CASE B: None of the corners of the hypercube is an eigenvector in the spectral representation of W.

Note: The synthesis of real as well as complex Hopfield type associative memories can be done using corresponding real / complex Hadamard matrices (based on say Sylvester construction). Detailed specification of algorithm is avoided for brevity.

5   Conclusion

In this research paper, it is reasoned that the complex valued associative memory proposed in [3] does not exist (i.e. network with desired "unitary" stable states does not exist) when the dimension of associated synaptic weight matrix is odd. Using Hadamard matrices, an algorithm to synthesize such a complex valued Hopfield neural network in even dimension is proposed. Also synthesis of real and complex Hopfield type associative memories with desired energy landscape is discussed

References

1. Hopfield, J.J.: Neural networks and physical systems with emergent collective computational abilities. Proceedings of National Academy of Sciences **79**, 2554–2558 (1982). USA
2. [MGZ] Muezzinoglu, M.K., Guzelis, C., Zurada, J.M.: A new design method for the complex-valued multistate hopfield associative memory. IEEE Transactions on Neural Networks **14**(4), July 2003
3. Rama Murthy, G., Praveen, D.: Complex-valued Neural Associative Memory on the Complex hypercube. IEEE Conference on Cybernetics and Intelligent Systems, Singapore, December 1–3, 2004
4. Rama Murthy, G.: Optimal Signal Design for Magnetic and Optical Recording Channels. Bellcore Technical Memorandum, TM-NWT-018026, April 1st, 1991
5. Rama Murthy, G., Nischal, B.: Hopfield-Amari Neural Network: Minimization of Quadratic forms. The 6th International Conference on Soft Computing and Intelligent Systems, Kobe Convention Center(Kobe Portopia Hotel), Kobe, Japan, November 20–24, 2012
6. Hirose, A.: Complex Valued Neural Networks: Theories and Applications. World scientific publishing Co, November 2003
7. Rama Murthy, G., Praveen, D.: A novel associative memory on the complex hypercube lattice. In: 16th European Symposium on Artificial Neural Networks, April 2008
8. Rama Murthy, G.: Some novel real/ complex valued neural network models. In: Advances in Soft Computing, Springer Series on Computational Intelligence: Theory and Applications, Proceedings of 9th Fuzzy days, Dortmund, Germany, September 18–20 2006
9. Jagadeesh, G., Praveen, D., Rama Murthy, G.: Heteroassociative memories on the complex hypercube. In: Proceedings of 20th IJCAI Workshop on Complex Valued Neural Networks, January 6–12, 2007
10. Sree Hari Rao, V., Rama Murthy, G.: Global dynamics of a class of complex valued neural networks. In: Special Issue on CVNNS of International Journal of Neural Systems, April 2008

11. Rama Murthy, G.: Infinite Population, Complex Valued State Neural Network on the Complex Hypercube. In: Proceedings of International Conference on Cognitive Science (2004)
12. Rama Murthy, G.: Multidimensional Neural Networks-Unified Theory. New Age International Publishers. New Delhi (2007)
13. Goles, E., Fogelman, F., Pellegrin, D.: Decreasing energy functions as a tool for studying threshold networks. Discrete Applied Mathematics **12**, 261–277 (1985)
14. Bruck, J.: On the Convergence Properties of the Hopfield Model. Proceeding of the IEEE **78**(10), October 1990
15. Zhu, X., Wei, W.: Fixed points of complex-valued bidirectional associative memory. Journal of Computational and Applied Mathematics **236**, 753–758 (2011)
16. Kosko, B.: Bidirectional Associative Memories. IEEE Transactions on Systems, Man and Cybernetics **18**(1), January/February 1988

On Acceleration of Incremental Learning in Chaotic Neural Network

Toshinori Deguchi[1(✉)], Toshiki Takahashi[1], and Naohiro Ishii[2]

[1] National Institute of Technology, Gifu College,
Gifu 501–0495, Japan
deguchi@gifu-nct.ac.jp
[2] Aichi Institute of Technology, Aichi 470–0392, Japan
ishii@aitech.ac.jp

Abstract. The incremental learning is a method to compose an associate memory using a chaotic neural network and provides larger capacity than correlative learning in compensation for a large amount of computation. A chaotic neuron has spatio-temporal sum in it and the temporal sum makes the learning stable to input noise. When there is no noise in input, the neuron may not need temporal sum. In this paper, to reduce the computations, a simplified network without temporal sum are introduced and investigated through the computer simulations comparing with the network as in the past. It turns out that the simplified network is able to learn input patterns quickly with the learning parameter varying.

1 Introduction

The incremental learning proposed by the authors is highly superior to the auto-correlative learning in the ability of pattern memorization[1,2].

The idea of the incremental learning is from the automatic learning[3]. In the incremental learning, the network keeps receiving the external inputs. If the network has already known an input pattern, it recalls the pattern. Otherwise, each neuron in it learns the pattern gradually. Therefore, the weak point of the learning is computational complexity. The network takes steps to learn input patterns.

The neurons used in this learning are the chaotic neurons, and their network is the chaotic neural network, which was developed by Aihara[4].

A chaotic neuron has spatio-temporal sum in it and the temporal sum makes the learning possible with noisy inputs. But, when inputs don't include any noises, the neuron can be more simple without the temporal sum. This simplification reduces the computational complexity.

In this paper, first, we explain the chaotic neural networks and the incremental learning, then simplify the network by eliminating the temporal sum from the chaotic neurons and examine the simplified network comparing with the usual one.

© Springer International Publishing Switzerland 2015
I. Rojas et al. (Eds.): IWANN 2015, Part II, LNCS 9095, pp. 370–379, 2015.
DOI: 10.1007/978-3-319-19222-2_31

2   Chaotic Neural Networks and Incremental Learning

The incremental learning was developed by using the chaotic neurons. The chaotic neurons and the chaotic neural networks were proposed by Aihara[4].

We presented the incremental learning that provides an associative memory[1]. The network type is an interconnected network, in which each neuron receives one external input, and is defined as follows[4]:

$$x_i(t + 1) = f\big(\xi_i(t + 1) + \eta_i(t + 1) + \zeta_i(t + 1)\big) , \tag{1}$$

$$\xi_i(t + 1) = k_s\xi_i(t) + vA_i(t) , \tag{2}$$

$$\eta_i(t + 1) = k_m\eta_i(t) + \sum_{j=1}^{n} w_{ij}x_j(t) , \tag{3}$$

$$\zeta_i(t + 1) = k_r\zeta_i(t) - \alpha x_i(t) - \theta_i(1 - k_r) , \tag{4}$$

where $x_i(t + 1)$ is the output of the i-th neuron at time $t + 1$, f is the output sigmoid function described below in (5), k_s, k_m, k_r are the time decay constants, $A_i(t)$ is the input to the i-th neuron at time t, v is the weight for external inputs, n is the size—the number of the neurons in the network, w_{ij} is the connection weight from the j-th neuron to the i-th neuron, and α is the parameter that specifies the relation between the neuron output and the refractoriness.

$$f(x) = \frac{2}{1 + \exp(\frac{-x}{\varepsilon})} - 1 . \tag{5}$$

In the incremental learning, each pattern is inputted to the network for some fixed steps before moving to the next. In this paper, this term is called "input period", and "one set" is defined as a period for which all the patterns are inputted. The patterns are inputted repeatedly for some fixed sets.

During the learning, a neuron which satisfies the condition of (6) changes the connection weights as in (7)[1].

$$\xi_i(t) \times (\eta_i(t) + \zeta_i(t)) < 0 . \tag{6}$$

$$w_{ij} = \begin{matrix} w_{ij} + \Delta w, \xi_i(t) \times x_j(t) > 0 \\ w_{ij} - \Delta w, \xi_i(t) \times x_j(t) \le 0 \end{matrix} \quad (i \ne j) , \tag{7}$$

where Δw is the learning parameter.

If the network has learned a currently inputted pattern, the mutual interaction $\eta_i(t)$ and the external input $\xi_i(t)$ are both positive or both negative at all the neurons. This means that if the external input and the mutual interaction have different signs at some neurons, a currently inputted pattern has not been learned completely. Therefore, a neuron in this condition changes its connection weights. To make the network memorize the patterns firmly, if the mutual interaction is less than the refractoriness $\zeta_i(t)$ in the absolute value, the neuron also changes its connection weights.

In this learning, the initial values of the connection weights can be 0, because some of the neurons' outputs are changed by their external inputs and this makes the condition establish in some neurons. Therefore, all initial values of the connection weights are set to be 0 in this paper. $\xi_i(0)$, $\eta_i(0)$, and $\zeta_i(0)$ are also set to be 0.

To confirm that the network has learned a pattern after the learning, the pattern is tested on the normal Hopfield's type network which has the same connection weights as the chaotic neural network. That the Hopfield's type network with the connection weights has the pattern in its memory has the same meaning as that the chaotic neural network recalls the pattern quickly when the pattern inputted. Therefore, it is the convenient way to use the Hopfield's type network to check the success of the learning.

3   Simulations with Simplified Network

3.1   Capacity

The simplified network is given by letting all the k-parameters be zero to eliminate the temporal sum, namely, $k_s = k_m = k_r = 0$.

In this paper, a set of patterns are generated and used to be learned by the networks. These patterns are random patterns generated with the method that all elements in a pattern are set to be -1 at first, then the half of the elements are chosen at random to turn to be 1.

For preliminary simulations, we checked how many patterns the networks can learn.

From the result of the former work[6], the parameters are assigned in Table 1 for the usual network. Both the input period and the number of sets are set to be 100.

Table 1. Parameters

$\Delta w = 3 \times 10^{-6}$ $\alpha = 6 \times 10^{-4}$,
$k_s = 0.95$,
$k_m = 0.1$,
$k_r = 0.95$,
$v = 2.0$,
$\theta_i = 0$,
$\varepsilon = 0.015$

The result of the simulation on the usual 100-neuron network is shown in Fig.1. From the result, until 162 patterns, all the inputted patterns are learned completely. In this paper, this number is called the capacity.

On the simplified network under the same condition, the network was not able to learn the same number of patterns as the usual network. Through some trial, it was found that setting $\Delta w = 3 \times 10^{-7}$ makes the simplified network learn

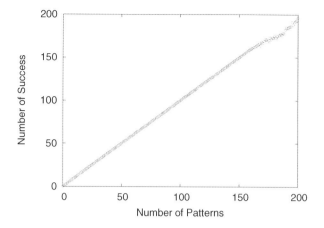

Fig. 1. Usual network

162 patterns as in Fig.2. Thus, the simplified network still learn the same number of patterns without the temporal sums, although the capacity is 158 patterns because when the number of inputs is 159, the network learns 158 patterns.

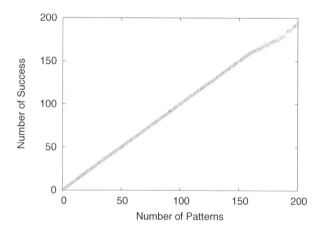

Fig. 2. Simplified network

3.2   Neurons Which Learn

In the incremental learning, not all the neurons change their weights at the same time but the neurons which learn in that step are decided by the learning condition. To investigate the differences between the simplified and the usual network, when and which neuron learns are inspected. Fig.3 shows the results. All the neurons are arranged vertically, and the horizontal axis shows the steps during the 1st pattern inputted in 10th set. The mark + indicates that the neuron learns at that step.

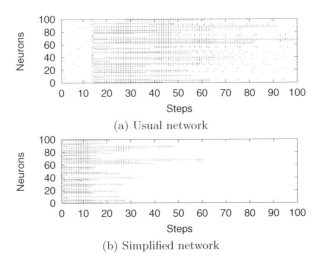

(a) Usual network

(b) Simplified network

Fig. 3. Neurons which learn

From Fig.3 (b), in the simplified network without the temporal sum, almost all the neurons learn from step 1, while, from Fig.3 (a), in the usual network, few neurons learn from step 1 and it was after step 14 that almost all the neurons learn. This is because the temporal sum is keeping the previous pattern information for some steps (for 13 steps in this case), and the new pattern information gets overwhelming at step 14 in the usual network. Although it would be unable for the simplified network to learn with noisy inputs without temporal sum which smooths noisy inputs[7], the simplified network may learn patterns faster.

3.3   Input Period

In the former simulations, the input period was kept to be 100 or 50. But, the simplified network may learn patterns with shorter input period. In the next simulations, to investigate the effect of the input period, the input period is changed from 1 to 100 and the number of successfully learned patterns is counted. The set of patterns is the same as in 3.1 and the number of input patterns is fixed to 162. The simulation results are shown in Fig.4 and Fig.5.

Fig.4 shows the results on the usual network. The horizontal axis is the input period and the vertical axis is the number of success. When the input period is short, no pattern is stored in the network, because the temporal sum is keeping the previous pattern information. With the input periods longer than 82 steps, the usual network was able to learn all the patterns.

Fig.5 shows the results on the simplified network. The number of success begins to rise from the input period of 5 steps, and reach 162 at the period of 52 which is 63% of that in the usual network. From these results, the simplified network can learn patterns faster than the usual netwrork.

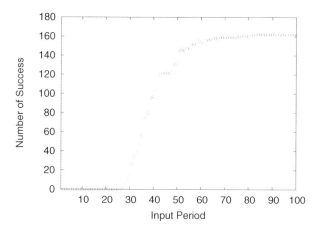

Fig. 4. Number of successfully learned patterns in usual network

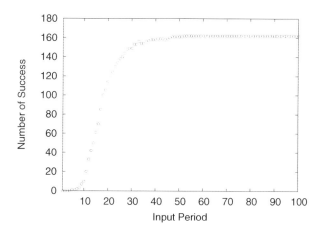

Fig. 5. Number of successfully learned patterns in simplified network

3.4   Noisy Inputs

The simplified network would loses the ability to learn in noisy inputs without the temporal sums. To verify the handling of noisy inputs, the following simulations investigate this ability.

In these simulations, to add noises, a fixed number of elements in an input pattern are chosen randomly every step and they reversed before they are inputted to the network. This fixed number is the number of noises. The results are shown in Fig.6.

The horizontal axis is the number of noises and the vertical axis is the capacity of the network, which is, in this paper, the maximum number of patterns when all the input patterns are stored in the network. As predicted previously, though the capacity of the usual network is above 100, that of the simplified

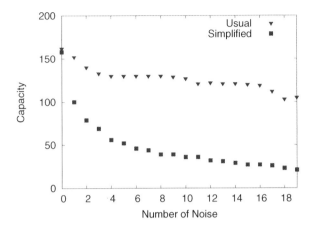

Fig. 6. Learning ability with noisy inputs

network became below 100 over 2 noises. Thus, the simplified network loses the ability to learn with noisy inputs.

3.5   Adjusted Network

The difference between the usual network and the simplified network is k-para-meters (and Δw). In this section, the other parameters are introduced to find the parameters with which the network can learn faster than the usual network and have more ability of learning in noisy inputs than the simplified network. Because the rage of the parameters are 4 dimensional including Δw, the parameters are changed with linear relation as follows:

$$k_s = 0.95r, \tag{8}$$
$$k_m = 0.1r, \tag{9}$$
$$k_r = 0.95r, \tag{10}$$
$$\Delta w = 2.7 \times 10^{-6}r + 3 \times 10^{-7}, \tag{11}$$

where r is the control parameter.

The simulations from $r = 0.1$ to 0.9 are carried out to adjust the parameters. From these simulations, a reasonable set of parameters is found at $r = 0.8$. In this paper, the network with these parameters is called the adjusted network.

Figure 7 shows the effect of the input period on the adjusted network. Although the network could not learn all the 162 patterns, curiously, the input period it needs is shorter than the simplified network. One of the available reasons is the effect of Δw. When it becomes larger, the connection weights change quickly, but are not set finely. Therefore, the learning is fast but not all the patterns are stored.

Figure 8 shows the ability for noisy inputs. As same as in Figure 6, the vertical axis is the capacity of the network. This network shows medium ability

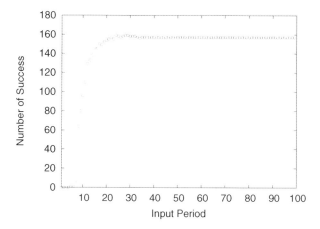

Fig. 7. Number of successfully learned patterns in adjusted network

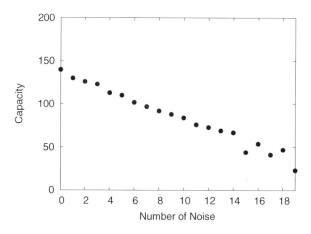

Fig. 8. Learning ability wity noisy inputs in adjusted network

between the other 2 networks. As same as the number of success in Figure 7, the capacity is smaller than the other 2 networks.

3.6   Varying the Learning Parameter

Described above, the adjusted network learns in shorter steps using Δw at $r = 0.8$. At the simplified network with the same Δw, there is a possibility that the network also learns in short steps.

Figure 9 shows the result of the simulations in which $\Delta w = 2.46 \times 10^{-6}$. This value makes the network learn more quickly but less patterns. The network learned only 66 patterns out of 162. When Δw is large, the network learns more quickly, and when Δw is small, the network learns more patterns.

Then, there arises the question that whether varying Δw from large value to small value during the learning makes the network learn all the patterns quickly. In these simulations, Δw is varied according to the equation (12),

$$\Delta w = 3 \times 10^{-6} - 2.7 \times 10^{-6} \times \frac{s-1}{99} \qquad (12)$$

where s is the number of learning set.

Figure 9 also shows this result. Although the number of success does not reach 162, it does 159 at 23 input period. The network learned almost all the patterns quickly.

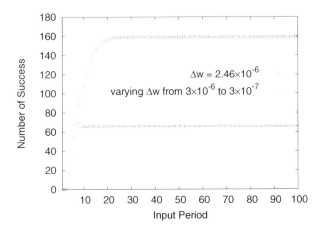

Fig. 9. Number of successfully learned patterns with different Δw

4   Conclusion

To reduce the amount of computation in the incremental learning, the simplified network was introduced and the behaviour of this network was investigated comparing with the usual network. The simplified network was able to learn patterns in shorter steps than the usual network and had almost the same capacity as the usual network. Furthermore, varying the learning parameter during the learning, the amount of steps for the learning can be reduced.

References

1. Asakawa, S., Deguchi, T., Ishii N.: On-demand learning in neural network. In: Proc. of the ACIS 2nd Intl. Conf. on Software Engineering, Artificial Intelligence, Networking & Parallel/Distributed Computing, pp. 84–89 (2001)
2. Deguchi, T., Ishii, N.: On refractory parameter of chaotic neurons in incremental learning. In: Negoita, M.G., Howlett, R.J., Jain, L.C. (eds.) KES 2004. LNCS (LNAI), vol. 3214, pp. 103–109. Springer, Heidelberg (2004)

3. Watanabe, M., Aihara, K., Kondo, S.: Automatic learning in chaotic neural networks. In: Proc. of 1994 IEEE Symposium on Emerging Technologies and Factory Automation, pp. 245–248 (1994)
4. Aihara, K., Tanabe, T., Toyoda, M.: Chaotic neural networks. Phys. Lett. A **144**(6,7), 333–340 (1990)
5. Deguchi, T., Matsuno, K., Ishii, N.: On capacity of memory in chaotic neural networks with incremental learning. In: Lovrek, I., Howlett, R.J., Jain, L.C. (eds.) KES 2008, Part II. LNCS (LNAI), vol. 5178, pp. 919–925. Springer, Heidelberg (2008)
6. Deguchi, T., Fukuta, J., Ishii, N.: On appropriate refractoriness and weight increment in incremental learning. In: Tomassini, M., Antonioni, A., Daolio, F., Buesser, P. (eds.) ICANNGA 2013. LNCS, vol. 7824, pp. 1–9. Springer, Heidelberg (2013)
7. Deguchi, T., Takahashi, T., Ishii, N.: On simplification of chaotic neural network on incremental learning. In: 15th IEEE/ACIS Intl. Conf. on Software Engineering, Artificial Intelligence, Networking and Parallel/Distributed Computing (SNPD 2014), pp. 1–4 (2014)

Comparing Optimization Methods, in Continuous Space, for Modelling with a Diffusion Process

Nuria Rico[✉], Maribel García Arenas, Desirée Romero,
J.M. Crespo, Pedro Castillo, and J.J. Merelo

Department of Architecture and Computer Technology, CITIC,
University of Granada (Spain), Granada, Spain
nrico@ugr.es

Abstract. Many probabilistic models are frequently used for natural growth-patterns modelling and their forecasting such as the diffusion processes. The maximum likelihood estimation of the parameters of a diffusion process requires a system of equations that, for some cases, has no explicit solution to be solved. Facing that situation, we can approximate the solution using an optimization method. In this paper we compare five optimization methods: an Iterative Method, an algorithm based on Newton-Raphson solver, a Variable Neighbourhood Search method, a Simulated Annealing algorithm and an Evolutionary Algorithm. We generate four data sets following a Gompertz-lognormal diffusion process using different noise level. The methods are applied with these data sets for estimating the parameters which are present into the diffusion process. Results show that bio-inspired methods gain suitable solutions for the problem every time, even when the noise level increase. On the other hand, some analytical methods as Newton-Raphson or the Iterative Method do not always solve the problem whether their scores depend on the starting point for initial solution or the noise level hinders the resolution of the problem. In these cases, the bio-inspired algorithms remain as a suitable and reliable approach.

Keywords: Diffusion process · Parameter estimation · Optimization methods

1 Introduction

Many probabilistic models are frequently used for natural growth-patterns modelling and their forecasting. Although there exists a lot of different growth-patterns, and many models to use, among the widely studied and useful models are diffusion processes. These kind os processes have shown that they are useful in many areas like biology [18,19], medicine [1,11], economy [4] or agriculture [17,20].

In this paper we focus our attention on the Gompertz-lognormal diffusion process [21]. It can be seen as a mixture between a Gompertz-type process

© Springer International Publishing Switzerland 2015
I. Rojas et al. (Eds.): IWANN 2015, Part II, LNCS 9095, pp. 380–390, 2015.
DOI: 10.1007/978-3-319-19222-2_32

[6] and a lognormal process [5] since the infinitesimal moments that define the Gompertz-lognormal diffusion process are:

$$A_1(x,t) = (m \cdot e^{-\beta t} + c)x$$
$$A_2(x,t) = \sigma^2 x^2,$$

with $m > \beta > 0$, $c \in \mathbb{R}$.

Taking $c = 0$ the process is Gompertz-type, and for a known fixed value of β the process is an homogeneous lognormal diffusion process. In other cases, the process has a shape with two inflection points. This makes the process useful for modelling situations with observed paths along time that turn from increasing to decreasing or from concave to convex.

To carry out the estimation, we consider the transition density function, that depends on m, β, c and σ^2 parameters. Estimating the set of parameters leads to the estimation of the main features of the process. The most widely used, specifically for predicting purposes, are the mean, mode and quantile functions. Their expressions depend on the set of parameters $\{m; \beta; c; \sigma^2\}$ and, in the case of a lognormal initial distribution being considered, on the $\{\mu_1; \sigma_1^2\}$ set too.

The problem we deal with is to find the maximum likelihood estimates (MLE) of the parameters of the model, from which the estimation of the parametric functions can be found. Thus, we assume x_{ij}, $i = 1, \cdots, d$, $j = n_1, \cdots, n_d$ a discrete sampling of d paths, observed in n_i times named t_{ij} $(i = 1, \cdots, d, j = 1, \cdots, n_i)$, not necessarily equal in each observed path, but $t_{i1} = t_1$, $i = 1, \cdots, d$. With the sample, we try to obtain MLE of the parameters maximising the likelihood function, or equivalently, maximising the log-likelihood without constants values. So the final function we have to maximise, using $b = e^{-\beta}$, and $k = \sum_{i=1}^{d} n_i$, is

$$L(m,b,c,\sigma^2) = \frac{d-k}{2} \ln \sigma^2 - \frac{1}{2} \sum_{i=1}^{d} \sum_{j=2}^{n_i} \frac{\left[\ln \frac{x_{ij}}{x_{ij-1}} - \frac{m}{\ln(b)} \left(b^{t_{ij}} - b^{t_{ij-1}} \right) + \left(\frac{\sigma^2}{2} - c \right) (t_{ij} - t_{ij-1}) \right]^2}{(t_{ij} - t_{ij-1})\sigma^2}$$

(1)

We differentiate (1) with respect to all the unknown parameters, and equalize the expressions to zero to obtain the system of equations to solve. In addition, we consider the case $t_{ij} - t_{ij-1} = h$, $i = 1, \ldots, d$ and rewrite the system as follows

$$A_{1,b} = \sum_{i=1}^{d} \sum_{j=2}^{n_i} b^{t_{ij}-1} \qquad A_{1,b}^* = \sum_{i=1}^{d} \sum_{j=2}^{n_i} t_{ij-1} b^{t_{ij}-1}$$

$$A_{2,b} = \sum_{i=1}^{d} \sum_{j=2}^{n_i} b^{2t_{ij}-1} \qquad A_{2,b}^* = \sum_{i=1}^{d} \sum_{j=2}^{n_i} t_{ij-1} b^{2t_{ij}-1}$$

$$A_{3,b} = \sum_{i=1}^{d} \sum_{j=2}^{n_i} b^{t_{ij}-1} \ln \frac{x_{ij}}{x_{ij-1}} \qquad A_{3,b}^* = \sum_{i=1}^{d} \sum_{j=2}^{n_i} t_{ij-1} b^{t_{ij}-1} \ln \frac{x_{ij}}{x_{ij-1}}$$

$$A_{4,b} = \sum_{i=1}^{d} \sum_{j=2}^{n_i} \ln^2 \frac{x_{ij}}{x_{ij-1}} \qquad A_{5,b} = \sum_{i=1}^{d} \sum_{j=2}^{n_i} \ln \frac{x_{ij}}{x_{ij-1}}$$

obtaining, after some operations, that all the estimations depend on b:

$$\sigma_b^2 = \frac{1}{h(k-d)} \frac{A_{4,b} A_{1,b}^* A_{2,b} - A_{4,b} A_{1,b} A_{2,b}^* + A_{1,b} A_{3,b}^* A_{3,b} - A_{1,b}^* A_{3,b}^2 + A_{3,b} A_{2,b}^* A_{5,b} - A_{3,b}^* A_{2,b} A_{5,b}}{A_{1,b}^* A_{2,b} - A_{1,b} A_{2,b}^*},$$

(2)

$$m_b = \frac{\ln(b)}{b^h - 1} \frac{A_{1,b}^* A_{3,b} - A_{1,b} A_{3,b}^*}{A_{1,b}^* A_{2,b} - A_{1,b} A_{2,b}^*} \quad \text{and} \quad c_b = \frac{\sigma^2}{2} - \frac{1}{h} \frac{A_{3,b} A_{2,b}^* - A_{3,b}^* A_{2,b}}{A_{1,b}^* A_{2,b} - A_{1,b} A_{2,b}^*} \quad \text{and, finally, the equa-}$$

tion

$$A_{5,b} - \frac{m_b}{\ln(b)}(b^h - 1)A_{1,b} + \left(\frac{\sigma_b^2}{2} - c_b\right) h(k - d) = 0, \tag{3}$$

that has no explicit solution.

This point take us to propose different approaches to find the estimates of the parameters, maximising the likelihood function via numerical approximations or finding the estimates with other strategy. We describe the proposed methods in Section 2. We compare afterwards, at Section 3 all the methods using four simulated data set with different noise levels. Finally, in Section 4 we establish the main conclusions of the study.

2   Methods

The aim of this work is to compare methods that allow us to adjust a Gompertz-lognormal diffusion process to sample paths by the estimation of its parameters. That estimated process is useful with predictive purpose, using the estimated mean function, i.e. in a lot of natural diffusion processes like diseases spreading.

We propose to use one deterministic method (Iterative Method), one quasi-deterministic method (Newton-Raphson method based algorithm), and three non-deterministic methods (Variable Neighbour Search [10,14]), Simulated Annealing [7] and Evolutionary Algorithm [2]). All of them search the estimates of the parameters but the way each method finds the estimations is different from one to the other. The Iterative Method and the Newton-Raphson provides the estimations by solving the equation (3) approximately and then they use the relations between the parameters displayed in (2) to estimate the complete set. Nevertheless, the non-deterministic methods, Variable Neighbourhood Search, Simulated Annealing and the Evolutionary Algorithm, acquire the estimations optimising the function (1) directly, with no additional knowledge.

2.1   Iterative Method

The Iterative Method, (from now, IM) uses relationships among the parameters and tries to set a solution using an iterative loop. This method includes three steps:

1. First, we fix an initial value for parameter β and the method uses it to get an initial set of solutions ($\{\hat{m}; \hat{\beta}; \hat{c}; \hat{\sigma}^2\}$) through assessing (2).
2. Second, the method calculates β' using the point of the path where the second inflection point is set, for improving the β value.
3. Third, the method goes to step 1 and sets β' as β. The process iterates until $|\beta - \beta'| < 10^{-8}$.

This method is useful when the second inflection point of the paths (used at second step) are clear, but sometimes it is not easy to calculate. Furthermore, if the method takes a wrong inflection point, the solution may not be precise enough.

2.2 Newton-Raphson Based Algorithm

Secondly, we provide estimations of the parameters using the Newton-Raphson method (NR from now) to find a root of the equation (3). In this case the objective function has to verify some analytical conditions to achieve its convergence to a correct solution, independently of the initial solution considered. In our case, the function defined by the equation does not usually fulfil the conditions, so the method provides a solution, but it depends on the starting point. This method works as follows:

1. It generates a random initial value in the search space and calculates the tangent line to the function at that point.
2. Then it uses the x-intercept of that line as the next approximation point.
3. The method iterates until stopping criteria is reached.

We have selected, as stopping criteria, a maximum tolerance level in the iterative process. In particular we consider a threshold of 10^{-8}.

2.3 Variable Neighbourhooh Search

Variable neighbourhood search (VNS) is a metaheuristics designed for solving optimization problems. It exploits systematically the idea of neighbourhood change, both in the descent to local minima and in the escape from the valleys which contain them. The basic steps of VNS metaheuristic are:

1. First, it selects a set of neighbourhood structures $\mathcal{N}_k$, $k = 1, \cdots, k_{max}$, and random distributions for the (2.b.i) step, that will be used in the search, and it finds an initial solution x.
2. It repeats the following sequence until a stopping condition is met:
   (a) Set $k \leftarrow 1$;
   (b) Repeat the following steps until $k > k_{max}$:
      i. Generate a point y randomly from the $k-$th neighbourhood of x ($y \in \mathcal{N}_k(x)$);
      ii. Apply some local search method with y as initial solution to obtain a local optimum given by y';
      iii. If the local optimum is better than the incumbent, move there ($x \leftarrow y'$), and continue the search with $\mathcal{N}_1(k \leftarrow 1)$; otherwise, set $k \leftarrow k+1$.

In this case, the local search (2.b.ii) is the Nelder-Mead algorithm [15] which does not carry out the gradient computation. The understanding for this choice is that if function (1) is not derivable, the gradient does not exist. Moreover, we repeat second step for each range, 10 times as maximum to assure the convergence to the optimum solution inside the current range. The stopping rule equals maximum tolerance to 10^{-8}.

2.4   Simulated Annealing

Although the Simulated Annealing (from now, SA) algorithm was proposed time ago, it is successfully used in many areas and is a relevant optimization method today. In a classical SA problem, we define a cost function (function (1), named f) and it is minimised meanwhile running. The algorithm search is as follows:

1. At the beginning it generates a random solution as starting point. The algorithm generates a new solution from the current one by means of a change operator (new state generator).
2. If the new solution is better, it is accepted. However, if it is worse, the algorithm accepts it using the probability of $p_a = e^{-\Delta f/T}$, where Δf is the increment of the cost function and T is the current value of temperature parameter. The temperature is then decreased using the temperature cold function T. The decrement rule used for this paper is $T_{n+1} = \alpha T_n$, where α is equal to 0.95 proposed in [8]. For each temperature value, the algorithm test 32 different solutions.
3. The algorithm starts with temperature equal to 10 and ends when the temperature decreases according to the f_T until 0.0001.

The algorithm matches with the version described in [13], except the new state generator, that is specific for this problem. The new state generator selects randomly one of the parameters we look for, and generates a new value close to the previous one, using a closed interval from the current one. This generator provides the steps the algorithm follows towards the final solution.

2.5   Evolutionary Algorithm

The Evolutionary Algorithm (EA from now) searches the values of the parameters including them in each individual of a population. The fitness function of the algorithm is the value of the function (1) used by the parameters each individual proposes by its chromosome. Each individual uses real codification with four values randomly generated at the beginning. Each value is randomly generated within a closed interval defined by an expert for this problem. The individuals are evolved with roulette-wheel selector using BLX-alpha crossover [12] with 0.8 rate and a normal distribution $N(0; 1)$ mutator to alter the individuals with 0.2 as application rate and 1% as mutation rate per gen.

The method simulates a generational evolution of 100 individuals along 120 iterations. The process is repeated until the number of generations is reached. This method takes much more time than IM, RN or VNS. In addition, this method provides a solution each time you run it and it does not diverge from the solution of the problem as Newton-Raphson based algorithm does. The reliability of the method is 100%, because the method always provides a suitable solution.

The values for the parameters have been tuning by experimentation running a set of previous experiment where the minimum number of generations and population size where fixed.

3   Simulated Paths Study

In order to compare the five different methods we carry out a simulation of the Gompertz-lognormal diffusion process. This is based on the algorithms derived from the numerical solution of stochastic differential equations [9,16].

The simulation allows us to find paths that validate the estimation procedures proposed. To this end, several different paths are generated using a recursive algorithm starting with initial value, x_0. This value is determined by the initial distribution being considered as either degenerate or lognormal. For the generation of paths, it is considered x_0 following a lognormal distribution $\Lambda(0; 1)$. The values of the parameters in the simulation are $m = 1$, $\beta = 0.2$ and $c = 0.013$. Four different values for σ^2 have been considered, taking the values 10^{-5}, 10^{-4}, 10^{-3} and 10^{-2}. For each combination, we simulate 100 paths with 500 data each one, from times $t_0 = 1$ to $t_{500} = 100$, being $t_i - t_{i-1} = 0.2$. Figure 1 shows the simulated paths.

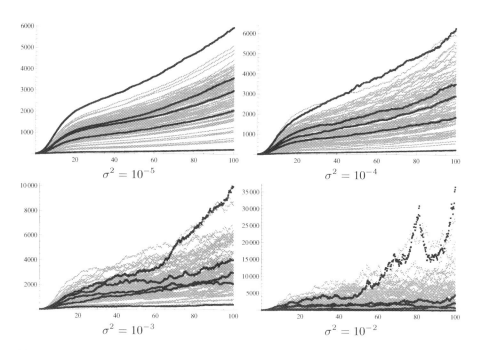

Fig. 1. Simulated paths for Gompertz-lognormal process from $t_0 = 1$ to $t_{500} = 100$, with values $m = 1$, $\beta = 0.2$, and $c = 0.013$ and different values of noise; $\sigma^2 = 10^{-5}$, $\sigma^2 = 10^{-4}$, $\sigma^2 = 10^{-3}$ and $\sigma^2 = 10^{-2}$. Black lines highlight the paths which last value $x_{i;500}$ are maximum, minimum and the three quartiles of the simulated paths range.

For each set of paths, we apply the five proposed methods 30 times, except IM, which is a deterministic approach. The averages of each estimated parameter, using just valid outputs, are calculated later the experimentation process.

We calculate the error measurements [3]: mean squared error (MSE), mean absolute error (MAE), mean absolute percentage error (MAPE), symmetric mean absolute percentage error (SMAPE) and mean relative absolute error (MRAE) in all paths and compute the average of them for each method and noise level. In the case of the MRAE, the error for each path is calculated between the estimated mean function and the path that a simple model gives. In this case a *naive* model is used; the predicted value is equal to the value before it.

We summarize the results of estimates and errors in Tables 1 and 2.

Table 1. Number of the successful solutions (*Succ.*); average and standard deviations of the estimates $(\overline{\hat{m}}, sd_{\hat{m}}, \overline{\hat{\beta}}, sd_{\hat{\beta}}, \overline{\hat{c}}, sd_{\hat{c}}, \overline{\hat{\sigma}^2}, sd_{\hat{\sigma}^2})$; average and standard deviation of the value of function (1) $(\overline{\hat{F}}, sd_{\hat{F}})$ and best found solution $(\hat{F}_{min})$. Paths simulated with $m = 1$, $\beta = 0.2$ and $c = 0.013$ and $\sigma^2 = \{10^{-5}, 10^{-4}, 10^{-3}, 10^{-2}\}$.

M	Succ.	$\overline{\hat{m}}$	$sd_{\hat{m}}$	$\overline{\hat{\beta}}$	$sd_{\hat{\beta}}$	$\overline{\hat{c}}$	$sd_{\hat{c}}$	$\overline{\hat{\sigma}^2}$	$sd_{\hat{\sigma}^2}$	$\overline{\hat{F}}$	$sd_{\hat{F}}$
					Noise Level $\sigma^2 = 10^{-5}$						
IM	1	1.0076	–	0.20034	–	0.01299	–	$10 \cdot 10^{-6}$	–	−205013.0	–
NR	12	1.0129	$7 \cdot 10^{-10}$	0.20273	$2 \cdot 10^{-8}$	0.01319	$3 \cdot 10^{-8}$	$12 \cdot 10^{-6}$	$7 \cdot 10^{-11}$	−256494.2	$2 \cdot 10^{-6}$
VNS	30	0.9972	$1 \cdot 10^{-2}$	0.20398	$3 \cdot 10^{-3}$	0.01363	$8 \cdot 10^{-4}$	$14 \cdot 10^{-6}$	$5 \cdot 10^{-6}$	−254626.4	2298.7
SA	30	1.0129	$3 \cdot 10^{-5}$	0.20273	$9 \cdot 10^{-6}$	0.01319	$3 \cdot 10^{-7}$	$12 \cdot 10^{-6}$	$4 \cdot 10^{-8}$	−256494.0	$1 \cdot 10^{-1}$
EA	30	1.0129	$2 \cdot 10^{-4}$	0.20273	$4 \cdot 10^{-6}$	0.01319	$1 \cdot 10^{-6}$	$12 \cdot 10^{-6}$	$4 \cdot 10^{-9}$	−256494.2	$1 \cdot 10^{-2}$
					Noise Level $\sigma^2 = 10^{-4}$						
IM	1	1.0144	–	0.19846	–	0.01338	–	$10 \cdot 10^{-5}$	–	−194477.8	–
NR	21	1.0395	$3 \cdot 10^{-8}$	0.20815	$3 \cdot 10^{-7}$	0.01354	$4 \cdot 10^{-7}$	$12 \cdot 10^{-5}$	$9 \cdot 10^{-10}$	−199315.7	$2 \cdot 10^{-5}$
VNS	30	0.9970	$1 \cdot 10^{-2}$	0.20854	$3 \cdot 10^{-3}$	0.01374	$1 \cdot 10^{-3}$	$12 \cdot 10^{-5}$	$7 \cdot 10^{-6}$	−199170.7	216.6
SA	30	1.0394	$1 \cdot 10^{-7}$	0.20815	$1 \cdot 10^{-5}$	0.01354	$1 \cdot 10^{-5}$	$12 \cdot 10^{-4}$	$7 \cdot 10^{-8}$	−199315.7	$1 \cdot 10^{-3}$
EA	30	1.0395	$1 \cdot 10^{-7}$	0.20815	$1 \cdot 10^{-5}$	0.01354	$1 \cdot 10^{-5}$	$12 \cdot 10^{-5}$	$7 \cdot 10^{-8}$	−199315.7	$8 \cdot 10^{-4}$
					Noise Level $\sigma^2 = 10^{-3}$						
IM	1	1.1366	–	0.22761	–	0.01478	–	$10 \cdot 10^{-4}$	–	−88350.2	–
NR	14	1.1366	$2 \cdot 10^{-8}$	0.22761	$9 \cdot 10^{-8}$	0.01478	$9 \cdot 10^{-8}$	$12 \cdot 10^{-4}$	$2 \cdot 10^{-12}$	−141799.8	$7 \cdot 10^{-8}$
VNS	30	0.9989	$1 \cdot 10^{-2}$	0.22737	$3 \cdot 10^{-3}$	0.01459	$8 \cdot 10^{-4}$	$13 \cdot 10^{-4}$	$1 \cdot 10^{-5}$	−141790.6	10.0
SA	30	1.1366	$2 \cdot 10^{-4}$	0.22759	$6 \cdot 10^{-5}$	0.01476	$2 \cdot 10^{-5}$	$12 \cdot 10^{-4}$	$6 \cdot 10^{-7}$	−141799.8	$8 \cdot 10^{-3}$
EA	30	1.1366	$3 \cdot 10^{-6}$	0.22761	$8 \cdot 10^{-7}$	0.01478	$1 \cdot 10^{-7}$	$12 \cdot 10^{-4}$	$5 \cdot 10^{-9}$	−141799.8	$4 \cdot 10^{-6}$
					Noise Level $\sigma^2 = 10^{-2}$						
IM	0	–	–	–	–	–	–	–	–	–	–
NR	16	1.5472	$2 \cdot 10^{-7}$	0.32957	$2 \cdot 10^{-7}$	0.02079	$2 \cdot 10^{-6}$	$12 \cdot 10^{-3}$	$4 \cdot 10^{-10}$	−84972.8	$2 \cdot 10^{-8}$
VNS	30	0.9371	$5 \cdot 10^{-3}$	0.33014	$2 \cdot 10^{-3}$	0.02081	$5 \cdot 10^{-4}$	$12 \cdot 10^{-3}$	$4 \cdot 10^{-5}$	−84972.3	0.75
SA	30	1.1366	$2 \cdot 10^{-4}$	0.22759	$6 \cdot 10^{-5}$	0.01476	$2 \cdot 10^{-5}$	$12 \cdot 10^{-3}$	$4 \cdot 10^{-5}$	−84965.0	$1 \cdot 10^{-3}$
EA	30	1.5391	$2 \cdot 10^{-3}$	0.32210	$6 \cdot 10^{-4}$	0.01560	$3 \cdot 10^{-6}$	$12 \cdot 10^{-3}$	$7 \cdot 10^{-6}$	−84965.1	$1 \cdot 10^{-3}$

Table 1 shows the average of the parameters estimations when the noise on the simulation is 10^{-5}, 10^{-4}, 10^{-3} and 10^{-2}. We also include the standard deviation because small changes in the value of the parameters could involve big alterations in objective function results.

The number of valid outputs, labelled *Succ.*, display that in spite of NR has been launched 30 runs, only some of their outputs, 12, 21, 14 and 16 times respectively, provides a successful solution.

As can be seen, all the methods provide a solution close to those used in the simulation for each parameter, and the closest is the obtained with the IM for every value of σ^2 but $\sigma^2 = 10^{-2}$, because there is no solution this time using IM. According to this table, all considered methods found good results for the estimations of the parameters in comparison to the values used for the simulation. This table also shows the average of the values with these estimations for the function (1).

Usually, researchers in bio-inspired algorithms present tables with the means of their algorithm scores. However, in this case, the averaged values in Table 1 are not useful for evaluating function (1) because the means of the scores are not really outputs of the methods. Due to the values of each parameter strongly influence the function value, we decide to select one of the estimated set of parameters, which is included in the left part of Table 2. We select the set of values $\{\hat{m}; \hat{\beta}; \hat{c}; \hat{\sigma}^2\}$ which minimise the function (1), labelled as $\{\hat{m}_{F_{min}}; \hat{\beta}_{F_{min}}; \hat{c}_{F_{min}}; \hat{\sigma}^2_{F_{min}}\}$, because the goal of this experimentation process is to minimise it.

Table 2 includes firstly the selected sets of estimations $\{\hat{m}_{F_{min}}; \hat{\beta}_{F_{min}}; \hat{c}_{F_{min}}; \hat{\sigma}^2_{F_{min}}\}$. Secondly, it shows $\hat{F}_{min}$, that is, the values of the function (1) that the selected sets of estimates provide. Finally, it presents the means of the error measurements MSE, MAE, MAPE, SMAPE and MRAE given by each method and the four sets of paths.

Table 2. Estimated sets of parameters that give the minimum values of function (1); values for function (1) that the sets provide and averages of the error measurements with the selected set of parameteres

M	$\hat{m}_{F_{min}}$	$\hat{\beta}_{F_{min}}$	$\hat{c}_{F_{min}}$	$\hat{\sigma}^2_{F_{min}}$	$\hat{F}_{min}$	MSE	MAE	MAPE	SMAPE	MRAE
				Noise Level $\sigma^2 = 10^{-5}$						
IM	1.008	0.200	0.013	$26 \cdot 10^{-5}$	-205013	2858	30.52	2.10	0.02	39.85
NR	1.013	0.203	0.013	$13 \cdot 10^{-6}$	-256494	2785	28.48	1.87	0.02	32.17
VNS	0.999	0.203	0.013	$13 \cdot 10^{-6}$	-256476	14052	94.33	5.74	0.06	71.20
SA	1.013	0.203	0.013	$12 \cdot 10^{-6}$	-256494	2791	28.51	1.87	0.02	32.18
EA	1.013	0.203	0.013	$13 \cdot 10^{-6}$	-256494	2785	28.48	1.87	0.02	32.17
				Noise Level $\sigma^2 = 10^{-4}$						
IM	1.017	0.198	0.013	$244 \cdot 10^{-5}$	-148757	51123	146.42	10.70	0.10	129.92
NR	1.039	0.208	0.014	$12 \cdot 10^{-5}$	-199316	25027	87.14	5.43	0.05	73.73
VNS	0.999	0.208	0.014	$13 \cdot 10^{-5}$	-199315	154407	300.77	17.88	0.20	232.38
SA	1.039	0.208	0.014	$12 \cdot 10^{-5}$	-199316	25025	87.13	5.43	0.05	73.73
EA	1.039	0.208	0.014	$12 \cdot 10^{-5}$	-199316	25027	87.14	5.43	0.05	73.73
				Noise Level $\sigma^2 = 10^{-3}$						
IM	1.035	0.186	0.022	$275 \cdot 10^{-4}$	-88517	16443800	2262.92	146.54	0.75	725.20
NR	1.137	0.228	0.015	$13 \cdot 10^{-4}$	-141800	341403	267.03	15.88	0.16	71.62
VNS	0.999	0.228	0.015	$13 \cdot 10^{-4}$	-141800	1461321	867.30	44.93	0.60	208.52
SA	1.136	0.228	0.015	$13 \cdot 10^{-4}$	-141800	340903	266.83	15.88	0.16	71.60
EA	1.137	0.228	0.015	$13 \cdot 10^{-4}$	-141800	341403	267.03	15.88	0.16	71.62
				Noise Level $\sigma^2 = 10^{-2}$						
IM	$-$	$-$	$-$	$-$	$-$	$-$	$-$	$-$	$-$	$-$
NR	1.547	0.330	0.021	$12 \cdot 10^{-3}$	-84973	5120383	955.6	79.18	0.54	139.8
VNS	0.939	0.330	0.021	$12 \cdot 10^{-3}$	-84973	9921164	1785.9	78.32	1.34	136.7
SA	1.540	0.322	0.016	$12 \cdot 10^{-3}$	-84965	5642929	1010.4	62.19	0.54	108.9
EA	1.539	0.322	0.016	$12 \cdot 10^{-3}$	-84965	5634217	1009.3	62.29	0.53	109.1

4   Conclusions

Table 1 shows that the IM provides an estimated set of parameters which is the closets to the used in simulation, excepting for $\sigma^2 = 10^{-2}$, where the method is not able to provide a solution. In spite of that, the values of function (1) to minimise, with those estimations, are the highest in all cases. Moreover, error measurements, shown in Table 2, for IM are not the lowest in any case. The main advantage of IM is it spends less time than the rest of the methods and it always gives the same solution.

Newton-Raphson method provides or not a solution depending on the starting point. When it is not good enough the method diverges, but when the initial value is close to the real one, the method provides a precise solution. In Table 1 appears data about the number of successfully runs (*Succ.*) for each noise level; 12, 21, 14 and 16 respectively in the simulation study. For the success cases, the means of the estimations obtained for the parameters are close to the values used in simulation, as well as results given by SA and EA methods. The function (1) evaluated with the estimations has, in mean, the lowest value, and the reached minimum is the lowest too, together with SA and EA methods. Despite the means and the minimum in these tree methods are close each other, NR is the one that presents lower standard deviation, thus the results are closer using NR than using SA or EA.

Table 2 shows that the better found solution (which minimise the function (1)) for NR is similar to the provided by EA and SA. The estimated set of parameters, value of function (1) reached and error measurements are very close for low noise and not much far for higher noise. Thus, the error measurements are not better nor worse than the results provided by EA or SA methods. The advantage of NR is that, if you have some knowledge about the expected solution, this method provides an exact solution in a few seconds. The drawback is that the method is not as reliable as the others.

Based on the values in Table 1, we note that VNS method gets competitive average values comparing them with the other methods, but the standard deviations obtained for all the estimations are high. It implies that the method varies in the solutions provided much more than the others methods. Hence, in Table 2 appears that VNS presents the highest error values almost always. Only for high noise level, IM gives worse results.

As we commented before, generally, SA and EA provide similar results to NR. For low noise cases, we can not appreciate any difference but, when data show high noise SA and EA provide different mean estimations for the parameters from NR and higher value of function (1). These bio-inspired methods do not need any previous knowledge about the objective function and they supply competitive solutions for all the runs. Nevertheless, SA and EA are slower than the other considered methods, and in Table 1 the main difference we appreciate between EA and SA is that the standard deviation for the $\overline{\hat{F}}$ is lower for EA. Otherwise, NR and EA obtain the same parameter estimations, which minimise function (1) and are very similar to EA ones. In Table 2 we can see similar behaviour. For hight noise level, EA and SA provide different results with respect to NR, and they have less average errors MAPE, SMAPE and MRAE than NR but higher errors MSE and MAE.

In short, the IM allows us, when no high noise is present, to estimate the parameters close to the real set, although it does not mean the function estimated looks like the observed path. In terms of errors and likelihood value, the NR method gives as good results as EA and SA, but only EA or SA provides always successful results. For high noise, bio-inspired algorithms give results more reliable than IM, NR or VNS give although they are less exact for high noise levels. VNS is not better than other methods in any case.

As future lines we proposes to apply the reliable methods, EA or SA, to model some processes of disease transmission, as influenza dissemination process.

Acknowledgments. The authors would like to thank the FEDER of European Union for financial support via project "Sistema de Información y Predicción de bajo coste y autónomo para conocer el Estado de las Carreteras en tiempo real mediante dispositivos distribuidos" (SIPEsCa) of the "Programa Operativo FEDER de Andalucía 2007-2013". We also thank all Agency of Public Works of Andalusia Regional Government staff and researchers for their dedication and their professionalism. This work has been supported in part by project ANYSELF (TIN2011-28627-C04-02), MICINN (Spain) MTM2011-28962, PRY142/14 (ha sido financiado íntegramente por la Fundación Pública Andaluza Centro de Estudios Andaluces en la IX Convocatoria de Proyectos de Investigación), PYR-2014-17 GENIL project, awarded by CEI-BIOTIC Granada and PETRA (SPIP2014-01437, funded by Dirección General de Tráfico).

References

1. Albano, G., Giorno, V., Romn-Romn, P., Romn-Romn, S., Torres-Ruiz, F.: Estimating and determining the effect of a therapy on tumor dynamics by means of a modified Gompertz diffusion process **364**, 206–219 (2015)
2. Bäck, T.: Evolutionary algorithms in theory and practice: evolution strategies, evolutionary programming, genetic algorithms. Oxford University Press (1996)
3. De Gooijer, J.G., Hyndman, R.J.: 25 years of time series forecasting. International Journal of Forecasting **22**, 443–473 (2006)
4. Gutiérrez, R., Román, P., Torres, F.: Inference and first-passage-times for the lognormal diffusion process with exogenous factors: application to modelling in economics. Applied Stochastic Models in Business and Industry **15**, 325–332 (1999)
5. Gutiérrez, R., Román, P., Romero, D., Torres, F.: Applications of the univariate lognormal diffusion process with exogenous factors in forecasting. Cybernetics and Systems **34**(8), 709–724 (2003)

6. Gutiérrez, R., Román, P., Romero, D., Serrano, J.J., Torres, F.: A new Gompertz-type diffusion process with application to random growth. Mathematical Biosciences **208**, 147–165 (2007)
7. Ingber, L.: Simulated annealing: practice versus theory. Mathematical and Computer Modelling **18**(11), 29–57 (1993)
8. Kirkpatrick, S., Gelatt Jr, C.D., Vecch, M.P.: Optimization by simulated annealing. Science **220**(4598), 671–680 (1983)
9. Kloeden, P.E., Platen, E., Schurz, H.: Numerical solution of SDE through computer experiments. Springer-Verlag (1994)
10. Lagarias, J.C., Reeds, J.A., Wright, M.H., Wright, P.E.: Convergence properties of the Nelde-Mead simplex method in low dimensions. SIAM Journal on Optimization **9**(1), 112–147 (1998)
11. Lo, C.F.: Stochastic Gompertz model of tumour cell growth. Journal of Theoretical Biology **248**(2), 317–321 (2007)
12. Lozano, M., Herrera, F., Krasnogor, N., Molina, D.: Real-coded memetic algorithms with crossover hill-climbing. Evolutionary Computation **12**(3), 273–302 (2004)
13. Michalewicz, Z.: Heuristic methods for evolutionary computation techniques. Heuristics **1**(2), 177–206 (1996)
14. Mladenovic, N., Hansen, P.: Variable neighborhood search. Computers and Operations Research **24**, 1097–1100 (1997)
15. Nelder, J.A., Mead, R.: A simplex method for function minimization. The Computer Journal **7**, 308–313 (1965)
16. Rao, N.J., Borwankar, J.D., Ramkrishna, D.: Numerical solution of Ito integral equations. Journal on Control and Optimization **12**, 124–139 (1974)
17. Rico, N., Romn-Romn, P., Romero, D., Torres-Ruiz, F.: Gompertz-lognormal diffusion process for modelling the accumulated nutrients dissolved in the growth of Capiscum Annuum. In: The 20th Annual Conference of The International Environmetrics Society. Book of abstracts, p. 90 (2009)
18. Romn-Romn, P., Romero, D., Torres-Ruiz, F.: A diffusion process to model generalized Von Bertalanffy growth patterns: fitting to real data. Journal of Theoretical Biology **263**(1), 59–69 (2010)
19. Romn-Romn, P., Torres-Ruiz, F.: Modelling logistic growth by a new diffusion process: Application to biological systems. Biosystems **110**(1), 9–21 (2012)
20. Romn-Romn, P., Torres-Ruiz, F.: Forecasting fruit size and caliber by means of diffusion processes. Application to "valencia late" oranges. Journal of Agricultural, Biological, and Environmental Statistics **19**(2), 292–313 (2014)
21. Romero, D., Rico, N., G-Arenas, M.: A new diffusion process to epidemic data. In: Moreno-Díaz, R., Pichler, F., Quesada-Arencibia, A. (eds.) EUROCAST. LNCS, vol. 8111, pp. 69–76. Springer, Heidelberg (2013)

Estimating Artificial Neural Networks with Generalized Method Moments

Alexandre Street de Aguiar[1] and João Marco Braga da Cunha[1,2(✉)]

[1] Electrical Engineering Department, Pontifical Catholic University of Rio de Janeiro, Rio de Janeiro, RJ 22451-900, Brazil
[2] Brazilian Development Bank (BNDES), Rio de Janeiro, RJ 20031-917, Brazil
jmarco@fgvmail.br

Abstract. In this article, we present a general framework for estimation of Artificial Neural Networks (ANN) parameters using the Generalized Method of Moments (GMM), as an alternative to the conventional *Quasi* Maximum Likelihood (QML). We used a simple generalization for nonlinear models of the usual orthogonality conditions from linear regression in addition to the moment conditions that replicate the QML estimation. Consequently the resultant models are overidentified. Monte Carlo simulations suggested that GMM can outperform QML in cases with small samples or elevated noise.

Keywords: Artificial Neural Networks · Feedforward perceptron · Generalized Method of Moments · *Quasi* Maximum Likelihood

1 Introduction

In the last few decades, nonlinear models have increased their relevance among econometricians. One of the most used techniques applies artificial neural networks due to its universal approximation of Borel-measurable functions capability ([1] and [2]). In particular, feedforward ANNs with a single hidden layer have received special attention since the late 80s. The seminal works [3] and [4] treated ANN learning as a parametric estimation problem and proposed the application of the *Quasi* Maximum Likelihood (QML) method. They also established identifiability conditions for the parameters and their asymptotic normality. A test for neglected nonlinearity was presented in [5]. It can be used to determine the ANNs architecture in an incremental procedure.

Since then, several learning algorithms and heuristics have been developed in the Computer Science and Artificial Intelligence literature. Nevertheless, QML is still the most widely used method for econometric purposes. This can be explained by two desirable features, consistency and asymptotic normality, which are not found in other methods.

In this paper, we introduce a new approach for estimating ANNs, using the Generalized Method of Moments. The GMM estimation maintains consistency and asymptotic normality under suitable conditions and can potentially generate

© Springer International Publishing Switzerland 2015
I. Rojas et al. (Eds.): IWANN 2015, Part II, LNCS 9095, pp. 391–399, 2015.
DOI: 10.1007/978-3-319-19222-2_33

more precise finite sample estimations. A more detailed explanation is presented in the following section. The third section presents Monte Carlo experiments and their results. The conclusions appear in the last section.

2   GMM for ANN

The goal of this paper is to present a general framework for estimating ANNs using overidentified GMMs. This method can be applied in a broad range of ANNs that use the supervised learning paradigm. Without loss of generality, we will focus our attention on the single-layer perceptron with logistic activation function and direct (linear) connection between the explicative variables (inputs) and conditional expected values (outputs), as it is done in [8]. So, we have:

$$G(\boldsymbol{x}_n; \psi) = \boldsymbol{\alpha}' \cdot \boldsymbol{x}_n + \sum_{m=1}^{M} \left\{ \lambda_m \cdot l(\boldsymbol{\omega}'_m \cdot \boldsymbol{x}_n) \right\}, \tag{1}$$

where $G(\cdot; \psi)$ is an ANN parametrized by $\psi = [\boldsymbol{\alpha}', \lambda_1, ..., \lambda_M, \boldsymbol{\omega}'_1, ..., \boldsymbol{\omega}'_M]$, $\boldsymbol{x}_n$ [1] are the explanatory variables (I of them, including a constant), M is the number of neurons in the hidden layer and $l(\cdot)$ is the logistic function. So, the econometric model is given by:

$$y_n = G(\boldsymbol{x}_n; \psi) + \varepsilon_n, \tag{2}$$

where y_n é is the dependent variable and ε_n is a zero-mean noise. In order to avoid multiple parametrizations of the same ANN, we imposed the following restrictions: $0 < \lambda_1 < \lambda_2 < ... < \lambda_M$.

Before we explain the particular case of ANNs' parameters estimation, we will provide an overview of the GMM.

2.1   GMM Overview

The GMM is an estimation method introduced in 1982 by [6]. It generalizes the traditional Method of Moments by allowing the number of moment conditions to be superior to the number of estimated parameters. In this case, we say that the model is overidentified. Given that it may be impossible to match all the sample moment conditions to their population counterparts, GMM proceeds by minimizing a quadratic form of their deviations. Formally, let $\{\boldsymbol{x}_n\}_{N \geq n \geq 1}$ be a sample from a J-variate stochastic process, $\psi^* \in \Psi \subset R^L$ a vector of unknown parameters, $W \in R^{K \times K}$ a positive semi-definite matrix, and $f : R^{J+L} \to R^K$ a function of moment conditions satisfying:

$$E[f(\boldsymbol{x}, \psi^*)] = 0. \tag{3}$$

Defining $c(\psi) = N^{-1} \sum_{n=1}^{N} f(\boldsymbol{x}_n, \psi)$, a GMM estimation for ψ^* is given by:

$$\hat{\psi}_{MGM} = \arg \min_{\psi \in \Psi} c(\psi)' \cdot W \cdot c(\psi). \tag{4}$$

---

[1] Vectors are defined as column matrices.

Although any positive semi-definite matrix W generates a consistent and asymptotically normal estimator, it is proven in [6] that GMM reaches the minimum asymptotic variance when:

$$W = (E[f(\boldsymbol{x}, \psi^*).f(\boldsymbol{x}, \psi^*)'])^{-1}. \tag{5}$$

In these cases, it's said that W is the optimal weighting matrix. An extensive discussion on different approaches for estimating W can be found in [7].

2.2 Moment Conditions for ANNs

The first step to construct the overidentified GMM is to show that the QML can be thought as an exactly identified GMM.

The QML is obtained by:

$$\hat{\psi}_{QML} = \arg\min_{\psi \in \Psi} F_{SQR}(\psi), \tag{6}$$

where $F_{SQR}(\psi) = (Y - G(X; \psi))' \cdot (Y - G(X; \psi))$, and the capital letters for variables denotes matrices with the observations stacked in the lines. It is convenient to name the residuals vector $U(\psi) = Y - G(X; \psi)$.

Consequently, the first order conditions of this optimization problem hold for $\hat{\psi}_{QML}$, so we have valid moment conditions given by:

$$\nabla_{\psi} F_{SQR}(\hat{\psi}) = \mathbf{0}'_{J \times 1}. \tag{7}$$

where ∇_{ψ} denotes the gradient vector with the partial derivatives relative to the elements of ψ. This is equivalent to:

$$2 \cdot U(\psi)' \cdot \nabla_{\psi} G(X; \psi) = \mathbf{0}'_{J \times 1}. \tag{8}$$

This means that residuals are expected to be orthogonal to the gradient of $G(\boldsymbol{x}_n; \psi)$.

Given that:

$$\nabla_{\psi} G(X; \psi) = [X, l(X \cdot \boldsymbol{\omega}_1), \ \dots \ , l(X \cdot \boldsymbol{\omega}_M),$$
$$\lambda_1 \cdot (l'(X \cdot \boldsymbol{\omega}_1) \otimes \mathbf{1}'_{I \times 1}) \circ X, \ \dots \ , \lambda_M \cdot (l'(X \cdot \boldsymbol{\omega}_M) \otimes \mathbf{1}'_{I \times 1}) \circ X], \tag{9}$$

where $l'(\cdot)$ is the derivative of the logistic function, and $\otimes$ and $\circ$ are the symbols for Kronecker and Hadamard products, respectively.

Now, defining:

$$D(X; \{\boldsymbol{\omega}_m\}_{M \geq m \geq 1}) = [X, l(X \cdot \boldsymbol{\omega}_1), \ \dots \ , l(X \cdot \boldsymbol{\omega}_M),$$
$$(l'(X \cdot \boldsymbol{\omega}_1) \otimes \mathbf{1}'_{1 \times I}) \circ X, \ \dots \ , (l'(X \cdot \boldsymbol{\omega}_M) \otimes \mathbf{1}'_{1 \times I}) \circ X], \tag{10}$$

that does not depend on $\boldsymbol{\lambda}$, and considering that the elements of $\boldsymbol{\lambda}$ are nonzero, we can derive convenient moment conditions that are equivalent to (8):

$$U(\psi)' \cdot D(X; \{\boldsymbol{\omega}_m\}_{M \geq m \geq 1}) = \mathbf{0}'_{J \times 1}. \tag{11}$$

Once we have the same number of first order conditions as we have parameters, QML is equivalent to an exactly identified GMM, where the moment conditions are the first order conditions of the minimization of the sum of squared residuals (SSR).

In order to overidentify the GMM, we need some extra moment conditions. In classical linear regressions, it is assumed that the error term is uncorrelated to any linear combination of the explanatory variables. This assumption leads to orthogonality (moment) conditions with the form:

$$E[\boldsymbol{\nu}' \cdot \boldsymbol{x} \cdot \varepsilon] = 0, \forall \boldsymbol{\nu} \in R^J. \tag{12}$$

In the case of nonlinear models, this assumption can be generalized. The error term is assumed to be uncorrelated to any measurable function $h(\cdot)$ of the explanatory variables. So, we have moment conditions with the form $E[h(\boldsymbol{x}) \cdot \varepsilon] = 0$.

Therefore, the only necessary ingredient to generate a valid moment condition is a measurable function. There are some natural candidates, such as polynomials, exponential of linear combinations, indicator functions and, of course, ANNs.

Our method exploits ANNs by taking the following steps. For each $\{\boldsymbol{\omega}_m\}_{M \geq m \geq 1}$ in a grid, we estimate the respective $\boldsymbol{\alpha}$ and $\boldsymbol{\lambda}$ by Ordinary Least Squares (OLS), and set $\boldsymbol{\alpha}$ to zero in order to avoid redundancy with the moment conditions generated from the first order conditions. Then we generate the principal components of the conditional expected value given by all the ANNs created in the previous steps and take the first components as extra moment. The matrix with this principal components is referred to as E.

3   Monte Carlo Simulation

In this section, we constructed some Monte Carlo experiments in order to compare the accuracy of the estimations generated by QML and GMM, under controlled setups.

3.1   Data Generation

The conditional expectation was generated by an ANN, avoiding approximation issues. It has two neurons on the hidden layer and one explanatory variable (besides the constant), drawn from an Uniform Distribution between -1 and 1. The parameters ψ^* are:

$$\alpha = \begin{bmatrix} -1.5927 \ 1.0000 \end{bmatrix}',$$
$$\lambda = \begin{bmatrix} 1.5927 \ 1.5927 \end{bmatrix}',$$
$$\omega_1 = \begin{bmatrix} 1.9520 \ -10.1690 \end{bmatrix}',$$
$$\omega_2 = \begin{bmatrix} 1.9520 \ 10.1690 \end{bmatrix}'.$$

This function approximates a Gaussian curve upon a 45 degree slope.

The noise term ε_n was drawn from a Gaussian Distribution, with five different levels of variance: 10%, 30%, 50%, 70% and 90% of the dependent variable's variance. Three sample sizes (50, 200 and 800) were used in the experiments. For each one of the 15 possible combinations of variance and sample size 1.000 repetitions were carried out.

It's important do notice that, given the Gaussian noise, QML coincides with the Maximum Likelihood Method and therefore it has minimum asymptotic variance. So any other estimation procedure can only beat it in small sample cases.

3.2   Competitor Methods

The QML is the benchmark method, and it was estimated with a two-step procedure. The initialization step consists of choosing from a grid of $\{\boldsymbol{\omega}_m\}_{M \geq m \geq 1}$ those which generate the minimal sum of squared residuals given that $\boldsymbol{\alpha}$ and $\boldsymbol{\lambda}$ were estimated by OLS. In the local optimization step, the sum of squared residuals are minimized as a function of the $\{\boldsymbol{\omega}_m\}_{M \geq m \geq 1}{}^2$, starting from the point obtained in the first step.

The alternative method, GMM, was tested in four different versions, GMM-10, GMM-12, GMM-14 and GMM-16, with 10, 12, 14 and 16 moment conditions, respectively. Once again, a two-step procedure was used, very similar to one performed under QLM. The first step entailed the use of a grid of $\{\boldsymbol{\omega}_m\}_{M \geq m \geq 1}$ to get an initial point in the parametric space and a weighting matrix, which was the optimal one considering the initial parameters. Then, in the local optimization step, the GMM objective function was minimized with the fixed weighting matrix. In this case, it is also possible to consider only the optimization over the $\{\boldsymbol{\omega}_m\}_{M \geq m \geq 1}$ and take $\boldsymbol{\alpha}$ and $\boldsymbol{\lambda}$ by Weighted Least Squares. Details are presented in the Appendix.

The same grid of $\{\boldsymbol{\omega}_m\}_{M \geq m \geq 1}$ was used in all cases, including the generation of extra moment conditions for GMM[3]. Similarly, the same optimization method was applied to all the local optimizations[4].

3.3   Comparison Criterion

Each estimated ANN is evaluated in terms of the root-mean-square deviation (RMSD) from the true one in a very tight grid[5]. We compared the 50% and 95% quantiles across the 1.000 simulations conducted for each setup.

---

[2] In both steps, it's useful write the sum of squared residual as a function of $\{\boldsymbol{\omega}_m\}_{M \geq m \geq 1}$ only, once that, given $\{\boldsymbol{\omega}_m\}_{M \geq m \geq 1}$, optimal $\boldsymbol{\alpha}$ and $\boldsymbol{\lambda}$ can estimated by OLS. See [8] for details.

[3] Details of the grid's generation are on the Appendix.

[4] The optimization was done using Matlab's functions *fminunc* and, in the cases it had errors, *fminsearch*.

[5] From -1 to 1 with step size of 10^{-3}.

3.4   Results

The results are separated by sample size. Table 1 presents the results for sample size 50.

The only case in which the QML was able to beat one of the GMMs, GMM-10, occurred with a 10% noise ratio. On the other hand, GMM-10 had a median RMSD almost 25% below that of the QML for the two largest noise ratios. The RMSE-advantage of the GMMs in percentage relative to the QML is positively correlated to the noise ratios. At the 90% noise ratio, all GMMs (at he median) are more than 20% below the QML. At the 95% quantile, RMSD reduction under GMM increases to over 80% in comparison to QML.

Table 2 shows that results for sample size 200 are more balanced than those for sample size 50. QML beats GMM on three results considering the median and on four considering the 95% quantile of RMSD. GMM-14 was the only one to beat QML at every noise ratio and under the median and the 95% percentile. GMMs delivered RMSD reductions of about 20% under both the median and the 95% quantile at the 90% noise ratio.

Table 1. Sample Size 50

Quantile	Noise Ratio	QML	GMM-10	GMM-12	GMM-14	GMM-16
	10%	0,106	0,108	0,098	0,095	0,097
	30%	0,227	0,200	0,200	0,203	0,201
50%	50%	0,352	0,286	0,293	0,296	0,291
	70%	0,539	0,408	0,424	0,425	0,425
	90%	1,057	0,795	0,821	0,820	0,813
	10%	0,176	0,185	0,166	0,169	0,173
	30%	0,356	0,311	0,308	0,309	0,309
95%	50%	0,723	0,459	0,466	0,458	0,459
	70%	1,927	0,687	0,699	0,713	0,680
	90%	7,709	1,383	1,367	1,369	1,372

Table 2. Sample Size 200

Quantile	Noise Ratio	QML	GMM-10	GMM-12	GMM-14	GMM-16
	10%	0,0451	0,0479	0,0450	0,0437	0,0451
	30%	0,0940	0,0955	0,0891	0,0850	0,0862
50%	50%	0,1557	0,1475	0,1394	0,1336	0,1377
	70%	0,2480	0,2161	0,2166	0,2150	0,2197
	90%	0,4996	0,3956	0,3922	0,3961	0,4000
	10%	0,0689	0,0784	0,0689	0,0671	0,0708
	30%	0,1518	0,1581	0,1387	0,1357	0,1393
95%	50%	0,2408	0,2288	0,2202	0,2154	0,2153
	70%	0,3789	0,3211	0,3247	0,3179	0,3217
	90%	0,7698	0,6262	0,6165	0,6008	0,6086

Table 3. Sample Size 800

Quantile	Noise Ratio	QML	GMM-10	GMM-12	GMM-14	GMM-16
	10%	0,0235	0,0264	0,0249	0,0250	0,0268
	30%	0,0444	0,0490	0,0470	0,0448	0,0446
50%	50%	0,0699	0,0750	0,0712	0,0666	0,0655
	70%	0,1108	0,1124	0,1085	0,1025	0,1009
	90%	0,2368	0,2148	0,2117	0,2088	0,2064
	10%	0,0334	0,0572	0,0377	0,0354	0,0375
	30%	0,0657	0,0930	0,0747	0,0664	0,0663
95%	50%	0,1058	0,1358	0,1146	0,1001	0,0987
	70%	0,1763	0,1922	0,1708	0,1641	0,1642
	90%	0,3485	0,3175	0,3079	0,3051	0,3036

For the largest sample size tested (table 3), the QML becomes more competitive, especially under lower noise ratio. Nevertheless, under the highest noise ratio, the GMMs surpass QML performance by 10% in both quantiles measured. Another pattern that can be noticed is that, apart from the 10% noise ratio cases, the performance of the GMMs increases monotonically as the number of moment conditions grows.

4   Conclusion

Three clear conclusions arise from the information gathered in the experiments.

Firstly, the higher the noise ratio, the more likely is that the QML will deliver poorer estimations vis a vis the GMMs. This is a solid result in favor of the GMM because the noise ratio is directly related to the complexity of the estimation problem. So, the GMM is more helpful to the econometrician in the more complicated cases. In addition, real world fields of application, such as finance and fraud detection, are known for producing data with elevated levels of noise.

Secondly, smaller samples are more favorable to the GMMs relative to QML. This was not an unexpected result, since it is known that QML is (asymptotically) efficient under Gaussian noise. GMMs are relatively better for smaller samples, which happen to be the more complicated cases for estimation. For example, in several applications of time series, particularly with annual data, there are no more than a few dozen observations. This can be attributed to lack of historical data and discountinuity.

Thirdly, the GMMs' performance is positively correlated with the number of moment conditions used[6]. This is an evidence that, at least for the range tested (until twice the number of parameters) the addition of moment conditions improves the estimations.

---

[6] In fact, the simple correlation between all GMM results reported, normalized (divided) by the mean across the four GMMs with the same noise ratios sample sizes, and the number of moment conditions is -0.4735.

In sum, the overall conclusion is that in the absence of a very large sample[7] and low noise levels, the estimation of ANNs should be performed (with all the statistically desirable properties given under QML) by applying a GMM.

References

1. Cybenko, G.: Approximation by Superpositions of a Sigmoidal Function. Mathematics of Control, Signals and Systems **2**(4), 303–314
2. Hornik, K.: Approximation Capabilities of Multilayer Feedforward Networks. Neural Networks **4**(2), 251–257 (1991)
3. White, H.: Learning in Artificial Neural Networks: A Statistical Perspective. Neural Computation **1**(4), 425–464
4. White, H.: Some Asymptotic Results for Learning in Single Hidden-Layer Feedforward Network Models. Journal of the American Statistical Association **84**(408), 1003–1013
5. White, H.: An additional hidden unit test for neglected nonlinearity in multilayer feedforward networks. In: International Joint Conference on Neural Networks, IJCNN. IEEE (1989)
6. Hansen, L.: Large Sample Properties of Generalized Method of Moments Estimators. Econometrica: Journal of the Econometric Society, 1029–1054
7. Hansen, L.P., Heaton, J., Yaron, A.: Finite-sample properties of some alternative GMM estimators. Journal of Business & Economic Statistics **14**(3), 262–280 (1996)
8. Medeiros, M., Tersvirta, T., Rech, G.: Building Neural Network Models for Time Series: a Statistical Approach. Journal of Forecasting **25**(1), 49–75 (2006)

Appendix A   Generating the Grid of $\{\omega_m\}_{M \geq m \geq 1}$

The idea behind the use of a grid is the same from randomly sampling candidates for initial $\{\omega_m\}_{M \geq m \geq 1}$[8]. The grid is nothing more than a (clever) discretization of the probability distribution. The first step is to define a sampling procedure.

The starting point are the biases[9] They are drawn from an Uniform Distribution between -25 and 0. In order to avoid redundant multiple parametrizations, the biases are sorted and assigned to the neurons in order according to m. Then, for each neuron, the other parameters are drawn from an Uniform Distribution between $-25 - \omega_{m,0}$ and $25 + \omega_{m,0}$. A set of extra restrictions must be applied:

- the interval $\left[min\left(\omega_{m,1}, \sum_{j>1} \omega_{m,j}\right), \max\left(\omega_{m,1}, \sum_{j>1} \omega_{m,j}\right)\right]$ must contain -3 or 3;

- and $\sum_{j>1} abs(\omega_{m,j})$ must be greater than 2.

---

[7] The number of observations that constitutes a large sample is strongly related to the number of estimated parameters.

[8] Explanatory variables rescaled between 0 and 1 (except for the constant).

[9] Biases are parameters $\omega_{m,1}$ that are multiplied by the constant one.

The first restriction prevents sample points from winding up far from zero, in the flat zones of the activation function. The second one avoids the concentration of points in a small part of the activation. If any restriction doesn't hold, the point is discarded and a new one is drawn.

Once the sampling procedure is defined, the following step consists of generating 1.000 initial points. After this, an iterative process is started. In each iteration, a new point is drawn. The closest point on the grid to the new one is taken and replaced by a convex combination of them. The weight of the new point should decrease to zero across the iterations. In the experiments, we performed 50.000 iterations. The weight of a new point in the kth iteration was given by $(2 + k \cdot 10^{-15})^{-1}$.

Appendix B   Details of the GMM Estimation

Applying the GMM formula for the case of ANNs, we have:

$$\hat{\psi}_{MGM} = \arg\min_{\psi \in \Psi} \ U(\psi) \cdot C \cdot W \cdot C' \cdot U(\psi)'c(\psi) \tag{13}$$

where $C = [D(X; \{\boldsymbol{\omega}_m\}_{M \geq m \geq 1}), E]$.

For the initial step, we used the objective function of the Continuously Updated GMM. In the case, the weighting matrix W is the optimal one given the parameters and the sample. Inserting the ANNs moment conditions into equation (5) and taking the sample counterpart of the expected value, we have:

$$W_{CUGMM} = (N^{-1} \cdot [C' \cdot U(\psi)' \cdot U(\psi) \cdot C])^{-1}. \tag{14}$$

Given that $E[U(\psi) \cdot C] = \mathbf{0}$, it follows that:

$$W_{CUGMM} = [C' \cdot C]^{-1} \cdot N^{-1} \cdot [U(\psi)' \cdot U(\psi)]^{-1}. \tag{15}$$

Now, calling $N^{-1} \cdot [U(\psi)' \cdot U(\psi)] = \sigma^2$ and applying (15) to (14), we get to:

$$\hat{\psi}_{CUMGM} = \arg\min_{\psi \in \Psi} \ (\sigma^{-1} \cdot U(\psi)) \cdot C \cdot [C' \cdot C]^{-1} \cdot C' \cdot (\sigma^{-1} \cdot U(\psi))'c(\psi). \tag{16}$$

So, basically, we have the residuals normalized by their standard deviation projected in the orthogonal space generated by the columns of E.

For the local optimization step, we use a fixed weighting matrix. So, for a given $\{\boldsymbol{\omega}_m\}_{M \geq m \geq 1}$, the block $C \cdot W \cdot C'$ is also given and we will call it W^+. We have to find the vectors $\boldsymbol{\alpha}$ and $\boldsymbol{\lambda}$ that minimize:

$$U(\psi)) \cdot W^+ \cdot U(\psi)'. \tag{17}$$

In this case, we can apply WLS' closed formula given by:

$$\begin{bmatrix} \boldsymbol{\alpha} \\ \boldsymbol{\lambda} \end{bmatrix} = (Z' \cdot W^+ \cdot Z)^{-1} \cdot Z' \cdot W^+ \cdot Y, \tag{18}$$

where $Z = [X, l(X \cdot \boldsymbol{\omega}_1), \ ... \ , l(X \cdot \boldsymbol{\omega}_M)]$.

An Hybrid Ensemble Method Based on Data Clustering and Weak Learners Reliabilities Estimated Through Neural Networks

Marco Vannucci$^{(\boxtimes)}$, Valentina Colla, and Silvia Cateni

TeCIP Institute, Scuola Superiore Sant'Anna, Via G. Moruzzi,
1, 56124 Pisa, Italy
{mvannucci,colla,s.cateni}@sssup.it

Abstract. In this paper a novel hybrid ensemble method aiming at the improvement of models accuracy in regression tasks is presented. The proposed ensemble is composed by a *strong* learner trained exploiting data belonging to the whole training dataset and a set of specialised *weak* learners trained by using data coming from limited regions of the input space determined by means of a Self Organising Map based clustering. In the simulation phase, the strong and weak learners operate alternatively according to their punctual self-estimated reliabilities so as to handle each specific sample by means of the most promising learner. The method has been tested both on literature and real world datasets achieving satisfactory results.

1 Introduction

Ensemble Methods (EM) are a widely used approach in machine learning, based on the idea of handling a learning task by means of the combination of a set of different models whose outputs are merged through a suitable algorithm in order to compute the final *ensemble* output [1]. Normally, the learners forming the ensemble are characterized by simple structures and each of them is trained by exploiting a different - but possibly overlapping - set of training data in order to promote their differentiation. Due to their low complexity, single learners are commonly named *weak learners*.

In the last years, EM have been employed for solving numerous real world problems belonging to different research and industrial areas achieving impressive results and often outperforming approaches based on the use of single models. To mention some of these works, EM have been recently applied in finance and economics for the forecasting of the trend of currency exchange rates [2] and for the prediction of market change [3]; in the industrial field EM have been successfully adopted for the classification of operating conditions of electrical power apparatuses [4]; in the medical field for images segmentation [5], and, in the field of internet-related applications, for web pages ranking and content classification [6].

One interesting aspect of EM is that, under specific conditions related to the differentiation of weak learners outputs corresponding to the same input, it

© Springer International Publishing Switzerland 2015
I. Rojas et al. (Eds.): IWANN 2015, Part II, LNCS 9095, pp. 400–411, 2015.
DOI: 10.1007/978-3-319-19222-2_34

has been demonstrated that the accuracy of the ensemble can arbitrary grow by increasing the number of learners within the ensemble [7]. Unfortunately this latter condition is often hard to obtain as a very high number of learners would be needed, requiring an unaffordable computational time, not in line with real world problems.

Unfortunately there are some issues which limit the efficiency of EM when facing real world problems. EM are in facts known to suffer the presence of noise or outliers [8] which are quite common in real world datasets. Moreover EM accuracy can degrade when they have to model different input-output relations occurring within the same dataset such as, for instance, sensibly different machine operating conditions in industrial tasks. This aspect is related to the way the ensemble output is calculated from the single weak learners outputs: this latter operation is performed, in most cases, by averaging or weighting the single learners contributions, without taking into account their effective reliability. When coping with outliers, noise or particular input patterns, this operation can compromise the EM performance since each weak learner - whose robustness and generalization capabilities are far lower with respect to standard approaches - is heavily affected by the nature of the presented input pattern. In these cases an output aggregation algorithm not based on single learners effective reliability may sensibly decrease the predictive performance of the ensemble.

Despite the prominence of this issue, in literature only few works try to overcome this drawback by involving a measure of weak learners reliability in the weighting of learners output. In [5] the EM output is calculated by weighting the weak learners contributions according to their overall accuracy (the more accurate the classifier is, the higher is the weight associated). Nevertheless, considering the overall performance of each classifier instead of a more specific and *local* measure of reliability, which accounts the particular nature of the pattern to be classified, does not solve completely the previously mentioned problems. In [11], an EM for classification which is able to self-estimate the punctual reliability of each weak classifier when coping with an arbitrary new sample is proposed. The estimated reliabilities are then used in the aggregation so as to assign a higher weight to the learners whose estimated reliability is higher on handling the specific pattern. According to the reported results, this latter method seems to overcome, for the classification tasks, the main encountered by EM.

In this paper a new hybrid EM (HyEM) with the aim of improving the predictive performance of standard models and EMs is proposed. The basic idea of this approach is to place side by side an EM and a standard model (a neural network, in this case) and to employ them alternatively, based on their reliability, when coping with a new pattern so as to exploit the strong point and avoid the criticalities of both approaches.

The paper is organised as follows: in section 2 a short review on most common EM and on the concept of reliability estimation is presented in order to point out their main characteristics; in section 3 the HyEM is described in details. The proposed method has been tested on a set of real world dataset coming from the industrial field: the results of these tests are presented in section 4 where they are compared to those achieved by an approach based on the use of a single standard *strong* learner and of the most widely used EM approaches. Final conclusions and an overview of the future work related to this novel approach are depicted in section 5.

2   Ensemble Methods and Models Reliability Estimation

The basic idea of ensemble methods (EM) is the achievement of higher performance than a standard predictive model by suitably combining the output of a finite set of different models: the so-called *weak learners* (WL). EM are used for both classification and regression purpose and there is no constraint on the nature of weak learners which can be indifferently decision trees, neural networks or any other kind of supervised learner: an ensemble can even be composed by heterogeneous learners.

The most interesting property of EM was pointed out in [7] where it is shown that, in particular conditions (i.e. performance of the learners and disagreement between them), the EM performance can be arbitrary high provided the number of involved learners is high enough.

On the basis of the above mentioned result, several methods have been developed for efficient EM development aiming at producing a set of weak-learners whose errors are uncorrelated, responses are diverse and whose single performance is the best as possible (or at least better than random guessing). The developed techniques can be classified according on how they operate for creating the suitable set of classifiers as follows:

- **Selection of input features:** each classifier is trained by using a subset of the available input variables in order to obtain diverse and uncorrelated learners but, on the other hand, the loss of informative content, can degrade the performance.
- **Adding randomness**: this technique aims at differentiating the learners by operating on some initial random parameters of the classifiers. This method is widely used when neural networks are adopted: in this case initial weights and biases of the networks are randomly set in order to promote their dissimilarity.
- **Manipulating the training sets**: in this approach learners are trained by using different patterns coming from the available training dataset. Most used ensemble methods - **Bagging** and **Boosting** - belong too this group of ensemble techniques. The approach of bagging is simple yet powerful: each classifier is trained with a subset of the whole dataset where single samples are drawn randomly with repetition so that single samples can be used in the training of more than one classifier [9] in order to improve dissimilar behaviour among the learners and does not imply the loss of informative content. Boosting algorithms, among which the AdaBoost [10] is the most known, sequentially build the ensemble by adding a new learner at each step. The new learner is trained with those patterns whose processing is characterized by lowest performance within the previous step ensemble. Boosting

techniques are scarcely used for regression tasks, which is the main concern of this paper, nevertheless boosting applied to Artificial Neural Networks (ANN) ensembles are investigated in [15] and [16].

As introduced in section 1, the approach presented in this paper puts into practice an aggregation based on the weak learners reliability in the processing of specific input samples. The concept of reliability is not new in the EM context since it is, for instances, handled by AdaBoost: in these cases the output returned for a new pattern is calculated as a weighted average of individual learners output where involved weights are proportional to each learner's performance on the dataset it was trained on. The major limit of this approach lies in the *global* nature of this reliability measure.

In order to overcome this limit, the proposed approach will exploit the work developed in [12] where artificial neural networks are used to estimate the performance of predictive models: the main element of novelty in this approach is the ability of providing such measure sample-wise. This characteristic allows to use this measure within the ensemble aggregation algorithm in order to efficiently exploit the contribution of the weak learners.

3   An Hybrid Ensemble Method Based on Learners Reliability Estimated Through ANNs

The method proposed in this paper aims at the improvement of the accuracy of a generic model when facing regression tasks by creating an hybrid ensemble (HyEM) which couples the original model to a set of more specific learners. According to the ensemble nomenclature the original model will be called SL whilst the other learners will be referred as WL. In this work, both the SL and WLs are implemented as artificial neural networks in order to exploit their well-known capabilities of universal approximation and robustness, nevertheless, the presented method does not depend on the nature of the adopted learners.

The basic idea behind this approach is to create a SL which is trained by exploiting the whole training dataset whilst each WL exploits for its training only a subset of the training data and, more precisely, only the samples belonging to a specific region of the input space determined according to a Self-Organizing-Map (SOM) [13] based clustering. Moreover, an ANN for reliability estimation is associated to each learner (namely the SL and all WLs) and trained according to the procedure described in [12]. The trained reliability network associated to each learner is able to estimate, for a generic input sample, the confidence of the learner on the output provided for the processed input sample. Finally, in the simulation phase, the output of the SL and of the weak learner corresponding to the cluster the handled sample belongs to, will be collected and the HyEM output will be the one whose estimated reliability is higher.

The main advantage of this approach is the achievement of a SL which is able to reproduce the general relation between input and output variables throughout the whole input domain on one hand, while, on the other hand, the WLs should be able to reproduce the peculiarities of the input-output relation in different and specific zones of the domain in order to model the (eventual) region-wise differences of the handled relation. In addition, when a new input sample is fed to HyEM, the reliability estimators provide an evaluation of the confidence on the output returned by the SL and by the WL corresponding to the cluster the sample belongs to: this information is used in the HyEM simulation phase in order to properly combine the system output to take maximum advantage from the generalization capabilities of the strong learner and the specialization of the selected weak learner. The use of the reliability estimators aims at overcoming the criticalities encountered by standard ensemble aggregation approaches mentioned in section 1.

In the following sections 3.1 and 3.2, the HyEM training and simulation procedures are described in detail.

3.1   HyEM Training Procedure

The training of the HyEM is based on the exploitation of an arbitrary dataset D which includes a set of input variables $InData$ and the corresponding target variables $TarData$. Available data samples are divided into training (65% of the whole D), validation (15% of the whole D) and test (20% of the whole D) sets. The result of this division are the following datasets:

- $InData_{TR}$, $TarData_{TR}$ from training data
- $InData_{VD}$, $TarData_{VD}$ from validation data
- $InData_{TS}$, $TarData_{TS}$ from test data

Training and validation data are used for the tuning of HyEM models and parameters while test data are exploited for the assessment of the whole HyEM system once trained. The various components of the HyEM system are trained as follows:

- The **SL training** procedure is straightforward and generates the model $MainSL$: $MainSL$ is a two-layers feed-forward neural network (FFNN) trained via the classical back-propagation algorithm by exploiting $InData_{TR}$ as input, $TarData_{TR}$ as target.
- The **reliability estimator** Rel_{SL} **associated to SL** exploits SL performance on validation data for its tuning. Validation data have been used instead of training data for this tuning stage in order to better assess the performance of the main model on data not used for its tuning. The output vector of $MainSL$ with respect to $InData_{VD}$ is obtained as follows:

$$SLOut_{VD} = MainSL(InData_{VD}) \tag{1}$$

and the corresponding sample-wise reliability vector is calculated:

$$SLRel_{VD} = -|TarData_{VD} - SLOut_{VD}| \tag{2}$$

Reliability is, in practice, the opposite of the discrepancy between predicted and actual target output so that the higher (the closer to zero) the better it is. Rel_{SL} is a two-layers FFNN trained exploiting $InData_{VD}$ as input, $SLRel_{VD}$ as target.

– The **WLs training** is based on the result of input data clustering provided by a SOM. $InData_{TR}$ is used to train the SOM whose dimension, in terms of number of clusters C, is a parameter of the HyEM. The neurons of the trained SOM act as centroids which partition the input space defined by $InData_{TR}$. Each WL $MainWL_i$ is a two-layers FFNN, thus, according to this clusterization of the input data, each WL_i ($i \in [1, C]$) is trained via back-propagation exploiting $InData_{TRi}$ and $TarData_{TRi}$ which represent the training input and target variables respectively for the cluster i.

– As for the strong learner, the **reliability estimator Rel_{WLi} associated to each weak learner** i is tuned on the basis of its performance on the validation data belonging to the corresponding cluster. This procedure, together with the training of the corresponding weak learner, is graphically depicted in figure Fig. 1. The output vector of a generic weak learner

$$WL_iOut_{VD} = MainWL_i(InData_{VDi}) \tag{3}$$

and the corresponding sample-wise reliability vector are calculated

$$WL_iRel_{VD} = -|TarData_{VDi} - WL_iOut_{VD}| \tag{4}$$

The neural network Rel_{WLi} is trained by exploiting $InData_{VDi}$ as input and WL_iRel_{VD} as target.

The learners (both the SL and the WLs) and the reliability estimators are implemented as two-layers FFNN where the number h of neurons in the hidden layer is set by adopting the following empirical rule:

$$h = \frac{ANN_{Params}}{k \cdot D_{samples}} \tag{5}$$

where ANN_{Params} is the number of free parameters of the neural network, $D_{samples}$ the number of training samples used for the network training and k is a parameter.

A SOM based approach has been used for data clusterization and preferred to other methods (i.e . k-means) in order to exploit its well known capabilities of partitioning the data space preserving its original topology and distribution [13].

3.2   HyEM Simulation Procedure

The simulation procedure of the HyEM system exploits the structure generated during its training composed by: the strong learner, the batch of weak learner together with the ANNs devoted to the associated reliability estimation and the

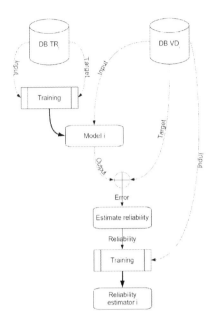

Fig. 1. Flow-chart depicting the data flow related to the training of reliability estimators for the learners involved in the ensemble

trained SOM which is used to determine the cluster associated to input patterns presented to the system. Simulation puts into practice the basic idea of selecting, for each specific input sample fed to the HyEM, the most promising predictor - according to the estimated reliability - between the SL and the weak learner WLi associated to the cluster the fed sample belongs.

Given an input sample P, the procedure, also depicted in figure 2, can be summarized as follows:

1. The output O_{SL} of the strong learner is calculated:

$$O_{SL} = MainSL(P) \tag{6}$$

2. The reliability of O_{SL} is estimated through Rel_{SL}:

$$R_{SL} = Rel_{SL}(P) \tag{7}$$

3. The trained SOM is used to determine the cluster i the pattern P belongs to. This operation is performed by identifying the neuron of the SOM closest to P with respect to euclidean distance.
4. The output of the weak learner WL$_i$ and the associated reliability are calculated:

$$O_{WLi} = MainWL_i(P) \tag{8}$$

$$R_{WLi} = Rel_{WLi}(P) \tag{9}$$

5. The output O_{HyEM} of the HyEM is calculated by aggregating the output returned by the strong learner and the selected weak learner on the basis of their reliabilities:

$$O_{HyEM} = \begin{cases} O_{SL}, & \text{if } R_{SL} \geq R_{WLi} \\ O_{WLi}, & \text{otherwise} \end{cases} \tag{10}$$

So doing, the output is selected which holds the highest estimated reliability.

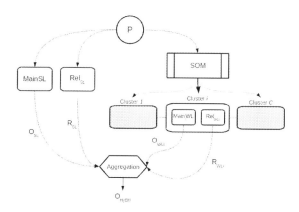

Fig. 2. Flow-chart depicting the data flow related to the simulation of the HyEM system when a new pattern P is provided

4   Test of the Proposed Approach

The proposed method has been tested on various datasets taken both from the UCI dataset repository [14] and from different industrial contexts. The features of the exploited datasets are summarized in table 1 in terms of number of samples and attributes. All these datasets concern regression tasks.

The UCI datasets included in these tests are:

- Combined Cycle Power Plant (CCPP)
- Physicochemical Properties of Protein Tertiary Structure (CASP)
- Concrete Compressive Strength (Concrete)
- Airfoil Self-Noise (Airfoil)
- Wine quality (Wine)

for which a complete description can be found on the UCI dataset repository website. On the other hand, the industrial datasets employed in this analysis can be shortly described as follows:

Obstruction (OBS) Includes data collected during the steel-making practice referring to the obstruction rate of nozzles from which the liquid steel flows during continuous casting. The target value to predict is the rate of nozzle obstructed area and varies between 0 and 5. Input variables include several process parameters and the processed steel chemical composition.

Mechanical properties gathers the results of laboratory tests performed on steel samples in order to determine the value of two different mechanical properties referred as A and B for confidentiality reason. A varies in the range [0;750], while B in the range [0; 620]. The input variables within the dataset to be used for A and B estimation include the steel chemical.

Plastic deformation (PD) Collects data regarding the measured pressure applied on steel products during their manufacturing. This pressure, which varies in the range [200;1000] and measured in MPa, is to be estimated by exploiting some process variables and product characteristics.

Train speed (TS) This dataset concerns the prediction of actual train speed based on the information provided by a set of sensors measuring train wheels angular speeds and frictions on the rail. The speed to be predicted varies between 0 and 250 km/h [17].

Table 1. Description of the datasets used for testing purpose within this work. The upper part of the table lists datasets taken from the UCI repository while lower part the industrial ones.

Dataset	Samples	Attributes
Concrete	1030	8
CCPP	9568	4
CASP	45730	9
Wine	4898	11
Airfoil	1503	5
PD	4935	7
Mech.Prop.	1995	12
OBS	7286	15
TS	94115	6

Within the test campaign, different configurations of HyEM have been assessed for each of the considered datasets by varying the number of weak learners included in the ensemble. As far ar the ANN composing the HyEM concerns, the number of neurons in the hidden layer has been calculated according to formula 5 where the parameter k was set to 5. The first layer activation function belongs to the *tansig* family whilst the output layer activation is linear. The performances of HyEM are compared, on one hand to those achieved by a ANN whose structure is the same of the strong learner embedded into the HyEM: this comparison puts into evidence the improvement due to the use of the WLs and the reliability estimators which support the SL within the HyEM architecture. On the other hand, HyEM results are compared to those achieved by the popular ensemble

Bagging method. In the tested bagging architectures, in order to pursue a fair comparison, the number of weak learners is set to be equal to the number of WLs forming the corresponding HyEM plus one, substituting the strong learner.

A representative selection of the results of the performed tests are reported in table 2, for all the datasets and all the tested methods, in terms of mean absolute prediction error with respect to the measure to be predicted through the different problems. In addition one column (#WL) specifies the number of employed weak learners (for bagging and HyEM) and the corresponding structure of the SOM used for the clustering. The last two columns are used to put into evidence the improvement achieved by the HyEM with respect to ANN and Bagging respectively: the value in these columns is the ratio (in percentage) between HyEM and other approaches prediction errors. The presented results have been obtained by using the 10-fold cross-validation technique.

Table 2. Performance, in terms of mean absolute error, of the tested approach on different datasets. Last two columns report the improvement achieved by the HyEM method with respect to ANN and Bagging respectively.

Dataset	#WL	ANN	Bagging	HyEM	Imp. ANN	Imp. Bag.
Airfoil	4 (2x2)	1.56	1.41	1.39	89.1%	98.6%
	9 (3x3)		1.52	1.52	97.4%	100%
CASP	9 (3x3)	3.60	3.33	3.33	92.5%	100%
	16 (4x4)		3.34	3.33	92.5%	99.7%
CCPP	9 (3x3)	3.08	3.05	2.92	94.8%	95.7%
	16 (4x4)		3.01	2.97	96.4%	98.7%
Concrete	4 (2x2)	4.31	4.30	4.0	92.8%	93.0%
	9 (3x3)		4.14	4.12	95.1%	100%
Wine	9 (3x3)	0.55	0.56	0.54	98.2%	96.4%
	16 (4x4)		0.57	0.55	100%	96.5%
OBS	9 (3x3)	0.12	0.12	0.09	75.0%	75.0%
	16 (4x4)		0.14	0.09	75.0%	64.3%
Mec. Prop. A	4 (2x2)	23.4	22.7	21.4	94.4%	96.5%
	9 (3x3)		22.9	22.1	91.5%	94.3%
Mec. Prop. B	4 (2x2)	20.1	19.4	18.8	93.5%	96.9%
	9 (3x3)		20.0	19.8	98.5%	99.0%
PD	4 (2x2)	18.2	16.9	15.8	86.8%	93.5%
	9 (3x3)		17.1	16.2	89.0%	94.7%
	16 (4x4)		17.8	16.0	87.9%	89.9%
TS	4 (2x2)	0.47	0.49	0.47	100%	95.9%
	9 (3x3)		0.51	0.43	91.5%	84.3%
	16 (4x4)		0.46	0.43	91.5%	93.5%

From the results shown in table 2 it stands out that the HyEM performs better than ANN and bagging: throughout all the tests the average improvement in terms of percent error reduction is 8% and 6.5% with respect to ANN and Bagging respectively. If for each problem the best performing HyEM architecture is considered, the improvement is 9.4% with respect to ANN and 8.3% with respect to Bagging. These figures point out the satisfactory performance of the proposed hybrid ensemble approach. Moreover, according to the performed tests, the performance of HyEM was never worse if compared to the other approaches involved in the test campaign (it does not happen for Bagging with respect to ANN, for instance).

These results show that the support of a set of specialized weak learners - of which only one is considered within the ensemble aggregation procedure on the basis of its location in the input space - combined to the use of a system for learners reliability estimation is effective.

The performance of different architectures of HyEM seems to be mainly affected by the dimension (in terms of data samples) of the dataset: in the case of small datasets, HyEMs with a low number of WLs perform better, probably due to the bigger number of samples available for the training of each WL; on the contrary, when datasets are formed by a high number of observations, the use of a higher number of WLs leads to better results, due to higher level of specialization reached by the use of more WLs trained with a sufficient number of data samples.

5   Conclusions and Future Work

This paper presents a novel hybrid ensemble method for regression tasks which couples the use of a strong learner and a set of weak learners specialized in handling patterns located in different regions of the input space. When a new pattern is fed to the HyEM, the SL and the associated WL operate alternatively on the basis of their estimated confidence in handling the specific input pattern. This approach is designed in order to exploit, for each input pattern, the best performing learner overcoming some of the criticalities encountered by standard EM.

The main elements of novelty of this approach are the use of locally specialized WLs, in the use of a SOM for the determination of the input space zones associated to each WL and the use of a reliability measure associated to the performance of the learner on each specific input pattern instead of a more global measure at WL level.

The HyEM has been tested on datasets with different characteristics coming both from the UCI repository and industrial contexts achieving satisfactory results by improving prediction accuracy with respect to the use of single ANN and a bagging ensemble.

In the future the basic concepts of the HyEM will be adopted, with opportune changes, for classification tasks and different aggregation strategies will be tested, for instance by combining the output of SL and WLs in different ways (but still related to estimated reliability) or including the contribution of more WLs.

References

1. Dietterich, T.G.: Ensemble methods in machine learning. In: Kittler, J., Roli, F. (eds.) MCS 2000. LNCS, vol. 1857, pp. 1–15. Springer, Heidelberg (2000)
2. Embrechts, M.J., Gatti, C.J., Linton, J., Gruber, T., Sick, B.: Forecasting exchange rates with ensemble neural networks and ensemble K-PLS: a case study for the US dollar per indian rupee. In: The 2012 International Joint Conference on Neural Networks (IJCNN), June 10–15, pp. 1–8 (2012)
3. Cheng, C., Xu, W., Wang, J.: A comparison of ensemble methods in financial market prediction. In: 2012 Fifth International Joint Conference on Computational Sciences and Optimization (CSO), June 23–26, pp. 755–759 (2012)
4. Hirose, H., Zaman, F.: More accurate diagnosis in electric power apparatus conditions using ensemble classification methods. IEEE Transactions on Dielectrics and Electrical Insulation **18**(5), 1584–1590 (2011)
5. Wei, W., Yaoyao, Z., Xiaolei, H., Lopresti, D., Zhiyun, X., Long, R., Antani, S., Thoma, G.: A classifier ensemble based on performance level estimation. In: IEEE International Symposium on Biomedical Imaging: From Nano to Macro, ISBI 2009, June 28-July 1, pp. 342–345 (2009)
6. Hashemi, H.B., Yazdani, N., Shakery, A., Naeini, M.P.: Application of ensemble models in web ranking. In: 2010 5th International Symposium on Telecommunications (IST), December 4–6, pp. 726–731 (2010)
7. Hansen, L.K., Salamon, P.: Neural Network Ensembles. IEEE Transactions on Pattern Analysis and Machine Intelligence **12**(10), 993–1001 (1990)
8. Opitz, D., Maclin, R.: Popular ensemble methods: an empirical study. Journal of Artificial Intelligence Research **11**, 169–198 (1999)
9. Breiman, L.: Bagging predictors. Machine Learning **24**(2), 123–140
10. Freund, Y., Schapire, R.: Experiments with a new boosting algorithm. In: Proc. of the 13th International Conference on Machine Learning, Bari, Italy, pp. 148–156 (1999)
11. Vannucci, M., Colla, V., Vannocci, M., Nastasi, G.: An ensemble classification method based on input clustering and classifiers expected reliability. In: Proc. of 6th European Modelling Symposium on Mathematical Modelling and Computer Simulation EMS2012, Malta, November 14–16 (2012)
12. Reyneri, L.M., Colla, V., Sgarbi, M., Vannucci, M.: Self-estimation of data and approximation reliability through neural networks. In: Cabestany, J., Sandoval, F., Prieto, A., Corchado, J.M. (eds.) IWANN 2009, Part I. LNCS, vol. 5517, pp. 89–96. Springer, Heidelberg (2009)
13. Haykin, S.: Neural networks - A comprehensive foundation - Chapter 9: Self-organizing maps. Prentice-Hall (1999). ISBN 0-13-908385-5
14. Bache, K., Lichman, M.: UCI Machine Learning Repository. School of Information and Computer Science. University of California, Irvine (2013). http://archive.ics.uci.edu/ml
15. Avnimelech, R., Intrator, N.: Boosting regression estimators. Neural Computation **11**, 499 (1999)
16. Karakoulas, G., Shawe Taylor, J.: Towards a strategy for boosting regressors. In: Smola, A., Brattlet, P., Scholkopf, B., Schuurmans, D. (eds.) Advances in Large Margin Classifiers, p. 247. MIT Press (2000)
17. Allotta, B., Colla, V., Malvezzi, M.: Train position and speed estimation using wheel velocity measurements. Proceedings of the Institution of Mechanical Engineers, Part F: Journal of Rail and Rapid Transit **216**(3), 207–225 (2002)

Conventional Prediction vs Beyond Data Range Prediction of Loss Coefficient for Quarter Circle Breakwater Using ANFIS

Arkal Vittal Hegde[(✉)] and Budime Raju

Department of Applied Mechanics and Hydraulics, National Institute of Technology,
Surathkal, Mangaluru, Karnataka, India
{arkalvittal,maniraju107}@gmail.com

Abstract. Protecting the lagoon area from the wave attack is one of the primary challenges in coastal engineering. Due to the scarcity of rubble and also to achieve economy, new types of breakwaters are being used in place of conventional rubble mound breakwaters. Emerged Perforated Quarter Circle Breakwaters (EPQCB) are artificial concrete breakwaters consisting of a curved perforated face fronting the waves with a vertical wall on rear side and a base slab resting on a low rubble mound base. The perforated curved front face has advantages like energy dissipation and good stability with less material as it is hollow inside. The estimation of hydrodynamic performance characteristics of EPQCB by physical model studies is complex, expensive and time consuming. Hence, computational intelligence (CI) methods are adopted for the evaluation of the performance characteristics like reflection, dissipation, transmission, run-up, rundown etc. A number of CI methods like Artificial Neural Network (ANN), Fuzzy logic, and hybrids such as ANFIS, ANN-PCO (particle swarm optimization), ANN-ACO etc., are available and are being used. The paper presents the work carried out to predict the dependent output variable of loss coefficient (K_l) beyond the range of values of one of the input variables i.e., wave period (T) adopted in present work, using the input data on variables of wave height (H), wave period (T), structure height (h_s), water depth (d), radius of the breakwater (R), spacing of perforations (S) and diameter of perforations (D) using ANFIS. For this purpose, both the conventional method of data segregation and also a new method called 'beyond data range' method are used for both training the ANFIS models and also to predict the dependent variable. Further, the input data was fed to the models in both dimensional and non-dimensional form in order to understand the effect of using non-dimensional data in place of dimensional parametric data. The performance of ANFIS models for all the four cases mentioned above was studied and it was found that prediction using conventional method with non-dimensional parameters performed better than other three methods. ANFIS models can be used to predict the performance characteristic K_l of EPQCB beyond the input data range of wave period T.

Keywords: ANN · ANFIS · Beyond data range prediction · Non-dimensional variable · Quarter circle breakwater

© Springer International Publishing Switzerland 2015
I. Rojas et al. (Eds.): IWANN 2015, Part II, LNCS 9095, pp. 412–421, 2015.
DOI: 10.1007/978-3-319-19222-2_35

1   Introduction

Coastal zone is a sensitive area which changes its profile dynamically. More than 40% of global population lives in the coastal area because of its socio-economic features. The protection of coastal area is a continuous challenge for the coastal engineers. Coastal protection works such as the seawalls, offshore breakwaters, groins and beach nourishment are being developed and installed to overcome the problem of erosion. As all of these structures are not successful in fulfilling their structural and functional requirements due to their improper design or location, the research is going on in the domain of coastal defense structures such as Berm Breakwater, floating pipe breakwater, perforated semi-circular and quarter-circle breakwaters and other artificial structures.

Quarter circle breakwater (QCB) is a pre-cast reinforced concrete caisson having perforations on its seaside circular face with a bottom slab resting on a low-mound rubble base. The concept of perforated seaside face was derived to absorb most of the wave energy and the vertical wall on rear side is to avoid the wave transmission. The paper discusses the loss coefficient of the breakwater which is a function of wave height (H_i), water depth (d), radius (R) and spacing-perforation ratio (S/D), height of structure (h_s) and wave period (T). Fig. 1 shows an emerged perforated quarter circular breakwater (EPQCB) with a free board, which means crest level of the structure is raised above the still water level and there is no overtopping of water over QCB crest.

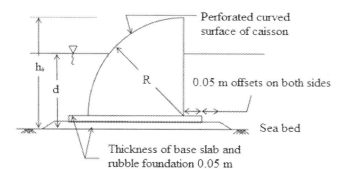

Fig. 1. Cross sectional view of EPQCB model used in present work

In the process of the evolution of these special type breakwaters, initially laborious time taking physical model tests are required in order to compute their hydrodynamic characteristics to decide their suitability to a particular location and design a stable breakwater section for that site. This process involves a lot many variables which affect the shape, strength, alignment, base stability and other aspects which are quite complex in nature. Hence, in view of saving time, effort and heavy cost of physical modeling, computational intelligence (CI) techniques are well suitable to predict the hydrodynamic performance of these structures for a different site, for different structural and wave conditions.

2   Methodology of Adaptive Neuro Fuzzy Interface System (ANFIS)

The application of computational intelligence in coastal engineering problems were initiated and successfully used for predicting the performance parameters for the past two decades. Although the invention of CI was initiated in late 1940s, due to the slow processing of these complex solution algorithms, research on CI got stagnated in 1980inspite of computer coding. However, Neuro fuzzy logics have advantages over programming languages that they can understand and modify the way data sets trains them. This unique feature brought CI into the necessary research domain with the fast processing computers of today. Computational Intelligence has became a superior choice for predicting and analysing complex issues which will save both time, money and routine laborious work. CI includes the methods like Artificial Neural Networks (ANN), Support Vector Machines (SVM), Adaptive Neuro Fuzzy Interface System (ANFIS) and their combination with the hybrid algorithms like Particle Swarm Optimization (PSO), Ant Colony Optimization (ACO), Genetic Algorithm (GA) etc.

The present work is carried out with one of hybrid CI model called the ANFIS (adoptive neuro fuzzy inference systems) model, based on a fusion of ideas from neural network and fuzzy controls, and possesses advantages of both the models. This model has high-level and computational power due to the merging of low-level learning neural networks with IF–THEN rule reasoning of fuzzy control systems. In brief, neural networks can improve their transparency, being closer to fuzzy control systems, while fuzzy control systems can self-adapt, being closer to neural networks. Generally, ANFIS is a graphical network representation of a Sugeno-type fuzzy system, endowed by neural learning capabilities. The network is a composition of nodes with specific functions, or duties, collected in layers with specific functions (Keskin et al., 2006).Neural fuzzy control system considered here is based on Tagaki-Sugeno-Kang (TSK) fuzzy rules. The TSK fuzzy rules are in the following forms:

$$R^j: \text{IF } x_1 \text{ is } A_1^{\ j} \text{ AND } x_2 \text{ is } A_2^{\ j} \text{ AND } \ldots \text{ AND } x_n \text{ is } A_n^{\ j}$$
$$\text{THEN } y = f_j = a_0^{\ j} + a_1^{\ j} x_1 + a_2^{\ j} x_2 + \ldots + a_n^{\ j} x_n \tag{1}$$

where x_i ($i = 1, 2, \ldots, n$) are input variables (five input parameters in this work), y is the output variable (loss coefficient K_l here), $A_i^{\ j}$ ($j=1, 2, \ldots, m$), are linguistic terms of the precondition part with membership functions $\mu_{A_i^{\ j}(X_i)}$, and $a_1^{\ j} \in R$ are coefficients of linear equations $f_i(x_1, x_2, \ldots, x_n)$. To simplify the discussion, it is necessary to focus on a specific neuro-fuzzy controller (NFC) referred to as an adaptive neural-based fuzzy inference system (ANFIS).Assume that the fuzzy control system under consideration has two inputs 'x_1' and 'x_2' and one output 'y' and then the rule base contains are given by two TSK fuzzy rules, as follows:

$$R^1: \text{IF } x_1 \text{ is } A_1^{\ 1} \text{ AND } x_2 \text{ is } A_2^{\ 1}, \text{THEN } y = y_1 = f_1 = a_0^{\ 1} + a_1^{\ 1} x_1 + a_2^{\ 1} x_2 \tag{2}$$

$$R^2: \text{IF } x_1 \text{ is } A_1^{\ 2} \text{ AND } x_2 \text{ is } A_2^{\ 2}, \text{THEN } y = y_2 = f_2 = a_0^{\ 2} + a_1^{\ 2} x_1 + a_2^{\ 2} x_2 \tag{3}$$

In fuzzy logic approaches, for given input values 'x_1' and 'x_2', the inferred output 'y*' is calculated by:

$$y^* = (\mu_1 f_1 + \mu_2 f_2) / (\mu_1 + \mu_2) \tag{4}$$

where, μ_j are firing strengths of R^j ($j = 1, 2$), and are given by:

$$\mu_j = \mu_{A1}{}^j (x_1) \times \mu_{A2}{}^j (x_2) \qquad j = 1, 2 \tag{5}$$

If product inference is used, the corresponding ANFIS architecture is shown in Fig. 2. ANFIS structure used in the present work has five inputs and one output as architecture, where following meanings can be attached to each layer:

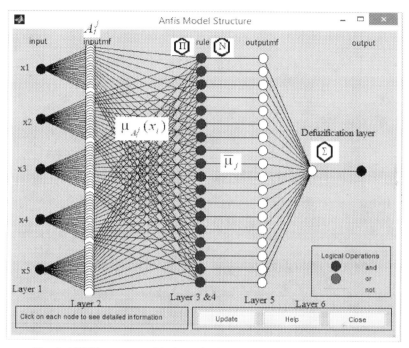

Fig. 2. ANFIS structure with theoretical explanation of working principle

Layer 1: Every node in this layer implies an input and it just passes external signals to the next layer.

Layer 2: Every node in this layer acts as a membership function $\mu_{Ai}{}^j (x_i)$, and its output specifies the degree to which the given x_i satisfies the quantifier $A_i{}^j$. Generally, $\mu_{Ai}{}^j (x_i)$ is selected as bell-shaped with a maximum equal to 1 and minimum equal to 0, such as:

$$y^* = \frac{\mu 1\, y1 + \mu 2\, y2}{\mu 1 + \mu 2} = \frac{\mu 1\, f1 + \mu 2\, f2}{\mu 1 + \mu 2} = \overline{\mu 1}\, f1 + \overline{\mu 2}\, f2 \tag{6}$$

and

$$\mu_{Ai}{}^j (x_i) = \exp\{-[((x_i - m_i{}^j)/\sigma_i{}^j)^2]^{bij}\} \tag{7}$$

where $\{m_i{}^j, \sigma_i{}^j, b_i{}^j\}$ is the parameter set to be tuned. In fact, continuous and piece wise differentiable functions, such as commonly used trapezoidal or triangular membership functions are also qualified candidates for node functions in this layer. Parameters in this layer are also called as precondition parameters.

Layer 3: Every node in this layer is labelled by Π and multiplies the incoming signals

$$\mu_j(x_1) = \mu_{A1}{}^j(x_1) * \mu_{A2}{}^j(x_2) \tag{8}$$

and sends the product out. The output of each node represents the firing strength of a rule.

Layer 4: Every node in this layer is labelled by N and calculates the normalized firing strength of a rule. That is the jth node calculates the ratio of the firing strength of the jth rule to that of all the rules as:

$$\overline{\mu j} = \mu j / \Sigma \mu i \tag{9}$$

Layer 5: Every node j in this layer calculates the weighted consequent value as

$$\overline{\mu j}(a_0{}^j + a_1{}^j x_1 + a_2{}^j x_2) \tag{10}$$

Where $\overline{\mu j}$ is the output of Layer 4 and $\{a_0{}^j, a_1{}^j, a_2{}^j\}$ is the set to be tuned. Parameters in this layer are referred to as consequent parameters.

Layer 6: This is the final layer containing only one node labelled as Σ, and it sums all incoming signals to obtain the final inferred result for the whole system (Keskin et al., 2006).

3   Data Collection and Usage

For the present work, experimental data of EPQCB comprising of 300 data points was collected from Balakrishna (2014). The physical model experiments on EPQCB were carried out by Balakrishna (2014) in the monochromatic wave flume of the Marine Structures Laboratory in the Department of Applied Mechanics and Hydraulics, NITK Surathkal, Mangaluru, India. Fig. 3 shows the experimental setup of EPQCB used by Balakrishna (2014). For the hydraulic model investigations, wave and site conditions of Mangaluru coast were considered.

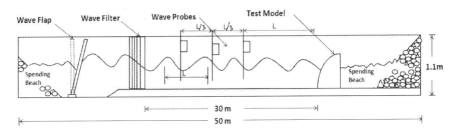

Fig. 3. Longitudinal section of wave flume used with experimental setup

In the present work, in addition to the conventional method of prediction (using the random data segregation of 73% and 27% for training and testing respectively),other objective was also to investigate the prediction capability of ANFIS beyond the data range of input variables used for training, referred hereby as 'beyond data range' prediction. This is essential because many a times, it is not possible to conduct wave flume experiments for larger values of input variables, which may demand a larger flume size, for example, a larger value of wave height, wave period, water depth etc.

Hence, it was decided in the present paper, to use ANFIS for predicting K_l for higher and lower wave period (T) values, than that used for training of the ANFIS model. Data points with wave periods T of 1.4 s (lowest of T values in the data set) and also 2.5 s (highest of T values in the data set) amounting to 27% (81 data points) of the total ensemble of 300 data points, was segregated as testing data. The remaining data set of 73% (i.e. 219 data points) for T=1.6 s, 1.8 s, 2.0 s and 2.2 s, was used for training. Further, in order to understand the effect of non-dimensional data, dimensional analysis was carried out using Buckingham's 'π' Theorem, and following non-dimensional parametric groups were obtained:-H_i/gT^2, d/gT^2, S/D, R/H_i & h_s/d and above mentioned procedure of data segregation, i.e. the conventional and the beyond data range methods were applied in this case too, for prediction of K_l.

The ANFIS algorithm was developed by using 'genfis2', which is a function that builds upon 'subclust' function to provide a quick, one-pass method that is used to take input-output training data and it generates a Sugeno-type fuzzy inference system that models the data behaviour on its own. Since no guideline is available in the literature regarding the number of clusters required for the given data set, subtractive clustering algorithm was employed for estimating the number of clusters and the cluster centres in the data set. Fundamentally, 'genfis1' with grid pattern and input membership functions of any kind of distribution and output parameter of linear or constant membership functions are a default in ANFIS. However, after running the model with 'genfis1',it was found to be unsuitable, owing to the fact that output parameter K_l is a highly complex and non-linear output variable. Hence, for all of the ANFIS models 'genfis2' with cluster radius of 0.1 was used, except for one case, where 'genfis2' with cluster radius of 0.58 was used (i.e. for the case of conventional method of prediction with non-dimensional parameters).

4   Results and Discussion

The input and target data was loaded into the MATLAB (R2013a) and 'genfis2' (sub-clustering method), 20 epochs, step size of 0.1 and usual default values were adopted for other variables. After running the code the network of ANFIS models were generated and the average Root Mean Square Error (RMSE) and Coefficient of Correlation (CC) values for training and testing data were obtained, for both the conventional and beyond data range prediction methods, as shown in Table 1. MATLAB output graphs are also presented in Figs. 4, 5, 6, 7 and 8 for comparing the conventional prediction and beyond data range prediction for both the testing and training data sets. Fig. 4 shows the agreement between predicted and actual (or experimental) K_l values for training data for the conventional non-dimensional method with high CC and low RMSE. Similarly, Fig. 5 depicts the similar agreement for testing data for same case. Hence, it is clear that the ANFIS models with conventional non-dimensional method predict the phenomenon of K_l very well. Now, looking at beyond data range prediction for non-dimensional method, Fig. 6 shows same for training performance with high CC and low RMSE. For testing performance, the beyond data range prediction has slightly lesser correlation coefficient compared to the conventional prediction, due to the over-estimation of K_l for a few data points as shown in Fig. 7 (some values are even beyond the physical limiting value of 1, i.e. $K_l > 1$). This problem however, can be solved by applying some filters/hybrid algorithms like: Early stopping and Regularization methods of ANN and will be solved in the near future.

Table 1. ANFIS training and testing results for prediction of K_l

Method of prediction	Dimensional analysis	RMSE of training	RMSE of testing	CC of training	CC of testing
Beyond data range prediction	Dimensional parameters	7.72E-7	0.0311	1.00	0.7230
	Non-dimensional parameters	3.6E-7	0.0897	1.00	0.7088
Convention method of prediction	Dimensional parameters	8.39E-7	0.0190	1.00	0.8204
	Non-dimensional parameters	1.69E-2	0.0174	0.889	0.8533

The difference between training and test RMSE and CC is large due to the manual segregation of the wave periods (T=1.4 sec and 2.5 sec) for the purpose of beyond the data range prediction. Hence the test data is highly differential in numbers compare to training data. Fig. 8 depicts the comparison of prediction of K_l for testing data in terms of CC for both the conventional non-dimensional method and beyond data range, non-dimensional case. The CC values are found to be above 70% in both the cases, and hence it clearly indicates the suitability of ANFIS models developed for both the conventional and beyond data range predictions. It is proposed to compare the results obtained using ANFIS in the present work, with results obtained by other soft computing techniques such as ANN and ANN-ACO etc..

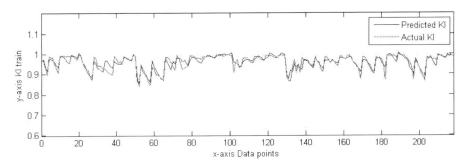

Fig. 4. Training performance of ANFIS model for conventional, non-dimensional method

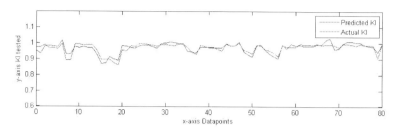

Fig. 5. Testing performance of ANFIS model for conventional, non-dimensional method

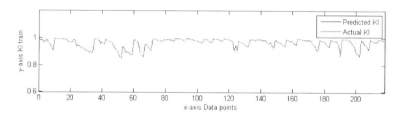

Fig. 6. Training performance of ANFIS model for beyond data range, non-dimensional method

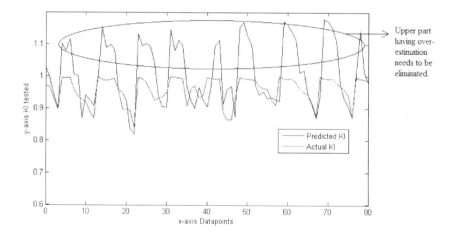

Fig. 7. Testing performance of ANFIS model for beyond data range, non-dimensional method

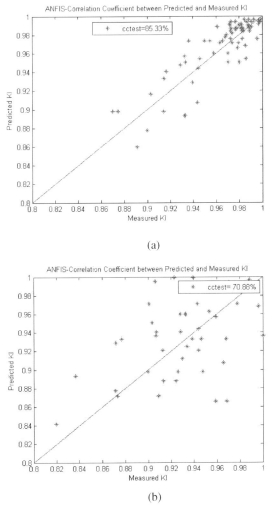

Fig. 8. Comparison of prediction of testing in terms of CC, a) conventional method non-dimensional, b) beyond data range, non-dimensional method

5   Conclusions

Based on the results obtained and discussion carried out, following conclusions have been drawn:

- Prediction using conventional method with non-dimensional parameters has better accuracy than other methods.
- ANFIS models can be used to predict the performance characteristic K_t of EPQCB beyond the input data range.
- For the case of dimensional parameters, conventional method of prediction gave better results compared to beyond the data range prediction.

References

1. Areerachakul, S.: Comparison of ANFIS and ANN for estimation of biochemical oxygen demand parameter in surface water. World Academy of Science, Engineering and Technology, vol. 6, April 26, 2012
2. Balakrishna, K.: Hydrodynamic performance characteristics of emerged perforated quarter circle breakwater. M. Tech Thesis, NITK, Surathkal, India (2014)
3. Dhinakaran, G., Sundar, V., Graw, K.: Effect of perforations and rubble mound height on wave transformation characteristics of surface piercing semicircular breakwaters. J. Ocean Eng. 36(15), 1182–1198 (2009)
4. ErolKeskin, M., Taylan, D., Terzi, O.: Adaptive neural-based fuzzy inference system (ANFIS) approach for modeling hydrological time series. Hydrological Sciences Journal 51(4), 588–598 (2006)
5. Londhe, S.N.: Soft computing approach for real-time estimation of missing wave heights. Ocean Engineering 35, 1080–1089 (2008)
6. Patil, S.G.: Computational Intelligence in prediction of wave transmission for horizontally interlaced multilayer moored floating pipe breakwater. Ph.D thesis, NITK, Surathkal, India (2012)
7. Qie, L., Zhang, X., Jiang, X., Qin, Y.: Research on partial coefficients for design of quarter circle caisson breakwater. J. Marine Science Applications 12, 65–71 (2013)
8. Rushil, G., Kiriti, S., Hegde, A.V.: Quarter circular breakwater: prediction of transmission using multiple regression and artificial neural network. Marine Technology Society Journal 48, 92–98 (2014)

Performance Evaluation of Least Squares SVR in Robust Dynamical System Identification

José Daniel A. Santos[1][(✉)], César Lincoln C. Mattos[2],
and Guilherme A. Barreto[2]

[1] Department of Industry, Federal Institute of Education,
Science and Technology of Ceará, Maracanaú, Ceará, Brazil
jdalencars@gmail.com
[2] Department of Teleinformatics Engineering, Center of Technology,
Federal University of Ceará, Campus of Pici, Fortaleza, Ceará, Brazil
cesarlincoln@terra.com.br, gbarreto@ufc.br

Abstract. Least Squares Support Vector Regression (LS-SVR) is a powerful kernel-based learning tool for regression problems. Nonlinear system identification is one of such problems where we aim at capturing the behavior in time of a dynamical system by building a black-box model from the measured input-output time series. Besides the difficulties involved in the specification a suitable model itself, most real-world systems are subject to the presence of outliers in the observations. Hence, robust methods that can handle outliers suitably are desirable. In this regard, despite the existence of a few previous works on robustifying the LS-SVR for regression applications with outliers, its use for dynamical system identification has not been fully evaluated yet. Bearing this in mind, in this paper we assess the performances of two existing robust LS-SVR variants, namely WLS-SVR and RLS-SVR, in nonlinear system identification tasks containing outliers. These robust approaches are compared with standard LS-SVR in experiments with three artificial datasets, whose outputs are contaminated with different amounts of outliers, and a real-world benchmarking dataset. The obtained results for infinite step ahead prediction confirm that the robust LS-SVR variants consistently outperforms the standard LS-SVR algorithm.

Keywords: Least Squares Support Vector Regression · Nonlinear dynamical system identification · NARX model · Outliers

1 Introduction

Least Squares Support Vector Machine (LS-SVM) is a widely used tool for classification [16] and regression [14,15] in the fields of pattern recognition and machine learning. Particularly in regression, when the model is called Least Squares Support Vector Regression (LS-SVR), it has found broad applicability in areas such as time series forecasting [18], control [6] and system identification [1,3].

Motivated by the theory of Support Vector Machines (SVMs) [19], LS-SVR uses a Sum-of-Squared-Error (SSE) cost function and equality constraints to

© Springer International Publishing Switzerland 2015
I. Rojas et al. (Eds.): IWANN 2015, Part II, LNCS 9095, pp. 422–435, 2015.
DOI: 10.1007/978-3-319-19222-2_36

replace the original convex Quadratic Programming (QP) optimization problem in SVM. Consequently, the global optimum is simpler to obtain by solving a set of linear equations. However, despite this computationally attractive feature, the use of a SSE-based cost function can lead to estimates which are too sensitive to the presence of outliers in the data or when the underlying assumption of Gaussian distribution for the error variables is not realistic. In this scenario, LS-SVR may present poor prediction performance on new data.

Despite the importance of robust methods for real-world applications, very few authors have developed learning strategies for LS-SVR models to handle outliers suitably during the training process. For example, Suykens et al. [17] were probably the first to propose a robust variant for the LS-SVR by introducing a weighted version of it based on M-Estimators [5]. Another approach worth mentioning is developed by Chuang et al. [20], who introduced an iterative method based on truncated least squares loss function, Concave-Convex Procedure (CCCP) and Newton algorithm.

It is worth noting that while the aforementioned robust LS-SVR variants were developed for standard regression problems, their applications to regression with time series data, especially to nonlinear system identification, is still an open issue. Nonlinear dynamical system identification is a complex problem which can be roughly understood as a set of well-defined nonlinear regression tools applied to modeling systems with memory (i.e. dynamics), aiming at describing the behavior in time of such systems, for control applications and for simulation purposes only.

Only recently, few authors has started to investigate LS-SVR robust versions for dynamical nonlinear black-box regression tasks [4,8,9]. However, these works evaluated their proposed models using results from one-step-ahead prediction tasks. For a more complete validation of a model in a dynamical system identification task, infinite steps ahead predictions (a.k.a free simulation) are strongly recommended, since one can judge correctly if the model has indeed captured in the long term the relevant dynamics of the system under study. As a consequence, it turns out that infinite steps ahead prediction is a problem far less trivial than one step ahead prediction.

From the exposed, the scope of this paper encompasses a comprehensive evaluation of the behavior of two robust LS-SVR variants, namely, Weighted Least Squares Support Vector Regression (WLS-SVR) [17] and Robust Least Squares Support Vector Regression (RLS-SVR) [20], in nonlinear system identification tasks. For this purpose, we use three synthetic datasets, whose outputs are deliberately contaminated with different amounts of outliers, and a real-world benchmarking dataset. The validation scenarios of the performance comparison to be carried out correspond to infinite steps ahead prediction tasks.

The remainder of this paper is organized as follows. In Section 2 we briefly discuss the robust nonlinear system identification with NARX models. In Section 3, LS-SVR, WLS-SVR and RLS-SVR models are described. In Section 4 the results of a comprehensive set of computer experiments are presented with the paper being concluded in Section 5.

2   Robust Nonlinear Dynamical System Identification

Given a dynamical system that could be explained by a nonlinear autoregressive with exogenous inputs (NARX) model, its i-th input vector $\boldsymbol{x}_i \in \mathbb{R}^P$ is obtained from L_y past observed outputs $y_i \in \mathbb{R}$ and L_u past control inputs $u_i \in \mathbb{R}$ [10]

$$y_i = m_i + \epsilon_i, \qquad m_i = g(\boldsymbol{x}_i), \qquad \epsilon_i \sim \mathcal{N}(\epsilon_i | 0, \sigma_n^2), \tag{1}$$

$$\boldsymbol{x}_i = [y_{i-1}, y_{i-2}, \cdots, y_{i-L_y}, u_{i-1}, u_{i-2}, \cdots, u_{i-L_u}]^T, \tag{2}$$

where i is the instant of observation, $m_i \in \mathbb{R}$ is the true (noiseless) output of the system, $g(\cdot)$ is an unknown nonlinear function and ϵ_i is a Gaussian distributed observation noise. After N instants, we have the dataset

$$\mathcal{D} = (\boldsymbol{x}_i, y_i)|_{i=1}^N = (\boldsymbol{X}, \boldsymbol{y}), \tag{3}$$

where $\boldsymbol{X} \in \mathbb{R}^{N \times P}$ is called *regressor matrix* and $\boldsymbol{y} \in \mathbb{R}^N$ is the vector of measured outputs. From the set $\mathcal{D}$, henceforth called the *estimation data*, we aim at building a model that explains well enough the dynamical behavior of the system under study.

After the estimation of a suitable model, it may be used to simulate the dynamical output of the identified system through iterative predictions. Given a new instant j, the prediction for *test data* follows

$$\hat{y}_j = f(\boldsymbol{x}_j) + \epsilon_j, \tag{4}$$

$$\boldsymbol{x}_j = [\hat{y}_{j-1}, \hat{y}_{j-2}, \cdots, \hat{y}_{j-L_y}, u_{j-1}, u_{j-2}, \cdots, u_{j-L_u}]^T, \tag{5}$$

where $\hat{y}_j$ is the j-th estimated noisy output. This procedure, in which past estimated outputs are used as regressors, is usually called *free simulation* or *infinite step ahead* prediction and will be adopted in this paper.

When the observed noise cannot be considered Gaussian, as in the presence of outliers, models obtained through Eq. (1) are not appropriated. In fact, the light tails of the Gaussian distribution are not able to justify the error deviations caused by the outliers.

In this paper we are interested in evaluating the performance of outlier-robust LS-SVR models in the identification of nonlinear dynamical system. We present these models in the next section.

3   Evaluated Models

Initially, let us consider the estimation dataset $\{(\boldsymbol{x}_1, y_1), \ldots, (\boldsymbol{x}_N, y_N)\}$, with the inputs $\boldsymbol{x}_i \in \mathbb{R}^p$ and correspondent outputs $y_i \in \mathbb{R}$. In a regression problem, the goal is to search for a function $f(\cdot)$ that approximates, with acceptable accuracy, the outputs y_i for all instances of the available data. For nonlinear case, f usually takes the form

$$f(\boldsymbol{x}) = \langle \boldsymbol{w}, \varphi(\boldsymbol{x}) \rangle + b, \quad \text{with} \quad \boldsymbol{w} \in \mathbb{R}^P, b \in \mathbb{R}, \tag{6}$$

where $\langle \cdot, \cdot \rangle$ denotes the dot-product in space of the input patterns, $\boldsymbol{w}$ is a vector of weights, b is a bias and $\varphi(\cdot)$ is a nonlinear map into some dot-product space $\mathcal{H}$, usually called feature space.

The formulation of the parameter estimation problem in LS-SVR leads to the minimization of the following functional [14,15]

$$J(\boldsymbol{w}, e) = \frac{1}{2}\|\boldsymbol{w}\|_2^2 + C\frac{1}{2}\sum_{i=1}^{N} e_i^2, \qquad (7)$$

subject to

$$y_i = \langle \boldsymbol{w}, \varphi(\boldsymbol{x}_i) \rangle + b + e_i, \quad i = 1, 2, \ldots, N \qquad (8)$$

where $e_i = y_i - f(\boldsymbol{x}_i)$ is the error due to the i-th input pattern and $C > 0$ is a regularization parameter.

The Lagrangian function of the optimization problem in Eqs. (7) and (8) is

$$L(\boldsymbol{w}, b, \boldsymbol{e}, \boldsymbol{\alpha}) = \frac{1}{2}\|\boldsymbol{w}\|_2^2 + C\frac{1}{2}\sum_{i=1}^{N} e_i^2 - \sum_{i=1}^{N} \alpha_i[\langle \boldsymbol{w}, \varphi(\boldsymbol{x}_i) \rangle + b + e_i - y_i], \qquad (9)$$

where α_i's are the Lagrange multipliers. The conditions for optimality are given by

$$\begin{cases} \frac{\partial L}{\partial \boldsymbol{w}} = 0 \implies \boldsymbol{w} = \sum_{i=1}^{N} \alpha_i \varphi(\boldsymbol{x}_i), \\ \frac{\partial L}{\partial e_i} = 0 \implies \sum_{i=1}^{N} \alpha_i = 0, \\ \frac{\partial L}{\partial b} = 0 \implies \alpha_i = Ce_i, \\ \frac{\partial L}{\partial \alpha_i} = 0 \implies \langle \boldsymbol{w}, \varphi(\boldsymbol{x}_i) \rangle + b + e_i - y_i = 0, \end{cases} \qquad (10)$$

for $i = 1, 2, \cdots, N$. After elimination of the variables e_i and $\boldsymbol{w}$, the optimal dual variables correspond to the solution of the following system of linear equations

$$\begin{bmatrix} 0 & \mathbf{1}^T \\ \mathbf{1} & \boldsymbol{\Omega} + C^{-1}I \end{bmatrix} \begin{bmatrix} b \\ \boldsymbol{\alpha} \end{bmatrix} = \begin{bmatrix} 0 \\ \boldsymbol{y} \end{bmatrix}, \qquad (11)$$

where $\boldsymbol{y} = [y_1, y_2, \ldots, y_n]^T$, $\mathbf{1} = [1, 1, \ldots, 1]^T$, $\boldsymbol{\alpha} = [\alpha_1, \alpha_2, \ldots, \alpha_n]^T$, $\boldsymbol{\Omega} \in \mathbb{R}^{N \times N}$ is the kernel matrix whose entries are $\Omega_{i,j} = k(\boldsymbol{x}_i, \boldsymbol{x}_j) = \langle \varphi(\boldsymbol{x}_i), \varphi(\boldsymbol{x}_j) \rangle$ where $k(\cdot, \cdot)$ is the chosen kernel function.

The resulting LS-SVR model for nonlinear regression is given by

$$f(\boldsymbol{x}) = \sum_{i=1}^{N} \alpha_i k(\boldsymbol{x}, \boldsymbol{x}_i) + b, \qquad (12)$$

where $\boldsymbol{\alpha}$ and b are the solution of the linear system in Eq. (11). The Gaussian kernel function $k(\boldsymbol{x}, \boldsymbol{x}_i) = \exp\left\{\frac{\|\boldsymbol{x} - \boldsymbol{x}_i\|_2^2}{2\sigma^2}\right\}$ was adopted in all the experiments in this paper.

3.1   The WLS-SVR Model

The Weighted Least Squares Support Vector Regression (WLS-SVR), developed by Suykens et. al. [17], is described by the minimization of the functional

$$J(\boldsymbol{w}, e) = \frac{1}{2}\|\boldsymbol{w}\|_2^2 + C\frac{1}{2}\sum_{i=1}^{N} v_i e_i^2, \tag{13}$$

subject to Eq. (8). $\boldsymbol{v} = [v_1, \ldots, v_N]^T$ is a vector of weights associated with the estimation data. If $v_k = 0$, one can delete the corresponding data sample from the model. In the same way of LS-SVR, the optimal dual variables are given by the solution of the following system of linear equations

$$\begin{bmatrix} 0 & \mathbf{1}^T \\ \mathbf{1} & \boldsymbol{\Omega} + C^{-1}\mathbf{V} \end{bmatrix} \begin{bmatrix} b \\ \boldsymbol{\alpha} \end{bmatrix} = \begin{bmatrix} 0 \\ \boldsymbol{y} \end{bmatrix}, \tag{14}$$

where the diagonal matrix $\mathbf{V} \in \mathbb{R}^{N \times N}$ is given by

$$\mathbf{V} = \mathrm{diag}\left\{ \frac{1}{v_1}, \ldots, \frac{1}{v_N} \right\}. \tag{15}$$

The weights v_i are determined based on the error variables $e_i = \alpha_i/C$ from the original LS-SVR approach in Eq. (11). In this paper, the robust estimates are obtained from *Hampel* weight function [13,17] as follows

$$v_i = \begin{cases} 1 & \text{if } |e_i/\hat{s}| \leq c_1, \\ \frac{c_2 - |e_i/\hat{s}|}{c_2 - c_1} & \text{if } c_1 < |e_i/\hat{s}| \leq c_2, \\ 10^{-4} & \text{otherwise}, \end{cases} \tag{16}$$

where $\hat{s} = \mathrm{IQR}/1.349$ is a robust estimate of the standard deviation of the LS-SVR error variables e_i. IQR stands for *Interquantile range*, which is the difference between the 75th percentile and 25th percentile and the constants c_1, c_2 are typically chosen as $c_1 = 2.5$ and $c_2 = 3.0$ [13].

3.2   The RLS-SVR Model

Let us consider again the WLS-SVR optimization problem in Eqs. (13) and (8) that is equivalent to the following unconstrained functional

$$J(\boldsymbol{w}, b) = \frac{1}{2}\|\boldsymbol{w}\|_2^2 + C\frac{1}{2}\sum_{i=1}^{N} v_i(y_i - (\langle \boldsymbol{w}, \varphi(\boldsymbol{x}_i) \rangle + b))^2. \tag{17}$$

In order to avoid setting the weights of the estimation samples, Yang et al. [20] developed a robust approach, called Robust Least Squares Support Vector Regression (RLS-SVR), through the minimization of another unconstrained functional given by

$$J(\boldsymbol{w}, b) = \frac{1}{2}\|\boldsymbol{w}\|_2^2 + C\frac{1}{2}\sum_{i=1}^{N} \text{robust}_2(\boldsymbol{w}, b, \boldsymbol{x}_i, y_i), \tag{18}$$

where $\text{robust}_2(\cdot)$ is a truncated least squares loss function

$$\text{robust}_2(\boldsymbol{w}, b, \boldsymbol{x}, y) = \min\{p, (y_i - (\langle \boldsymbol{w}, \varphi(\boldsymbol{x}_i)\rangle + b))^2\}, \tag{19}$$

where $p \geq 0$ is the truncation parameter, which controls the errors reducing the effects of the outliers. Let the error of the estimation sample be $e_i = y_i - (\langle \boldsymbol{w}, \varphi(\boldsymbol{x}_i)\rangle + b)$, then the loss function of the optimization problem in Eq. (18) can be rewritten by

$$\text{robust}_2(p, e_i) = \min\{p, e_i^2\} = \begin{cases} e_i^2, & \text{if } |e_i| \leq \sqrt{p}, \\ p, & \text{if } |e_i| > \sqrt{p}. \end{cases} \tag{20}$$

It may be seen easily in Eq. (20) that when p is large enough , the solution of RLS-SVR is the same as LS-SVR. In this paper, we set $0 \leq p \leq 1$ in all the experiments.

The function $\text{robust}_2(\cdot)$ is neither differentiable nor convex. In order to overcome that difficulty, the RLS-SVR approach firstly performs a smoothing procedure to make the loss function $\text{robust}_2(\cdot)$ [2]. Then, considering $z_i = \langle \boldsymbol{w}, \varphi(\boldsymbol{x}_i)\rangle + b$, we can write

$$\text{robust}_2(\boldsymbol{w}, b, \boldsymbol{x}_i, y_i) = \min\{p, (y_i - z_i)^2\} = (y_i - z_i)^2 + h(z_i), \tag{21}$$

where

$$h(z_i) = \begin{cases} 0, & y_i - \sqrt{p} \leq z_i \leq y_i + \sqrt{p}, \\ p - (y_i - z_i)^2, & \text{otherwise.} \end{cases} \tag{22}$$

It is possible to note that $h(\cdot)$ is a non-smooth function. In order to solve that problem, $h(\cdot)$ is replaced with another smoothing function $h^*(\cdot)$ given by

$$h^*(z_i) = \begin{cases} p - (y_i - z_i)^2, & z_i < y_i - \sqrt{p} - h \ \text{ or } \ z_i > y_i + \sqrt{p} + h \\ -\frac{(h+2\sqrt{p})(y_i+h-\sqrt{p}-z_i)^2}{4h}, & |z_i - y_i + \sqrt{p}| \leq h \\ 0, & y_i + h - \sqrt{p} < z_i < y_i - h + \sqrt{p} \\ \frac{-(h+2\sqrt{p})(y_i-h+\sqrt{p}-z_i)^2}{4h}, & |z_i - y_i - \sqrt{p}| \leq h, \end{cases} \tag{23}$$

where h is the smoothing parameter, typically taking its values between 0.001 and 0.5. Now the function $h^*(\cdot)$ is continuous and twice-differentiable. Then, the functional in Eq. (18) can rewritten as

$$J_{rob}(\boldsymbol{w}, b) = \frac{1}{2}\|\boldsymbol{w}\|_2^2 + C\frac{1}{2}\sum_{i=1}^{N}(y_i - z_i)^2 + C\frac{1}{2}\sum_{i=1}^{N} h^*(z_i), \tag{24}$$

where

$$J_{vex}(\boldsymbol{w}, b) = \frac{1}{2}\|\boldsymbol{w}\|_2^2 + C\frac{1}{2}\sum_{i=1}^{N}(y_i - z_i)^2, \tag{25}$$

$$J_{cav}(\boldsymbol{w}, b) = C \frac{1}{2} \sum_{i=1}^{N} h^*(z_i). \tag{26}$$

Since J_{cav} is non-convex, it is difficult to minimize the functional J_{rob} by classical convex optimization algorithms. The next step is using concave-convex procedure (CCCP) [21] to transform a concave-convex optimization problem into a iteratively series of convex optimization problems. Finally, it is applied the Newton algorithm [2] to solve the series of convex optimization problems. The above steps are detailed in [2,20,21]. Due to the lack of space, only the final iterative formula to obtain the Lagrange multipliers and the bias is shown below

$$\begin{bmatrix} b^{t+1} \\ \boldsymbol{\alpha}^{t+1} \end{bmatrix} = - \begin{bmatrix} 0 & \mathbf{1}^T \\ \mathbf{1} & \mathbf{I} + C\boldsymbol{\Omega} \end{bmatrix} \begin{bmatrix} 0 \\ \boldsymbol{\lambda}^t - C\boldsymbol{y} \end{bmatrix}, \tag{27}$$

where the vector $\boldsymbol{\lambda}$ is iteratively calculated by

$$\lambda_i^t = \begin{cases} C(y_i - z_i), & z_i < y_i - \sqrt{p} - h \text{ or } z_i > y_i + \sqrt{p} + h \\ \frac{C(h+2\sqrt{p})(y_i+h-\sqrt{p}-z_i)}{4h}, & |z_i - y_i + \sqrt{p}| \le h \\ 0, & y_i + h - \sqrt{p} < z_i < y_i - h + \sqrt{p} \\ \frac{C(h+2\sqrt{p})(y_i-h+\sqrt{p}-z_i)}{4h}, & |z_i - y_i - \sqrt{p}| \le h. \end{cases} \tag{28}$$

Given a tolerance parameter ε, b^{t+1} and $\boldsymbol{\alpha}^{t+1}$ must be calculated using the Eqs. (27) and (28). It must check if $||(b^{t+1}, \boldsymbol{\alpha}^{t+1}) - (b^t, \boldsymbol{\alpha}^t)|| < \varepsilon$ holds to end the process. Otherwise, do $t = t + 1$ and repeat the algorithm.

4   Simulations and Discussion

In order to evaluate the performances of the previously described models in nonlinear system identification under infinite step ahead prediction scenarios, we will carry out computer experiments with three artificial datasets, whose outputs are contaminated with outliers, and a real-world benchmarking dataset.

The first dataset, labeled *Artificial 1*, was generated according to [7] and is giving by

$$y_i = y_{i-1} - 0.5 \tanh(y_{i-1} + u_{i-1}^3), \tag{29}$$

$$u_i \sim \mathcal{N}(u_i|0, 1), \quad -1 \le u_i \le 1, \text{for both estimation and test data.} \tag{30}$$

The dataset contains 150 samples for estimation and 150 for test. The estimation data was corrupted with additive Gaussian noise with zero mean and variance 0.0025.

The following artificial datasets were generated according the seminal work by Narendra and Parthasarathy [12]. The first one, labeled *Artificial 2* is given by

$$y_i = \frac{y_{i-1} y_{i-2}(y_{i-1} + 2.5)}{1 + y_{i-1}^2 + y_{i-2}^2} + u_{i-1} \tag{31}$$

$$u_i = \begin{cases} U(-2, 2), & \text{for estimation data} \\ \sin(2\pi i/25), & \text{for test data} \end{cases}, \tag{32}$$

Table 1. RMSE values from free simulation results with and without outliers in artificial datasets

	Artificial 1			
outliers %	0%	5%	10%	20%
LS-SVR	**0.0309**	0.0654	0.1395	0.1518
WLS-SVR	0.0329	**0.0626**	0.1288	0.1174
RLS-SVR	0.0322	0.0682	**0.1083**	**0.1134**
	Artificial 2			
outliers %	0%	5%	10%	20%
LS-SVR	0.2805	**0.3808**	0.4937	0.8804
WLS-SVR	**0.2363**	0.3869	0.4634	0.6209
RLS-SVR	0.2854	0.4530	**0.4036**	**0.6089**
	Artificial 3			
outliers %	0%	5%	10%	20%
LS-SVR	0.2993	0.3150	0.6756	0.6467
WLS-SVR	**0.2890**	**0.2915**	0.2976	**0.3973**
RLS-SVR	0.2891	0.2921	**0.2823**	0.4606

where $U(-2,2)$ is a random number uniformly distributed between -2 and 2. The dataset contains 300 samples for estimation and 100 samples for test. The estimation data was corrupted with additive Gaussian noise with zero mean and variance 0.29.

The last artificial dataset, labeled *Artificial 3*, is given by

$$y_i = \frac{y_{i-1}}{1 + y_{i-1}^2} + u_{i-1}^3 \tag{33}$$

$$u_i = \begin{cases} U(-2,2), & \text{for estimation data} \\ \sin(2\pi i/25) + \sin(2\pi i/10), & \text{for test data} \end{cases} \tag{34}$$

Once again the dataset contains 300 samples for estimation and 100 test samples. The estimation data was corrupted with additive Gaussian noise with zero mean and variance 0.65.

The real-word dataset is called wing flutter and is available at DaISy repository of *Katholieke Universiteit Leuven*[1]. This dataset corresponds to a mechanical SISO (Single Input Single Output) system with 1024 samples of each sequence u_i and y_i, in which 512 samples were used for estimation and the other 512 for test.

All the artificial datasets were progressively corrupted with a number of outliers equal to 5%, 10% and 20% of the estimation samples. To each randomly chosen sample was added a uniformly distributed value $U(-M_y, +M_y)$, where M_y is the maximum absolute output value. Such outliers contamination methodology is similar to the one performed in [11].

The orders L_u and L_y chosen for the regressors of each artificial dataset were set to their largest delays according to Eqs. (29), (31) and (33). For the

---

[1] http://homes.esat.kuleuven.be/smc/daisy/daisydata.html

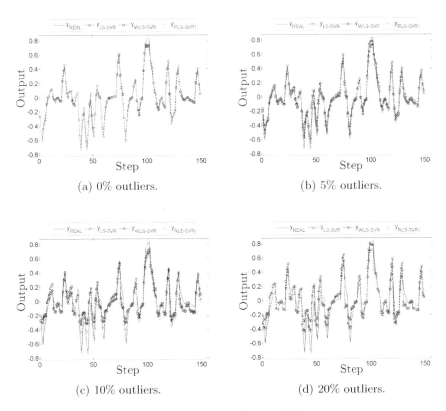

(a) 0% outliers.   (b) 5% outliers.

(c) 10% outliers.   (d) 20% outliers.

Fig. 1. Free simulation with Artificial 1 dataset

Table 2. Free simulation RMSE with wing flutter dataset

Real dataset		
LS-SVR	WLS-SVR	RLS-SVR
0.6433	**0.4709**	0.6433

wing flutter dataset the orders $L_u, L_y \in \{1, 2, 3, 4, 5\}$ were set after execution of a 5-fold cross validation strategy. The same strategy was used to set the hyperparameters $C \in \{2^{-5}, 2^{-4}, \ldots, 2^{20}\}$ and $\sigma \in \{2^{-10}, 2^{-9}, \ldots, 2^{10}\}$ in the search for their optimal values. Furthermore, for the RLS-SVR model, a new 5-fold cross validation is then performed to search for the optimal values of truncation parameter $p \in \{0.1, 0.2, \ldots, 1.0\}$ and the smoothing parameter $h \in \{0.01, 0.05, 0.10, 0.15, 0.20, 0.25, \ldots, 0.50\}$. The chosen value for the tolerance parameter was $\varepsilon = 0.001$. All algorithms were written in Matlab R2013a and the simulations ran on a HP ProBook notebook with 2.30Ghz Intel Core i5 processor, 4GB of RAM memory and Windows 7 Professional operational system.

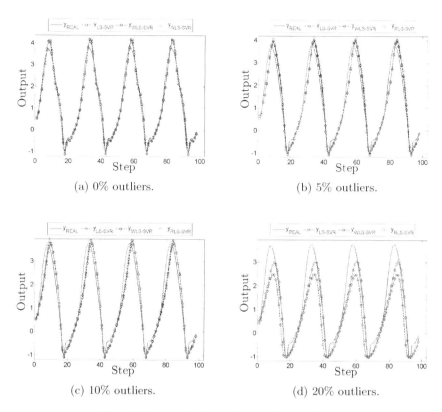

(a) 0% outliers.

(b) 5% outliers.

(c) 10% outliers.

(d) 20% outliers.

Fig. 2. Free simulation with Artificial 2 dataset

The obtained Root Mean Square Error (RMSE) values for the artificial datasets during the free simulation phase are reported in Table 1. In almost all cases contaminated with outliers the robust approaches consistently outperformed the traditional LS-SVR, except for the *Artificial 2* dataset with 5% of outliers. In the scenarios without outliers, the robust models achieved performance closer to the traditional LS-SVR, except for the *Artificial 2* dataset where the RMSE value for the WLS-SVR model was significantly lower that those achieved by the other methods.

Comparing the two robust models, the WLS-SVR presented in general smaller RMSE values than RLS-SVR models in scenarios without outliers and with 5% of contamination. The only exception was in *Artificial 1* dataset without outliers. The RLS-SVR performed better in scenarios with 10% e 20% of outliers in almost all datasets, except for the *Artificial 3* with 20% of outliers.

It is important to note that, despite the fact that the WLS-SVR and the RLS-SVR are outlier-robust methods, this does not mean they are fully insensitive to outliers. For the *Artificial 1* and *2* datasets, the RMSE obtained increased

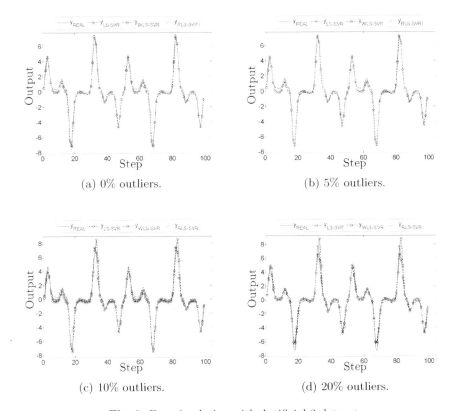

(a) 0% outliers.

(b) 5% outliers.

(c) 10% outliers.

(d) 20% outliers.

Fig. 3. Free simulation with Artificial 3 dataset

significantly when the contamination also increased. A good resilience to outliers was achieved only for *Artificial 3* dataset, where WLS-SVR and RLS-SVR were less affected for the cases up to 10% of contamination.

The predicted outputs for the artificial test data are illustrated in Figs. 1 to 3, where the effect of the incremental addition of outliers to the estimation data can be more easily perceived. Note that in the scenarios with higher rates of contamination, especially for the *Artificial 1* and *2* datasets (Figs. 1c, 1d, 2c and 2d), the predicted outputs can become very different from the real ones. It should be also observed that, as expected, the performance of the conventional LS-SVR model deteriorates faster than those of the robust approaches.

The obtained RMSE for the wing flutter dataset during the free simulation experiments are shown in Table 2. The RMSE value of the WLS-SVR model was considerable lower than ones with other approaches. For this dataset, LS-SVR and RLS-SVR outputs had the same behavior, because the RLS-SVR model stopped after a single iteration. The simulated test outputs are illustrated in Figs. 4a-4c, where we can see in Fig. 4b that the predicted outputs of the

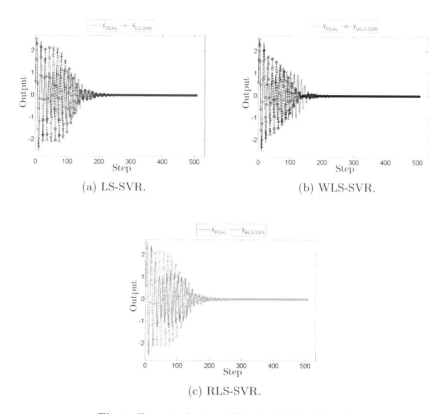

(a) LS-SVR.   (b) WLS-SVR.

(c) RLS-SVR.

Fig. 4. Free simulation with wing flutter dataset

WLS-SVR model followed the dynamical behavior of the system better than LS-SVR and RLS-SVR models, as can be seen in Figs. 4a and 4c, respectively.

As a final remark, we can write down some words on the computational complexity of the algorithms. It was observed in the experiments that the training times of the RLS-SVR model were usually longer than the ones observed for the LS-SVR and WLS-SVR models. In fact, this was somewhat expected since the computational complexities of the LS-SVR, WLS-SVR and RLS-SVR models are $O((N+1)^3)$, $O(2(N+1)^3)$ and $O(T_r(N+1)^3)$, respectively, where T_r is the total number of iterations of the RLS-SVR model.

5   Conclusions

In this paper we carried out a comprehensive performance evaluation of two robust LS-SVR variants, namely the WLS-SVR and RLS-SVR models, applied to nonlinear dynamical system identification tasks under free simulation scenarios and in the presence of outliers. None of these models were evaluated in such difficult scenarios before.

The experiments were conducted on three artificial datasets, contaminated with different rates of outliers in the outputs of the estimation data, and a real-world dataset, available in a benchmark identification database.

Generally, the obtained results in free simulation with the robust models presented lower values of RMSE than ones obtained with traditional LS-SVR. Moreover, between the robust models, the RLS-SVR achieved in general the best results in the scenarios with higher rates of outliers. For scenarios without outliers and the real-world dataset, the WLS-SVR model outperformed the other methods.

However, the results also showed that the WLS-SVR and the RLS-SVR models are not fully insensitive to the presence of outliers. Both methods presented considerable differences between some outlier-free and outlier corrupted scenarios. Depending on the application, such differences can make these approaches unfeasible to use in practice.

In the future, we will continue to study and develop techniques to improve the robustness of LS-SVR algorithms applied to nonlinear dynamical system identification, besides to investigate some features of the computational complexity in RLS-SVR method.

Acknowledgments. The authors thank the financial support of FUNCAP (Fundação Cearense de Apoio ao Desenvolvimento Científico e Tecnológico), IFCE (Instituto Federal de Educação, Ciência e Tecnologia do Ceará) and NUTEC (Núcleo de Tecnologia Industrial do Ceará).

References

1. Cai, Y., Wang, H., Ye, X., Fan, Q.: A multiple-kernel lssvr method for separable nonlinear system identification. Journal of Control Theory and Applications **11**(4), 651–655 (2013)
2. Chapelle, O.: Training a support vector machine in the primal. Neural Computation **19**(5), 1155–1178 (2007)
3. Falck, T., Dreesen, P., De Brabanter, K., Pelckmans, K., De Moor, B., Suykens, J.A.: Least-squares support vector machines for the identification of wiener-hammerstein systems. Control Engineering Practice **20**(11), 1165–1174 (2012)
4. Falck, T., Suykens, J.A., De Moor, B.: Robustness analysis for least squares kernel based regression: an optimization approach. In: Proceedings of the 48th IEEE Conference on Decision and Control, 2009 Held Jointly with the 2009 28th Chinese Control Conference, CDC/CCC 2009, pp. 6774–6779. IEEE (2009)
5. Huber, P.J., et al.: Robust estimation of a location parameter. The Annals of Mathematical Statistics **35**(1), 73–101 (1964)
6. Khalil, H.M., El-Bardini, M.: Implementation of speed controller for rotary hydraulic motor based on LS-SVM. Expert Systems with Applications **38**(11), 14249–14256 (2011)
7. Kocijan, J., Girard, A., Banko, B., Murray-Smith, R.: Dynamic systems identification with Gaussian processes. Mathematical and Computer Modelling of Dynamical Systems **11**(4), 411–424 (2005)

8. Liu, Y., Chen, J.: Correntropy-based kernel learning for nonlinear system identification with unknown noise: an industrial case study. In: 2013 10th India International Symposium on Dynamics and Control of Proccess Systems, pp. 361–366 (2013)
9. Liu, Y., Chen, J.: Correntropy kernel learning for nonlinear system identification with outliers. Industrial and Enginnering Chemistry Research pp. 1–13 (2013)
10. Ljung, L.: System Identification Theory for the User. 2nd edn. (1999)
11. Majhi, B., Panda, G.: Robust identification of nonlinear complex systems using low complexity ANN and particle swarm optimization technique. Expert Systems with Applications **38**(1), 321–333 (2011)
12. Narendra, K.S., Parthasarathy, K.: Identification and control of dynamical systems using neural networks. IEEE Transactions on Neural Networks **1**(1), 4–27 (1990)
13. Rousseeum, P.J., Leroy, A.M.: Robust Regression and Outlier Detection. 1st edn. (1987)
14. Saunders, C., Gammerman, A., Vovk, V.: Ridge regression learning algorithm in dual variables. In: Proceedings of the 15th International Conference on Machine Learning, ICML 1998, pp. 515–521. Morgan Kaufmann (1998)
15. Suykens, J.A.K., Van Gestel, T., De Brabanter, J., De Moor, B., Vandewalle, J.: Least Squares Support Vector Machines, 1st edn. World Scientific Publishing (2002)
16. Suykens, J.A.K., Vandewalle, J.: Least squares support vector machine classifiers. Neural Processing Letters **9**(3), 293–300 (1999)
17. Suykens, J.A., De Brabanter, J., Lukas, L., Vandewalle, J.: Weighted least squares support vector machines: robustness and sparse approximation. Neurocomputing **48**(1), 85–105 (2002)
18. Van Gestel, T., Suykens, J.A., Baestaens, D.E., Lambrechts, A., Lanckriet, G., Vandaele, B., De Moor, B., Vandewalle, J.: Financial time series prediction using least squares support vector machines within the evidence framework. IEEE Transactions on Neural Networks **12**(4), 809–821 (2001)
19. Vapnik, V.N.: The Nature of Statistical Learning Theory. Springer (1995)
20. Yang, X., Tan, L., He, L.: A robust least squares support vector machine for regression and classification with noise. Neurocomputing **140**, 41–52 (2014)
21. Yuille, A.L., Rangarajan, A.: The concave-convex procedure. Neural Computation **15**(4), 915–936 (2003)

On the Generalization of the Uninorm Morphological Gradient

Manuel González-Hidalgo, Sebastia Massanet$^{(\boxtimes)}$, Arnau Mir,
and Daniel Ruiz-Aguilera

Department of Mathematics and Computer Science, University of the Balearic
Islands, E-07122 Palma de Mallorca, Spain
{manuel.gonzalez,s.massanet,arnau.mir,daniel.ruiz}@uib.es

Abstract. The morphological gradient is a widely used edge detector
for grey-level images in many applications. In this communication, we
generalize the definition of the morphological gradient of the fuzzy mathematical morphology based on uninorms. Concretely, instead of defining
the morphological gradient from the usual definitions of fuzzy dilation
and erosion, where the minimum and the maximum are used, we define
it from the generalized fuzzy dilation and erosion, where we consider a
general t-norm and t-conorm, respectively. Once the generalized morphological gradient is defined, we determine which t-norm and t-conorm have
to be considered in order to obtain a high performance edge detector.
Some t-norms and their dual t-conorms are taken into account and the
experimental results conclude that the t-norms of the Schweizer-Sklar
family generate a morphological gradient which outperforms notably the
classical morphological gradient based on uninorms.

Keywords: Fuzzy mathematical morphology · Edge detection ·
Uninorms · Fuzzy implications · Hysteresis

1 Introduction

Edge detection is a fundamental low level operation in image processing which
is essential for developing high-level operations related with fields such as computer vision. Its performance is crucial for the final results of image processing
methods. In recent decades, a great number of edge detection algorithms has
been developed. There are different approaches from the classical ones [1] based
on the use of a set of convolution masks, to the new techniques based on fuzzy
sets and their extensions [2].

Among the fuzzy approaches, the fuzzy mathematical morphology which generalizes the binary morphology [3] using concepts and techniques of the theory
of fuzzy sets [4,5] can be highlighted. This theory allows a better processing
and a representation with higher flexibility of the uncertainty and the ambiguity
present in each level in an image. The morphological operators are the basic
tools of this theory. A morphological operator P converts an input image A in
a new image $P(A,B)$ using a structuring element B. The four basic morphological operations are dilation, erosion, closing and opening. Because the grey

© Springer International Publishing Switzerland 2015
I. Rojas et al. (Eds.): IWANN 2015, Part II, LNCS 9095, pp. 436–449, 2015.
DOI: 10.1007/978-3-319-19222-2_37

level images can be viewed as fuzzy sets, morphological fuzzy operators can be defined using fuzzy tools. Therefore, conjunctions (usually continuous t-norms and uninorms, see [6]) and their residuals implications have been used. Recently, a fuzzy mathematical morphology based on discrete t-norms has been introduced with good results in edge detection [7].

A general framework based on fuzzy conjunctions and fuzzy implications to define the morphological operators was established by De Baets in [8] and [9]. De Baets studied the case of t-norms concluding that the pair given by the Lukasiewicz t-norm and its residual implication is the representative of the unique family of t-norms satisfying all the desirable algebraic properties. Thus, the previous pair $(T_{\mathrm{LK}}, I_{\mathrm{LK}})$ is often used to implement an edge detector based on t-norms. This edge detector is known as the fuzzy morphological gradient, defined as the residual between the fuzzy dilation and the fuzzy erosion. Later, in [10] and [11] the fuzzy morphology based on uninorms was studied. More recently this fuzzy morphology has been applied to image processing, providing remarkable results, especially in edge detection and noise removal [6,12–14]. In this theory, the number of uninorms and implications that can be used to define a uninorm based morphological gradient with a notable performance (see [15] and also [16]) is higher than in the case of t-norms. More specifically, the configuration (U^{N_C}, I_{N_C}), where U^{N_C} is the idempotent uninorm generated from the classical negation and I_{N_C} is its residual implication, is the best configuration according to the performance measures used in [15,16], obtaining the best results with respect to other configurations.

In this work, a generalization of the fuzzy erosion and dilation is proposed to define a morphological gradient with a better performance than the classical morphological gradient based on uninorms. In this way, the erosion and dilation of the fuzzy mathematical morphology can be generalized by changing the minimum and the maximum in their expressions. The maximum can be considered as a particular case of a t-conorm and the minimum, as a particular case of a t-norm and therefore, they can be changed by a general operator of these families of aggregation functions. Because the maximum is the smallest t-conorm and the minimum, the largest t-norm, the new morphological gradient should be able to detect more edges in the image. Therefore, the next step will be to compare the results obtained by the two approaches, both from the visual and the quantitative point of view.

To perform a comparison of the results, several performance measures will be used, namely the measure proposed by Pratt *FoM* (see Chapter 15 of [1]), the ρ-coefficient [17] and the F-measure [18]. To use these measures, the edge image must be binary and the width of the edges has to be of one pixel, consistent with the restrictions imposed by Canny in [19]. Therefore, once the fuzzy edge image is obtained, a *thinning* algorithm as *Non-Maxima Suppression* (NMS) introduced by Canny, will be implemented. After that, the non-supervised algorithm of hysteresis based on the determination of the "instability zone" in the image histogram, proposed in [20], will be performed to binarize the image.

Table 1. Considered t-norms

Name	Expression
Łukasiewicz	$T_{LK}(x,y) = \max\{x + y - 1, 0\}$
Minimum	$T_M(x,y) = \min\{x,y\}$
Product	$T_P(x,y) = xy$
Nilpotent Minimum	$T_{nM}(x,y) = \begin{cases} 0 & \text{if } x + y \leq 1, \\ \min\{x,y\} & \text{otherwise.} \end{cases}$
Drastic	$T_D(x,y) = \begin{cases} 0 & \text{if } x,y \in [0,1), \\ \min\{x,y\} & \text{otherwise.} \end{cases}$
Schweizer-Sklar	$T_\lambda^{SS}(x,y) = \begin{cases} T_M(x,y) & \text{if } \lambda = -\infty, \\ T_P(x,y) & \text{if } \lambda = 0, \\ T_D(x,y) & \text{if } \lambda = +\infty, \\ (\max\{x^\lambda + y^\lambda - 1, 0\})^{\frac{1}{\lambda}} & \text{if } \lambda \in \mathbb{R} \setminus \{0\}. \end{cases}$

The communication is organized as follows. In Section 2, the definitions of the classical morphological operators and the fuzzy operators that define them are introduced. In Section 3, the generalized dilation and erosion, as well as the morphological gradient derived from them are defined. In the next section, the comparison of both edge detectors is performed, comparing the results both from the visual and the quantitative point of view. Finally, some conclusions and future work are exposed.

2   Preliminaries

Fuzzy morphological operators are defined using fuzzy operators such as fuzzy conjunctions (t-norms and conjunctive uninorms) and fuzzy implications. For more details on these connectives, see [21], [22] and [23], respectively.

Definition 1. *A* t-norm *T (t-conorm S) is a commutative, associative and increasing function from $[0,1]^2$ to $[0,1]$ with 1 (0) as neutral element.*

Let us remember that t-norms and t-conorms are dual operators. Given a t-norm T, its dual t-conorm T^* is defined as $T^*(x,y) = 1 - T(1-x, 1-y)$ for all $x, y \in [0,1]$ and vice-versa. The t-norms that we will use throughout the paper have been listed in Table 1. Let us note that the t-norms T_λ^{SS} belong to the parametric family of Schweizer-Sklar and are strict if $\lambda \in [0, +\infty)$ and nilpotent if $\lambda \notin [0, +\infty)$. The t-conorms considered in this work are their duals. Moreover, for any t-norm T and t-conorm S it is satisfied that $T \leq T_M$ and $S_M \leq S$, with $S_M = T_M^*$.

The associativity of a t-norm T (t-conorm S) allows us to extend it to an n-ary operator using recursion, defining for each n-tuple $(x_1, \ldots, x_n) \in [0,1]^n$:

$$\mathop{\mathsf{T}}_{i=1}^{n} x_i = T\left(\mathop{\mathsf{T}}_{i=1}^{n-1} x_i, x_n\right) = T(x_1, x_2, \ldots, x_n)$$

$$\left(\mathop{\mathsf{S}}_{i=1}^{n} x_i = S\left(\mathop{\mathsf{S}}_{i=1}^{n-1} x_i, x_n\right) = S(x_1, x_2, \ldots, x_n)\right).$$

Now, in order to define the morphological operators based on uninorms, let us introduce this class of fuzzy conjunctions.

Definition 2. *A* uninorm *is a commutative, associative, non-decreasing function $U : [0,1]^2 \to [0,1]$ with neutral element $e \in (0,1)$, i.e., $U(e,x) = U(x,e) = x$ for all $x \in [0,1]$.*

It is known that $U(0,1) \in \{0,1\}$. A uninorm U such that $U(0,1) = 0$ is called *conjunctive* and if $U(0,1) = 1$, then it is called *disjunctive*.

Definition 3. *A binary operator $I : [0,1]^2 \to [0,1]$ is a* fuzzy implication *if it is decreasing in the first variable, increasing in the second one and it satisfies $I(0,0) = I(1,1) = 1$ and $I(1,0) = 0$.*

In this framework, the basic fuzzy morphological operators are dilation and erosion. From now on, we will follow this notation: I will denote a fuzzy implication; U a conjunctive uninorm; A a grey-level image and B a grey-level structuring element, both modelled as mappings $A : d_A \to [0,1]$ and $B : d_B \to [0,1]$ where $d_A, d_B \subseteq \mathbb{Z}^2$ both finite and $T_v(A)$ will denote the translation of a fuzzy set A by $v \in \mathbb{Z}^2$ defined as $T_v(A)(x) = A(x - v)$.

Definition 4. *The* fuzzy dilation $D_U(A,B)$ *and the* fuzzy erosion $E_I(A,B)$ *of A by B are the grey-level images defined as*

$$D_U(A,B)(y) = \max_{x \in d_A \cap T_y(d_B)} U(B(x-y), A(x)),$$

$$E_I(A,B)(y) = \min_{x \in d_A \cap T_y(d_B)} I(B(x-y), A(x)).$$

With some few properties, the following proposition ensures the extensivity of the fuzzy dilation and the antiextensivity of the fuzzy erosion.

Proposition 1. *Let U be a conjunctive uninorm with neutral element $e \in (0,1)$, I an implication that satisfies the neutrality principle for implications derived from uninorms (NP$_e$), i.e., $I(e,y) = y$ for all $y \in [0,1]$ and B a grey-level structuring element such that $B(0) = e$. Then the following inclusions hold: $E_I(A,B) \subseteq A \subseteq D_U(A,B)$.*

Therefore, as in the classical morphology, the difference between the fuzzy dilation and the fuzzy erosion in a grey-level image, $\delta_{U,I}(A,B) = D_U(A,B) \setminus E_I(A,B)$, called the *fuzzy morphological gradient*, can be used in edge detection.

3   Generalization of the Morphological Gradient

In this section, the main goal will be to generalize the definitions of the fuzzy dilation and erosion given in Definition 4.

Definition 5. *Let $\hat{S}$ be a t-conorm and $\hat{T}$ be a t-norm. Let A and B be grey level images. For every $y \in d_A$, consider the finite set with cardinal n_y given by $K_y = d_A \cap T_y(d_B) = \{x_1, \ldots, x_{n_y}\}$. The generalized fuzzy dilation $\hat{D}_{\hat{S},U}(A,B)$ and the generalized fuzzy erosion $\hat{E}_{\hat{T},I}(A,B)$ of A by B are the grey level images defined as:*

$$\hat{D}_{\hat{S},U}(A,B)(y) = \overset{n_y}{\underset{i=1}{\hat{S}}}\, U(B(x_i - y), A(x_i)),$$

$$\hat{E}_{\hat{T},I}(A,B)(y) = \overset{n_y}{\underset{i=1}{\hat{T}}}\, I(B(x_i - y), A(x_i)).$$

Remark 1. Note that the previous definitions generalize the classical fuzzy dilation and erosion due to the fact that $\hat{E}_{T_{\mathrm{M}},I}(A,B) = E_I(A,B)$ and $\hat{D}_{S_{\mathrm{M}},U}(A,B) = D_U(A,B)$.

The properties of t-conorms and t-norms allow us to prove the next result straightforwardly.

Proposition 2. *Let $\hat{S}$ be a t-conorm and $\hat{T}$ be a t-norm. Let U and I be a conjunctive uninorm and a fuzzy implication satisfying the conditions of Proposition 1. Then the generalized fuzzy dilation and erosion of an image A by a structuring element B satisfy: $\hat{E}_{\hat{T},I}(A,B) \subseteq E_I(A,B) \subseteq A \subseteq D_U(A,B) \subseteq \hat{D}_{\hat{S},U}(A,B)$.*

Proof. Since the following inequalities hold $\hat{T} \leq T_{\mathrm{M}}$ and $S_{\mathrm{M}} \leq \hat{S}$, we have that:

$$\hat{E}_{\hat{T},I}(A,B) \subseteq \hat{E}_{T_{\mathrm{M}},I}(A,B) = E_I(A,B), \quad D_U(A,B) = \hat{D}_{S_{\mathrm{M}},U}(A,B) \subseteq \hat{D}_{\hat{S},U}(A,B).$$

Using Proposition 1, we get:

$$\hat{E}_{\hat{T},I}(A,B) \subseteq E_I(A,B) \subseteq A \subseteq D_U(A,B) \subseteq \hat{D}_{\hat{S},U}(A,B).$$

Therefore, the definition of the generalized morphological gradient can be derived directly from the previous proposition:

$$\delta_{\hat{S},\hat{T},U,I}(A,B) = \hat{D}_{\hat{S},U}(A,B) \setminus \hat{E}_{\hat{T},I}(A,B).$$

As it has been already said in the introduction, the generalized morphological gradient extends the usual morphological gradient being able to detect more edges of the image.

Corollary 1. *Let $\hat{S}$ be a t-conorm and $\hat{T}$ be a t-norm. Let U and I be a conjunctive uninorm and a fuzzy implication satisfying the conditions of Proposition 1. Then the following inequality holds: $\delta_{U,I}(A,B) \subseteq \delta_{\hat{S},\hat{T},U,I}(A,B)$.*

3.1 Edge Detector

Fuzzy methods of edge detection, the framework where morphological gradients belong to, generate an image where the value of a pixel determines the membership degree of that pixel to the set of edges. This idea contradicts the restrictions given by Canny in [19]. There, a representation of the edge image as a binary image with edges of one pixel width is recommended. Hence, the fuzzy edge image must be thinned and binarized. Indeed, the fuzzy edge image will contain large values where there is a strong image gradient, but to identify edges the broad regions present in areas where the slope is large must be thinned so that only the magnitudes at those points which are local maxima remain. Non Maxima Supremum (NMS), an algorithm proposed by Canny, performs this by suppressing all values along the line of the gradient that are not peak values [19]. NMS has been performed using P. Kovesis' implementation in MATLAB [24].

Finally, to binarize the image, we have implemented an automatic non-supervised hysteresis based on the determination of the instability zone of the histogram to find the threshold values [20]. Hysteresis allows to choose which pixels are relevant in order to be selected as edges, using their membership values. Two threshold values T_1, T_2 with $T_1 \leq T_2$ are used. All the pixels with a membership value greater than T_2 are considered as edges, while those which are lower to T_1 are discarded. Those pixels whose membership value is between the two values are selected if, and only if, they are connected with other pixels above T_2. The method needs some initial set of candidates for the threshold values. In this case, the set $\{0.01, \ldots, 0.25\}$ has been introduced, the same one which is used in [20]. In Figure 1, we display the block diagram of the edge detector algorithm proposed in this section and in Figure 2, the intermediate images which are being obtained in each step.

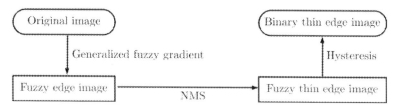

Fig. 1. Block diagram of the proposed edge detector

3.2 Objective Comparison Method

Nowadays, it is well-established in the literature that the visual inspection of the edge images obtained by several edge detectors can not be the unique criterion with the aim of proving the superiority of one edge detector with respect to the others. This is because each expert has different criteria and preferences and consequently, the reviews given by two experts can differ substantially. For this reason, when we obtain the binary edge image with edges of one pixel width

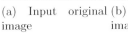

(a) Input original image | (b) Fuzzy edge image obtained with the fuzzy gradient | (c) NMS | (d) Output binary thin edge image

Fig. 2. Sequence of the proposed edge detector

(DE) corresponding to the edges detected by the method, some objective performance measure is needed. The use of objective performance measures on edge detection is growing in popularity to compare the results obtained by different edge detection algorithms. There are several measures of performance for edge detection in the literature, see [25] and [26]. These measures require, in addition to the DE image obtained by the edge detector we want to evaluate, a reference edge image or *ground truth* edge image (GT) which is a binary edge image with edges of one pixel width containing the real edges of the original image. In this work, we will use the following objective measures to evaluate the similarity between DE and GT:

1. The measure proposed by Pratt [1], *Pratt's figure of merit*, defined as FoM $= \dfrac{1}{\max\{card\{DE\}, card\{GT\}\}} \cdot \displaystyle\sum_{x \in DE} \dfrac{1}{1 + ad^2}$, where $card$ is the number of edge pixels of the image, a is a scaling constant and d is the separation distance between an obtained edge pixel with respect to an ideal one. In this paper, we will consider $a = 1$ and the Euclidean distance d.

2. The ρ-coefficient [17], given by $\rho = \dfrac{card(E)}{card(E) + card(E_{FN}) + card(E_{FP})}$, where E is the set of well detected edge pixels, E_{FN} is the set of edges of the GT which have not been detected by the considered edge detector and E_{FP} is the set of edge pixels which have been detected but without any correspondence in the GT.

3. The F-measure [18] which is given by the weighted harmonic mean of the precision PR and recall RE, i.e., $F = \dfrac{2 \cdot PR \cdot RE}{PR + RE}$, where $PR = \dfrac{\#(E)}{\#(E) + \#(E_{FP})}$ and $RE = \dfrac{\#(E)}{\#(E) + \#(E_{FN})}$ and $\#(A)$ means the cardinality of A.

Larger values of *FoM*, ρ and F ($0 \leq FoM, \rho, F \leq 1$) are indicators of a better capability to detect edges.

4   Experimental Results and Analysis

In this section we will show some preliminary results to show the potential of the generalized morphological gradient, the edge detector generated from the generalized morphological operators. The performance of this approach will be objectively evaluated and compared with the usual morphological gradient based on uninorms, using some images of the dataset of the University of South Florida[1] ([27]). Concretely, the first 15 images of the dataset and their edge specifications have been used. In [27], the details about the ground truth edge images and their use for the comparison of edge detectors are specified.

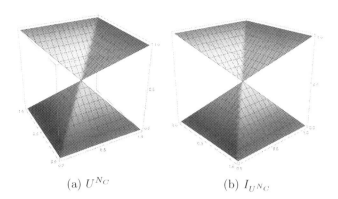

(a) U^{N_C} (b) $I_{U^{N_C}}$

Fig. 3. Best pair uninorm-implication for the usual morphologial gradient based on uninorms

The results included in this section have been obtained using the following isotropic structuring element

$$B = e \cdot \begin{pmatrix} 0.86\ 0.86\ 0.86 \\ 0.86\ \ 1\ \ 0.86 \\ 0.86\ 0.86\ 0.86 \end{pmatrix}.$$

This structuring element was already used in [5] and it provides notable results in practise. As internal operators U and I into both the generalized and usual morphological operators, we have considered the idempotent uninorm U^{N_C} and its residual implication $I_{U^{N_C}}$ (the graphs of U^{N_C} and $I_{U^{N_C}}$ are shown in Figure 3), whose expressions are given by:

$$U^{N_C}(x,y) = \begin{cases} \min\{x,y\}, & \text{if } y \le 1-x, \\ \max\{x,y\}, & \text{if } y > 1-x, \end{cases} \quad I_{U^{N_C}}(x,y) = \begin{cases} \max\{1-x,y\}, & \text{if } x \le y, \\ \min\{1-x,y\}, & \text{if } x > y. \end{cases}$$

Note that the pair $(U^{N_C}, I_{U^{N_C}})$ is the best configuration of the usual morphological gradient derived from uninorms for edge detection purposes (see [15]). Finally, as external operators, t-norm $\hat{T}$ and t-conorm $\hat{S}$, we have considered the t-norms of Table 1 except the drastic t-norm whose expression is not adequate to detect edges and their dual t-conorms.

---

[1]This image dataset can be downloaded from ftp://figment.csee.usf.edu/pub/ROC/edge_comparison_dataset.tar.gz

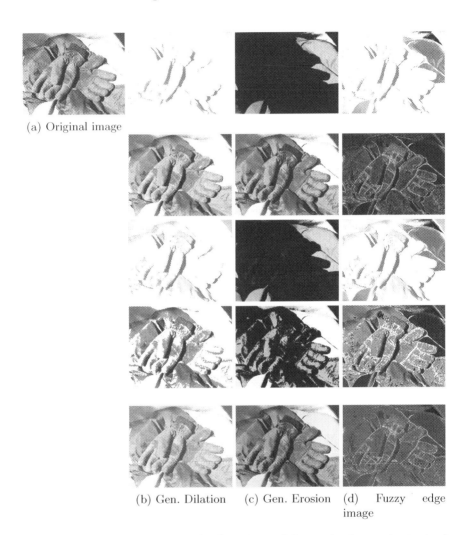

(a) Original image

(b) Gen. Dilation   (c) Gen. Erosion   (d)   Fuzzy   edge
image

Fig. 4. Generalized dilation, generalized erosion and fuzzy edge image obtained using $\delta_{\hat{S},\hat{T},U^{N_C},I_{U^{N_C}}}$ considering as, from top to bottom, $\hat{T}$ the t-norms T_{LK}, T_{M}, T_{P}, T_{nM} and T_{-10}^{SS} and as t-conorms $\hat{S}$, the corresponding dual t-conorms

First of all, in Figure 4, we show the generalized fuzzy dilation and erosion and the fuzzy edge image, obtained by the generalized morphological gradient using the external t-norms and t-conorms enumerated above for some images. We can see how the fuzzy edge images obtained using T_P and specially, T_{LK} contain high edge membership values in regions where no significant edge is present. This low performance is due to the behaviour of the generalized erosion and dilation with these operators. On the other hand, the nilpotent minimum and the Schweizer-Sklar t-norms obtain interesting results. Furthermore, the

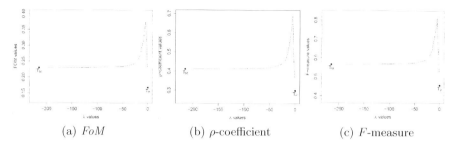

(a) *FoM*   (b) *ρ*-coefficient   (c) *F*-measure

Fig. 5. Evolution of the means of the values of a measure obtained by $\hat{T} = T^{SS}_\lambda$ depending on the values of λ including also $T^{SS}_{-\infty} = T_M$ and $T^{SS}_0 = T_P$

Table 2. Mean and standard deviation of some configurations of the generalized morphological gradient according to the considered objective measures

Conf.	FoM		ρ		F	
	Mean	Std.	Mean	Std.	Mean	Std.
$\delta_{S_M,T_M,U^{N_C},I_{U^{N_C}}}$	0.2300	0.0871	0.4112	0.1577	0.5660	0.1623
$\delta_{S_P,T_P,U^{N_C},I_{U^{N_C}}}$	0.1680	0.0483	0.2989	0.0768	0.4551	0.0926
$\delta_{S_{nM},T_{nM},U^{N_C},I_{U^{N_C}}}$	0.2370	0.0458	0.4665	0.0880	0.6315	0.0842
$\delta_{S^{SS}_{-5},T^{SS}_{-5},U^{N_C},I_{U^{N_C}}}$	0.3758	0.0711	0.6905	0.1173	0.8113	0.0860

Schweizer-Sklar family of t-norms depends on the value of the parameter λ whose role on the performance on the resulting edge detector deserves to be studied.

Remark 2. Note that the use of external operators such as a t-norm $\hat{T} \neq T_M$ and a t-conorm $\hat{S} \neq S_M$ can imply that the fuzzy edge image contains pixels with edge membership values greater than zero in plain regions of the original image. Although this is an undesired behaviour, these pixels usually have the lowest edge membership values and the thinning and hysteresis algorithms are capable of discarding them as final edges in the binary edge image with edges of one pixel width.

The bad behaviour of T_{LK} observed in the first experiment, which occurs also with the remaining images, allows us to discard this t-norm in the second experiment. At this point, let us check the performance of the different t-norms $\hat{T}$ and t-conorms $\hat{S}$ to generate a generalized morphological gradient which improves the results of the usual morphological gradient. Thus, we have computed the mean and the standard deviation of the 15 values of the three considered measures obtained by each configuration of the generalized morphological gradient, applied to the considered images of the dataset. In particular, in addition to T_M (which generates the usual morphological gradient), T_P and T_{nM}, we have considered the

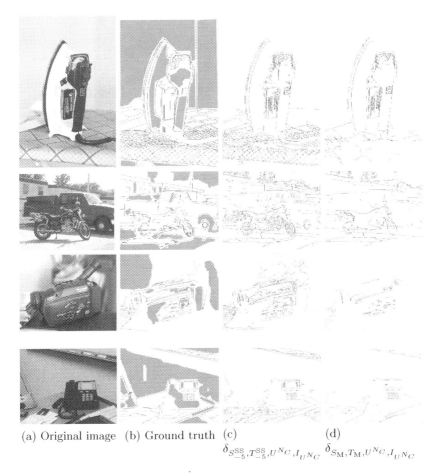

(a) Original image (b) Ground truth (c)   (d)

$$\delta_{S^{SS}_{-5},T^{SS}_{-5},U^{NC},I_{U^{NC}}} \quad \delta_{S_{M},T_{M},U^{NC},I_{U^{NC}}}$$

Fig. 6. Original image, ground truth edge image and the results obtained by the best configuration of the generalized morphological gradient and the usual one for several images

t-norms T^{SS}_{λ} taking $\lambda \in \{-1, -2, -3, \ldots, -15, -20, -25, \ldots, -195, -200\}$. The considered λ values have been chosen according to the following two remarks:

1. Since $T^{SS}_{\lambda} > T^{SS}_{0} = T_{P}$ for $\lambda > 0$ and T_{P} already provides questionable results, we have only considered negative values of λ. In addition, since $T^{SS}_{\lambda} \to T_{M}$ when $\lambda \to -\infty$ and the results of T^{SS}_{-200} are almost similar to the ones obtained by T_{M}, we have only reached this value of λ.
2. From the results, we can observe that the best ones are obtained when $\lambda \in [-15, -1]$ and consequently, we have refined the step of the values in this range.

In Figure 5, the evolution of the means of the values of each measure for each T^{SS}_{λ} is displayed. From these figures, the best t-norm of this family is

T_{-5}^{SS} according to the three measures. Note that the evolution of the measures according to the values of λ does not depend of the measure. In the figures, we have included the mean values obtained by T_M and T_P as limiting cases of the family when $\lambda \in (-\infty, 0)$.

Some conclusions emerge from the previous figures. Note that the graphs of the mean values of the measure seem to be smooth with respect to λ and therefore, the curve approaches to the mean values of T_M and T_P. These curves present a global maximum with respect to the considered λ values at $\lambda = -5$ which means that the pair T_{-5}^{SS} and its dual t-conorm improve drastically the results obtained using T_M. Consequently, the generalized morphological gradient generated from this pair outperforms the usual morphological gradient. In Table 2, the mean and standard deviation of the values of the measures obtained by each configuration of the generalized morphological gradient are collected. As it can be observed, the configuration $\delta_{S_{-5}^{SS}, T_{-5}^{SS}, U^{N_C}, I_{U^{N_C}}}$ obtains a higher mean with a lower standard deviation, deriving in a more robust edge detector than the usual morphological gradient. On the other hand, neither the configuration from T_P nor the one from T_{nM} and their dual t-conorms improve the usual morphological gradient.

To support the previous claim, we have performed a Wilcoxon test and a t-test for the configuration $\delta_{S_{-5}^{SS}, T_{-5}^{SS}, U^{N_C}, I_{U^{N_C}}}$ over the configuration $\delta_{S_M, T_M, U^{N_C}, I_{U^{N_C}}}$. The results show that the first configuration is statistically better than the usual morphological gradient obtaining a p-value less than 10^{-5}. Although the configuration derived from T_{-5}^{SS} is the one with a higher mean value from the Schweizer-Sklar family of t-norms, the t-norms T_{-6}^{SS} and T_{-7}^{SS} are also statistically similar, and they obtain also notable results. In Fig. 6 we can observe some of the results obtained by the best configuration of the generalized morphological gradient and the usual one. Note that the visual results agree with the quantitative results.

5   Conclusions and Future Work

In this article, we have proposed a generalization of the morphological operators based on uninorms in order to define a generalized morphological gradient capable of detecting a greater number of edges of an image. This generalization is based on considering a general t-conorm and t-norm into the definitions of erosion and dilation instead of the usual maximum and minimum. The preliminary obtained results show the potential of this generalization as long as the considered t-norm and t-conorm are of the Schweizer-Sklar families. In the experiments carried out in this paper, we have proved that the configuration $\delta_{S_{-5}^{SS}, T_{-5}^{SS}, U^{N_C}, I_{U^{N_C}}}$ outperforms severely the usual morphological gradient. Other operators of the family, such as the ones with $\lambda \in \{-7, -6\}$ can be also used with great results.

As future work, we want to extend the comparison started in this work to all the images of the considered dataset. This comparison experiment will provide further evidences of the superiority of the generalized morphological gradient over the classical one. Furthermore, note that the study made in this communication uses as internal operators the best ones for the usual morphological gradient based on uninorms. However, it is possible that other internal operators could be more suitable for the generalized morphological gradient and consequently, the results could be further improved. Once the study of the generalized morphological gradient is finished, comparison experiments with some classical edge detectors should be addressed. In addition, it would be also worth to study other possible applications of these generalized operators in image processing such as segmentation, contrast adjustment and noise removal.

Acknowledgments. This paper has been supported by the Spanish Grant TIN2013-42795-P.

References

1. Pratt, W.K.: Digital Image Processing, 4th edn. Wiley-Interscience (2007)
2. Bustince, H., Barrenechea, E., Pagola, M., Fernandez, J.: Interval-valued fuzzy sets constructed from matrices: Application to edge detection. Fuzzy Sets and Systems **160**(13), 1819–1840 (2009)
3. Serra, J.: Image analysis and mathematical morphology, vols. 1, 2. Academic Press, London (1982, 1988)
4. Bloch, I., Maître, H.: Fuzzy mathematical morphologies: a comparative study. Pattern Recognition **28**, 1341–1387 (1995)
5. Nachtegael, M., Kerre, E.E.: Classical and fuzzy approaches towards mathematical morphology. In: Kerre, E.E., Nachtegael, M. (eds.) Fuzzy Techniques in Image Processing. Studies in Fuzziness and Soft Computing, ch. 1, vol. 52, pp. 3–57. Physica-Verlag, New York (2000)
6. González-Hidalgo, M., Mir-Torres, A., Ruiz-Aguilera, D., Torrens, J.: Image analysis applications of morphological operators based on uninorms. In: Proceedings of the IFSA-EUSFLAT 2009 Conference, pp. 630–635, Lisbon, Portugal (2009)
7. González-Hidalgo, M., Massanet, S., Torrens, J.: Discrete t-norms in a fuzzy mathematical morphology: Algebraic properties and experimental results. In: Proceedings of WCCI-FUZZ-IEEE, Barcelona, Spain, pp. 1194–1201 (2010)
8. De Baets, B.: Fuzzy morphology: A logical approach. In: Ayyub, B.M., Gupta, M.M. (eds.) Uncertainty Analysis in Engineering and Science: Fuzzy Logic. Statistics, and Neural Network Approach, pp. 53–68. Kluwer Academic Publishers, Norwell (1997)
9. De Baets, B.: Generalized idempotence in fuzzy mathematical morphology. In: Kerre, E.E., Nachtegael, M. (eds.) Fuzzy Techniques in Image Processing. Studies in Fuzziness and Soft Computing, ch. 2, vol. 52, pp. 58–75. Physica-Verlag, New York (2000)
10. De Baets, B., Kwasnikowska, N., Kerre, E.: Fuzzy morphology based on uninorms. In: Proceedings of the seventh IFSA World Congress, Prague, pp. 215–220 (1997)
11. González, M., Ruiz-Aguilera, D., Torrens, J.: Algebraic properties of fuzzy morphological operators based on uninorms. In: Artificial Intelligence Research and Development. Frontiers in Artificial Intelligence and Applications, Amsterdam, vol. 100, pp. 27–38. IOS Press (2003)

12. González-Hidalgo, M., Mir-Torres, A., Ruiz-Aguilera, D.: A robust edge detection algorithm based on a fuzzy mathematical morphology using uninorms (ϕMM-U morphology). In: Manuel, J., Tavares, R.S., Natal Jorge, R.M. (eds.) Computational Vision and Medical Image Processing: VipIMAGE 2009, Bristol, PA, USA, pp. 630–635. CRC Press, Taylor & Francis Inc. (2009)

13. González-Hidalgo, M., Mir-Torres, A.: and J. Torrens Sastre. Noise reduction using alternate filters generated by fuzzy mathematical operators using uninorms (ϕMM-U morphology). In: Burillo, P., Bustince, H., De Baets, B., Fodor, J. (eds.) EUROFUSE WORKSHOP 2009. Preference Modelling and Decision Analysis, pp. 233–238. Public University of Navarra, Pamplona (2009)

14. González-Hidalgo, M., Mir Torres, A., Torrens Sastre, J.: Noisy image edge detection using an uninorm fuzzy morphological gradient. International Conference on Intelligent Systems Design and Applications (ISDA 2009), pp. 1335–1340 (2009)

15. González-Hidalgo, M., Massanet, S., Mir, A., Ruiz-Aguilera, D.: On the choice of the pair conjunction-implication into the fuzzy morphological edge detector. IEEE Transactions on Fuzzy Systems (2014). doi:10.1109/TFUZZ.2014.2333060

16. Gonzlez-Hidalgo, M., Massanet, S., Mir, A., Ruiz-Aguilera, D.: On the pair uninorm-implication in the morphological gradient. In: Madani, K., Correia, A.D., Rosa, A., Filipe, J. (eds.) Computational Intelligence. Studies in Computational Intelligence, vol. 577, pp. 183–197. Springer (2015)

17. Grigorescu, C., Petkov, N., Westenberg, M.A.: Contour detection based on nonclassical receptive field inhibition. IEEE Transactions on Image Processing **12**(7), 729–739 (2003)

18. Rijsbergen, C.J.V.: Information retrieval. Butterworths (1979)

19. Canny, J.: A computational approach to edge detection. IEEE Trans. Pattern Anal. Mach. Intell. **8**(6), 679–698 (1986)

20. Medina-Carnicer, R., Muñoz-Salinas, R., Yeguas-Bolivar, E., Diaz-Mas, L.: A novel method to look for the hysteresis thresholds for the Canny edge detector. Pattern Recognition **44**(6), 1201–1211 (2011)

21. Klement, E.P., Mesiar, R., Pap, E.: Triangular norms. Kluwer Academic Publishers, London (2000)

22. Fodor, J.C., Yager, R.R., Rybalov, A.: Structure of uninorms. Int. J. Uncertainty, Fuzziness. Knowledge-Based Systems **5**, 411–427 (1997)

23. Baczyński, M., Jayaram, B.: Fuzzy Implications. Studies in Fuzziness and Soft Computing, vol. 231. Springer, Heidelberg (2008)

24. Kovesi, P.D.: MATLAB and Octave functions for computer vision and image processing. Centre for Exploration Targeting, School of Earth and Environment, The University of Western Australia http://www.csse.uwa.edu.au/~pk/research/matlabfns/

25. Lopez-Molina, C., De Baets, B., Bustince, H.: Quantitative error measures for edge detection. Pattern Recognition **46**(4), 1125–1139 (2013)

26. Papari, G., Petkov, N.: Edge and line oriented contour detection: State of the art. Image and Vision Computing **29**(2–3), 79–103 (2011)

27. Bowyer, K., Kranenburg, C., Dougherty, S.: Edge detector evaluation using empirical ROC curves. In: IEEE Conf. on Computer Vision and Pattern Recognition (CVPR 1999), vol. 1, pp. 354–359 (1999)

Revisiting Image Vignetting Correction by Constrained Minimization of Log-Intensity Entropy

Laura Lopez-Fuentes, Gabriel Oliver, and Sebastia Massanet[✉]

Department Mathematics and Computer Science, University of the Balearic Islands,
Crta. Valldemossa km. 7,5, E-07122 Palma de Mallorca, Spain
llopezfuentes@hotmail.com, goliver@uib.eu, s.massanet@uib.es

Abstract. The correction of the vignetting effect in digital images is a key pre-processing step in several computer vision applications. In this paper, some corrections and improvements to the image vignetting correction algorithm based on the minimization of the log-intensity entropy of the image are proposed. In particular, the new algorithm is able to deal with images with a vignetting that is not in the center of the image through the search of the optical center of the image. The experimental results show that this new version outperforms notably the original algorithm both from the qualitative and the quantitative point of view. The quantitative measures are obtained using an image database with images to which artificial vignetting has been added.

Keywords: Vignetting · Entropy · Optical center · Gain function

1 Introduction

Vignetting is an undesirable effect in digital images which needs to be corrected as a pre-processing step in computer vision applications. This effect is based on the radial fall off of brightness away from the optical center of the image. Specially, those applications which rely on consistent intensity measurements of a scene are highly affected by the existence of vignetting. For example, image mosaicking, stereo matching and image segmentation, among many others, are applications where the results are notably improved if some previous vignetting correction algorithm is applied to the original image.

There are three types of vignetting according to their cause [1]: natural vignetting, pixel vignetting and mechanical vignetting. Natural vignetting is caused by light reaching different locations on the camera sensor at different angles and is most significant with wide angle lenses. On the other hand, pixel vignetting only affects digital cameras and is caused by angle-dependence of the digital sensors. Mechanical vignetting is an abrupt vignetting which only appears in the corners of the image and is caused by the occlusion by objects entering the camera field of view. Most of the proposed vignetting correction algorithms

© Springer International Publishing Switzerland 2015
I. Rojas et al. (Eds.): IWANN 2015, Part II, LNCS 9095, pp. 450–463, 2015.
DOI: 10.1007/978-3-319-19222-2_38

are designed to correct the gradual and continuous natural and pixel vignettings since the mechanical one can often be eliminated by using a longer focal length.

There exist basically two different approaches to correct vignetting. The first one lies on the use of multiple images of the same scene at different angles which need to be stitched into a single one, panoramic image. If vignetting affects these images, when the border of an image is superposed on the center of the neighbouring image, the differences of intensities are significant obtaining a poor panoramic image. Some vignetting correction algorithms handle this problem (see [2,3]). However, there exists a second approach which is the vignetting correction based on a single input image from a non-predetermined camera. This problem setting leads to a research that is more useful in practice because it does not require an explicit calibration step or an image sequence as input. Nevertheless, single-image vignetting correction is a challenging problem to solve since the resulting algorithm needs to differentiate the global brightness variations caused by vignetting from those caused by textures or lighting. This second approach has only been recently studied and some single-image vignetting correction methods have been proposed in the literature. First, Zheng et al. in [4] proposed a method based on a segmentation of the image into regions and a posterior fitting of vignetting functions to the image intensities within these regions. After that, again Zheng et al. in [5] presented a new method that uses the observed symmetry of radial gradient distributions in natural images. Another method, which is designed for microscopy images, was proposed by Leong et al. in [6]. There, the authors smooth the whole image with a Gaussian kernel to eliminate image structure and retain only the low-frequency intensity vignetting field. Other vignetting correction methods can be seen in [7–10].

In addition to the previous algorithms, in an unpublished article [11], a novel method for single-image vignetting correction through the constrained minimization of log-intensity entropy was proposed. The author applies the concept of information minimization to vignetting correction and uses a constrained radial polynomial vignetting function to correct the effect. The method proves to be faster and more accurate than Zheng's method based on the symmetry of radial gradient distributions [5], which is often used as the comparison benchmark. However, from our point of view, the method presents some mathematical inaccuracies which cause some undesired results and jeopardize the validity of the method. In spite of these technical problems, the underlying idea of the method is interesting and therefore, in this article, we will propose a new version of this algorithm fixing the technical problems of the original version. Furthermore, we will also add an improvement based on the search of the optical center of the image before applying the vignetting correction algorithm. In [11] as well as in most of the vignetting correction methods, the optical center of the image is supposed to be at the center of the image. However, in practice the optical center may lie at a considerable distance from the image center [12] and in such cases, the actual image optical center needs to be determined for accurate removal of vignetting effects.

The paper is organized as follows. First, in Section 2, the actual single-image vignetting correction based on the minimization of the log-intensity entropy will be recalled. In the next section, we will present the corrected and improved version of the algorithm. In Section 4, some comparison experiments will be carried out to check the performance of our proposal with respect to the original algorithm and Zheng's algorithm [5]. Finally, we end the paper with a section devoted to conclusions and the future work we want to develop.

2   Log-Intensity Entropy Minimization Algorithm

As we have already commented in the introduction, T. Rohlfing proposed in [11] a vignetting correction algorithm based on the constrained minimization of the log-intensity entropy. The algorithm relies on a method developed in [13] where the authors proved that the entropy of an image can be considered as an adequate optimization criterion for the model-free correction of shading artifacts and for microscopy image shading correction. The main hypothesis of that method is that the homogeneous objects have homogeneous intensities which correspond to a single peak and low entropy histogram. On the other hand, if there exists some gradual change of the intensity, as the one caused by vignetting, the peak of the histogram would not be as sharp as before and additional information would be added with an increment of the entropy. Thus, in order to reduce and correct this spatially varying intensity bias the minimization of the entropy becomes necessary.

At this point, for the sake of completeness and in order to fully understand the corrections and improvements that will be made to this algorithm in Section 3, let us recall the main steps of the original version. We refer the reader to [11] for further details. To correct vignetting from the image, Rohlfing's algorithm uses a sixth grade polynomial gain function g which depends on the distance from the image center to the pixel being treated. Specifically,

$$g_{a,b,c}(r) = 1 + ar^2 + br^4 + cr^6$$

where

$$r(i,j) = \frac{\sqrt{(i - \bar{i})^2 + (j - \bar{j})^2}}{\sqrt{\bar{i}^2 + \bar{j}^2}},$$

(i, j) is whichever pixel from the image and $(\bar{i}, \bar{j})$ is the image center. Therefore we have a function which depends on three real parameters a, b and c, verifying that $g(0)$ corresponds to the image center and $g(1)$ corresponds to the image corners. Once the gain function has been calculated it is multiplied by the original function, obtaining in this way an image with reduced vignetting. Namely,

$$\text{Final image}(i,j) = \text{Original image}(i,j)g_{a,b,c}(r).$$

Obviously, not all the values of the parameters a, b and c generate suitable vignetting correction functions. The vignetting effect is increasing as we move to the borders of the image and consequently, this function g must be strictly increasing for all $0 < r < 1$. In [11], the author determines that the function g will be strictly increasing if, and only if, the parameters satisfy one of the following conditions

$$
\begin{aligned}
C_1 &= (c \geq 0 \wedge 4b^2 - 12ac < 0), \\
C_2 &= (c \geq 0 \wedge 4b^2 - 12ac \geq 0 \wedge q_- \leq 0 \wedge q_+ \leq 0), \\
C_3 &= (c \geq 0 \wedge 4b^2 - 12ac \geq 0 \wedge q_- \geq 0 \wedge q_+ \geq 0), \\
C_4 &= (c < 0 \wedge q_- \leq 0 \wedge q_+ \geq 0), \\
C_5 &= (c < 0 \wedge q_- \geq 0 \wedge q_+ \leq 0),
\end{aligned}
$$

where q_+ and q_- are defined as

$$
q_+ = \frac{-2b + \sqrt{4b^2 - 12ac}}{6c}, \quad q_- = \frac{-2b - \sqrt{4b^2 - 12ac}}{6c}. \tag{1}
$$

The criterion for the determination of the optimal values of a, b and c among the ones which satisfy one of the above conditions was established as the minimization of the log-intensity entropy. Let us recall this concept. First, the luminance values $L = \{0, \ldots, 255\}$ are mapped to N histogram bins i using the function $i : L \to \mathbb{R}^+$ given by

$$
i(L) = (N - 1)\frac{\log(1 + L)}{\log 256}.
$$

Then the histogram bins n_k are computed using the following formula:

$$
n_k = \sum_{\substack{x,y: \\ \lfloor i(L(x,y)) \rfloor = k}} (1 + k - i(L(x,y))) + \sum_{\substack{x,y: \\ \lceil i(L(x,y)) \rceil = k}} (k - i(L(x,y))) \tag{2}
$$

At this point, to account for gaps in the image intensity distribution that can appear when scaling quantized data, the histogram is smoothed using a Gaussian kernel G_σ with standard deviation σ: $\hat{n} = n \star G_\sigma$. At this point, the discrete entropy is computed as

$$
H = \sum_k \hat{p}_k \log \hat{p}_k \tag{3}
$$

where $\hat{p}_k = \dfrac{\hat{n}_k}{\sum_j \hat{n}_j}$.

Finally, a hill climbing optimization algorithm is implemented to search in $\mathbb{R}^3$ among those triplets satisfying one of the conditions C_1-C_5 the optimal values (a, b, c) which minimize the log-intensity entropy H. Starting from $(0, 0, 0)$, each one of the parameters is increased and diminished independently by $\delta > 0$. From these 6 triplets, we compute the entropy of the cases satisfying one of the conditions and we update the optimal (a, b, c) to the one with the lowest entropy value. When none improves the entropy, δ is reduced by a factor $0 < k < 1$ and the process is repeated. If a new minimum value is achieved, δ is reset to its initial value. Otherwise, we continue reducing δ until we reach a prefixed value δ_0.

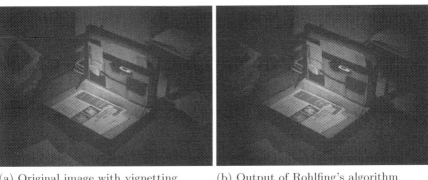

(a) Original image with vignetting
$H = 2.0643$

(b) Output of Rohlfing's algorithm
$H = 1.9127$

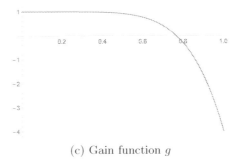

(c) Gain function g

Fig. 1. Undesired behaviour of the original algorithm based on the constrained minimization of the log-intensity [11]

3   A Corrected and Improved Algorithm

In this section, we will take a closer look to Rohlfing's algorithm and we will prove that it suffers from some mathematical inaccuracies that cause it to provide some undesired results with some images. As a matter of example, consider the triplet $a = b = 0$ and $c = -5$ which generates the gain function $g(r) = 1 - 5r^6$ which is clearly a strictly decreasing function as can be seen in Figure 1-(c). Note that this function satisfies conditions C_4 and C_5 but although it may reduce the entropy when it is applied to an image with high vignetting as in Figure 1-(a), it increases in fact the vignetting, darkening the whole image.

In addition to the previous problem, an exhaustive analysis of the conditions C_1-C_5 allows to deduce that some cases have been omitted and some others are not well-defined. Note that conditions C_2 and C_3 have non sense when $c = 0$ since q_+ and q_- cannot be defined. Therefore and with the aim to fully determine all the cases leading to suitable gain functions, let us correct these mistakes. Consider the gain function $g_{a,b,c}(r) = 1 + ar^2 + br^4 + cr^6$. To obtain a strictly increasing function, the first derivative must satisfy

$$g'(r) = 2ar + 4br^3 + 6cr^5 > 0$$

for all $0 < r < 1$. Since $r > 0$ and changing $r^2 = q$ we obtain the following inequality

$$a + 2bq + 3cq^2 > 0$$

for all $0 < q < 1$. Thus, the problem is reduced to study this general polynomial. It can be easily checked that this inequality holds if, and only if, one of the following nine conditions on the parameters holds:

[Horizontal positive line]	$C_1 = (a > 0 \wedge b = c = 0)$,
[Increasing line with non-positive root]	$C_2 = (a \geq 0 \wedge b > 0 \wedge c = 0)$,
[Decreasing line with root ≥ 1]	$C_3 = (c = 0 \wedge b < 0 \wedge -a \leq 2b)$,
[Convex parab. without roots]	$C_4 = (c > 0 \wedge b^2 < 3ac)$,
[Convex parab., only one non-positive root]	$C_5 = (c > 0 \wedge b^2 = 3ac \wedge b \geq 0)$,
[Convex parab., only one root ≥ 1]	$C_6 = (c > 0 \wedge b^2 = 3ac \wedge -b \geq 3c)$,
[Convex parab., non-positive highest root]	$C_7 = (c > 0 \wedge b^2 > 3ac \wedge q_+ \leq 0)$,
[Convex parab., lowest root ≥ 1]	$C_8 = (c > 0 \wedge b^2 > 3ac \wedge q_- \geq 1)$,
[Concave parab., lowest root ≤ 0,	$C_9 = (c < 0 \wedge b^2 > 3ac \wedge q_+ \geq 1 \wedge q_- \leq 0)$,
highest root ≥ 1]	

$$(4)$$

where q_+ and q_- are the roots of the polynomial given by Equations (1).

Once the conditions have been corrected, the algorithm does not obtain undesired results. However, as the original version of the algorithm, the method supposes the optical center of the image to be at the center of the image. This assumption leads to applying the gain function g from the center of the image $r = 0$ to the borders $r = 1$. However, it was shown in [12] that the optical center may not coincide with the center of the image and in fact, it could be quite displaced. Thus, it would be necessary to look for the optical center in order to compute the radial distance r of a pixel from the optical center rather than from the center of the image. Several techniques have been proposed for optical center estimation. Some estimate the optical center by locating the radial lens distortion [14] or the vanishing point [15], among other strategies. In this article, we will use a simple approach based on two steps. First, a low-pass Gaussian filter is applied to the image in order to extract the luminance pattern of the image. If the standard deviation of the Gaussian filter is very high, the objects and the structures of the image are removed or smoothed and the obtained image provides an estimation of the shading and luminance pattern of the original image. After that, we compute the center of mass CM of the image I using the following formula:

$$CM = \left(\frac{\sum\limits_{\substack{1 \leq i \leq N \\ 1 \leq j \leq M}} i \cdot I(i,j)}{\sum\limits_{\substack{1 \leq i \leq N \\ 1 \leq j \leq M}} I(i,j)} , \frac{\sum\limits_{\substack{1 \leq i \leq N \\ 1 \leq j \leq M}} j \cdot I(i,j)}{\sum\limits_{\substack{1 \leq i \leq N \\ 1 \leq j \leq M}} I(i,j)} \right) \qquad (5)$$

where I is a $N \times M$ image. In Figure 2, we can see how this method is able to detect the optical center of the image when it is not located at the center of the image. Now, we can modify the computation of the radial distance r to account for the optical center located in $CM = (CM_1, CM_2)$ by

$$r(i,j) = \frac{\sqrt{(i - CM_1)^2 + (j - CM_2)^2}}{\max\limits_{(v_1, v_2) \text{ vertex}} \sqrt{(v_1 - CM_1)^2 + (v_2 - CM_2)^2}}. \qquad (6)$$

Fig. 2. Computation of the center of mass as the optical center of the image. Red: optical center; Blue: center of the image.

To sum up, in Algorithm 1, we have included for a quick view the pseudo-code of the proposed algorithm, where the log-intensity entropy is computed using Equation (3).

Algorithm 1. Proposed vignetting correction algorithm.

Input: I image with vignetting
Output: F corrected image

1 Initial values $(a, b, c) = (0, 0, 0)$, $\delta = 8$, $H_{\min} = \text{log-entropy}(I)$, $F = I$;
2 Compute CM using Equation (5) and r using Equation (6);
3 **while** $\delta > \frac{1}{256}$ **do**
4 $\quad\big|\quad v_1 = (a + \delta, b, c), v_2 = (a - \delta, b, c), v_3 = (a, b + \delta, c)$;
5 $\quad\big|\quad v_4 = (a, b - \delta, c), v_5 = (a, b, c + \delta), v_6 = (a, b, c - \delta)$;
6 $\quad\big|\quad H = \min_{vi \text{ satisfies } C_1 \cup \ldots \cup C_9} \text{log-entropy}(I \cdot g_{a,b,c}(r))$;
7 $\quad\big|\quad$ **if** $H < H_{\min}$ **then** $H_{\min} = H$, $\quad F = I \cdot g_{a,b,c}(r)$ for the corresponding v_i;
8 $\quad\big|\quad \delta = 8$, $\quad (a, b, c) = v_i$ found;
9 $\quad\big|\quad$ **else**
10 $\quad\big|\quad\big\lfloor\ \delta = \frac{\delta}{2}$;

In Figure 3, we include a graphical example of how the minimization of the log-intensity entropy helps to correct the vignetting present in an image. While the presence of vignetting reduces the peak of the histogram, the algorithm is able to correct it and get an image, quite similar to the original one.

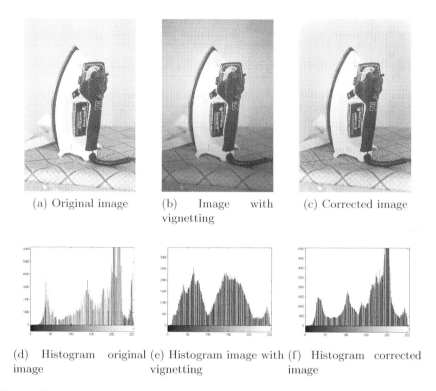

(a) Original image   (b) Image with vignetting   (c) Corrected image

(d) Histogram original image   (e) Histogram image with vignetting   (f) Histogram corrected image

Fig. 3. Correction of the histogram of the image with vignetting after applying the proposed algorithm

To end this section, in Figure 4, we include a visual example of the intermediate images obtained by the algorithm including the values of the log-intensity entropy and the triplet (a, b, c) used to correct it. Note that the image is progressively improved while the log-intensity entropy values decrease.

4   Experimental Results

In this section, we will check the performance of the vignetting correction algorithm proposed in Section 3 with respect to:

1. Rohlfing's algorithm [11], including the corrections on the conditions but without the optical center search,
2. Zheng's algorithm [5], which is often used as a benchmark algorithm in vignetting correction.

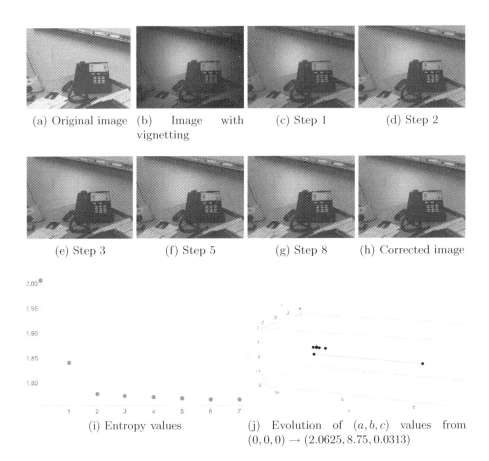

(a) Original image   (b)   Image   with   (c) Step 1   (d) Step 2
                               vignetting

(e) Step 3   (f) Step 5   (g) Step 8   (h) Corrected image

(i) Entropy values   (j) Evolution of (a, b, c) values from
                            $(0, 0, 0) \rightarrow (2.0625, 8.75, 0.0313)$

Fig. 4. Intermediate obtained images using the proposed algorithm and the evolution of the log-intensity entropy values and (a, b, c) triplets

All the computations has been carried out using MATLAB R2014a. We have used the implementation of the Zheng's algorithm available in [16], made by the same author and using the default parameter values.

The comparison have been made both from the visual point of view of the results and using objective quantitative performance measures. Nowadays, it is well-established in the literature that the visual inspection of the images obtained by several vignetting correction algorithms can not be the unique criterion with the aim of proving the superiority of one method with respect to the others. This is because each expert has different criteria and preferences and consequently, the reviews given by two experts can differ substantially. For this reason, the use of objective performance measures is growing in popularity to compare the results obtained by different vignetting correction algorithms. However, in order to be able to use some measures, we need an image database that includes for each image with vignetting, the corresponding original image without vignetting.

4.1   Image Database

The use of a quantitative measure to compare the output images of different methods implies the availability of a reference image to which compare the results obtained by the different algorithms. This reference image is considered as the perfect image that the algorithms should obtain. For vignetting correction purposes, the reference image must be an image without vignetting. For this reason, the addition of artificial vignetting to natural images with no visible vignetting is a reasonable way to obtain a suitable image database to use in this comparison setting.

We have chosen the first 15 images of the dataset of the University of South Florida[1] ([17]). The image dataset contains indoor and outdoor scenes, and natural and man-made objects. Each of the images contains a single object approximately centred in the image, appearing unoccluded and set against a natural background for the object. Some images have highlights, reflections or low resolution and a visual inspection has been performed to check the absence of vignetting. From these images, we have generated a new database adding artificial vignetting to these images using a simplified version of the Kang-Weiss model (see [18]). Let $A(r)$ be the off-axis illumination factor which is given by

$$A(r) = \frac{1}{\left(1 + \left(\frac{r}{f}\right)^2\right)^2}$$

where f is the effective focal length of the camera and r is the radial distance to the origin point where the factor is applied. Using this factor, the image with vignetting V from an image I is obtained by $V(i,j) = I(i,j) \cdot A(r)$. As the origin point where the factor is applied, we have randomly chosen a point located in the subimage from $\frac{N}{4}$ to $\frac{3N}{4}$ of width and from $\frac{M}{4}$ to $\frac{3M}{4}$ of height of the $N \times M$ image. Thus, the optical center of the image and the center of the image differ in general. Different values of $f = \{200, 300, 400, 500, 1000\}$ have been considered for the 15 images, generating 75 vignetting images.

4.2   Objective Measures

In addition to the visual comparison of the corrected images obtained by the algorithms, the performance will be quantitatively measured by two widely used performance objective measures, namely PSNR and SSIM. Let I_1 and I_2 be two images of dimensions $N \times M$. In the following, we suppose that I_1 is the original vignetting-free image and I_2 is the restored image for which some vignetting correction algorithm has been applied. The peak signal-to-noise ratio (PSNR) is defined by

$$PSNR(I_2, I_1) = 10 \log_{10}\left(\frac{R^2}{MSE(I_2, I_1)}\right) \tag{7}$$

---

[1] This image dataset can be downloaded from ftp://figment.csee.usf.edu/pub/ROC/edge_comparison_dataset.tar.gz

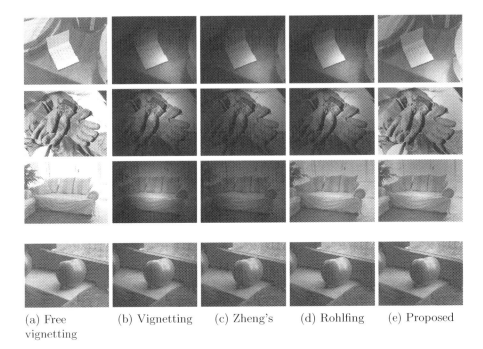

(a) Free   (b) Vignetting   (c) Zheng's   (d) Rohlfing   (e) Proposed
vignetting

Fig. 5. Comparison of the outputs of the different vignetting correction algorithms for several images

where R is the maximum fluctuation in the input image and MSE is the mean-squared error computed using the following expression:

$$MSE(I_2, I_1) = \frac{\sum_{\substack{1 \leq i \leq N \\ 1 \leq j \leq M}} (I_2(i,j) - I_1(i,j))^2}{N \times M}. \tag{8}$$

On the other hand, the structural similarity index measure (SSIM) was introduced in [19] under the assumption that human visual perception is highly adapted for extracting structural information from a scene. The measure is defined as follows:

$$\text{SSIM}(I_2, I_1) = \frac{(2\mu_1\mu_2 + C_1)}{(\mu_1^2 + \mu_2^2 + C_1)} \cdot \frac{(2\sigma_{12} + C_2)}{(\sigma_1^2 + \sigma_2^2 + C_2)}, \tag{9}$$

where for $k = 1, 2$, μ_k and σ_k^2 are the mean and the variance of each image, σ_{12} is the covariance between the two images, $C_1 = (0.01 \cdot 255)^2$ and $C_2 = (0.03 \cdot 255)^2$. Larger values of PSNR and SSIM ($0 \leq SSIM \leq 1$) are indicators of better capabilities for vignetting correction.

4.3   Comparison Results

We have applied the three considered vignetting correction algorithms to each vignetting image and we have obtained the values for the two considered objective

Table 1. Mean and standard deviation of the PSNR values obtained by the vignetting correction methods according to the value of f

f	PSNR Results							
	Vignetting		Zheng		Rohlfing		Proposed	
	Mean	σ	Mean	σ	Mean	σ	Mean	σ
200	8.0352	2.1118	7.9633	2.0659	9.5656	2.8170	12.2968	3.1512
300	10.0883	2.2567	10.0994	2.1933	10.9610	2.6213	13.0706	3.2286
400	12.8669	2.3230	13.0135	2.2496	15.5541	3.2876	17.2203	4.0243
500	14.3902	2.7940	14.5138	2.8246	16.5148	3.9082	16.9943	3.3770
1000	23.4953	3.1797	22.5514	3.1464	24.6718	3.3896	25.2753	3.6489

Table 2. Mean and standard deviation of the SSIM values obtained by the vignetting correction methods according to the value of f

f	SSIM Results							
	Vignetting		Zheng		Rohlfing		Proposed	
	Mean	σ	Mean	σ	Mean	σ	Mean	σ
200	0.3736	0.0558	0.3479	0.0609	0.5108	0.1541	0.7034	0.1018
300	0.5693	0.0497	0.5621	0.0526	0.6252	0.0636	0.7350	0.0868
400	0.7305	0.0520	0.7352	0.0539	0.8019	0.0692	0.8499	0.0481
500	0.8034	0.0364	0.8072	0.0382	0.8437	0.0451	0.8628	0.0291
1000	0.9158	0.0184	0.9120	0.0170	0.9194	0.0171	0.9212	0.0175

measures comparing the output to the original free-vignetting image. The results grouped according to the focal length f used to generate the image with vignetting are shown in Tables 1 and 2. As it can be seen, the proposed method outperforms the other methods severely, specially for low values of f which correspond to high levels of vignetting. Note also that the proposed algorithm and Rohlfing's one improve always the images from the quantitative point of view, obtaining better values of the measures with respect to the images with vignetting.

In Figure 5, some results are displayed comparing the output of the different algorithms. The proposed algorithm is able to reduce the vignetting even if the original image is corrupted by high amounts of it. Note that Zheng's algorithm gives results which although reducing somewhat the amount of vignetting, it changes completely the tone of the image, darkening it, specially for higher amounts of vignetting. Another conclusion which emerges from the images in the first two rows, in which vignetting is very off-centred, is the capability of the step included in the proposed algorithm to detect the optical center of the image. While the proposed algorithm applies the gaining function with origin at the optical center, Rohlfing's algorithm applies it at the center of the image being unable to reduce the vignetting uniformly in the whole image.

5 Conclusions and Future Work

Single-image vignetting correction is a useful technique to avoid the need of the calibration of the camera or the disposal of several images to correct vignetting. In this paper, we have deeply analysed the vignetting correction algorithm based on the minimization of the log-intensity entropy of the image, that was proposed in [11]. We have proved that the algorithm had several mathematical inaccuracies which caused to perform inappropriately in several images. Here, we have corrected these mistakes and we have proposed a revised version of the algorithm adding the improvement of the search of the optical center of the image to handle adequately images with off-centred vignetting. The comparison results ensure the potential of this algorithm from both the visual and the quantitative point of view.

As a future work, we want to generalize our algorithm to deal with color images and with images having more than one focus of vignetting. These last images correspond to images obtained using several illumination focuses.

Acknowledgments. This paper has been partially supported by the Spanish Grants TIN2013-42795-P and DPI2011-27977-C03-3.

References

1. Ray, S.F.: Applied Photographic Optics: Lenses and Optical Systems for Photography, Film, Video. Electronic and Digital Imaging, Focal (2002)
2. Goldman, D.B.: Vignette and exposure calibration and compensation. IEEE Transactions on Pattern Analysis and Machine Intelligence **32**(12), 2276–2288 (2010)
3. Jia, J., Tang, C.-K.: Tensor voting for image correction by global and local intensity alignment. IEEE Transactions on Pattern Analysis and Machine Intelligence **27**, 36–50 (2005)
4. Zheng, Y., Lin, S., Kambhamettu, C., Yu, J., Kang, S.B.: Single-image vignetting correction. IEEE Transactions on Pattern Analysis and Machine Intelligence **31**(12), 2243–2256 (2009)
5. Zheng, Y., Yu, J., Kang, S.B., Lin, S., Kambhamettu, C.: Single-image vignetting correction using radial gradient symmetry. In: IEEE Computer Society Conference on Computer Vision and Pattern Recognition (CVPR 2008), June 24–26, Anchorage, Alaska, USA, pp. 1–8. IEEE Computer Society (2008)
6. Leong, F.J. W-M., Brady, M., McGee, J.O'D.: Correction of uneven illumination (vignetting) in digital microscopy images. Journal of Clinical Pathology 56(8), 619–621 (2003)
7. Zheng, Y., Lin, S., Kang, S.B., Xiao, R., Gee, J.C., Kambhamettu, C.: Single-image vignetting correction from gradient distribution symmetries. IEEE Trans. Pattern Anal. Mach. Intell. **35**(6), 1480–1494 (2013)
8. Yu, W., Chung, Y., Soh, J.: Vignetting distortion correction method for high quality digital imaging. In: Proceedings of the 17th International Conference on Pattern Recognition, ICPR 2004, vol. 3, pp. 666–669 (2004)

9. He, K., Tang, P.-F., Liang, R.: Vignetting image correction based on gaussian quadrics fitting. In: Fifth International Conference on Natural Computation, ICNC 2009, vol. 5, pp. 158–161 (2009)
10. Cho, H., Lee, H., Lee, S.: Radial Bright Channel Prior for Single Image Vignetting Correction. In: Fleet, D., Pajdla, T., Schiele, B., Tuytelaars, T. (eds.) ECCV 2014, Part II. LNCS, vol. 8690, pp. 189–202. Springer, Heidelberg (2014)
11. Rohlfing, T.: Single-image vignetting correction by constrained minimization of log-intensity entropy (2012) (unpublished)
12. Lenz, R.K., Tsai, R.Y.: Techniques for calibration of the scale factor and image center for high accuracy 3-D machine vision metrology. IEEE Transactions on Pattern Analysis and Machine Intelligence **10**(5), 713–720 (1988)
13. Likar, B., Maintz, J.B., Viergever, M.A., Pernus, F.: Retrospective shading correction based on entropy minimization. Journal of Microscopy **197**(3), 285–295 (2000)
14. Willson, R.: Modeling and Calibration of Automated Zoom Lenses. PhD thesis, Robotics Institute, Carnegie Mellon University, Pittsburgh, PA (1994)
15. Wang, L.-L., Tsai, W.-H.: Computing camera parameters using vanishing-line information from a rectangular parallelepiped. Machine Vision and Applications **3**(3), 129–141 (1990)
16. Zheng, Y.: Matlab Central - nu_corrector (May 2010). http://www.mathworks.com/matlabcentral/fileexchange/27315-nu-corrector
17. Bowyer, K., Kranenburg, C., Dougherty, S.: Edge detector evaluation using empirical ROC curves. In: IEEE Conf. on Computer Vision and Pattern Recognition (CVPR 1999), vol. 1, pp. 354–359 (1999)
18. Kang, S.B., Weiss, R.: Can We Calibrate a Camera Using an Image of a Flat, Textureless Lambertian Surface? In: Vernon, D. (ed.) ECCV 2000. LNCS, vol. 1843, pp. 640–653. Springer, Heidelberg (2000)
19. Wang, Z., Bovik, A.C., Sheikh, H.R., Simoncelli, E.P.: Image quality assessment: From error visibility to structural similarity. IEEE Transactions on Image Processing **13**(4), 600–612 (2004)

Hybrid Dynamic Learning Systems for Regression

Kaushala Dias[✉] and Terry Windeatt

Centre for Vision Speech and Signal Processing, University of Surrey, Guildford, UK
{k.dias,t.windeatt}@surrey.ac.uk

Abstract. Methods of introducing diversity into ensemble learning predictors for regression problems are presented. Two methods are proposed in this paper, one involving pruning and the other a hybrid approach. In these ensemble learning approaches, diversity is introduced while simultaneously training, as part of the same learning process. Here not all members of the ensemble are trained in the same manner, but selectively trained, resulting in a diverse selection of ensemble members that have strengths in different parts of the training set. The result is that the prediction accuracy and generalization ability of the trained ensemble is enhanced. Pruning and hybrid heuristics attempt to combine accurate yet complementary members; therefore these methods enhance the performance by dynamically modifying the pruned aggregation through distributing the ensemble member selection over the entire dataset. A comparison is drawn with Negative Correlation Learning and a static ensemble pruning approach used in regression to highlight the performance improvement yielded by the dynamic methods. Experimental comparison is made using Multiple Layer Perceptron predictors on benchmark datasets, and on a signal calibration application.

Keywords: Ensemble methods · Ensemble pruning · Ensemble learning · Neural networks

1 Introduction

It is recognized in the context of ensemble methods that combined outputs of several predictors generally give improved accuracy compared to a single predictor [1]. Further performance improvements have also been shown by selecting ensemble members that are complementary [1]. The selection of classifiers, rather than regressors, has previously received more attention and given rise to many different approaches to pruning [3]. Some of these methods have been adapted to the regression problem [3]. The selection of ensemble members, has the potential advantage of both reduced ensemble size as well as improved accuracy.

The first proposed dynamic method, Ensemble Learning with Dynamic Ordered Pruning (ELDOP) for regression, uses the Reduced Error (RE) pruning method without back fitting (Section 3.1) for selecting the diverse members in the ensemble and only these are used for training [5]. To enhance the diversity, the selection and training of ensemble members are performed for every pattern in the training set.

The second method, Hybrid Ensemble Learning with Dynamic Ordered Selection (HELDOS), also uses the same RE method to select diverse members, but the ensemble is split into two sub ensembles, of which one contains the most diverse members

© Springer International Publishing Switzerland 2015
I. Rojas et al. (Eds.): IWANN 2015, Part II, LNCS 9095, pp. 464–476, 2015.
DOI: 10.1007/978-3-319-19222-2_39

and other less diverse. The most diverse members are then trained using Negative Correlation Learning (NCL) method [6] while the less diverse members are trained independently. Here again the selection and training of ensemble members are performed for every pattern in the training set.

By dynamic, we mean that the subset of predictors is chosen differently depending on its performance on the test sample. In ELDOP, given that only selected members of the ensemble are allowed to train for a given training pattern, the assumption is made that only a subset of the ensemble will perform well on a test sample. Therefore the method aims to automatically harness the ensemble diversity as a part of ensemble training. ELDOP is novel, since pruning occurs with training, and unlike [9], in the test phase there is no need to search for the closest training pattern. HELDOS harnesses diversity by training the ensemble members differently for a given training pattern. Training a subset of the ensemble using NCL, which relies on the correlation among the members of the sub-ensemble, and the rest independently, introduces variation into the way the members are trained. It is also assumed that a subset of the ensemble will perform well on a test sample. The novelty in HELDOS is due to its simultaneous training with varied training methods.

In contrast to ELDOP, HELDOS trains all members in the ensemble for a given training pattern. Therefore, for an epoch of training, a member is trained with varying number of times by the two methods. All members in the HELDOS method are trained equal number of times since all the members are trained for all the patterns in the training set. Whereas the number of times a member is trained in the ELDOP method, can vary from 1 to all pattern in an epoch. This is due to its member selection by ordering the members based on the RE method and selecting a certain percentage of the members for training. With this a member can be considered a strong or a weak predictor in different parts of the problem. This can contribute to the generalization ability of the member predictors. Therefore diversity by means of the training rate is introduced in this paper.

2   Related Research

The main objective of using ensemble methods in regression problems is to harness the complementarity and diversity of individual ensemble member predictions [1]. In [2] ordered aggregation pruning using Walsh coefficient has been suggested. In Negative Correlation Learning, diversity of the predictors is introduced by simultaneously training a collection of predictors using a cost function that includes a correlation penalty term [6]; thereby collectively enhancing the performance of the entire ensemble. Empirical evidence shows that this approach tends to over-fit, but with an additional regularization term, Multi-objective Regularized Negative Correlation Learning tackles over-fitting for noisy data. By weighting the outputs of the ensemble members before aggregating, an optimal set of weights is obtained in [8] by minimizing a function that estimates the generalization error of the ensemble; this optimization being achieved using genetic algorithms. With this approach, predictors with weights below a certain level are removed from the ensemble. In [7] a dynamic ensemble selection approach is described, in which many ensembles that perform well on an optimization set or a validation set are searched from a pool of over-produced ensembles and from this the best ensemble is selected using a selection function for computing the final output for the test sample. In [9], for ordered aggregation, dynamically selecting the ensemble order that has been defined by the ensemble member performance on the training set has shown to improve prediction accuracy; here the ensemble order of the training   pattern closest to the test pattern is searched and selected for the prediction phase. Here scaling factors come into effect when searching large training sets. Through instance selection [4], the training set is reduced by removing redundant or non-useful instances which improve prediction accuracy. The techniques used in instance selection can also potentially be useful in pruning to design ensembles with improved diversity [5].

3   Pruning and Hybrid Learning

3.1   Reduced Error Pruning

Reduced Error Pruning without back fitting method (RE) [3], modified for regression problems, is used to establish the order of predictors in the ensemble that produces a minimum in the ensemble training error. Starting with the predictor that produces the lowest training error, the remaining predictors are subsequently incorporated one at a time into the ensemble to achieve a minimum ensemble error. The sub ensemble S_u is constructed by incorporating to S_{u-1} the predictor that minimizes

$$S_u = arg_k \min u^{-1}\left(\sum_{i=1}^{u-1} C_{S_i} + C_k\right) \qquad (1)$$

Where, for M predictors $k \in (1,...,M)\backslash\{S_1, S_2,...,S_{u-1}\}$ and $\{S_1, S_2,...,S_{u-1}\}$ label predictors that have been incorporated in the pruned ensemble at iteration u-1. For the proposed methods C_i is calculated per individual training pattern and expressed as

$$C_i = f_i(x_n) - y_n \qquad (2)$$

where $i = 1,2,...,M$, the function $f_i(x)$ is the output of the ith predictor and (x_n, y_n) is the training data where n = $(1,2,...,N)$ patterns. Therefore the information required for the ordering of the training error is contained in the vector C. Therefore ordering is achieved using RE in both methods, ELDOP and HELDOS.

In ELDOP pruning is carried out by selecting a certain percentage of the predictors with the lowest error that are ordered in C to be trained using the Back Propagation update rule. This can be expressed as the error term used for the ith predictor for the nth pattern

$$E_{ni} = \mu_{ni}(f_i(x_n) - y_n) \qquad (3)$$

$$\text{where} \quad \mu_{ni} = 1 \quad \text{if } i \in P_n$$

$$\mu_{ni} = 0 \quad \text{otherwise}$$

P_n is defined as 50% of predictors with the lowest error, according to RE.

3.2 Hybrid Learning (HELDOS)

Negative Correlation Learning (NCL) attempts to train and combine individual pre-dictors in the same learning process. In NCL all the individual outputs are trained simultaneously through the correlation penalty terms in their error functions. The error function of an individual predictor contains the empirical risk function, which is the general error of the predictor, and the correlation penalty term. In NCL [6] the error term for the ith predictor for the nth pattern is expressed as

$$E_{ni} = (1 - \lambda)(f_i(x_n) - y_n) + \lambda(f(x_n) - y_n) \tag{4}$$

where $f(x_n)$ is the mean output of the selected predictors of the ensemble contribut-ing towards NCL and λ is the parameter $0 < \lambda < 1$ that is used to adjust the strength for the penalty term. The value of λ in this paper has been set to 1, so that equation (4) becomes

$$E_{ni} = (f(x_n) - y_n) \tag{5}$$

In the hybrid method after ordering the predictors using the RE method described in section 3.1, the ensemble members are split into two sub-ensembles. One sub-ensemble is trained using NCL, while the second is trained using the Back Propa-gation update rule. To accommodate the selection and training with the respective methods, equation (5) is modified to the following

$$E_{ni} = \mu_{ni}(f(x_n) - y_n) + (1 - \mu_{ni})(f_i(x_n) - y_n) \tag{6}$$

$$\text{where} \quad \mu_{ni} = 1 \quad \text{if } i \in P_n$$

$$\mu_{ni} = 0 \quad \text{otherwise}$$

P_n is defined as 50% of predictors with the lowest error, according to RE.

4 Methods

4.1 ELDOP Method

Dynamic selection of ensemble members provides an ensemble tailored to the specific test instance. The method described here is for a regression problem where the en-semble members are simultaneously ordered and trained on a pattern by pattern basis. The ordering of ensemble members is based on the method of RE and only the first 50% of the ordered members for a given training pattern are used for learning. Therefore diversity is encouraged by training half of the ensemble members that per-form well. The training continues until a pre-determined number of epochs of the training set are completed.

Training data $D = (x_n, y_n)$, where $n = (1,2,..,N)$ and f_m is an ensemble member, where $m = (1,2,..,M)$. S is a vector with max index of m.

1. **For** $n = 1....N$
2. $S \leftarrow$ empty vector
3.   **For** $m = 1...M$
4.   Evaluate $C_m = f_m(x_n) - y_n$
5.   **End for**
6.   **For** $u = 1...M$
7.   min $\leftarrow +\infty$
8.   **For** k in $(1,...,M)\backslash\{S_1, S_2,...,S_u\}$
9.   Evaluate $z = u^{-1}\left(\sum_{i=1}^{u-1} C_{S_i} + C_k\right)$
10.   **If** $z < $ min
11. S_u $\leftarrow k$
12.   min $\leftarrow z$
13.   **End if**
14.   **End for**
15.   **End for**
16. $E_{ni} = \mu_{ni}(f_i(x_n) - y_n)$
            where $\mu_{ni} = 1$ if $i \in P_n$
                    $\mu_{ni} = 0$ otherwise
            P_n is 50% of predictors with the lowest error.
17. **End for**

Fig. 1. Pseudo-code implementing the training process with ordered ensemble pruning per training pattern for the ELDOP method

The implementation of ELDOP method consists of two stages. First the base ensemble members M are ordered and trained on a pattern by pattern basis. As shown in the pseudo-code in figure 1, this is achieved by building a series of nested ensembles in which the ensemble of size u contains the ensemble of size u-1. Taking a single pattern of the training set, the method starts with an empty ensemble S, in step 2, and builds the ensemble order, in steps 6 to 15, by evaluating the training error of each predictor in M. The predictor that increases the ensemble training error least is iteratively added to S. This is achieved by minimizing z in step 9. Then the update rule is applied to the first 50% of the ordered ensemble member in S (step 16). Therefore in one epoch of training, the Back Propagation update rule would be applied a different number of times for each predictor, the more effective predictors being trained the most.

In the second stage, the ensemble output for each test pattern is evaluated. The assumption is made that the outputs of ensemble members that perform well for a test pattern would cluster together. Therefore the second stage starts by clustering the ensemble outputs into two clusters, which is shown in step 1 in figure 2. In step 2, the mean and the standard deviation are calculated. Taking the ensemble member outputs of each cluster, the outputs that are within one standard deviation from the mean are selected for the sub-cluster of each original cluster. This is denoted by Sk. Finally the mean of each of these sub-clusters is calculated as the outputs of the original clusters (step 10).

Ensemble member output for a test pattern (x_n, y_n) is f_m, where $m = (1,2,...,M)$.
f_j are ensemble member outputs in cluster C_k, $j = 1,2,...,J$ number of members.
μ_k, σ_k are the mean and the standard deviation of C_k
S_k is the sub-cluster in C_k
1.   Using K-means (K = 2) separate f_m into two clusters C_1, C_2
2.   Find mean and standard deviation of the two clusters; μ_1, σ_1, μ_2, σ_2,
3.   Calculate cluster mean as follows for each of the two clusters C_1, C_2:
4.   **For** $k = 1,2$
5.   **For** $j = 1....J$
6.   **If** $(\mu_k + \sigma_k) > f_j > (\mu_k - \sigma_k)$
7.   **Then** $S_k \leftarrow f_j$
8.   **End if**
9.   **End for**
10.   Evaluate   the mean of S_k ; $\bar{\bar{\mu}}_k$(This is the cluster output for comparison)

Fig. 2. Pseudo-code implementing the ensemble output evaluation for test pattern

4.2   HELDOS Method

HELDOS also provides an ensemble tailored to the specific test instance, for the regression problem. The ensemble members are simultaneously ordered and trained on a pattern by pattern basis. The ordering of ensemble members is based on the method of RE and the first 50% of the ordered members are trained using the NCL method and the rest are trained independently. Therefore diversity is encouraged by training subsets of the ensemble members with differing methods. The training continues until a pre-determined number of epochs of the training set is completed.

The implementation of HELDOS method also consists of two stages where the base ensemble members M are ordered and trained on a pattern by pattern basis. This is achieved by building a series of nested ensembles in which the ensemble of size u contains the ensemble of size u-1. This is shown in the pseudo code in figure 3, from steps 2 to 15. Here the predictor that increases the ensemble training error least is iteratively added to S. Taking a single pattern of the training set the ensemble order of M predictors is thus built and using this order the ensemble is split into two equal sub-ensembles (step 16). The first 50% of the ensemble members in S is trained using the Negative Correlation Learning method using equation (6). The second 50% of the ensemble members are trained independently using the Back Propagation update rule on their individual errors. Therefore in one epoch of training, the Negative Correlation Learning and Back Propagation update rule would be applied a different number of times for each predictor.

Training data $D = (x_n, y_n)$, where $n = (1,2,...,N)$ and f_m is an ensemble member, where $m = (1,2,...,M)$. S is a vector with max index of m.

1. **For** $n = 1....N$
2. $S \leftarrow$ empty vector
3.   **For** $m = 1...M$
4.   Evaluate $C_m = f_m(x_n) - y_n$
5.   **End for**
6.   **For** $u = 1...M$
7.   min $\leftarrow +\infty$
8.   **For** k in $(1,...,M)\backslash\{S_1, S_2,...,S_u\}$
9.   Evaluate $z = u^{-1}\left(\sum_{i=1}^{u-1} C_{S_i} + C_k\right)$
10.   **If** $z < $ min
11. S_u $\leftarrow k$
12.   min $\leftarrow z$
13.   **End if**
14.   **End for**
15.   **End for**
16. $E_{ni} = \mu_{ni}(f(x_n) - y_n) + (1 - \mu_{ni})(f_i(x_n) - y_n)$
            where $\mu_{ni} = 1$ if $i \in P_n$
                    $\mu_{ni} = 0$ otherwise
            P_n is 50% of predictors with the lowest error.
17. **End for**

Fig. 3. Pseudo-code implementing the training process with ordered ensemble pruning per training pattern for the HELDOS method

The second stage of this method follows the same process described in ELDOP. The ensemble output for each test pattern is evaluated and based on the assumption that outputs of the ensemble members that perform well cluster together, the second stage starts by clustering the ensemble outputs into two clusters. This is shown in step 1 in figure 2. In step 2, the mean and the standard deviation are calculated. The ensemble member outputs that are within one standard deviation from the mean are selected for the sub-cluster of each original cluster. This is denoted by S_k. Finally the mean of each of these sub-clusters is calculated as the outputs of the original clusters, as shown in step 10.

5   Results

Neural Networks with Multi-Layer Perceptron (MLP) architecture with 5 nodes in the hidden layer as described in [3] has been selected in this experiment. The training and test pattern split is 70% and 30% respectively. In this experiment 32 base predictors are trained with identical training samples. This is to avoid the diversity being introduced by random selections of the training patterns. The Mean Squared Error (MSE) is used as the performance indicator for both training and test sets, and averaged over 10 iterations. Training is stopped after fifty epochs.

Table 1 shows MSE performance comparison of Negative Correlation Learning (NCL) [6], Ordered Aggregation (OA) [3], Dynamic Ensemble Selection and Instantaneous Pruning (DESIP) [9] and the two proposed methods of Ensemble Learning with Dynamic Ordered Pruning (ELDOP) and Hybrid Ensemble Learning with Dynamic Ordered Selection (HELDOS). In table 1, grayed results indicate the minimum MSE over the five methods for the datasets, shown in table 2. It is observed that the lowest MSE values have been achieved by ELDOP or HELDOS. Figures 4 and 5 show the comparison of the training and test error plots with ensemble size for NCL, ELDOP and HELDOS. It is observed that ensembles with ELDOP and HELDOS are more accurate with fewer members than the other methods.

Table 1. Averaged MSE with Standard Deviation for 10 iterations for NCL, OA, DESIP, ELDOP and HELDOS

Dataset	Multiplier	NCL	OA	DESIP	ELDOP	HELDOS
Servo	10^0	0.25±0.49	1.35±1.69	0.14±0.24	0.10±0.14	0.21±0.46
Wisconsin	10^1	2.89±7.63	2.82±6.81	2.37±5.21	0.64±1.71	1.70±5.38
Concrete Slump	10^1	4.39±6.69	4.81±7.37	4.03±5.99	1.15±1.62	2.86±5.31
Auto93	10^2	0.52±1.57	1.02±2.73	0.72±1.92	0.45±1.50	0.35±1.10
Body Fat	10^1	0.10±0.34	3.66±4.62	0.09±0.32	0.29±0.52	0.06±0.23
Bolts	10^2	0.94±1.71	2.71±2.27	0.79±1.22	0.66±0.76	0.54±0.60
Pollution	10^3	1.99±3.38	3.57±5.56	1.70±2.68	2.14±3.19	1.39±2.74
Auto Price	10^7	0.36±0.71	1.09±1.84	0.46±0.70	0.20±0.32	0.17±0.47

Table 2. Benchmark Datasets

Dataset	Instances	Attributes	Source
Servo	167	5	UCI-Repository
Wisconsin	198	36	UCI-Repository
Concrete Slump	103	8	UCI-Repository
Auto93	82	20	WEKA
Body Fat	252	15	WEKA
Bolts	40	8	WEKA
Pollution	60	16	WEKA
Auto Price	159	16	WEKA

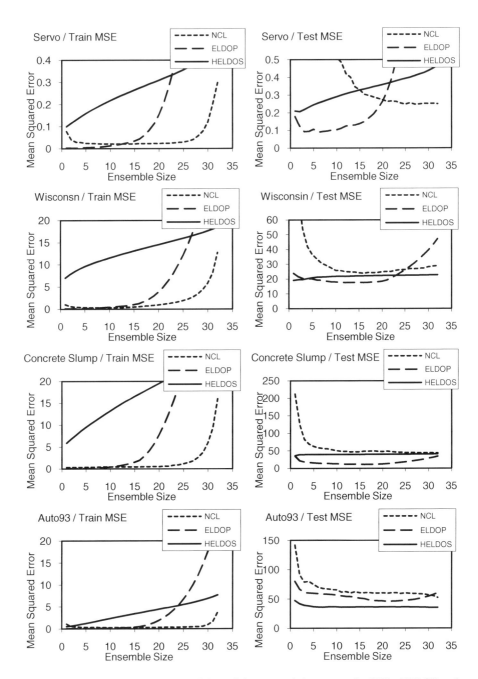

Fig. 4. Comparison of the MSE plots of the training set and the test set for NCL, ELDOP and HELDOS

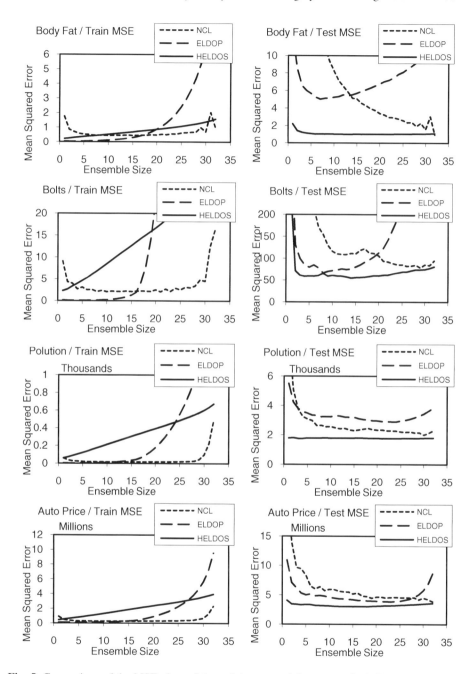

Fig. 5. Comparison of the MSE plots of the training set and the test set for NCL, ELDOP and HELDOS

6   Signal Calibration Application

In this application the output of a Radio Frequency Source Device used in a military environment is calibrated to give out a signal with a variable amplitude level, where the amplitude level is specified by the user of the device. A learning based system is trained with measurements taken from the un-calibrated device, which is then used to predict values for the intermediate levels. These predicted values are then used as correction values in the calibration process. Here the accuracy of the predictions is important in order to meet tight specification requirements, since a large error in the prediction would produce undesirable behavior by the device, rendering the device unsuitable for its intended use. Therefore in the real world this learning based application would be required to operate with its own tight specification requirements.

Due to the compression characteristics of the Radio Frequency amplifiers in the source device, the output shows nonlinear behavior. This behavior is compounded by the influence of temperature on the amplifiers. Therefore such devices are calibrated using a multi-dimensional look-up table that contains the necessary information to correct the nonlinearity. However the resolution of the output can be limited due to the size of the look-up table. A learning based system can be beneficial in this situation, since the resolution of the output is not limited by the size of memory used by the predictors. Predictors are able to interpolate values with any resolution given that they have adequate training data. However the accuracy of the output can be limited by the size of memory used by the predictors and the method of training. Therefore care needs to be taken when designing predictors for increased accuracy; the number of nodes, the number of layers and the training algorithm in a neural network can affect efficiency as well as accuracy.

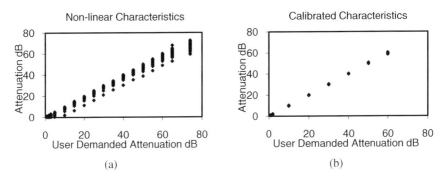

(a)   (b)

Fig. 6. Non-linear characteristics and the calibrated characteristics of the Radio Frequency Source Device

The learning base system with neural networks is used to learn the non-linear characteristics of the Radio Frequency Source Device. In figure 6(a) the non-linear characteristics of the device's attenuation is shown. Here each point in the graph represents the measured attenuation level for a given frequency, for an attenuation demand and the temperature at which the amplifiers are operating. The spread of the points in the vertical direction shows the level of non-linearity in the measured data.

The application then utilizes the frequency, user demanded attenuation level and the temperature of the amplifiers as input features to the neural networks and the device output attenuation level as the outputs of the neural network to train an ensemble of neural networks. The predictors thus produced are then utilized to correct for the output of the device. Figure 6(b) shows the calibrated output of the Radio Frequency Source Device.

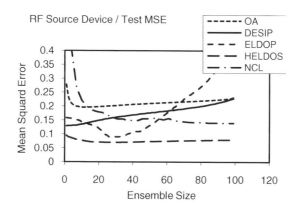

Fig. 7. Calibration application MSE performance of NCL, OA, DESIP, ELDOP and HELDOS

For this application a neural network with 10 hidden layer nodes was used. The data set consisted of 660 training instances and 630 test instances. Here ensemble methods NCL, OA, DESIP, ELDOP and HELDOP have been applied to obtain the performance differences and the results follow a similar trend to the performance on the benchmark data. This is shown in figure 7 and the minimum MSE is shown in table 3.

Table 3. Average MSE with standard deviation for 10 iterations

NCL	OA	DESIP	ELDOP	HELDOS
0.14±0.06	0.20±0.07	0.13±0.03	0.09±0.05	0.07±0.03

7   Conclusion

Dynamic pruning utilizes a distributed approach to ensemble selection and is an active area of research for both classification and regression problems. Unlike static methods, dynamic methods provide a solution tailored to a specific instance. In this paper two novel dynamic methods are introduced, one which combines ensemble learning with dynamic pruning of regression ensembles and the other that combines learning approaches for regression problems. Experimental results show that test error has been reduced in both methods by introducing pruning in the training phase of ensembles as well as combining two learning methods in the training phase. The motivation for the proposed methods is based on the introduction of diversity through varied learning methods [5].

In DESIP [9] the ensemble selection for a test pattern is based on the closest training instance and therefore a search is necessary to determine the pruned ensemble, while in ELDOP the ensemble is trained with the pruned selection, therefore eliminating the need to search. In NCL, OA, DESIP and HELDOS the entire ensemble is utilized in training, while ELDOP trains only the selected members of the ensemble, with a commensurate reduction in training time. For the data sets used in this experiment, HELDOS or ELDOP achieves best results, but which of the two methods performs best is data dependent.

As further work, varying the percentage of split for the ensemble member selection and varying the value of K in K-means clustering will be investigated. Bias/Variance and time complexity analysis should also help to understand the performance relative to other ensemble methods with similar complexity.

References

1. Tsoumakas, G., Partalas, I., Vlahavas, I.: An Ensemble Pruning Primer. In: Okun, O., Valentini, G. (eds.) Applications of Supervised and Unsupervised Ensemble Methods. SCI, vol. 245, pp. 1–13. Springer, Heidelberg (2009)
2. Windeatt, T., Zor, C.: Ensemble Pruning using Spectral Coefficients. IEEE Trans. Neural Network. Learning Syst. **24**(4), 673–678 (2013)
3. Hernández-Lobato, D., Martínez-Muñoz, G.: Suárez A, Empirical Analysis and Evaluation of Approximate Techniques for Pruning Regression Bagging Ensembles. Neurocomputing **74**, 2250–2264 (2011)
4. Olvera-Lopez J., Carrasco-Ochao J., Martinez-Trinidad J., Kittler J.: A review of instance selection methods. Artificial Intelligence Review **34**(2) 133–143 (2010)
5. Brown, G., Wyatt, J., Harris, R., Yao, X.: Diversity creation methods: a survey and categorization. Information Fusion **6**(1), 5–20 (2005)
6. Chen H., Yao X.: Multiobjective Neural Network Ensembles Based on Regularized Negative Correlation Learning. IEEE Trans. Knowledge and Data Engineering **22**(12), 1738–1751 (2010)
7. Dos Santos, E.M., Sabourin, R., Maupin, P.: A Dynamic Overproduce-and-choose Strategy for the selection of Classifier Ensembles. Pattern Recognition **41**, 2993–3009 (2008)
8. Zhau, Z.-H., Wu, J., Tang, W.: Ensembling Neural Networks: many could be better than all. Artificial Intelligence **137**, 239–263 (2002)
9. Dias K., Windeatt T.: Dynamic Ensemble Selection and Instantaneous Pruning for Regression. In: European Symposium on Artificial Neural Networks, Computational Intelligence and Machine Learning, ESANN 2014, pp. 643–648 (2014)

A Novel Algorithm to Train Multilayer Hardlimit Neural Networks Based on a Mixed Integer Linear Program Model

Jose B. da Fonseca[✉]

Faculty of Sciences and Technology, New University of Lisbon,
2829-615 Caparica, Portugal
jbfo@fct.unl.pt
http://www.dee.fct.unl.pt

Abstract. In a previous work we showed that hardlimit multilayer neural networks have more computational power than sigmoidal multilayer neural networks [1]. In 1962 Minsky and Papert showed the limitations of a single perceptron which can only solve linearly separable classification problems and since at that time there was no algorithm to find the weights of a multilayer hardlimit perceptron research on neural networks stagnated until the early eighties with the invention of the Backpropagation algorithm [2]. Nevertheless since the sixties there have arisen some proposals of algorithms to implement logical functions with threshold elements or hardlimit neurons that could have been adapted to classification problems with multilayer hardlimit perceptrons and this way the stagnation of research on neural networks could have been avoided. Although the problem of training a hardlimit neural network is NP-Complete, our algorithm based on mathematical programming, a mixed integer linear model (MILP), takes few seconds to train the two input XOR function and a simple logical function of three variables with two minterms. Since any linearly separable logical function can be implemented by a perceptron with integer weights, varying them between -1 and 1 we found all the 10 possible solutions for the implementation of the two input XOR function and all the 14 and 18 possible solutions for the implementation of two logical functions of three variables, respectively, with a two layer architecture, with two neurons in the first layer. We describe our MILP model and show why it consumes a lot of computational resources, even a small hardlimit neural network translates into a MILP model greater than 1G, implying the use of a more powerful computer than a common 32 bits PC. We consider the reduction of computational resources as the near future work main objective to improve our novel MILP model and we will also try a nonlinear version of our algorithm based on a MINLP model that will consume less memory.

Keywords: Hardlimit neural networks · Mixed integer linear programming · Training a hardlimit neural network with a MILP model · Solving a MILP model with the cplex solver

© Springer International Publishing Switzerland 2015
I. Rojas et al. (Eds.): IWANN 2015, Part II, LNCS 9095, pp. 477–487, 2015.
DOI: 10.1007/978-3-319-19222-2_40

1   Introduction

In their famous book Perceptrons [2] Minsky and Papert showed that a single perceptron can only solve linearly separable classification problems and to solve a nonlinearly separable classification problem they would need a multilayer perceptron. Since at that time, 1962, there was no algorithm to find the weights of a multilayer perceptron, research on neural networks stagnated until the early eighties with the invention of the Backpropagation algorithm. Nevertheless since the early sixties there appeared many algorithms to train hardlimit neural networks or multilayer threshold gates to implement logical functions that could have been adapted to solve nonlinearly separable classification problems. The first published work on the synthesis of multilayer threshold gates to implement logical functions seems to be [3]. So why to create one more algorithm? Because it is shown that the training of a multilayer hardlimit perceptron is NP-Complete [4,5] and we will show that our solution is more efficient than previous algorithms in terms of runtime, although it consumes a lot of memory that increases exponentially with the network size. Nevertheless our MILP model is simple and beautiful with only two constraints and given a set of possible integer values of weights it always finds all the possible solutions with a given architecture. We could have used a nonlinear model that would consume much less memory since the weights could be, in this case, represented by a simple array of integers, since it is shown that any logical function can be implemented by a multilayer hardlimit neural network with integer weights [4]. Nevertheless the nonlinear model could be trapped in a local optimum and the nonlinear model needs an initial feasible solution. On the contrary our MILP model always finds the global optimum since it is a linear model. We have already done some experiments with a nonlinear MNILP model but the results are very poor: the model did not converge to any solution using all possible nonlinear solvers.

2   What is Mathematical Programming? What is a MILP Model?

A mathematical programming model is a set of constraints over the variables of the model subjected to a maximization or minimization of an objective variable also defined in terms of the model variables. A linear mathematical program model is a mathematical program where all constraints are linear and also the expression that defines the objective variable. A mixed integer linear program model (MILP) is a linear mathematical programming model with integer and binary variables. A linear model can neither have the multiplication or division of two variables nor a nonlinear operation over a model variable. To solve the MILP model we use a black box algorithm, the solver, normally the Cplex solver [6], through a modelling language, the GAMS software, that makes transparent the use of the solver by a non specialist.

3   Description of a MILP Model to Train Hardlimit Multilayer Neural Networks

The main idea behind our work is expressed by the following theorem:

Theorem 1. Any linearly separable logical function of n variables can be computed by a Rosenblatt neuron or threshold gate with integer weights such that

$$|w_i| \le \frac{(n+1)^{(n+1)/2}}{2^n}, i = 0, ..., n \tag{1}$$

For $n=3$, this theorem guarantees that all linearly separable logical functions of three variables can be implemented by a single Rosenblatt perceptron or threshold gate with integer weights varying between -2 and 2. You can find a demonstration of theorem 1 in [4]. Since there are only 104 linearly separable functions over 256 possible functions of three variables, we considered an architecture with two layers, two hidden neurons in the first layer and an output neuron, to implement a two input XOR and a three variable function. Preliminary results seem to show that we can implement almost all logical functions of three variables with this architecture with the integer weights varying only between -1 and 1 to reduce the model size, but we found one counter example $f=m_1+m_2+m_3+m_4$ where the MILP did not find any solution for this architecture. Varying the integer weights between -2 and 2 we exhausted the memory because the model takes a dimension of 1.8G and the GAMS software on a 32 bits machine only uses 2G RAM. In a near future we plan to install GAMS on a 64 bits machine that will not have this memory limit.

The *brain* of our MILP model is an indexed binary variable, $weights(w_i, bias_j)$, that assumes the value 1 for all combinations of integer weights that are considered solutions of the implementation of the logical function. For the case of a two input XOR defined by (2) the solutions are generated by the set of linear constraints (3).

$$xor(b_0, b_1) \equiv b_0 \ne b_1 \equiv b_0 \bar{b}_1 + \bar{b}_0 b_1 \tag{2}$$

$$\forall w_{ij}, bias_i, b_0, b_1 : \tag{3}$$
$$weights(w_{ij}) f(w_{ij}, bias_i, b_0, b_1) = xor\_2in(b_0, b_1) weights(w_{ij}, bias_i)$$

The set of *linear* constraints (3) are equivalent to the simpler set of *nonlinear* constraints (4), since we make a restriction based on a logical operation over a model variable. In this sense (3) results from a linearization of (4).

$$\forall w_{ij}, bias_i, b_0, b_1 \setminus \{weights(w_{ij}, bias_i) = 0\} : \tag{4}$$
$$f(w_{ij}, bias_i, b_0, b_1) = xor(b_0, b_1)$$

Since we want to obtain all possible solutions of the XOR with a given architecture and integer weights varying between -1 and 1, we must maximize the number of solutions, the objective variable which is defined by (5).

$$obj = \sum_{w_{ij},bias_i} weights(w_{ij}, bias_i) \qquad (5)$$

For the case of a XOR function (5) is implemented by the following line of GAMS code: calc_obj.. sum((w11, w12, w13, w14, w21,w22, bias1, bias2, bias3), weights(w11, w12, w13, w14, w21, w22, bias1, bias2, bias3))=e=obj;

So we linearized a very nonlinear set of constraints. But the price we paid is very high: since the binary variable weights and the constraint calc_ws1 are indexed by all the weights and input variables in this simple example we will have a search space with dimension 39x4=78732, assuming only three possible values for the weights -1, 0, 1 which generates a model with 18M. But if we increase the domain of variation of weights to -2,-1,0,1,2 we will have a search space with dimension 59x4= 7812500 which corresponds to a model with 1.3G and we got an out of memory message in a machine with 2G RAM. The set of linear constraints (4) is implemented by the following GAMS code:

calc_ws1(w11,w12,w13,w14,w21,w22,bias1,bias2,bias3,b0,b1).. weights(w11,w12, w13, w14, w21, w22, bias1, bias2, bias3) * f(w11, w12, w13, w14, w21, w22, bias1, bias2, bias3, b0, b1) =e= xor_2in(b0,b1) * weights(w11, w12, w13, w14, w21, w22, bias1, bias2, bias3);

When weights(w11,w12,w13,w14,w21,w22,bias1,bias2,bias3)=0 the constraint is relaxed and equivalent to 0=e=0 and when weights(w11, w12, w13, w14, w21, w22, bias1, bias2, bias3)=1 the constraint imposes that the output of the neural network, f(w11, w12, w13, w14, w21, w22, bias1, bias2, bias3, b0, b1), must be equal to the desired output xor_2in(b0, b1) for all combinations of (b0, b1), i.e. (w11, w12, w13, w14, w21, w22, bias1, bias2, bias3) is a solution of the weights that implement a two input xor. We have already developed a first version of a nonlinear model to solve this problem but although the model got smaller, about 5M, all nonlinear solvers we have access did not reach any solution.

Function f, the output of the multilayer perceptron, is defined as a parameter given by (6).

$$f = ((\qquad (6)$$
$$(((b_0 - 1)(w_{11} - 2) + (b_1 - 1)(w_{12} - 2) + bias_1 - 2 \geq 0)(w_{21} - 2) +$$
$$(b_0 - 1)(w_{13} - 2) + (b_1 - 1)(w_{14} - 2) + bias_2 - 2 \geq 0)(w_{22} - 2) +$$
$$bias_3 - 2)) \geq 0$$

Note that the subtraction of the weights by 2 is only correct for weights varying in {-1,0,1}. For a greater set of variation we must increase this constant. Equation (6) is implemented by the single line of GAMS code:
f(w11,w12,w13,w14, w21,w22, bias1, bias2, bias3, b0, b1)= (((((((ord(b0)-1) * (ord(w11)-2) + (ord(b1)-1)*(ord(w12)-2) + ord(bias1)-2) ge 0) * (ord(w21)-2) + (((ord(b0)-1) * (ord(w13)-2) + (ord(b1)-1)*(ord(w14)-2) + ord(bias2)-2) ge 0) * (ord(w22)-2) + ord(bias3)-2)) ge 0);

3.1   Implementing a Three Input Logical Function

As a second example we will show how to obtain the weights of a two layer hardlimit perceptron, with two neurons in the first layer, to implement a three input logical function defined by (7) and the weights varying between -1 and 1 to reduce the combinatorial explosion. We use two layers of neurons because from the 256 possible functions of three variables only 106 are linearly separable and so the remaining cannot be implemented by a single neuron. Augmenting the number of neurons in the first layer and the weight variation domain to -2..2 we got an out of memory with a 2G RAM PC and so there are lots of three input logical functions that still cannot be implemented with this architecture since to implement all the nonlinearly separable logical functions of three variables we will need more neurons in the first layer and to augment the variation domain to -2..2 to guarantee the implementation of minterms in the first layer following theorem 1. The sets of constraints (8) impose that for all solutions the output of the network must be equal to the desired function.

$$f(b_0, b_1, b_2) = m_1 + m_4 = \bar{b_2}\bar{b_1}b_0 + b_2\bar{b_1}\bar{b_0} \tag{7}$$

$$\forall w_{ij}, bias_i, b_0, b_1, b_2: \tag{8}$$
$$weights(w_{ij}, bias_i)f(w_{ij}, bias_i, b_0, b_1, b_2) =$$
$$weights(w_{ij}, bias_i)f3in(w_{ij}, bias_i, b_0, b_1, b_2)$$

4   Conclusions and Future Work

Although there exist many published algorithms to train multilayer threshold gates or hardlimit neural networks to implement logical functions, we showed that our algorithm, to our knowledge the first in the literature to be based on a MILP model and solved by the Cplex solver, has better runtimes and is simpler than previous works and can also solve classification problems. In a near future we plan to implement a nonlinear MINLP model that will consume much less memory, although it will have the drawbacks of the possibility of getting trapped in a local optimum and the necessity of an initial feasible suboptimal solution.

References

1. Barahona da Fonseca, J.: Are Rosenblatt multilayer perceptrons more powerfull than sigmoidal multilayer perceptrons? From a counter example to a general result. In: Proceedings of ESANN 2013, Ciaco, Belgium, pp. 345–350 (2013)
2. Minsky, M., Papert, S.: Perceptrons. MIT Press, Massachusetts (1965)
3. Gonzalez, R., Lawler, E.L.: Two-level threshold minimization. In: Proceedings of the Sixth Annual Symposium on Switching Circuit Theory and Logical Design, 41–4 (1965)

4. Orponen, P.: Computational Complexity of Neural Networks: a Survey. NC-TR-94-010 Technical Report, Department of Computer Science, University of London, England (1994)
5. Blum, A.L., Rivest, R.L.: Training a 3-node neural network is NP-complete. Lecture Notes in Computer Science, 661 (1993)
6. ILOG: Cplex Solver (2005)

A GAMS code of the First MILP Model to Implement a two input XOR with a two layer Threshold Gate

```
* XOR 2in

sets b0 /0*1/;
alias(b1,b0);

sets w11 /1*3/;
alias( w12, w13, w14, w21, w22, bias1, bias2, bias3, w11);


Parameter xor_2in(b0,b1), f(w11,w12,w13,w14, w21,w22, bias1, bias2, bias3, b0, b1);

*Definition of a 2 Input XOR:
xor_2in(b0,b1)=( ord(b0) ne ord(b1));

* Output of the Neural Network:
f(w11,w12,w13,w14, w21,w22, bias1, bias2, bias3, b0, b1)=
( ((
( ( ( ord(b0)-1 ) * (ord(w11)-2) + (ord(b1)-1)*(ord(w12)-2) +
ord(bias1)-2 )  ge 0) * (ord(w21)-2) +
( ( ( ord(b0)-1 ) * (ord(w13)-2) + (ord(b1)-1)*(ord(w14)-2) + ord(bias2)-2 )  ge 0) * (ord(w22)-2) +
ord(bias3)-2 ) ) ge 0 ) ;

variable obj;

binary variable
weights(w11,w12,w13,w14, w21,w22, bias1, bias2, bias3);

**CONSTRAINTS**

calc_ws1(w11,w12,w13,w14,w21,w22,bias1,bias2,bias3,b0,b1)..

calc_ws1(w11,w12,w13,w14,w21,w22,bias1,bias2,bias3,b0,b1)..
weights(w11,w12,w13,w14,w21,w22,bias1,bias2,bias3)*
f(w11,w12,w13,w14, w21,w22, bias1, bias2, bias3, b0, b1)=e=
xor_2in(b0,b1)*weights(w11,w12,w13,w14,w21,w22,bias1,bias2,bias3);

calc_obj.. obj=e=
sum( (w11,w12,w13,w14, w21,w22, bias1, bias2, bias3),
weights(w11,w12,w13,w14, w21,w22, bias1, bias2, bias3) );

Model Xor2in /all/;

Solve Xor2in using mip maximizing obj;

display weights.l, obj.l;
```

B Output of a Run of the First MILP Model

```
----      49 VARIABLE weights.L

INDEX 1 = 1  INDEX 2 = 1  INDEX 3 = 1  INDEX 4 = 1  INDEX 5 = 1  INDEX 6 = 3
             1
2.3      1.000
INDEX 1 = 1  INDEX 2 = 1  INDEX 3 = 1  INDEX 4 = 1  INDEX 5 = 3  INDEX 6 = 1
             1
3.2      1.000
INDEX 1 = 1  INDEX 2 = 3  INDEX 3 = 1  INDEX 4 = 3  INDEX 5 = 1  INDEX 6 = 3
             2
2.1      1.000
INDEX 1 = 1  INDEX 2 = 3  INDEX 3 = 1  INDEX 4 = 3  INDEX 5 = 3  INDEX 6 = 1
             2
```

```
1.2      1.000
INDEX 1 = 1  INDEX 2 = 3  INDEX 3 = 3  INDEX 4 = 1  INDEX 5 = 1  INDEX 6 = 1
                3
2.2      1.000
INDEX 1 = 1  INDEX 2 = 3  INDEX 3 = 3  INDEX 4 = 1  INDEX 5 = 3  INDEX 6 = 3
                1
1.1      1.000
INDEX 1 = 3  INDEX 2 = 1  INDEX 3 = 1  INDEX 4 = 3  INDEX 5 = 1  INDEX 6 = 1
                3
2.2      1.000
INDEX 1 = 3  INDEX 2 = 1  INDEX 3 = 1  INDEX 4 = 3  INDEX 5 = 3  INDEX 6 = 3
                1
1.1      1.000
INDEX 1 = 3  INDEX 2 = 1  INDEX 3 = 3  INDEX 4 = 3  INDEX 5 = 1  INDEX 6 = 3
                2
2.1      1.000
INDEX 1 = 3  INDEX 2 = 1  INDEX 3 = 3  INDEX 4 = 1  INDEX 5 = 3  INDEX 6 = 1
                2
1.2      1.000
----     49 VARIABLE obj.L               =        10.000
```

These 10 solutions correspond to the following sets of weight values, respectively:

```
w11= -1  w12= -1  w13= -1  w14= -1  w21= -1  w22= 1
bias1=0 bias2=1 bias3=-1

w11= -1  w12= -1  w13= -1  w14= -1  w21= 1  w22= -1
bias1=1 bias2=1 bias3=-1

w11= -1  w12= 1  w13= -1  w14= 1  w21= -1  w22= 1
bias1=0 bias2=-1 bias3=0

w11= -1  w12= 1  w13= -1  w14= 1  w21= 1  w22= -1
bias1=-1 bias2=0 bias3=0

w11= -1  w12= 1  w13= 1  w14= -1  w21= -1  w22= -1
bias1=0 bias2=0 bias3=1

w11= -1  w12= 1  w13= 1  w14= -1  w21= 1  w22= 1
bias1=-1 bias2=-1 bias3=-1
```
This solution corresponds to the first direct implementation of the XOR, the two neurons in the first layer implement the two minterms, m2 and m1 respectively, and the third neuron implement the OR of these minterms resulting in the two input xor

```
w11= 1  w12= -1  w13= -1  w14= 1  w21= -1  w22= -1
bias1=0 bias2=0 bias3=1

w11= 1  w12= -1  w13= -1  w14= 1  w21= 1  w22= 1
bias1=-1 bias2=-1 bias3=-1
```
This solution also corresponds to the second direct implementation of the XOR, the two neurons in the first layer implement the two minterms, m1 and m2 respectively, and the third neuron implement the OR of these minterms. This solution is equivalent to the previous since it is irrelevant which neuron in the first layer implements m1 or m2.

```
w11= 1  w12= -1  w13= 1  w14= -1  w21= -1  w22= 1
bias1=0 bias2=-1 bias3=0

w11= 1  w12= -1  w13= 1  w14= -1  w21= 1  w22= -1
bias1=-1 bias2=0 bias3=0
```

C   GAMS code of the Second MILP Model to Implement a Three Variable Logical Function, with a single Perceptron

```
* F 3in=m6

sets b0 /0*1/;
alias(b1,b2,b0);

sets w11 /1*5/;
alias( w12, w13, w14, w15, w16,w21, w22, bias1, bias2, bias3, w11);

Parameter f_3in_d(b0,b1,b2), f(w11,w12,w13, bias1, b0, b1,b2);

f_3in(b0,b1,b2)=(ord(b0)=1) * (ord(b1)=2) * (ord(b2)=2);

f(w11,w12,w13, bias1, b0, b1,b2)=
( ( ( ord(b0)-1 ) * (ord(w11)-3) + (ord(b1)-1)*(ord(w12)-3) + (ord(b2)-1)*(ord(w13)-3)
+ ord(bias1)-3 )  ge 0) ;

variable obj;
```

```
binary variable weights(w11,w12,w13, bias1);

calc_ws1(w11,w12,w13, bias1,b0,b1,b2)..
weights(w11,w12,w13, bias1)*
f(w11,w12,w13, bias1,b0,b1,b2)=e= f_3in(b0,b1,b2)*weights(w11,w12,w13, bias1);

calc_obj.. obj=e=sum( (w11,w12,w13, bias1), weights(w11,w12,w13, bias1));

Model f3in /all/;

Solve f3in using mip maximizing obj;

display weights.l, obj.l, f_3in_d;
```

D   Output of the Second MILP Model to Implement a Three Variable Logical Function, f =m6, with a Single Perceptron with Weights between -2 and 2

```
----    48 VARIABLE weights.L
INDEX 1 = 1
                 1
4.4      1.000
INDEX 1 = 2
                 1
4.4      1.000
----    48 VARIABLE obj.L          =        2.000
----    48 PARAMETER f_3in
                 1

0.1      1.000
```

These two solutions correspond to the following weights that implement m6= :

```
w1=-2, w2=1, w3=1, bias=-2
w1=-1, w2=1, w3=1, bias=-2
```

E   GAMS code of the Third MILP Model to Implement a Three Variable Logical Function with a Two Layer Perceptron

```
* F 3in
sets b0 /0*1/;
alias(b1,b2,b0);
sets w11 /1*3/;
alias( w12, w13, w14, w15, w16,w21, w22, bias1, bias2, bias3, w11);

Parameter f_3in_d(b0,b1,b2),
f(w11,w12,w13,w14,w15,w16, w21,w22, bias1, bias2, bias3, b0, b1,b2);

*f_3in_d=m1 + m4
f_3in_d(b0,b1,b2)=(ord(b0)=1) * (ord(b1)=1) * (ord(b2)=2) +
(ord(b0)=2) * (ord(b1)=1) * (ord(b2)=1);

* Output of the Neural Network, f():

f(w11,w12,w13,w14,w15,w16, w21,w22, bias1, bias2, bias3, b0, b1, b2)=
( ((  ( ( ( ord(b0)-1 ) * (ord(w11)-2) + (ord(b1)-1)*(ord(w12)-2) + (ord(b2)-1)*(ord(w13)-2) + ord(bias1)-2 )  ge 0) * (ord(w21)
-2) + ( ( ( ord(b0)-1 ) * (ord(w14)-2) +
(ord(b1)-1)*(ord(w15)-2)+ (ord(b2)-1)*(ord(w16)-2) + ord(bias2)-2 )  ge 0) * (ord(w22)-2) +
 ord(bias3)-2 ) )  ge 0) ;

variable obj;

binary variable
weights (w11,w12,w13,w14,w15,w16, w21,w22, bias1, bias2, bias3);

**CONSTRAINTS**
calc_ws1(w11,w12,w13,w14,w15,w16,w21,w22, bias1, bias2, bias3, b0,b1,b2)..
weights(w11,w12,w13,w14,w15,w16, w21,w22, bias1, bias2, bias3)*
f(w11,w12,w13,w14,w15,w16, w21,w22, bias1, bias2, bias3,b0,b1,b2)=e=
```

```
weights(w11,w12,w13,w14,w15,w16, w21,w22, bias1, bias2, bias3)*
 f_3in(b0,b1,b2);

calc_obj.. obj=e=sum(
(w11,w12,w13,w14,w15,w16, w21,w22, bias1, bias2, bias3),
weights( w11,w12,w13,w14,w15,w16, w21,w22, bias1, bias2, bias3) );

Model f3in /all/;

Solve f3in using mip maximizing obj;
```

F   Output of a Run of the Second MILP Model for $f=m_1+m_4$

```
----     47 VARIABLE weights.L

INDEX 1 = 1  INDEX 2 = 1  INDEX 3 = 1  INDEX 4 = 1  INDEX 5 = 2  INDEX 6 = 1   INDEX 7 = 3  INDEX 8 = 1

                 1
3.2     1.000
INDEX 1 = 1  INDEX 2 = 1  INDEX 3 = 1  INDEX 4 = 1  INDEX 5 = 3  INDEX 6 = 1   INDEX 7 = 3  INDEX 8 = 1
                 1
3.2     1.000
INDEX 1 = 1  INDEX 2 = 1  INDEX 3 = 3  INDEX 4 = 1  INDEX 5 = 3  INDEX 6 = 3   INDEX 7 = 3  INDEX 8 = 1
                 2
1.2     1.000
INDEX 1 = 1  INDEX 2 = 1  INDEX 3 = 3  INDEX 4 = 3  INDEX 5 = 1  INDEX 6 = 1  INDEX 7 = 3  INDEX 8 = 3
                 1
1.1     1.000
INDEX 1 = 1  INDEX 2 = 2  INDEX 3 = 1  INDEX 4 = 1  INDEX 5 = 1  INDEX 6 = 1  INDEX 7 = 1  INDEX 8 = 3
                 1
2.3     1.000
INDEX 1 = 1  INDEX 2 = 2  INDEX 3 = 1  INDEX 4 = 1  INDEX 5 = 3  INDEX 6 = 1   INDEX 7 = 3  INDEX 8 = 1
                 1
3.2     1.000
INDEX 1 = 1  INDEX 2 = 3  INDEX 3 = 1  INDEX 4 = 1  INDEX 5 = 1  INDEX 6 = 1   INDEX 7 = 1  INDEX 8 = 3
                 1
2.3     1.000
INDEX 1 = 1  INDEX 2 = 3  INDEX 3 = 1  INDEX 4 = 1  INDEX 5 = 2  INDEX 6 = 1   INDEX 7 = 1  INDEX 8 = 3
                 1
2.3     1.000
INDEX 1 = 1  INDEX 2 = 3  INDEX 3 = 3  INDEX 4 = 1  INDEX 5 = 1  INDEX 6 = 3   INDEX 7 = 1  INDEX 8 = 3
                 2
2.1     1.000
INDEX 1 = 1  INDEX 2 = 3  INDEX 3 = 3  INDEX 4 = 3  INDEX 5 = 3  INDEX 6 = 3   INDEX 7 = 1  INDEX 8 = 1
                 3
2.2     1.000
INDEX 1 = 3  INDEX 2 = 1  INDEX 3 = 1  INDEX 4 = 1  INDEX 5 = 1  INDEX 6 = 3   INDEX 7 = 3  INDEX 8 = 3
                 1
1.1     1.000
INDEX 1 = 3  INDEX 2 = 1  INDEX 3 = 1  INDEX 4 = 3  INDEX 5 = 3  INDEX 6 = 1   INDEX 7 = 3  INDEX 8 = 1
                 2
1.2     1.000
INDEX 1 = 3  INDEX 2 = 3  INDEX 3 = 1  INDEX 4 = 1  INDEX 5 = 3  INDEX 6 = 3   INDEX 7 = 1  INDEX 8 = 1
                 3
2.2     1.000
INDEX 1 = 3  INDEX 2 = 3  INDEX 3 = 1  INDEX 4 = 3  INDEX 5 = 1  INDEX 6 = 1  INDEX 7 = 1  INDEX 8 = 3
                 2
2.1     1.000

----     47 VARIABLE obj.L              =        14.000

----     47 PARAMETER f_3in_d
                 0            1

0.0                    1.000
1.0     1.000
```

These 14 solutions correspond to the following sets of weight values, respectively:

```
1. w11= -1  w12= -1  w13= -1  w14= -1 w15= 0  w16= -1  w21= 1  w22= -1
bias1=1 bias2=0 bias3=-1
2. w11= -1  w12= -1  w13= -1  w14= -1 w15= 1  w16= -1  w21= 1  w22= -1
bias1=1 bias2=0 bias3=-1
3. w11= -1  w12= -1  w13= 1  w14= -1 w15= 1  w16= 1  w21= 1  w22= -1
bias1=-1 bias2=0 bias3=0
4. w11= -1  w12= -1  w13= 1  w14= 1 w15= -1  w16= -1  w21= 1  w22= 1
bias1=-1 bias2=-1 bias3=-1**
```
This solution corresponds to the direct implementation of the two minterms by the two neurons of the first layer and the OR of the two minterms by the third neuron

```
5. w11= -1  w12= 0  w13= -1  w14= -1 w15= -1  w16= -1  w21= -1  w22= 1
bias1=0 bias2=1 bias3=-1
```

6. w11= -1 w12= 0 w13= -1 w14= -1 w15= 1 w16= -1   w21= 1 w22= -1
bias1=1 bias2=0 bias3=-1
7. w11= -1 w12= 1 w13= -1 w14= -1 w15= -1 w16= -1   w21= -1 w22= 1
bias1=0 bias2=1 bias3=-1
8. w11= -1 w12= 1 w13= -1 w14= -1 w15= 0 w16= -1   w21= -1 w22= 1
bias1=0 bias2=1 bias3=-1
9. w11= -1 w12= 1 w13= 1 w14= -1 w15= -1 w16= 1   w21= -1 w22= 1
bias1=0 bias2=-1 bias3=0
10. w11= -1 w12= 1 w13= 1 w14= 1 w15= 1 w16= -1   w21= -1 w22= -1
bias1=0 bias2=0 bias3=1
11. w11= 1 w12= -1 w13= -1 w14= -1 w15= -1 w16= 1   w21= 1 w22= 1
bias1=-1 bias2=-1 bias3=-1
This solution also corresponds to the direct implementation of the two minterms by the two neurons of the first layer and the OR
of the two minterms by the third neuron
12. w11= 1 w12= -1 w13= -1 w14= 1 w15= 1 w16= -1   w21= 1 w22= -1
bias1=-1 bias2=0 bias3=0
13. w11= 1 w12= 1 w13= -1 w14= -1 w15= -1 w16= 1   w21= -1 w22= -1
bias1=0 bias2=0 bias3=1
14. w11= 1 w12= 1 w13= -1 w14= 1 w15= -1 w16= -1   w21= -1 w22= 1
bias1=0 bias2=-1 bias3=0

G   of a Run of the Second MILP Model for $f=m_1+m_7$

```
----    48 VARIABLE weights.L

INDEX 1 = 1  INDEX 2 = 1  INDEX 3 = 3  INDEX 4 = 1  INDEX 5 = 3  INDEX 6 = 1
  INDEX 7 = 1  INDEX 8 = 1
                2
2.2 1.000
INDEX 1 = 1  INDEX 2 = 1  INDEX 3 = 3  INDEX 4 = 2  INDEX 5 = 1  INDEX 6 = 3
  INDEX 7 = 1  INDEX 8 = 3
                1
2.2 1.000
INDEX 1 = 1  INDEX 2 = 1  INDEX 3 = 3  INDEX 4 = 2  INDEX 5 = 3  INDEX 6 = 1
  INDEX 7 = 1  INDEX 8 = 1
                2
2.1 1.000
INDEX 1 = 1  INDEX 2 = 1  INDEX 3 = 3  INDEX 4 = 3  INDEX 5 = 1  INDEX 6 = 3
  INDEX 7 = 1  INDEX 8 = 3
                1
2.1 1.000
INDEX 1 = 1  INDEX 2 = 3  INDEX 3 = 1  INDEX 4 = 1  INDEX 5 = 1  INDEX 6 = 3
  INDEX 7 = 1  INDEX 8 = 1
                2
2.2 1.000
INDEX 1 = 1  INDEX 2 = 3  INDEX 3 = 1  INDEX 4 = 2  INDEX 5 = 1  INDEX 6 = 3
  INDEX 7 = 1  INDEX 8 = 1
                2
2.1 1.000
INDEX 1 = 1  INDEX 2 = 3  INDEX 3 = 1  INDEX 4 = 2  INDEX 5 = 3  INDEX 6 = 1
  INDEX 7 = 1  INDEX 8 = 3
                1
2.2 1.000
INDEX 1 = 1  INDEX 2 = 3  INDEX 3 = 1  INDEX 4 = 3  INDEX 5 = 3  INDEX 6 = 1
  INDEX 7 = 1  INDEX 8 = 3
                1
2.1 1.000
INDEX 1 = 2  INDEX 2 = 1  INDEX 3 = 3  INDEX 4 = 1  INDEX 5 = 1  INDEX 6 = 3
  INDEX 7 = 3  INDEX 8 = 1
                1
2.2 1.000
INDEX 1 = 2  INDEX 2 = 1  INDEX 3 = 3  INDEX 4 = 1  INDEX 5 = 3  INDEX 6 = 1
  INDEX 7 = 1  INDEX 8 = 1
                2
1.2 1.000
INDEX 1 = 2  INDEX 2 = 1  INDEX 3 = 3  INDEX 4 = 3  INDEX 5 = 1  INDEX 6 = 3
  INDEX 7 = 1  INDEX 8 = 3
                1
1.1 1.000
INDEX 1 = 2  INDEX 2 = 3  INDEX 3 = 1  INDEX 4 = 1  INDEX 5 = 1  INDEX 6 = 3
  INDEX 7 = 1  INDEX 8 = 1
                2
1.2 1.000
INDEX 1 = 2  INDEX 2 = 3  INDEX 3 = 1  INDEX 4 = 1  INDEX 5 = 3  INDEX 6 = 1
  INDEX 7 = 3  INDEX 8 = 1
                1
2.3 1.000
INDEX 1 = 2  INDEX 2 = 3  INDEX 3 = 1  INDEX 4 = 3  INDEX 5 = 3  INDEX 6 = 1
  INDEX 7 = 1  INDEX 8 = 3
                1
1.1 1.000
INDEX 1 = 3  INDEX 2 = 1  INDEX 3 = 3  INDEX 4 = 1  INDEX 5 = 1  INDEX 6 = 3
  INDEX 7 = 3  INDEX 8 = 1
                1
```

```
1.2 1.000
INDEX 1 = 3  INDEX 2 = 1  INDEX 3 = 3  INDEX 4 = 2  INDEX 5 = 1  INDEX 6 = 3
  INDEX 7 = 3  INDEX 8 = 1
               1
1.1 1.000
INDEX 1 = 3  INDEX 2 = 3  INDEX 3 = 1  INDEX 4 = 1  INDEX 5 = 3  INDEX 6 = 1
  INDEX 7 = 3  INDEX 8 = 1
               1
1.2 1.000
INDEX 1 = 3  INDEX 2 = 3  INDEX 3 = 1  INDEX 4 = 2  INDEX 5 = 3  INDEX 6 = 1
  INDEX 7 = 3  INDEX 8 = 1
               1
1.1 1.000

----    48 VARIABLE obj.L          =       18.000

----    48 PARAMETER f_3in_d

            0        1

1.0    1.000
1.1             1.000
```

Now does not exist direct solutions, i.e. where the first two neurons implement m1 and m7 and the third neuron implementing the OR
of these two minterms, because the
implementation of m7 by a perceptron of the first layer with integer weights would imply a bias of -3 and weights equal to 1, i.e.
a three input AND, and the weights only vary between -2 and 2.

These 18 solutions correspond to the following sets of values of weights, respectively:

```
1.  w11= -1  w12= -1  w13= 1  w14= -1 w15= 1  w16= -1   w21= 1  w22= -1
bias1=0 bias2=0 bias3=0
2.  w11= -1  w12= -1  w13= 1  w14= 0 w15= -1  w16= 1    w21= -1  w22= 1
bias1=0 bias2=0 bias3=-1
3.  w11= -1  w12= -1  w13= 1  w14= 0 w15= 1  w16= -1    w21= -1  w22= -1
bias1=0 bias2=-1 bias3=0
4.  w11= -1  w12= -1  w13= 1  w14= 1 w15= -1  w16= 1    w21= -1  w22= 1
bias1=0 bias2=-1 bias3=-1
5.  w11= -1  w12= 1  w13= -1  w14= -1 w15= -1  w16= 1   w21= -1  w22= -1
bias1=0 bias2=0 bias3=0
6.  w11= -1  w12= 1  w13= -1  w14= 0 w15= -1  w16= 1    w21= -1  w22= -1
bias1=0 bias2=-1 bias3=0
7.  w11= -1  w12= 1  w13= -1  w14= 0 w15= 1  w16= -1    w21= -1  w22= 1
bias1=0 bias2=0 bias3=-1
8.  w11= -1  w12= 1  w13= -1  w14= 1 w15= 1  w16= -1    w21= -1  w22= 1
bias1=0 bias2=-1 bias3=-1
9.  w11= 0  w12= -1  w13= 1  w14= -1 w15= -1  w16= 1    w21= 1  w22= -1
bias1=0 bias2=0 bias3=0
10. w11= 0  w12= -1  w13= 1  w14= -1 w15= 1  w16= -1    w21= -1  w22= -1
bias1=-1 bias2=0 bias3=0
11. w11= 0  w12= -1  w13= 1  w14= 1 w15= -1  w16= 1     w21= -1  w22= 1
bias1=-1 bias2=-1 bias3=-1
12. w11= 0  w12= 1  w13= -1  w14= -1 w15= -1  w16= 1    w21= -1  w22= -1
bias1=-1 bias2=0 bias3=0
13. w11= 0  w12= 1  w13= -1  w14= -1 w15= 1  w16= -1    w21= 1  w22= -1
bias1=0 bias2=1 bias3=-1
14. w11= 0  w12= 1  w13= -1  w14= 1 w15= 1  w16= -1     w21= -1  w22= 1
bias1=-1 bias2=-1 bias3=-1
15. w11= 1  w12= -1  w13= 1  w14= -1 w15= -1  w16= 1    w21= 1  w22= -1
bias1=-1 bias2=0 bias3=-1
16. w11= 1  w12= -1  w13= 1  w14= 0 w15= -1  w16= 1     w21= 1  w22= -1
bias1=-1 bias2=-1 bias3=-1
17. w11= 1  w12= 1  w13= -1  w14= -1 w15= 1  w16= -1    w21= 1  w22= -1
bias1=-1 bias2=0 bias3=-1
18. w11= 1  w12= 1  w13= -1  w14= 0 w15= 1  w16= -1     w21= 1  w22= -1
bias1=-1 bias2=-1 bias3=-1
```

On Member Labelling in Social Networks

Rafael Corchuelo$^{(\boxtimes)}$, Antonia M. Reina Quintero, and Patricia Jiménez

ETSI Informática, Avda. Reina Mercedes s/n, E-41012 Sevilla, Spain
{corchu,reinaqu,patriciajimenez}@us.es

Abstract. Software agents are increasingly used to search for experts, recommend resources, assess opinions, and other similar tasks in the context of social networks, which requires to have accurate information that describes the features of the members of the network. Unfortunately, many member profiles are incomplete, which has motivated many authors to work on automatic member labelling, that is, on techniques that can infer the null features of a member from his or her neighbourhood. Current proposals are based on local or global approaches; the former compute predictors from local neighbourhoods, whereas the latter analyse social networks as a whole. Their main problem is that they tend to be inefficient and their effectiveness degrades significantly as the percentage of null labels increases. In this paper, we present Katz, which is a novel hybrid proposal to solve the member labelling problem using neural networks. Our experiments prove that it outperforms other proposals in the literature in terms of both effectiveness and efficiency.

Keywords: Social networks · Member labelling · Hybrid approach · Neural networks

1 Introduction

On-line social media have sprouted out during the last decade. They have paved the way for on-line social networks whose members typically interact to share or to retrieve information from one another. Never before has it been easier to find information about individuals, their demographics, their likes, their dislikes, the activities in which they engage, their opinions, their thoughts, and so on. And something that is even more important: their relationships.

Software agents are being used in tasks such as searching for experts regarding a given topic, recommending resources (posts, videos, music, and the like), assessing opinions, targeting advertisements, sociological studies, and so on. For these agents to succeed in producing accurate information, it is very important that the information in a member's profile be as complete as possible. Unfortunately, it is not uncommon that many members do not complete their profiles [11], which makes it very difficult for software agents to work well.

Many authors have paid attention to a problem that is commonly referred to as member labelling (aka. member classification, node classification, link-based classification, or collective classification). Simply put, the idea is to infer the

© Springer International Publishing Switzerland 2015
I. Rojas et al. (Eds.): IWANN 2015, Part II, LNCS 9095, pp. 488–499, 2015.
DOI: 10.1007/978-3-319-19222-2_41

features of a member of a social network as accurately as possible using solely the features available from members with whom he or she has a relationship [1]. This has proven to work well because social networks have a property that is known as homophily [19], according to which members who have similar features tend to have stronger relationships than members that have very dissimilar features. The current proposals in the literature are based on local or global methods. The former learn a predictor from the features of the members of a social network, including some neighbours; the latter tackle the problem from a global perspective and attempt to analyse social networks as a whole. The main problem with current proposals is that they have proven to be inefficient and ineffective as the size of a social network or the number of null features increases.

This motivated us to work on Katz, which is a novel hybrid proposal to solve the member labelling problem. It is based on neural networks, which are used to infer a predictor for each member feature using the information provided by an unbounded neighbourhood. It starts analysing each member's profile in isolation, and then explores his or her neighbourhood searching for the relationships and features that contribute the most to producing a better predictor. It is not a local method since it explores an unbounded context and selects the most interesting features and neighbours to learn a predictor; neither is it a global method because it does not attempt to analyse social networks as a whole; that is the reason why we refer to Katz as a hybrid proposal. Our experiments on quite a large real-world social network prove that it outperforms other proposals in the literature in terms of both effectiveness and efficiency.

The rest of the paper is organised as follows: Section 2 describes our proposal; Section 3 reports on the results of our experiments; Section 4 surveys the related work and compares it to ours; finally, Section 5 presents our conclusions.

2   Our Proposal

Katz works on a social network that is represented as a graph in which a node represents all of the features of a member profile and an edge represents a relationship to another member. It analyses the network and returns a map in which each feature is associated with a set of neural networks that can be used to label a new member regarding that feature. Note that each feature is predicted by means of a set of neural networks that are learnt from different partitions of the social network; the goal, which has been confirmed empirically, is to decrease the error rate by using an ensamble-predictor approach instead of the single-predictor approach that is common in the literature. In the following subsections, we first present the main procedure of Katz and then an ancillary procedure that is used to extend a neural network to the most appropriate neighbourhood.

Main Procedure: Figure 1 shows the main procedure of Katz. It works on a graph (N, E) that represents a social network. N is a collection of vectors of the form $(m, f_1, f_2, \ldots, f_n)$, where m is the unique identifier of a member of the social network and f_i are the values of its features $(i = 1 \ldots n)$; features can be

```
 1: Katz(N, E)
 2:    m = ∅
 3:    for each feature f used in N do
 4:       ns = ∅
 5:       repeat β times
 6:          t = select nodes in N with a value for f
 7:          ts = create a training set with ⌈γ|t|⌉ nodes from t
 8:          vs = t \ ts
 9:          n = null
10:          do
11:             (n', ts', vs') = expandNeuralNetwork(n, ts, vs, N, E)
12:          exit when n = n'
13:             (n, ts, vs) = (n', ts', vs')
14:          end
15:          w = 1/error(n, vs)
16:          ns = ns ∪ {(n, w)}
17:       end
18:       m = m ∪ {(f, ns)}
19:    end
20: return m
```

Fig. 1. Main procedure of Katz

either numeric (e.g., age, salary, or opinion polarity about a topic) or categorical (e.g., nationality, gender, or dislikes). E is a collection of vectors of the form (m_1, m_2, k, w), where m_1 and m_2 are the identifiers of two members of the social network, k denotes a kind of relationship between them, and w is the weight of that relationship. The relationships include any kind of interaction between any two members of a social network (e.g., replies to posts, post forwards, friendship requests, message exchanges, and so on). Thus, the weight of edge (m_1, m_2, k, w) is computed as the number of actual interactions of type k that have occurred between members m_1 and m_2.

The result of the main procedure is computed in variable m, which is a map that associates every feature in the social network with a collection of tuples of the form (n, w), where n is a neural network, which acts as a regressor or a classifier for the corresponding feature, and w is its weight, which is the inverse of the error rate; that is, the smaller the error rate, the more important the neural network and the larger the error rate, the less important the neural network. Katz returns β rules for every feature, where β is a user-provided parameter. To label a new member regarding a given feature, the neural networks are applied one after the other. In the case of numeric features, the values predicted by each rule are weighted according to their normalised error rate and then averaged; in the case of categoric features, the results are weighted according to the normalised error rate and the most voted one is returned.

The main procedure basically iterates over the set of features in the social network; in each iteration, it repeats the following procedure β times: it first selects the subset of nodes that have a value for the feature being analysed and then splits it into a training set and a validation set. The size of the training set is controlled by means of γ, which is a user-provided parameter; the remaining

```
1: expandNeuralNetwork(n, ts, vs, N, E)
2:    if n = null then
3:        n = learn network from ts
4:    else
5:        c = expand the neighbourhood of ts and vs using (N, E)
6:        for each (u, v) in c do
7:            n′ = learn a network from u
8:            ts′ = u
9:            vs′ = v
10:           if error(n′, vs′) < error(n, vs) then
11:               (n, ts, vs) = (n′, ts′, vs′)
12:           end
13:       end
14:   end
15: return (n, ts, vs)
```

Fig. 2. Procedure to expand a neural network

X0 = member	age(X0)	gender(X0)	group(X0)	school(X0)
m1	23	male	student	physics
m2	22	male	student	arts
m3	23	female	lecturer	physics
m4	24	female	staff	physics

X0 = member	age(X0)	gender(X0)	group(X0)	school(X0)	X1 = sends-message(X0)	age(X1)	gender(X1)	group(X1)	school(X1)	weight(X1)
m1	23	male	student	physics	m5	24	female	lecturer	arts	12
m1	23	male	student	physics	m6	32	male	student	physics	18
m2	22	male	student	arts	m7	22	male	student	arts	2
m2	22	male	student	arts	m8	24	female	lecturer	arts	1
m2	22	male	student	arts	m9	32	female	lecturer	physics	23
m3	23	female	lecturer	physics	m4	24	female	staff	physics	18
m3	23	female	lecturer	physics	m9	32	female	lecturer	physics	90
m3	23	female	lecturer	physics	m1	23	male	student	physics	12
m4	24	female	staff	physics	null	null	null	null	null	null

Fig. 3. Excerpt of an expanded training set

nodes are used for validation purposes. It then initialises a neural network n to a null network that does nothing, and then repeatedly expands it until no further expansion is possible. Expanding a neural network consists of extending it to some neighbours as long as this helps to reduce the error rate. We provide additional details in the following subsection.

Expanding Neural Networks: Figure 2 shows the procedure to expand a neural network. It works on a neural network n, a training set ts, a validation set vs, and a social network (N, E); it returns a new neural network, the training set from which it was learnt, and the validation set on which it was validated.

The procedure first checks if the input neural network is null, in which case it simply learns a neural network from the training set and returns it. Otherwise, it first expands the neighbourhood of the training and the validation sets using the information provided by the social network. Expanding the neighbourhood of a dataset means that its vectors are expanded with additional components that represent the features of a kind of neighbour. For instance, Figure 3 shows

Table 1. Experimental results

Age Nullif.	NJ E	T	LG E	T	MP E	T	B E	T	Katz E	T
5.00%					5.55	7.01	9.70	12.39	8.85	6.75
10.00%					9.39	8.22	14.30	12.74	12.55	6.69
15.00%					16.45	9.18	10.70	15.60	10.88	4.36
20.00%					6.49	10.61	11.16	16.30	11.75	4.71
25.00%					20.43	11.37	20.00	14.55	18.83	4.70
30.00%					24.08	10.94	31.86	16.23	15.62	4.68
35.00%					35.55	13.56	21.17	15.63	10.59	5.05
40.00%					23.39	12.46	15.31	17.88	17.86	4.66
45.00%					50.13	12.92	36.26	21.65	13.18	6.38
50.00%					40.23	15.81	17.44	20.75	10.48	5.59
Mean					23.17	11.21	18.79	16.37	13.06	5.36

Nation. Nullif.	NJ E	T	LG E	T	MP E	T	B E	T	Katz E	T
5.00%	19.12	5.56	19.77	5.60	3.44	11.77	7.94	13.51	5.40	4.10
10.00%	26.85	6.79	21.28	6.95	6.81	12.27	8.33	15.06	9.09	3.76
15.00%	27.55	7.33	22.01	7.83	9.00	12.64	14.96	18.77	13.55	3.74
20.00%	22.76	9.04	22.75	6.96	10.33	15.55	8.94	20.13	17.36	3.23
25.00%	34.96	7.25	23.56	7.39	19.24	16.57	8.59	20.62	25.13	4.40
30.00%	48.85	7.99	24.25	7.27	7.88	12.54	17.79	22.72	12.34	5.07
35.00%	23.34	8.59	25.05	8.13	22.35	13.76	35.32	26.06	20.84	4.76
40.00%	22.85	7.19	35.82	8.14	24.67	16.75	27.80	23.97	20.46	2.44
45.00%	54.20	7.53	26.50	8.65	19.85	16.79	19.85	23.39	21.07	3.33
50.00%	49.13	7.86	44.71	10.52	52.53	17.77	17.79	27.51	44.04	3.57
Mean	29.96	7.51	25.57	7.74	17.61	14.64	16.73	21.17	17.93	3.84

Gender Nullif.	NJ E	T	LG E	T	MP E	T	B E	T	Katz E	T
5.00%	5.40	6.35	10.25	6.54	5.62	5.40	9.73	4.65	8.01	4.08
10.00%	5.72	9.33	13.54	6.90	14.50	5.61	16.40	4.86	12.73	4.14
15.00%	6.52	10.71	22.70	6.04	12.84	6.42	11.45	5.97	11.17	4.44
20.00%	14.85	12.21	26.83	6.96	24.15	6.08	17.67	6.67	15.25	4.76
25.00%	15.29	13.68	31.03	7.16	5.59	7.49	23.57	5.72	17.17	3.22
30.00%	7.71	15.19	20.18	8.41	35.44	5.93	32.64	6.49	10.59	4.14
35.00%	25.13	16.59	21.76	9.59	6.35	4.73	23.68	6.95	16.38	3.55
40.00%	18.76	18.08	23.57	7.91	20.12	3.62	24.36	7.03	14.36	4.06
45.00%	13.36	19.68	29.06	7.91	27.63	3.21	14.86	5.82	13.36	4.09
50.00%	28.18	20.87	26.63	9.50	27.29	2.88	42.70	6.43	15.29	3.83
Mean	12.09	14.27	22.16	7.69	17.95	5.14	21.71	6.06	13.43	4.03

School Nullif.	NJ E	T	LG E	T	MP E	T	B E	T	Katz E	T
5.00%	21.75	5.50	16.52	7.54	4.06	5.81	9.85	4.55	6.46	5.41
10.00%	23.86	6.59	18.81	9.03	11.93	5.69	15.55	4.98	5.48	4.04
15.00%	27.77	7.92	19.95	9.47	10.85	5.89	10.73	5.32	8.01	5.27
20.00%	30.22	8.19	21.11	7.34	11.14	6.30	8.69	5.33	7.12	5.24
25.00%	35.98	9.64	22.26	7.92	8.18	6.55	22.06	5.59	21.34	5.30
30.00%	24.09	9.59	23.47	9.40	25.70	6.88	19.83	5.69	28.90	2.09
35.00%	40.91	11.79	24.63	8.97	14.68	7.94	35.23	6.84	28.30	2.07
40.00%	53.60	9.75	25.74	9.76	21.84	9.44	24.57	7.36	25.61	1.57
45.00%	30.26	7.60	30.94	9.48	30.13	9.62	31.53	8.74	29.31	2.28
50.00%	32.96	7.78	28.06	11.67	21.03	12.00	19.98	10.05	32.13	2.29
Mean	32.14	8.43	22.75	9.06	15.95	7.61	19.00	6.44	17.27	3.56

Group Nullif.	NJ E	T	LG E	T	MP E	T	B E	T	Katz E	T
5.00%	14.36	5.01	18.56	6.07	8.53	9.52	9.17	4.13	8.33	5.40
10.00%	23.36	9.27	23.09	9.35	14.53	9.32	13.18	9.30	11.34	5.06
15.00%	27.86	10.79	25.28	10.75	17.54	10.82	15.18	10.74	12.83	3.85
20.00%	32.36	12.26	27.57	12.30	20.53	12.23	17.17	12.27	14.33	4.26
25.00%	36.86	13.84	29.75	13.64	23.53	13.76	19.17	13.82	15.84	4.19
30.00%	41.36	15.22	32.08	15.31	26.53	15.22	21.18	15.13	17.33	5.77
35.00%	45.86	16.70	34.43	16.54	29.53	16.56	23.17	16.60	18.84	5.14
40.00%	50.36	18.02	36.70	18.25	32.53	18.33	25.18	18.06	20.33	2.97
45.00%	54.86	19.72	38.72	19.74	35.53	19.76	27.18	19.64	21.83	3.82
50.00%	59.36	21.27	45.05	21.01	38.53	21.32	29.17	21.35	23.33	3.41
Mean	38.66	14.21	30.72	14.29	24.73	14.68	19.98	14.10	16.43	4.39

Lik./Dis. Nullif.	NJ E	T	LG E	T	MP E	T	B E	T	Katz E	T
5.00%	19.49	5.49	13.50	7.91	10.34	5.19	15.19	10.84	10.75	6.26
10.00%	24.50	9.33	20.25	9.28	18.35	9.26	20.20	9.31	13.75	5.62
15.00%	27.00	10.78	21.12	10.80	17.84	10.83	22.70	10.83	15.25	9.25
20.00%	29.49	12.18	21.95	12.17	20.34	12.25	25.20	12.20	16.75	7.68
25.00%	31.99	13.62	22.85	13.69	22.85	13.74	27.70	13.75	18.25	6.99
30.00%	34.49	15.10	23.63	15.12	25.35	15.06	30.19	15.14	19.75	9.72
35.00%	37.00	16.73	24.50	16.64	27.84	16.63	32.70	16.51	21.25	9.48
40.00%	39.50	17.95	25.42	18.02	30.35	18.12	35.20	18.09	22.76	16.02
45.00%	42.00	19.64	32.22	19.50	32.85	19.51	37.69	19.46	24.26	19.39
50.00%	44.50	21.34	37.04	20.85	35.35	20.91	40.19	21.19	25.75	13.73
Mean	33.00	14.22	22.65	14.40	24.15	14.15	28.70	14.73	18.85	10.41

an excerpt of an initial training set on the left; on the right, that training set has been expanded with the features of the neighbours regarding the 'sends-message' relationship; note that the weight of the relationship is added as an additional feature to the vector.

Then, the procedure iterates through the set of expansions of the training set and learns a new neural network from each one. It returns the expanded neural network that achieves the smallest error rate together with the training set from which it was learnt and validation set on which the error rate was computed.

3 Experimental Results

We conducted a series of experiments to analyse how Katz performs in practice. The experiments were carried out using a Java 1.7 implementation that was run

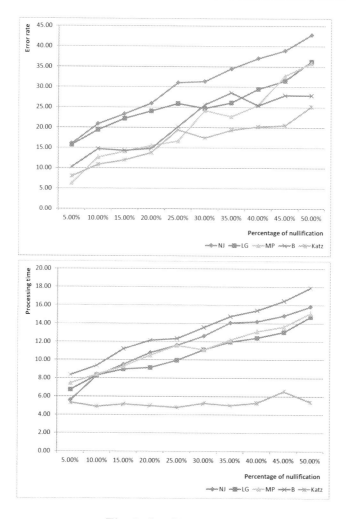

Fig. 4. Graphic summary

on a four-threaded Intel Core i7 computer that ran at 2.93 GHz, had 16 GiB of RAM, Windows 7 Pro 64-bit, Oracle's Java Development Kit 1.7.9_02, and Weka 3.6.8.

We implemented the general framework by Neville and Jensen [17] (NJ) and then the specific proposals by Lu and Getoor [12] (LG), Macskassy and Provost [13] (MP), and Bhagat et al. [3] (B). Regarding the previous proposals, we considered that a labelling was estable when no more than 5% of the features changed in an iteration of the method. Regarding Katz, we experimented with several combinations of parameters and kinds of neural networks. We found out that the following values for the parameters work quite well: $\beta = 10$, that is, 10 neural networks learnt for each feature, and $\gamma = 0.25$, that is, 25% of the nodes available in the social network are used for training purposes and the remaining for validation purposes. Regarding the learning technique, we found out that RBFN networks [4] are the best performing in this context.

The experiments were performed on a dataset that consisted in a dump of our university social network. This network has $56,431$ members, each of which is characterised by a profile that includes the following features: age (a natural value), gender (male, female), group (student, lecturer, staff), nationality (Spanish, French, Italian, and so on), school (Computer-Science, Mathematics, Physics, Philology, and so on), likes, and dislikes; other features like name, address, national id or passport were discarded to keep the data anonymous; neither was it very interesting to attempt to predict them. The likes and dislikes are sets of key words that are selected by the members from a list that is computed automatically from the messages post by the members of the network; to deal with them in our experiments, we selected the top 50 key words and created binary features of the form likes_X or dislikes_Y, where X or Y represents key words. The relationships between the members of the network are the following: posts-to-wall, replies-to-post, forwards-post, sends-message, follows-member, requests-friendship. This dataset was particularly useful because almost every profile has accurate features that are set automatically using the students' registration data or the lecturers' and staff's work contracts, and the likes and dislikes are also selected from sets of pre-computed key words. That is, we had quite a large correctly labelled dataset on which could conduct quite a precise validation.

To evaluate our proposal and compare it to others, we created several datasets from the previous one. They were versions of the original dataset in which we nullified the features of 5% up to 50% nodes that were chosen randomly. This helped us to evaluate how our proposal works and compare it to others in terms of error rate (E) and processing time (T). The error rate was computed as the percentage of wrong predictions; in the case of numeric features a $\pm 10\%$ tolerance threshold was established to consider a prediction wrong. The processing time was measured in CPU plus IO hours, since these timings are far more reliable and stable than user times.

Table 1 shows our results and Figure 4 summarises them using a couple of charts. (The columns that correspond to proposals NJ and LG regarding feature 'age' are empty because these methods cannot be applied to numeric features.) Regarding effectiveness, the first conclusion is that the error rate increases steadily as the percentage of nullification increases, but Katz keeps the smallest global mean in the majority of cases, where global mean refers to the computed mean for each feature for a given nullification percentage, cf. the upper part of Figure 4. To compare the results more precisely, we have computed the tendency lines for each proposal according to the percentage of nullification, which is denoted as N:

Proposal	Error rate tendency	R^2
NJ	$2.78N + 14.72$	0.99
LG	$1.88N + 15.17$	0.93
MP	$2.99N + 4.16$	0.95
B	$2.15N + 9.17$	0.88
Katz	$1.69N + 7.37$	0.92

Note that the R^2 coefficient is very good in every case, which means that there is a clear linear tendency in the results. The smallest slope corresponds to Katz, which means that it is the proposal whose error rate increases at the lowest pace as the percentage of nullification increases; it is followed by LG, but note that the error rate of this proposal is roughly double as Katz's.

Regarding efficiency, the first conclusion is that Katz seems to have a behaviour that is very stable, whereas the other proposals seem to require more processing time as the percentage of nullification increases. To confirm this idea, we have also computed the tendency lines for each proposal, namely:

Proposal	Processing time tendency	R^2
NJ	$1.04N + 5.62$	0.98
LG	$0.66N + 6.08$	0.98
MP	$0.78N + 6.93$	0.97
B	$1.00N + 7.63$	0.98
Katz	$0.08N + 4.81$	0.95

Note that the R^2 coefficient is again very good in every case. The smallest slope corresponds again to Katz, which means that it is the proposal whose processing time increases at a lower pace as the percentage of nullification increases. Note that it is very close to 0.00, which means that the processing time remains almost constant; the reason is that the size of the training sets decrease as the percentage of nullification increases, which makes learning neural networks easier; unfortunately, as the percentage of nullification increases, the number of neighbours that must be explored to keep as a low error rate as possible increases. Katz is followed by the other proposals, which require considerably more processing time since they have to iterate until the labelling is stable enough, which is more and more difficult as the percentage of nullification increases.

4   Related Work

There are two mainstream approaches to the member labelling problem [9], namely: local and global methods. They both work on a graph-based representation of the social network being analysed, where the nodes store member features and the edges keep track of their interactions, but differ in that the former focus on learning local predictors from every member and his or her local neighbourhood, whereas the latter analyse the social network as a whole.

Below, we report on both approaches and discuss on how our proposal improves on them from a conceptual point of view.

Local Methods: These methods can be further classified into instantiations of the Iterative Classification Algorithm by Neville and Jensen [17] or instantiations of the Gibbs Sampling Algorithm by Geman and Geman [8].

The methods that are based on the Iterative Classification Algorithm [17] transform a social network into a dataset of vectors, each of which provides the features of a member's profile plus some aggregated features that correspond to the members in his or her neighbourhood. They analyse each feature in isolation as follows: they first learn a local predictor from the members whose profiles provide a non-null value for that feature. (Informally, this is commonly referred to as "the member is labelled".) The predictor is either a regressor or a classifier depending on whether the feature being analysed is numeric or categorical. It is then used to compute the label of the unlabelled members, as long as they have at least a labelled neighbour. Note that labelling a member will likely change the values of the aggregated features in the neighbourhood, so the labelling process needs to be repeated iteratively until the labels do not change dramatically or do not change at all. The previous idea has been instantiated many times in the literature, the difference being the kind of predictor used: Neville and Jensen [17] used Naive-Bayes predictors, Lu and Getoor [12] used logistic regression, Macskassy and Provost [13] used a voting approach, and Bhagat et al. [3] and McDowell et al. [16] used k-nearest neighbours. Recently, Cataltepe et al. [6] have used different types of predictors for member features and neighbourhood features, which are then combined to produce an ensamble predictor.

The methods that are based on the Gibbs Sampling Algorithm [8] work in four phases, namely: bootstrapping, burn-in, collecting, and labelling. In the bootstrapping phase, they learn a predictor in a way that is very similar to the methods that are based on the Iterative Classification Algorithm, and then use it to label the unlabelled members. Then, the burn-in phase is repeated a number of times; in each repetition, the members that were initially unlabelled are randomly ordered and then new labels are computed using a predictor, which can be the same that was used in the bootstrapping phase or a new one [14]. In the sample collection phase, the process is repeated a pre-defined number of times and the count of labels assigned to each member is computed. Finally, in the labelling phase, the members that were initially unlabelled are assigned the most likely label according to the counts that were computed in the previous phase. Both McDowell et al. [15] and Macskassy and Provost [14] have instantiated this idea; the former used Naive-Bayes and k-nearest neighbours and the latter used different combinations of predictors.

Global Methods: The most common methods in this category are based on random walks and optimisation.

A random walk on a graph is a very special case of a Markov chain. The core idea was introduced by Zhu et al. [23]: they rely on a transition matrix P that encodes the probability that a random walk proceeds between any two members of a social network using their relationships. Given an unlabelled member, the method assigns it the most common label out of the members that can be reached from it using random walks. That is, it requires to compute an approximation to the closure of P; in practice, the procedure stops when the closure is stable enough, that is, when the probabilities do not change dramatically or do not change at all. This general idea has been instantiated many times in the literature, with some variations, namely: Szummer and Jaakkola [20] start their walks from unlabelled nodes and consider only labelled nodes that can be reached in a pre-defined number of steps; contrarily, Callut et al. [5] consider only walks that start and end in a labelled node regarding a given feature, but do not go through labelled nodes regarding the same feature in the intermediate steps.

The methods that are based on optimisation map the problem into a number of constraints plus an objective function for which the global maximum or the minimum needs to be found. The main problem with this approach is that it typically results in an optimisation problem with a number of variables that very typically exceeds the limits of current solvers [2]. Thus, the authors who have instantiated this idea have focused on approximating the results as efficiently as possible [7,10,18,21,22].

Discussion: The main difference amongst Katz and the other methods in the literature is that it does not analyse a pre-defined neighbourhood, neither treats it all of the members of a social network, all of the features, and all of their relationships equally. It first learns a predictor using the member's features only and then tries to improve it by searching the most adequate neighbours and features using the error rate as the only search heuristic. It is then a hybrid approach since it does not focus on a local or a global neighbourhood, but finds the most appropriate for each feature. This, in turn, leads to a method that needs not be applied repeatedly in order to label a social network very well and very efficiently, as our experimental results prove.

A key feature is that Katz does not rely on a single predictor, but on several predictors that are learnt from different parts of the social network in order to adapt better to its peculiarities and reduce the error rate. The methods that are based on the Gibbs Sampling Algorithm also label a member using several predictions, but the difference is that Katz makes predictions using several predictors, not the different predictions that are computed using a single predictor from randomising the order in which it is applied to the members of a network. This has resulted in a method that has proven experimentally to achieve a very low error rate, even in cases in which there are many null labels, which have proven to be very difficult to deal with using other proposals.

Another strong point is that Katz relies on using neural networks, which have been proven to learn good models from complex data like ours [4]. This allows it to be applied to any kind of features, whereas some methods in the literature can only be applied to a kind of features. For instance, the proposals by Neville and Jensen [17] and Lu and Getoor [12] can only be used with categorical features because, unfortunately, the technique on which they rely does not deal with numeric features.

Finally, a common weak feature of the existing methods is that they tend to be inefficient. Iterative methods typically require an unbounded number of iterations for a labelling to become stable; Macskassy and Provost [14] proposed to limit the number of iterations of their proposal and they experimentally proved that roughly 2,200 iterations was enough to achieve good results, but it is not completely clear if this figure works in general. Global methods typically lead to problems that are not computationaly tractable and then can only be approximated. Our experiments prove that Katz is very efficient in practice and more scalable than other proposals in the literature.

5   Conclusions

In this paper, we have introduced Katz, which is a new hybrid proposal to solve the problem of labelling the members of a social network, that is, to infer the values of a member's profile missing features using his or her neighbourhood. It is based on neural-network predictors that are computed from a member's profile and an unbounded neighbourhood, which makes it very effective; furthermore, it does not require to iterate multiple times until the labelling converges, which makes it very efficient. Our experiments on a real-world university social network prove that it outperforms other proposals in the literature.

Acknowledgments. Our work was supported by the European Commission (FEDER), the Spanish and the Andalusian R&D&I programmes (grants TIN2007-64119, P07-TIC-2602, P08-TIC-4100, TIN2008-04718-E, TIN2010-21744, TIN2010-09809-E, TIN2010-10811-E, TIN2010-09988-E, TIN2011-15497-E, and TIN2013-40848-R). We are grateful to our support staff for their help to obtain a dump of our university social network. We also thank Opileak.com for sharing their social media analysis platform with us.

References

1. Aggarwal, C.C.: Social Network Data Analytics. Springer (2011)
2. Bhagat, S., Cormode, G., Muthukrishnan, S.: Node classification in social networks. In: Social Network Data Analytics, pp. 115–148. Springer (2011)
3. Bhagat, S., Cormode, G., Rozenbaum, I.: Applying Link-Based Classification to Label Blogs. In: Zhang, H., Spiliopoulou, M., Mobasher, B., Giles, C.L., McCallum, A., Nasraoui, O., Srivastava, J., Yen, J. (eds.) WebKDD 2007. LNCS, vol. 5439, pp. 97–117. Springer, Heidelberg (2009)
4. Bianchini, M., Maggini, M., Jain, L.C.: Handbook on Neural Information Processing, vol. 49. Springer (2013)
5. Callut, J., Françoisse, K., Saerens, M., Dupont, P.E.: Semi-supervised Classification from Discriminative Random Walks. In: Daelemans, W., Goethals, B., Morik, K. (eds.) ECML PKDD 2008, Part I. LNCS (LNAI), vol. 5211, pp. 162–177. Springer, Heidelberg (2008)
6. Cataltepe, Z., Sonmez, A., Baglioglu, K., Erzan, A.: Collective Classification Using Heterogeneous Classifiers. In: Perner, P. (ed.) MLDM 2011. LNCS, vol. 6871, pp. 155–169. Springer, Heidelberg (2011)

7. Chakrabarti, S., Dom, B., Indyk, P.: Enhanced hypertext categorization using hyperlinks. In: SIGMOD Conference, pp. 307–318 (1998)
8. Geman, S., Geman, D.: Stochastic relaxation, Gibbs distributions, and the Bayesian restoration of images. IEEE Trans. Pattern Anal. Mach. Intell. **6**(6), 721–741 (1984)
9. Kazienko, P., Kajdanowicz, T.: Collective classification, structural features. In: Encyclopedia of Social Network Analysis and Mining, pp. 156–168 (2014)
10. Kleinberg, J.M., Tardos, É.: Approximation algorithms for classification problems with pairwise relationships: metric labeling and Markov random fields. J. ACM **49**(5), 616–639 (2002)
11. Lenhart, A., Madden, M.: Teens, privacy and online social networks. Tech. rep., Pew Internet (2007)
12. Lu, Q., Getoor, L.: Link-based classification. In: ICML, pp. 496–503 (2003)
13. Macskassy, S.A., Provost, F.: A simple relational classifier. In: Proceedings of the SIGKDD 2003 2nd Workshop on Multi-Relational Data Mining (2003)
14. Macskassy, S.A., Provost, F.J.: Classification in networked data: a toolkit and a univariate case study. Journal of Machine Learning Research **8**, 935–983 (2007)
15. McDowell, L., Gupta, K.M., Aha, D.W.: Cautious inference in collective classification. In: AAAI. pp. 596–601 (2007)
16. McDowell, L., Gupta, K.M., Aha, D.W.: Cautious collective classification. Journal of Machine Learning Research **10**, 2777–2836 (2009)
17. Neville, J., Jensen, D.: Iterative clasification in relational data. In: AAAI 2000 Workshop on Learning Statistical Models from Relational Data, pp. 42–49 (2000)
18. Neville, J., Jensen, D.: Dependency networks for relational data. In: ICDM, pp. 170–177 (2004)
19. Singla, P., Richardson, M.: Yes, there is a correlation: from social networks to personal behavior on the Web. In: WWW, pp. 655–664 (2008)
20. Szummer, M., Jaakkola, T.: Partially labeled classification with Markov random walks. In: NIPS, pp. 945–952 (2001)
21. Taskar, B., Abbeel, P., Koller, D.: Discriminative probabilistic models for relational data. In: UAI, pp. 485–492 (2002)
22. Yedidia, J.S., Freeman, W.T., Weiss, Y.: Generalized belief propagation. In: NIPS, pp. 689–695 (2000)
23. Zhu, X., Ghahramani, Z., Lafferty, J.D.: Semi-supervised learning using Gaussian fields and harmonic functions. In: ICML, pp. 912–919 (2003)

Applications of Computational Intelligence

Deconvolution of X-ray Diffraction Profiles Using Genetic Algorithms and Differential Evolution

Sidolina P. Santos[1], Juan A. Gomez-Pulido[2]([✉]),
and Florentino Sanchez-Bajo[3]

[1] High Technical School, Polytechnic Institute of Leiria, Leiria, Portugal
[2] Polytechnic School, University of Extremadura, Caceres, Spain
[3] Industrial Engineering School, University of Extremadura, Badajoz, Spain
sidolina.santos@ipleiria.pt, {jangomez,fsanbajo}@unex.es

Abstract. Some optimization problems arise when X-ray diffraction profiles are used to determine the microcrystalline characteristics of materials, like the detection of diffraction peaks and the deconvolution process necessary to obtain the pure diffraction profile. After applying the genetic algorithms to solve satisfactorily the first problem, in this work we propose two evolutionary algorithms to solve the deconvolution problem. This optimization problem targets the objective of obtaining the profile that contains the microstructural characteristics of a material from the experimental data and instrumental effects. This is a complex problem, ill-conditioned, since not only there are many possible solutions, but also some of them lack physical sense. In order to avoid such circumstance, the regularization techniques are used, where the optimization of some of their parameters by means of intelligent computing permits to obtain the optimal solutions of the problem.

Keywords: X-ray · Diffraction profiles · Deconvolution · Genetic algorithms · Differential evolution

1 Introduction

X-ray diffraction is an usual technique in physics that allows to obtain information about the microstructural nature of materials, such as the crystalline structure of a material (size, shape, orientation), the average distance between the layers of atoms, etc [1]. The electromagnetic waves, when collide against the material particles, cause diffractions whose angles and wave intensities are measured to plot the diffraction profile, where the Y axis represents the intensity of the radiation (*counts*) and the X axis is the diffraction angle (2θ, expressed in degrees). This profile has some peaks corresponding with the angles of the diffracted rays of high intensity.

The parameters that cause each peak in the profile give us information about the internal structure of the material. An optimization problem involves to find

© Springer International Publishing Switzerland 2015
I. Rojas et al. (Eds.): IWANN 2015, Part II, LNCS 9095, pp. 503–514, 2015.
DOI: 10.1007/978-3-319-19222-2_42

the value of these parameters and to know how many peaks are in the profile, whereas a second optimization problem tackles the optimal deconvolution that allows us to have the pure diffraction profile of a material.

2   Deconvolution Problem

The relationship among the diffraction profiles of materials is defined by the Fredholm's linear integral one-dimension first-class equation (1), where: x and y are profile samples (diffraction angles); $\mathbf{f}$ is the pure diffraction profile, that represents the microstructural characteristics of the material; $\mathbf{h}$ is the experimental diffraction profile (experimental data); and $\mathbf{g}$ is the equation kernel or instrumental diffraction profile (due to the instrumental and spectral effects, and it represents a "standard" material free of microstructural effects).

$$\int_{-\infty}^{+\infty} g(x - y) f(y) \, dy = h(x) \tag{1}$$

The goal is to obtain $\mathbf{f}$ once $\mathbf{h}$ and $\mathbf{g}$ are known (usually, in discrete terms, where there are N samples y for $\mathbf{f}$ and M samples x for $\mathbf{h}$). Nevertheless, this is an ill-conditioned problem, because there is not an only solution $\mathbf{f}$ but several sets of $\mathbf{f}$ that solve the equation, and because a small change in $\mathbf{h}$ produces a great change in $\mathbf{f}$. If an usual deconvolution is done, we can obtain a $\mathbf{f}$ without physical sense; therefore, we must use regularization methods in order to find an appropriate $\mathbf{f}$, which minimizes certain function composed of a basic function (physical deconvolution) plus a regularization term (to give a physical sense).

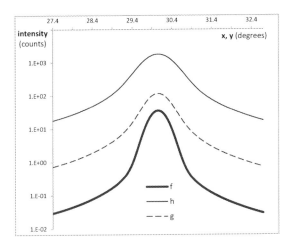

Fig. 1. Difraction profiles: pure ($\mathbf{f}$), experimental ($\mathbf{h}$) and instrumental ($\mathbf{g}$)

2.1   Discretization of the Convolution

The convolution equation (1) can be discretized through the quadrature rule (2), where w_j are N weights calculated using the Simpson's rule. This way, the discrete convolution function becomes (3), expressed in matrix notation in (4), where the **K** matrix is built from **g**.

$$\int_{-\infty}^{+\infty} f(y)\, dy = \sum_{j=1}^{N} w_j f(y_j) \tag{2}$$

$$\sum_{j=1}^{N} w_{ij} g(x_i - y_j) f(y_j) = h(x_i) \rightarrow \sum_{j=1}^{N} k_{ij} f_j = h_i \tag{3}$$

$$\mathbf{Kf = h} \tag{4}$$

The discrete convolution equation is (5), where x_i and y_j are 2θ samples; $M \geq N$; $h_i = h(x_i), i = \{1, ..., M\}$ (known values); $f_j = h(y_j), j = \{1, ..., N\}$ (unknown values); and $k_{ij} = w_{ij}\, g(x_i - y_j)$ (known values).

$$\begin{pmatrix} k_{11} & k_{12} & ... & k_{1N} \\ k_{21} & k_{22} & ... & k_{2N} \\ ... & ... & ... & ... \\ k_{M1} & k_{M2} & ... & k_{MN} \end{pmatrix} \begin{pmatrix} f_1 \\ f_2 \\ ... \\ f_N \end{pmatrix} = \begin{pmatrix} h_1 \\ h_2 \\ ... \\ h_M \end{pmatrix} \tag{5}$$

The w_{ij} weights are really the w_j weights, because the integral is in y: $w_{ij} = w_j$ and $k_{ij} = w_j\, g(x_i - y_j)$. Thus, the matrix **K** turns into (6).

$$K = \begin{pmatrix} w_1 g(x_1 - y_1) & ... & w_N g(x_1 - y_N) \\ w_1 g(x_2 - y_1) & ... & w_N g(x_2 - y_N) \\ ... & ... & ... \\ w_1 g(x_M - y_1) & ... & w_N g(x_M - y_N) \end{pmatrix} \tag{6}$$

The weights w_j are the same than in the extended Simpson's rule (7), where $h = \frac{y_N - y_1}{2}$, $f_j = f(y_j)$ and $E(f) = -\frac{nh^5}{90} f^{(4)}(e)$ is the global error.

$$\int_{y_1}^{y_N} f(y)\, dy = \frac{h}{3}[f_1 + 4(f_2 + f_4 + ... + f_{N-1}) +$$

$$2(f_3 + f_5 + ... + f_{N-2}) + f_N] - E(f) = \sum_{j=1}^{N} w_j\, f(y_j) \tag{7}$$

In this rule, the factors are 1/3 (first), 2/3 (evens), 4/3 (odds), and 1/3 (last) [4]. The matrix **K** expressed in (6) is built calculating the differences $x_i - y_j$, so $x_1 > y_N$. The distance from x_1 to y_N represents the data interval, and go from an initial angle to a final one: $x_1 = 2\theta_1$, $x_2 = 2\theta_2$, $x_3 = 2\theta_3$, etc. The experimental (h) and instrumental (g) peaks have the same angular range, x_i. This way, the difference $x_i - y_j$ can be taken as the difference with regard to the angle value for which the intensity is maximum. If this value is $x_{max} = 2\theta_{max}$, for example, we can consider that 0 and $x_i - y_j < 0$ are values of the instrumental profile to the left of the value of maximum intensity, and $x_i - y_j > 0$ are to its right. The values out of $\pm N/2$ (approximately) would be null (the profile takes null values outside the range). The profiles generated have N even, so there are $N/2$ values under the maximum, the maximum and $N/2$ above the maximum.

2.2   Solving the Deconvolution Problem

The deconvolution problem is similar to solve a system of algebraic equations (8) where $\mathbf{A}$ is ill-conditioned due to the dependence almost linear between adjacent rows and columns.

$$\mathbf{Ax} = \mathbf{b} \tag{8}$$

Let (9) be the system of M linear equations with N unknown quantities, where: $i = \{1, ..., M\}$ and $j = \{1, ..., N\}$; $\mathbf{A}$ = matrix $M \times N (M \geq N)$, where a_{ij} are the system coefficients, whose norm is $||\mathbf{A}|| = \sqrt{\sum_{i=1}^{M} \sum_{j=1}^{N} |a_{ij}|^2}$; $\mathbf{b}$ = vector of size M, where b_j are the independent terms of the system; and $\mathbf{x}$ = vector of size N, where x_j are the system unknowns.

$$\begin{pmatrix} a_{11} & ... & a_{1N} \\ a_{21} & ... & a_{2N} \\ ... & ... & ... \\ a_{M1} & ... & a_{MN} \end{pmatrix} \begin{pmatrix} x_1 \\ x_2 \\ ... \\ x_N \end{pmatrix} = \begin{pmatrix} b_1 \\ b_2 \\ ... \\ b_M \end{pmatrix} \rightarrow \begin{matrix} a_{11}x_1 + ... + a_{1N}x_N = b_1 \\ a_{21}x_1 + ... + a_{2N}x_N = b_2 \\ ... \\ a_{M1}x_1 + ... + a_{MN}x_N = b_M \end{matrix} \tag{9}$$

A vector $(s_1, s_2, ..., s_N)$ is a solution if the system becomes like (10).

$$\begin{matrix} a_{11}s_1 + a_{12}s_2 + ... + a_{1N}s_N = b_1 \\ a_{21}s_1 + a_{22}s_2 + ... + a_{2N}s_N = b_2 \\ ... \\ a_{M1}s_1 + a_{M2}s_2 + ... + a_{MN}s_N = b_M \end{matrix} \tag{10}$$

The system of equations can be incompatible (if $rank(\mathbf{A}) < rank(\mathbf{A/b})$, there is not solution) or compatible (if $rank(\mathbf{A}) = rank(\mathbf{A/b})$). In this last case, there is a solution, that can be determined (only one solution) or indeterminate (more than one solution). The deconvolution of X-ray diffraction profiles is indeterminate: there are many possible solutions $\mathbf{x}$; for each solution the norm is calculated according to (11). Therefore, the physical deconvolution problem is to minimize $\mathbf{A(f)}$, in other words, to find the $\mathbf{f}$ with minimum norm.

$$A(\mathbf{f}) = ||\mathbf{Kf\text{-}h}||^2 \tag{11}$$

Nevertheless, the solution with minimum norm does not depend continuously of $\mathbf{h}$ (a small change in $\mathbf{h}$ produces a great random change in $\mathbf{f}$, due to the non-singular nature of $\mathbf{K}^T\mathbf{K}$). The instrumental limitations produce to lose details of **f**, because the data noise (**h**) generates oscillatory terms in the solution that have not physical sense. As a result of this instability of the solution (degenerate kernel), a function **f** that minimizes **A(f)** could not be necessary close to the "true" solution. In general, the area of possible solutions is restricted according to additional properties for the functions, such as smoothness, non-negativity, etc., that allow to select the best solution within a certain class of functions.

2.3   Regularization Methods

There are several deconvolution methods, like Stokes, series spread of **f**, regularization of order P (or LWL), and iterative. We consider in this work the regularization methods, that convert the ill-conditioned problem into a stable minimization problem. Thus, the physical deconvolution is to minimize **A(f)** (11). Nevertheless, in order to give a physical sense to the deconvolution, a regularization term is added, so the deconvolution must minimize the quadratic function (12), where $\mathbf{B}(\mathbf{L}, \mathbf{f}) = ||\mathbf{Lf}||^2$ is the stability term, **L** is the linear operator and α is the stabilizer parameter $(0 < \alpha < \infty)$.

$$\mathbf{C}(\mathbf{L}, \mathbf{f}, \alpha) = ||\mathbf{Kf} - \mathbf{h}||^2 + \alpha||\mathbf{Lf}||^2 = \mathbf{A}(\mathbf{f}) + \alpha \mathbf{B}(\mathbf{L}, \mathbf{f})$$

Summarizing: **A(f)** is the residue that measures the data fitting to the model (the sole minimization of **A(f)** gives a good tradeoff but with unstable solutions); **B(L,f)** is the stability term that measures the stabilization or "smoothness" degree of the solution and represents "a priori" knowledge of it; and α is the stabilizer parameter that controls the regularization "strength": if α is increased, **A(f)** becomes higher, decreasing **B(L,f)**, and if it decreases, **A(f)** and **B(L,f)** have the opposite behaviour.

Given a **L** operator, for each α we obtain an only $\mathbf{f}_\alpha$ that minimizes $\mathbf{C}(\mathbf{L},\mathbf{f},\alpha)$. Therefore, given **L** and α, the system of linear equations to be solved is not (4), but (12), where **H** is the stability matrix (positive). Solving such system of equations the solution $\mathbf{f}_\alpha$ is obtained with physical sense (regularized solution). Within the set of regularized solutions $\mathbf{f}_\alpha$, we must select the best one, for that a balance between the residue value and the stability term exists, as Fig. 2 shows.

$$(\mathbf{K}^T\mathbf{K} + \alpha\mathbf{H})\mathbf{f}_\alpha = \mathbf{K}^T\mathbf{h} \qquad (12)$$

A possible initial selection for α could be the pointed out in (13), that makes comparable the contribution of **A(f)** and **B(L,f)**.

$$\alpha_0 = \frac{tr(\mathbf{K^T K})}{tr(\mathbf{H})} \qquad (13)$$

The regularization methods are easy to apply, but they must select the optimal values for **L** and α. In the chosen method (regularization of order P, or LWL method), $\mathbf{Lf} = \mathbf{f}^p$, where $\mathbf{f}^p$ represents the order-p derivative of **f**. We have considered the regularizations of order 0 and 1 in the experiments:

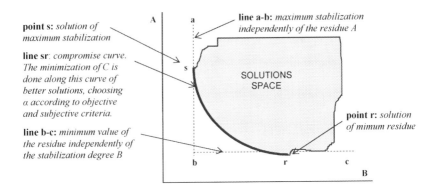

Fig. 2. Space of possible solutions

- Order 0: $\mathbf{Lf} = \mathbf{f}$. We minimize $\mathbf{C}(\mathbf{f}, \alpha) = \mathbf{A}(\mathbf{f}) + \alpha ||\mathbf{f}||^2$. The system of equations is $(\mathbf{K}^T\mathbf{K} + \alpha\mathbf{I})\mathbf{f}_\alpha = \mathbf{K}^T\mathbf{h}$, where $\mathbf{I}$ is the Identity matrix.
- Order 1: $\mathbf{Lf} = \mathbf{f}'$. We minimize $\mathbf{C}(\mathbf{f}, \alpha) = \mathbf{A}(\mathbf{f}) + \alpha ||\mathbf{f}'||^2$. The system of equations is $(\mathbf{K}^T\mathbf{K} + \alpha\mathbf{M})\mathbf{f}_\alpha = \mathbf{K}^T\mathbf{h}$, where $\mathbf{M}$ is represented in (14).

$$M = \begin{pmatrix} 1 & -1 & 0 & ... \\ -1 & 2 & -1 & ... \\ 0 & -1 & 2 & ... \\ ... & ... & ... & ... \end{pmatrix} \tag{14}$$

3   Optimization Problem

The deconvolution is a complex problem that tries to find the profile where the microstructure effects in a material are taken into account.

In the field of X-ray diffraction, there are procedures to obtain microstructure parameters without doing the deconvolution of the pure profile. In these cases, it is usual to consider some hypothesis about a determined analytic shape of the profiles. Nevertheless, the deconvolution problem has not an analytic solution, because it would suppose to analyze the profiles in terms of analytic functions, and that is not the case, not only for the pure profile, but also for the experimental and instrumental profiles.

The deconvolution problem could be tackled through linear algebra, solving a matrix equation. In this case, the solution is function of the regularization parameter. We can try different values of that parameter (with an approximate initial estimation) until finding an adequate solution (a solution where, for example, the oscillations in the profile tail are minimized). Since there is not an optimal parameter at the start, a procedure based on evolutionary algorithms could be a good choice to find the optimal parameter. In addition, it is not possible to use traditional optimization methods because the solution of minimum norm is the no-regularized solution, in other works, without stability term.

After formulating the deconvolution of profiles as a non-deterministic optimization problem, the Evolutionary Algorithms (EAs) [5] are a good option to obtain optimal solutions. With this purpose, we have considered Genetic Algorithms (GAs) and Differential Evolution (DE).

When considering GA [6], it is important to define an adequate structure of the individual in the population, the fitness function and the role of the selection, cross and mutation operators. Each population individual is represented by a real value for each regularization method. Since this representation is real-valued (continuous) instead of binary (discrete), the crossover between individuals is realized by the SBXCrossover operator [7]. The selection of the individuals to cross is done using a Binary Tournament selection operator [8]. The mutation operator applied is a Polynomial Mutation operator [9], again because of the continuous value representation

The algorithm DE [10] selects vectors of individuals where some arithmetic and genetic operators are applied to obtain the best individual in each generation. DE uses some typical strategies that allow changing parameters of the genetic operators. The main advantages of DE are the speed and robustness, due to its simple structure and easiness of use. The structure of the individual in the population, fitness function and stop criterion are the same than in GA.

There are some papers where inverse problems of great complexity have been tackled by evolutionary techniques, like [11], where a problem that has not analytic solution is solved by EAs.

4   Experimental Results

Given a L determined by the regularization method chosen, GA and DE minimize the fitness function $\mathbf{C}$ (12), obtaining the optimal α, from which the profile $\mathbf{f}_\alpha$ is obtained solving the system of linear equations (12). This way, when the EA ends its execution, it returns the optimal solution given by the pair $\{\alpha, C\}$.

4.1   Experimental Data

We have generated four datasets from the simulation of instrumental $\mathbf{g}$ and pure $\mathbf{f}$ profiles as Voight functions [12], so their convolution (experimental profile $\mathbf{h}$) is also another Voight function. Each dataset was generated having in mind different values for their characteristics, but sharing the same values for the peak maximum height in the instrumental and pure profiles (20), the samples step ($0.01°$) and the position of the peak maximum ($30°$), as well as truncating data to the 1% of the peak maximum intensity and assuring a maximum intensity value for the experimental profile between 1500 and 7000 counts.

On the other hand, a simulated statistical noise $\mathbf{rh}$ was added to the experimental profile $\mathbf{h}$, generating an experimental profile with noise $\mathbf{hr}$. This way, the deconvolution process uses $\mathbf{hr}$ and $\mathbf{g}$ to obtain a "pure" profile that can be compared with the original one $\mathbf{f}$. Figure 3 shows the four datasets, where the three profiles of the deconvolution and the statistical noise can be observed.

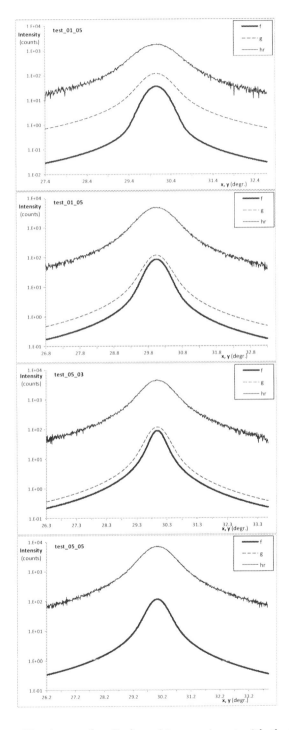

Fig. 3. The dataset (profiles) used to experiment with the EAs

4.2   Experimental Framework

Figure 4 shows a detail of the framework developed in Java to perform experiments in any platform, make easier their configurations and see the results. The parameters that define the experiment are stored in the plain-text file "PARAMETERS.txt", which is loaded in the framework before starting the execution.

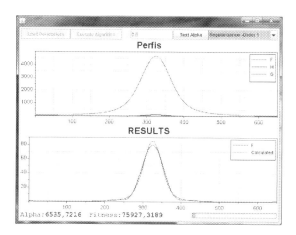

Fig. 4. Experimental framework developed in Java to find the optimal deconvolution of diffraction profiles using evolutionary algorithms. It shows, among other features, a comparison between the deconvoluted profile and the original simulated.

4.3   Experiments

Table 1 shows the values of the main parameters that configure the EAs.

Table 1. Main configurations of the EAs

Parameter	AG	ED
Population size	100	100
Iterations	4.000	4.000
Cross probability	0,5	0,7
Mutation probability	0,7	N/A
Differential weight	N/A	0,5

We have performed 7 experiments for the profile *test_03_05_v2*, that considers a step of 0.05 (131 samples), four (*Exp1* to *Exp4*) with GA and three (*Exp5* to *Exp7*) with DE. Each experiment consists of 25 runs of the algorithm, where each of them returns the minimum fitness found together with the corresponding optimal value of α, from which the optimal profile **f** can be generated.

Table 2 shows the values of the main parameters that configure each experiment according to the EA considered. In some experiments, the fitness is penalized by the amount *penalty* if negative values appear in the profile **f** generated. The way to penalize is determined by the *penalty type* variable: if it is equal to zero, then the value to penalize is always the same; if it equals 1, then the value assigned to the negative value of **f** is multiplied by *penalty*, ie, the amount of the penalty is proportional to the negative values of **f**.

Table 3 shows the results of the minimum fitness **C** found, together with their corresponding statistics generated from the 25 runs of each experiment (average fitness and its standard deviation). On this table, particularly in the DE algorithm experiments, a large average and standard deviation is presented. These values reflect a premature convergence of this algorithm, specifically convergence rates of 72% in exp5, 48% on exp6 and 36% on exp7. These rates were calculated according to (15). Nevertheless, we can observe in the second plot of Fig. 4 an example of how the obtained F fits well to the desired F.

$$Rate = \frac{Premature\ convergence\ results}{Total\ number\ of\ results} \tag{15}$$

Table 4 shows the optimal values of α obtained in each run. A wide search range for possible values of α, between 10^{-5} and 10^8, has been established for all the experiments. Nevertheless, the optimal values of α obtained in each run are in a smaller range, as we can see in the columns α_{min} and α_{max} of the table.

Table 2. Experiment configuration

Experiment	Type	Regularization order	Penalization type	Penalization value
Exp1	N/A	0	0	10^8
Exp2	N/A	1	0	10^8
Exp3	N/A	1	1	10^5
Exp4	N/A	0	1	10^5
Exp5	best/1/exp	1	0	10^8
Exp6	rand/1/bin	1	0	10^8
Exp7	best/1/exp	1	1	10^4

Finally, an additional experiment *Exp8* was performed; it is *Exp3* but considering a higher number of experimental samples, after reducing the step from 0.05 to 0.01. The purpose of this experiment was to check the influence of the number of samples in the precision of the results and the computing time.

4.4   Computing Time

The computing time for an AE run varies between few seconds and several minutes depending on the computer, profile size (number of samples) and the AE configuration. Nevertheless, the number of samples is the parameter with

Table 3. Fitness results

Experiment	Best fitness	Fitness average	Standard deviation
Exp1	1,318,250.55	1,324,110.88	6,048.71
Exp2	15,707.18	15,791.61	110.47
Exp3	15,481.04	15,494.17	11.53
Exp4	318,677.53	318,764.16	116.94
Exp5	15,754.93	17,988,409.30	11,207,830.29
Exp6	15,735.52	11,997,528.88	12,470,963.46
Exp7	15,480.00	3,500,395.69	4,646,554.24

Table 4. Values of α

Experiment	optimal α	α_{min}	α_{max}
Exp1	23,25	23,25	23,59
Exp2	149,79	149,79	154,25
Exp3	144,99	144,38	145,71
Exp4	2,14	2,06	2,16
Exp5	150,21	150,21	152,47
Exp6	149,91	149,91	153,26
Exp7	144,97	144,96	144.98

more influence on the computing time. For example, using an Intel Core i7-2600@3.7GHz, an AG run for a profile of 651 samples (step 0.01) takes 6 minutes, whereas if we consider a 0.05 step (131 samples) the time decreases to 30 seconds.

The computing time can be reduced, even for smaller steps, using multicore computers or designing specific-purpose coprocessors, already used with success in the optimization problem of detecting X-ray diffraction peaks [13].

4.5   Results Discussion

After studying the experimental results, we can say that using AEs allows a good deconvolution, because the profiles obtained were very close to the previously simulated. In addition, the results were better using regularization of order 1 than order 0; this move us to follow the strategy of a mono-objective optimization (the order of the regularization is not already an objective) where the only objective to minimize is **C** (in order to obtain an optimal α), setting **L** as the identity matrix. Finally, the results do not improve significantly when considering a smaller sample step (and therefore, a bigger number of profile samples processed), whereas the computing time increases a lot.

Similar experiments have been performed with the other profiles of the dataset, with similar results and conclusions.

5   Conclusions and Future Work

The evolutionary algorithms considered in this work have been useful to find optimal solutions in the deconvolution of X-ray diffraction profiles. These algorithms were programmed to implement regularization methods that avoid to find solutions apparently optimal but without physical sense. The good results obtained encourage us to continue deepening in this research line.

This is a problem with many possibilities to explore yet as future works. Thus, we can consider higher regularization orders, parallelize the fitness to shorten the computation time and include new evolutionary algorithms to improve the results. In addition, it would be interesting to compare the results with other deconvolution techniques, like Stokes, iterative, Ergun or Mencik methods.

References

1. Waseda, Y., Matsubara, E., Shinoda, K.: X-Ray Diffraction Crystallography. Springer (2011)
2. Enzo, S., et al.: A profile-fitting procedure for analysis of broadened x-ray diffraction peaks. Journal of Applied Crystallography **21**, 536–542 (1988)
3. Pereira, S., Gómez, J.A., Vega, M.A., Sánchez, J.M., Sánchez, F.: Aplicación de los Algoritmos Genéticos y la Evolución Diferencial para la Optimización de Perfiles de Difracción de Rayos X. MAEB 2009, Malaga, Spain, February, 11–13, pp. 9–1 (2009)
4. Abramowitz, M., Stegun, I.A.: Handbook of Mathematical Functions. Dover, New York (1964)
5. Fogel, D.B., Back, T., Michalewicz, Z.: Evolutionary Computation 1. Basic Algorithms and Operators. IOP, Philadelphia (2000)
6. Golberg, D.E.: Genetic Algorithms. Addison-Wesley (1988)
7. Deb, K., Agrawal, R.B.: Simulated binary crossover for continuous search space. Complex Systems **9**(3), 1–15 (1994)
8. Goldberg, D.E., Deb, K.: A comparative analysis of selection schemes used in genetic algorithms. Foundations of Genetic Algorithms **1**, 69–93 (1991)
9. Deb, K.: Multi-objective optimization using evolutionary algorithms. Foundations of genetic algorithms. John Wiley & Sons (2001)
10. Storn, R., Price, K.: Differential evolution. A simple and efficient heuristic for global optimization over continuous spaces. J. Global Optimization **11**, 341–359 (1997)
11. Macías, D., Olague, G., Méndez, E.R.: Inverse Scattering with Far-field Intensity Data: Random Surfaces that Belong to a Well-defined Statistical Class. Waves in Random and Complex Media **16**(4), 545–560 (2006)
12. Sánchez-Bajo, F., Cumbrera, F.L.: The use of the pseudo-Voigt function in the variance method of x-ray line-broadening analysis. Journal of Applied Crystallography **30**(4), 427–430 (1997)
13. Gomez, J., Sanchez, F., Pereira, S., Vega, M., Sanchez, J.: Custom Hardware Processor to Compute a Figure of Merit for the Fit of X-Ray Diffraction Peaks. X-Ray Optics and Instrumentation **2008**, 1–7 (2008)

Using ANN in Financial Markets
Micro-Structure Analysis

Brayan S. Reyes Daza[1] and Octavio J. Salcedo Parra[2(✉)]

[1] Universidad Distrital Francisco José de Caldas, Bogotá D.C., Colombia
bsreyesd@correo.udistrital.edu.co
[2] Universidad Nacional de Colombia, Bogotá D.C., Colombia
osalcedo@udistrital.edu.co

Abstract. The present document presents/displays a model of Neuronal Networks Artificial RNA for the prognosis of the rate of nominal change in Colombia, including flow orders and the differential of the interest rates like variables of entrance to the model. Additionally methodological conclusions from the traditional treatment of the series of time were extracted.

Keywords: Type of nominal change · Artificial prognosis · Not-linear model · Neuronal networks · Micro-structure of the financial markets

1   Introduction

The analysis for modeling and forecasting the exchange rate of one currency against other has had several important stages. In this sense, the first models developed have, as its starting point, the balance of flows between countries, the raised of the rate could be determined by supply and demand functions for foreign exchange each of the countries, including models are part of this line of development are those of Meade (1951) and Mundell - Fleming (1963) mentioned by Manrique (2001).

Subsequently with the dynamic acquired by the financial markets and the collapse of the Bretton Woods, several models appeared in which the international trade, international flows of trade, the price of exports, domestic goods and international portfolio of assets, were taken as variables relevant to the determination of the exchange rate. This approach is called the balance of stocks or asset market, which is divided into monetary models (flexible price, sticky prices) and balance models, which attempt to explain the fluctuations in the exchange rate using a process similar to that Prices are subject to other financial assets. Currently in Colombia the literature dealing with microeconomic models and the exchange rate is not very wide, in contrast to the literature on real rates, Cárdenas (1997) mentioned the work of Wiesner (1978), Urrutia (1981), Lopez (1987) and Steiner (1987) on the process of crawling peg in Colombia's crawling peg system. The work of Cardenas (1997), analyzes the determinants of nominal exchange rate in the period 1985-1986, under the two regimes, crawling peg and band's[65] system, through: a simple monetary model, the fixed-price monetary model and portfolio balance model, the conclusion about the determinants suggests that the monetary model with flexible prices is set in a good way the

© Springer International Publishing Switzerland 2015
I. Rojas et al. (Eds.): IWANN 2015, Part II, LNCS 9095, pp. 515–523, 2015.
DOI: 10.1007/978-3-319-19222-2_43

behavior of the Colombian exchange rate. According to the exchange rate model responds to changes in the money supply and interest rates. In this context, based on the microstructure and RNA, and taking into account the criticism entirely macro models of Evans and Lyons (1999) and expressed by Barkoulas, Baum, Caglayan and Chakraborty (2001) on the processes and martingale type long-term memory, this paper examines the behavior of a forecasting model of the TCN of the peso against the U.S. introduce a variable dollar of market microstructure (order flow) in a system of daily observations with macroeconomic variables (interest rates), under a non-linear modeling of RNA, daily time-series of one (1) year seeking to measure the predictive power and behavior modeling using the root mean square error (RMSE) and absolute average percentage error (MAPE) [67].

For this item is divided into four sections: The first section covered the introduction. Section 2 reviews the related to the RNA and the microstructure of financial markets. Section 3 describes the data, analyzes and presents the results. Section 4 is for conclusions and recommendations.

2   Artificial Neural Networks -ANN- And Financial Markets Micro-structure

2.1   Artificial Neural Networks

An Artificial Neural Network (ANN) is an attempt to perform a computer simulation of the behavior of the human brain through small-scale replica of the patterns that it plays in shaping results from the events received. More formally the ANN is nonlinear statistical models used mainly for classification and prediction of data and variables, inspired by biological nervous systems, which try to simulate the human learning process in the belief that having been created by the selection process natural mechanism to be efficient. (Montenegro, 2001).

The structure of an artificial neuron is an emulation of a biological neuron so that it could do the following parallel with the biological neuron:

- The inputs Xi represent discrete pulses from other neurons and are absorbed by the dendrites (UTP, 2000).
- The weights Wi represent the intensity of the synapse connecting two neurons. 0, is the threshold value the neuron must overcome, to produce the biological process within the cell when activated, analogy can be seen in Figure 1.

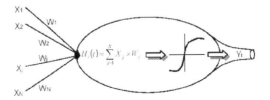

Fig. 1. ANN structure

Where

- *H(t)*: Potential synaptic neuron *i* at time t.
- *X_j*: input data from the information source *j*.
- *W_j*: The synaptic weight associated with the input X_j

In this project the Multilayer Perceptron-MLP-learning ANN architecture was used, with the back propagation technique, in which the topology can have an input layer with n neurons, for this case study, we used a two neurons layers for the first model (the order-flow-ODF and the difference between the DTF and the Libor-DDL-with the following setbacks: ODF_{t-1}, ODF_{t-2}, ODF_{t-9}, ODF_{t-13}, ODF_{t-36}, DDL_n, DDL_{t-5}, DDL_{t-6}, DDL_{t-8}, DDL_{t-9}, DDL_{t-17}), and five neurons (TRM_{t-1}, TRM_{t-2}, TRM_{t-3}; TRM_{t-4} and TRM_{t-6}) for the second model, at least one hidden layer (with four and eight neurons, for first and second ANN models respectively) also with n neurons and an output layer with m neurons, for these models are only used one output neuron-TRM-in RNA synthesis considered has an architecture with an input layer, a hidden layer and output layer, so it can be expressed as RNA (I, H, O).

The functional architecture of the network is:

$$Y_t = g(\sum_{h=1}^{H} c_h g(\sum_{i=1}^{I} x_{tt} w_{ik} + \theta_h) + d) \tag{1}$$

Where ch are the weights that connect the neuron h in the hidden layer neuron to the output layer and d the threshold of the neuron in the output layer, weights (w, c) and thresholds (0, d) are adjusted during the training of the network. The formula for adjusting the weights of the network depends on the position of the weights connecting layers, particularly if the weights are in the hidden layer or output layer.

The base model of the artificial neural network worked is:

$$Yt = g(\sum_{h=1}^{H} c_h g(\sum_{i=1}^{I} x_{tt} w_{ih} + \theta_h) + d) \tag{2}$$

2.2   Financial Market Microstructure and Exchange Rate

The microstructure of financial markets is the study of the processes and outcomes that occur in exchanging assets under explicit trading rules (Marín and Rubio, 2001). This microstructure focuses on the interaction between the mechanisms of the negotiation process and its results in terms of prices and quantities traded. It is recognized that specific rules under which the negotiation process occurs directly affect the outcome of such processes, i.e. the behavior of agents in the game of supply and demand determine the price and transaction volumes.

The support of the microstructures for the TCN study is summarized in order flow and spread, the first concept concerns the volume of transactions and the same meaning, i.e. the volume to settle and if bought or sold, which is treated as excess demand, any time you perform an operation does not necessarily imply a zero-sum balance, the latter in turn is conditional on the price as it enters the information asymmetries in financial markets will. It is also important to mention that the approach of modeling TCN microstructures at the following address:

$$P_t = f(X, I, Z) + g(i) + \varepsilon_t \qquad (3)$$

3   Empirical Analysis

3.1   Available Data

The availability of detailed databases on the intraday and daily activity in financial markets has opened the possibility to econometric research on the functioning of these markets, therefore the flow volume order ODF, TRM and the differential rate DDL Libor interest from April 16, 2003 to April 16, 2004. The set of training artificial neural network is defined according to a measure of error between the data generated and the training set (actual data), this value usually varies, whereas small values.

In the training process is likely that if the RNA undergoes a process of overtraining, which causes the RNA from a loop or has a sub training process, then both processes, the RNA lose their adjustment capacity, prognosis and generalization (Buitrago and Alcala, 1998).   In regard to the variable interest rate differential (DDL) was taken, the libor ninety 90 days macroeconomic component of the U.S. economy, the source of the data is the system REUTER and Colombia is ninety took the DTF 90 This data comes from the website of CORFINSURA. The foreign exchange market in the country is relatively new 1991. Finally, taking into account the definition given before order flow, this is calculated as the difference between all the purchase transactions initiated (T) and all transactions initiated on sale (P) on the same day, so if OD> 0 is more intent to purchase and therefore upward pressure on exchange rate - which is not necessarily explained by macro variables of the market - and when OD <0 the reverse process occurs.

3.2   Results Analysis

Figure 2 shows the variables in levels. Hopefully, a high volatility characteristic of financial time series. Using the ADF and KPSS tests determined the stationary of the series.

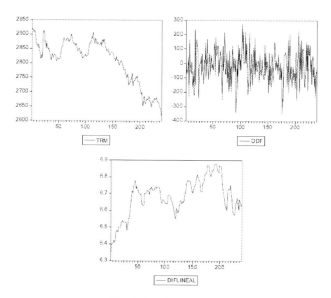

Fig. 2. Level Series

Table 1. Unit root tests

	ADF	**Critic Value 1%**	**KPSS**	**Critic Value 1%**	**Decision**
TRM (1)	-1.829393	-2.574714	1.550905	0.739	KD
ODF(2)	-15.03944	-3.457747	0.173559	0.739	1(0)
DIFERENCIAL	-2.66635	-3.457865	0.851827	0.739	KD

- (1) $Q(36) = 44.25(0.163)$
- (2) $Q(36) = 43.19(0.191)$
- (3) $Q(36) = 22.937(0.955)$

The results show that at a significance level of 1%, and TRM series of interest rate differential is not stationary. ODF Series is stationary. Additionally we present the Ljung-Box test for the residue of the auxiliary regression in each case and in brackets the p-value associated with the test.In the case of parametric models should be considered the property of stationary. Below are the results of Johansen co integration test and univariate models for forecasting TRM worked with the first difference in natural logarithm (which is the return compounded continuously).

3.3   Artificial Neural Network Models - ARN

As explained previously, RNA did not have a specific parametric model, which was done, was to vary the input vector, independent variables, order flow, differential interest rate (DTF-Libor) at time t and lag in t.

For the first model to implement the 8 - test, we obtained the following most significant setbacks for the model: (ODF_{t-1}, ODF_{t-2}, ODF_{t-9}, ODF_{t-3}, ODF_{t-36}, DDL_{t-1}, DDL_{t-5}, DDL_{t-6}, DDL_{t-8}, DDL_{t-9}, DDL_{t-17})

For the second model considers the TRM t = 1, like an AR (p), in this case, gives significance to the lags TRM_{t-1}; TRM_{t-2}; TRM_{t-3}; TRM_{t-14}; TRM_{t-16}

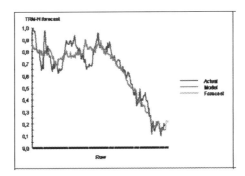

Fig. 3. ANN First Model Forecast h = 10

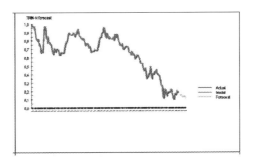

Fig. 4. ANN Second Model Forecast h = 10

Finally, we see that as time goes by (days) in second model of RNA loses predictability, however, their test statistics (MAPE, RMSE and approximately R2) behave appropriate and acceptable values, which makes RNA techniques to forecast the TRM a viable tool in the manner presented.

Table 2. Statistical Models AR (p) and RNA - One year

	Forecast 10 datos		
	RMSE	**MAPE**	**R²**
RNA First model	0.0775	1.054.870	93.54
RNA Second model	0,036	5.671	973.662

3.4   Time Series Model Multivariate Analysis

The following table presents the Johansen co-integration test. In all cases occur in the trace test. Results were obtained using 2 lags. In all cases there is one co integrating vector. However, as discussed above, one of the variables is 1 (0), which may affect this conclusion, in particular, this variable may be the one that forms the co integrating vector. Using the Schwartz criterion, the model was first considered appropriate. In estimating the VEC was obtained the following results:

Table 3. Johansen co-integration test

Vectors	First model	Critic Value 1%	Second model	Critic Value 1%	Third model	Critic Value 1%	Fourth model	Critic Value 1%
None	99.38	29.75	104.03	41.07	100.82	35.65	106.29	48.45
At least one	7.90(*)	16.31	12.36(*)	24.6	9.59(*)	20.04	14.94(*)	30.45
At least two	2.42	6.51	2.54	12.97	0.00	6.65	3.52	16.26

Table 4. Alpha Matrix and Co Integration Vector

	TRM	ODF	DIFERENTIAL
Coefficient	1	32.33773	446.6811
Statistical t		[10.4216]	[16.9962]
Alpha Matrix			
Coefficient	-0.00268	0.024488	-4.18E-07
Statistical t	[-9.72161]	[5.37546]	[-0.61896]

Table 5. Granger causality and Portmanteau Test

	Exelude	Test	P-value
Caused: TRM	D(ODF)	96.86	0.00
	D(DIFLINEAL)	2.43	0.30
Caused: ODF	D(TRM)	2.32	0.31
	D(DIFLINEAL)	1.04	0.59
Caused: DIFERENTIAL	D(TRM)	1.120	0.571
	D(ODF)	1.163	0.559
	Portamentau(40)	699.439	0.6377

3.5   Time Series Model Univariate Analysis

It was considered an alternative model of the first difference in the rate of change. A GARCH (1.1).

Table 6. Garch(1,1) Model

	Coefficient	z-	P-	
c	-0.05	-2.05	0.04	
D21	1.52	9.95		0
D25	-1.48	-3.9.4		0
D203	0.87	6.13		0
AR(1)	-0.34	-1.9.3	0.22	
MA(1)	0.55	2.31	0.02	
Variance				
C	0.03	1.68	0.09	
ARC(1)	0.27	2.97		0
GARCH(1,1)	0.45	2.19	0.02	

	Test	P-
Ljung-Box	34.74	0.17
Jarque-Bera	0.31	0.85
Arch(4) Model	1.62	0.80

Table 7. Forecast Evaluation With Models

	Garch(1,1)	Vec	First model ANN	Second model ANN
Rmse	22.90	20.66	31.98	17.24
Mape	17.94	15.72	26.93	12.78

The forecast evaluation results show that the neural network second model has a better prognosis than the parametric models introduced.

4   Conclusions and Recommendations

This work moves in the direction of the recent work of Evans and Lyons of the microstructure and exchange rate, making a first approach to the case of Colombia, together with a new non-linear approach to model-ANN-raised in the case of Mass inflation, Lopez and Cherub (2002). This approach allows the analysis of how signals are perceived differently by actors and somehow reflected in the microstructure of the information, being, however, asymmetries of the forex market.

The use of ANN, which is typically non-linear statistical models, which can be expressed as a generic model forecasts allowed approximation of functions acceptably good for a model of TRM according to the RMSE, with the inclusion of the variable microeconomic order to flow.

The explanatory power of macroeconomic variables, as mentioned by Evans and Lyons (1999), behavior exchange rate in the short term, daily-is very significant, since for the developed models behave as non-significant in this sense the financial market with speculative behavior, explained largely by the order flow is a determinant dollar price in the short term, as the daily trading volume far exceeds the volume of bids and actual demand of the economy, which largely explains the power of financial markets in determining price them.

The structure of the market is in itself is a determinant of prices and the determination of the nominal exchange rate, specifically the order flow as an indicator of the mechanisms of negotiation, agreement and settlement, it affects the price behavior. As a recommendation on methodology, it would be appropriate to initiate the study of research that merging different techniques of bio-inspired computing such as genetic algorithms and fuzzy logic in order to thus obtain predictive and optimization methods more effective and powerful as they try to simulate human behavior in ways different from the traditional linear models and computational.

References

1. Baillie, R., Bollerslev, T.: Common stochastic trends in a system of exchange rates. Journal of Finance **44**, 167–181 (1989)
2. Barkoulas, T., Baum, C., Caglayan, M., Chakraborty, A.: ``Persistent dependence in foreign exchange rates? A reexamination''. Department of Economics and Accounting. University of Liverpool (2000)
3. Buitrago, A., Alcalá, J.: Análisis, Diseño e Implementación de un Prototipo de Sistema Neuronal Para Pronóstico de Series de Tiempo Económicas. Departamento de Ingeniería de Sistemas. Universidad nacional (1998)
4. Cárdenas, M.: La tasa de cambio en Colombia. Editores Tercer Mundo S.A. Primera Edición, Bogotá (1997)
5. Diebold, F., Nason, J.: Nonparametric exchange rate prediction? Journal of International Economics **28**, 315–332 (1990)
6. Dominguez, K.: Are foreign exchange forecasts rational? New evidence from survey data. Economic Letters 21, 277–281 (1986)
7. Dornbusch, R.: Equilibrium and disequilibrium exchange rates, Zeitschrift fur Wirtschafts und Sozialwissenschaften, vol. 102, 573–799. MIT Press, Cambridge (1982)
8. Grossman, Rogoff, K. (eds.) Handbook of International Economics. Elsevier Science, Amsterdam 1689–1729
9. Flood, R., Rose, A.: Fixing exchange rates: A virtual quest for fundamentáis. Journal of Monetary Economics **36**, 3–37 (1995)
10. Gradojevic, N., Yang, J.: ``The Application of Artificial Networks To Exchange Rate Forecasting: The Role of Market Microstructure Variables'' Bank of Cañada December 2000, working paper 2000-23
11. Kuan, C-M., Liu, T.: Forecasting exchange rates using feedforward and recurrent neural networks. Journal of Applied Econometrics 10 347–364 (1995)
12. National Center for Biotechnology Information. http://www.ncbi.nlm.nih.govJalil, M.Y., Meló, L.F.: Una relación no lineal entre la inflación y los medios de pago. Borradores de Economía Banco de la República (1999)

Cluster Analysis of Finger-to-nose Test for Spinocerebellar Ataxia Assessment

Michel Velázquez-Mariño[1]([⊠]), Miguel Atencia[2], Rodolfo García-Bermúdez[1,3],
Daniel Pupo-Ricardo[1], Roberto Becerra-García[1], Luis Velázquez Pérez[4],
and Francisco Sandoval[5]

[1] Grupo de Procesamiento de Datos Biomédicos (GPDB),
Universidad de Holguín, Holguín, Cuba
{mvelazquez,depupor,idertator}@facinf.uho.edu.cu
[2] Campus de Excelencia Internacional Andalucía Tech, Departamento de
Matemática Aplicada, Universidad de Málaga, Málaga, España
matencia@ctima.uma.es
[3] Facultad de Ciencias Informáticas,
Universidad Laica Eloy Alfaro de Manabí, Manta, Ecuador
rodolfo.garcia@live.uleam.edu.ec
[4] Centro para la Investigación y Rehabilitación de Ataxias Hereditarias,
Holguín, Cuba
velazq63@gmail.com
[5] Departamento de Tecnología Electrónica, Universidad de Málaga, Málaga, España
sandoval@dte.uma.es

Abstract. The Finger-to-nose test (FNT) is an accepted neurological
evaluation to study the coordination conditions. In this work, a methodology for the analysis of data from FNT is proposed, aimed at assessing
the evolution of the condition of Spinocerebellar Ataxia type 2 (SCA2)
patients. First of all, test results obtained from both patients and healthy
individuals are processed through principal component analysis in order
to reduce data dimensionality. Next, data were grouped in order to determine classes of typical responses. The Mean Shift algorithm was used
to perform an unsupervised clustering with no previous assumption on
the number of clusters, whereas the k-means method provided an independent validation on the optimal cluster number. Experimental results
showed the highest internal evaluation for distribution into three clusters, which could be identified as the responses of healthy subjects, SCA2
patients with medium incoordination level, and patients with severe incoordination. A membership function is defined, which allows to establish
the subjects' condition based on the classification of their responses. The
results support that these protocols and the implemented clustering procedure can be used to accurately evaluate the incoordination stages of
healthy subjects and SCA2 patients, thus offering a method to assess the
impact of therapies and the progression of incoordination.

Keywords: Finger-to-nose test · Cluster analysis · Incoordination
assessment · NeuroScreening Coordination

© Springer International Publishing Switzerland 2015
I. Rojas et al. (Eds.): IWANN 2015, Part II, LNCS 9095, pp. 524–535, 2015.
DOI: 10.1007/978-3-319-19222-2_44

1   Introduction

Spinocerebellar Ataxia type 2 (SCA2) has the highest world prevalence rate in Holguín, Cuba [23]. Its main clinical manifestations include ataxia gait, dysdiadochokinesia and dysmetria of the upper and lower limbs [7]. The dysmetria is a condition associated to coordination problems and its measurement can be useful to check disease progression and evaluate the impact of treatments [6,8]. The Finger-to-nose test (FNT) is a neurological test to evaluate the upper limbs coordination state. There are different variants of this test, but mainly they all have a common goal, namely assessing dysmetria and time execution. The test consists in executing rapid and alternative movements: the subjects are asked to touch their nose tip and a target, alternatively, with their index finger [6,14].

The FNT is commonly used to explore the coordination condition in many diseases [4,12], and it is included in the Scale for the Assessment and Rating of Ataxia, which is validated for spinocerebellar ataxias, Friedreich's ataxia and sporadic ataxias [18,20]. Reports have been published on the use and helpfulness of this test combined with others in some studies of SCA2 [11,22]. However, the analysis of the FNT results proved challenging even for experienced therapists [19], thus an automated analysis could help experts in the assessment of the disease progression. Apart from the conventional analysis of variance to obtain differences between healthy subjects and ataxia patients, we are aware of only a rather limited number of contributions aimed at a quantitative analysis of FNT results [13], and no contribution has ever addressed, to the best of our knowledge, clustering and classification of FNT data. Therefore, the introduction of computational intelligence techniques is a promising direction in order to enhance the quantitative processing of FNT results.

Cluster analysis allows to make unsupervised classifications of features [9,24], and it has been applied in biomedical research to find groups in large medical data [10], or process movement signals [15]. The purpose of the present study is to describe and apply a clustering procedure to find significant differences in the performance results from FNT test protocols. The obtained clusters are validated by means of the *silhouettes* technique [17]. A significant contribution is the definition of a membership value, based in the assignation of test performances to the clusters found, in order to rate how much each individual belongs to each of the found clusters. The ultimate goal is to aid in distinguishing healthy individuals from symptomatic SCA2 patients, as well as measuring the disease evolution among SCA2 patients.

In Section 2, the proposed methodology is described, after a brief presentation of the experimental setting. The obtained results are presented in Section 3, with a discussion of the obtained classification. Finally, Section 4 provides some conclusions and directions for further research.

2   Clustering Procedure for Finger-to-nose Data

The proposed clustering methodology, depicted in Figure 1, comprises three steps: dataset construction, data processing, and interpretation of results. First

Fig. 1. Clustering procedure

of all, distance and time data were obtained by means of the standard FNT test from both healthy individuals and SCA2 patients. The captured data were processed by a set of machine learning methods, progressing through three stages: dimensionality reduction so that the most significant features are selected; data clustering in order to group individuals into classes of similar behaviour, in terms of performance in the FNT test; and clustering evaluation, which provides a validation of the number and sizes of obtained clusters. Finally, the degree of membership of data to each cluster is measured by a suitably defined membership score, thus allowing for interpreting the meaning of each cluster. This methodology is described with further detail in the following three subsections.

2.1 Dataset Construction

Ten symptomatic patients with SCA2 and ten healthy subjects were included in this study. They performed three test protocols for seven tests in total, each test with different stimuli. The protocols were executed twice, first with the dominant hand, yielding a total of 280 instances. Data were recorded using the NeuroScreening Coordination software [21] in a touchscreen device GEMINI Tablet of 7 diagonal screen inches, ARMv7 Processor up to 1GHz frequency, 512 MB of RAM and Android Operating System 4.0.4. All patients signed an informed consent before of the study execution. The dysmetria, hypermetria, hypometria and metria were computed as distances between stimulus and response, but the latter three were only recorded when responses were beyond, before and over the horizontal line of the stimulus respectively. For each magnitude, the statistical descriptors (minimum, maximum, mean, standard deviation, coefficient of variation and total) and the sum of occurrences were calculated. The time responses were registered as the time between one and the next responses, also computing their statistical descriptors. As a result, 33 features were extracted using the report routines of the NeuroScreening Coordination. Each instance was labelled with the subject state (either healthy or SCA2) to be used as reference for an external cluster evaluation, but it must be emphasized that these labels were not used at all during clustering, which was genuinely unsupervised.

2.2 Data Processing

Since the number of instances is rather limited, compared to the number of variables, data were preprocessed through Principal Component Analysis (PCA) in order to reduce data dimensionality. Next, clustering was performed with the Mean Shift (MS) algorithm, which has the advantage that it does not require to specify the number of clusters in advance. Also, an independent clustering procedure has been performed with the k-means algorithm, by choosing different values for the number of clusters. The last stage of data processing consists in validating the obtained number of clusters with the methods of silhouettes [17], which amounts to an internal validation of the results. All data were processed using the *scikit-learn* module for Python language [16], and the choice of methods and parameters follows the recommendations of the software.

Dimensionality Reduction. First, we applied the Principal Component Analysis algorithm to explore the data and select the most relevant linear combinations of features to reduce the data dimensionality. PCA tries to reduce the dimension of the data by searching for a few orthogonal linear combinations of the original variables with the largest variance [5].

Clustering. At a second stage, we obtained clusters of data by means of the MS algorithm, which is a non-parametric clustering method that does not require knowledge about the number of clusters beforehand. It was a method originally designed to estimate the mode of a probability distribution, that was adapted to compute a—rather limited—number of local modes, or *centroids* of corresponding "blobs" in a smooth density of samples, i.e. regions with an accumulation of data points. The algorithm works by updating every candidate x_i for centroids at every iteration t, according to the equation:

$$x_i^{t+1} = x_i^t + m(x_i^t) \tag{1}$$

by means of the *mean shift* vector m given by

$$m(x_i) = \frac{\displaystyle\sum_{x_j \in N(x_i)} K(x_j - x_i)\, x_j}{\displaystyle\sum_{x_j \in N(x_i)} K(x_j - x_i)} \tag{2}$$

where $N(x_i)$ is the neighborhood of samples at a given distance around x_i and $K(x)$ is a kernel function [2,3]. There is no need no choose a fixed number of centroids, because initially every data point is considered a candidate x_i^0, and all these points converge, under the update law (2), to one of a limited number of centroids. In this work, we have chosen the flat kernel implemented in *scikit-learn*:

$$K(x) = \begin{cases} 1 & \text{if } \|x\| \leq \lambda \\ 0 & \text{if } \|x\| > \lambda \end{cases} \tag{3}$$

where λ is an heuristic parameter.

In order to reassert the results provided by the MS algorithm, the k-means algorithm was also used to obtain clusters and compare the results between both strategies. The number of categories k must be previously specified in the k-means method, and the problem can be simply stated as minimizing the within-cluster distance, or *inertia*:

$$\sum_{i=1}^{k} \sum_{x_j \in C_i} (\|x_j - \mu_i\|^2) \tag{4}$$

where μ_i is the mean of the cluster C_i. One of the advantages of the k-means algorithm is that the inertia defined in Equation (4) provides a measure of the internal coherence of clusters [1]. In this work we implement the solution of the k-means problem with the standard Lloyd's algorithm, defining the number of clusters in the range from $k = \max(1, c - 3)$ to $k = c + 3$, where c is the number of clusters found by the MS method.

Clustering Evaluation. In a third data processing step, an internal evaluation was performed by means of the silhouette values [17], defined by:

$$s_i = \frac{b_i - a_i}{\max(a_i, b_i)} \qquad \begin{aligned} a_i &= d(x_i, A) \qquad x_i \in A \\ b_i &= \min_{x_i \notin C} d(x_i, C) \end{aligned} \tag{5}$$

where $a_i = d(x_i, A)$ is the average distance of the data point x_i to all other samples of its cluster A, and b_i is the minimum of the averages of the distances of x_i to all samples of the remaining clusters. The silhouette score is the mean of all silhouette values, which is easily proved to lie in the interval $(-1, 1)$, and it provides an evaluation of the clustering results, since a higher score denotes denser, more separated clusters, thus the clustering is more meaningful. As explained below, a silhouette plot can be constructed by means of this score in order to aid to interpret the clustering results, so that the best strategy (either mean shift or k-means) can be selected.

2.3   Results Interpretation

Since clustering was applied to subjects' responses without any knowledge about the subjects health state, we propose a simple membership function to evaluate to which extent the subject belongs to each found cluster, thus classifying their incoordination. This membership function is based on the total of responses of an individual that are classified per cluster C_n:

$$F = \sum_{n=1}^{k} \sum_{x_i \in C_n} \frac{n}{10} R_i \tag{6}$$

where k is the number of clusters and R_i is the response of individual x_i. The coefficient $\dfrac{n}{10}$ is an incremental weight for each cluster. Clusters are sorted in ascending order by the values that each element adopts, so that higher values imply that the subjects' responses are worse, and hence more incoordination of movements.

Besides, we establish boundary values to classify subjects into the clusters according to the weights assigned to each cluster. For a cluster C_n, vc_{C_n} is its center, vl_{K_n} is the lower boundary and vu_{C_n} is the upper boundary:

$$vc_{C_n} = \frac{1}{10} n\, M_n$$

$$vl_{C_n} = \begin{cases} \frac{1}{10}\left(n - \frac{1}{2}\right) M_n & \text{if } n > 1 \\ vc_{C_n} & \text{if } n = 1 \end{cases} \qquad (7)$$

$$vu_{C_n} = \begin{cases} \frac{1}{10}\left(n + \frac{1}{2}\right) M_n & \text{if } n < k \\ vc_{C_n} & \text{if } n = k \end{cases}$$

where $M_n = \max_{x_i \in C_n} R_i$ is the maximum of all possible subjects' responses classified in the cluster C_n. The rationale for the extreme cases C_1, C_k, is that all responses from the subjects with the best coordination state are classified in C_1 and all responses from the subjects with the worse coordination state are classified in C_k.

3   Results

According to the procedure explained above, first of all the PCA algorithm was applied to the recorded data from the experiment. The explained variance ratio for the most relevant linear combinations obtained is shown in Table 1, where it can be seen that the first combination is enough to explain most of the data variability, so it is chosen as the only feature.

Table 1. Results of PCA algorithm, most relevant values are showed

Num.	Explained variance ratio
1	0.98275
2	0.01362
3	0.00342
4	0.00008
⋮	⋮
33	0.00000

Next, data was analyzed by means of the MS algorithm, which detected three clusters in the dataset. The computation of the silhouette score for this cluster assignment yielded the value 0.617. In parallel, following the clustering procedure, the k-means method was executed four times, varying the number of cluster from two to five. The silhouettes scores for these clustering allocations are shown in Table 2, which highlights that the maximal score was attained with three clusters. It is remarkable that this was precisely the number of clusters that resulted from the MS method, thus providing a strong support for choosing three clusters as the suitable number k of categories. Besides, for the class centers defined by k-means with $k = 3$, the silhouette score is higher than the obtained by the MS algorithm, suggesting that k-means produces a more accurate clustering than MS. This comparisong is graphically shown in Figure 2(A).

Table 2. Results of k-means executions with different number of clusters

Num. of clusters	Silhouette score
2	0.594
3	0.637
4	0.604
5	0.549

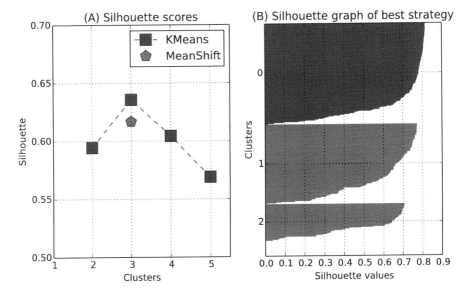

Fig. 2. Silhouette. (A) Silhouette scores for each clustering strategy. (B) Silhouette values graph of the best clustering strategy.

The silhouette graph of the best clustering strategy, namely k-means with three clusters, is shown in Figure 2(B). The cluster 0 shows the widest and highest silhouette, hence it is the strongest cluster and it comprises the most samples (130), probably because SCA2 patients in good coordination condition were included. The clusters 1 and 2 are a little narrower in comparison to the cluster 0 and they include less samples (103 and 47 respectively). A large value of the silhouette for a given element indicates that there is no doubt in the assignment of the element to the cluster it belongs. On the contrary, a low silhouette score suggests an element belonging to the intersection of two clusters.

Figure 3 shows the data organization and the cluster centroids. The cluster 0 is more compact and distances from the elements to the cluster centre are smaller. The clusters 1 and 2 are more scattered and distances from the elements to the respective cluster centres are larger, which means that the variances of the responses are higher. The clusters are not well separated, but three coordination stages are clearly shown. This is a significant finding, because the original classification from therapists only distinguishes two categories: healthy individuals and SCA2 patients at the first stage. This result shows that it is possible to find more specific classifications for the subjects' coordination by means of machine learning techniques. As a result, the subjects' coordination can be classified by means of the membership function, which uses this cluster assignment.

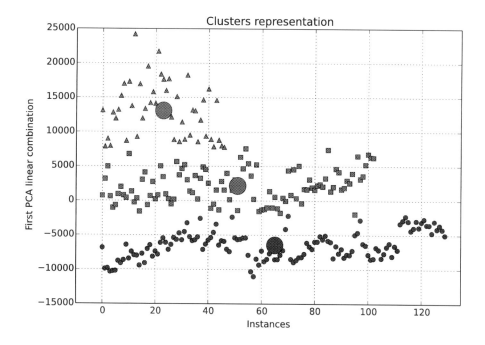

Fig. 3. Cluster representation of the reduced features from the PCA algorithm

In order to perform an external evaluation of the results, the cluster assignment was compared to the evaluation by a human expert. The comparison disclosed that 91.42% of healthy subjects' responses were classified in the cluster 0 and the remaining 8.57% were classified in cluster 1. Moreover 98.57% of SCA2 patients' responses were classified in the clusters 1 and 2 (65.00% and 33.57%, respectively) whereas only 1.43% of symptomatic responses were classified in the cluster 0. This results reveals that the designed protocols are useful to get information about the coordination condition of SCA2 patients and establish differences between different degrees of disease progression, as well as distinguishing asymptomatic individuals. Based on this results, it is meaningful to label cluster 0 as incoordination low level, cluster 1 as incoordination medium level, and cluster 2 as incoordination high level.

The construction of clusters is useful for classifying the subjects' responses, but it is also interesting to have available a measurement of the extent to which a subject belongs to the categories. This is the purpose of the membership function defined in Equation (6). Figure 4 shows the results of computing such function for each individual of our experimental setting. The membership function reported correctly 9 healthy subjects into the incoordination low level and 1 healthy subject into the lower bound of incoordination middle level. Six SCA2 patients were also reported into the incoordination middle level, 2 just in the boundary between incoordination middle and high level, and 2 into the incoordination high level. Therefore, the experts' assessment is consistent with the evaluation provided by the proposed methodology, which provides a remarkable support to the algorithmic design.

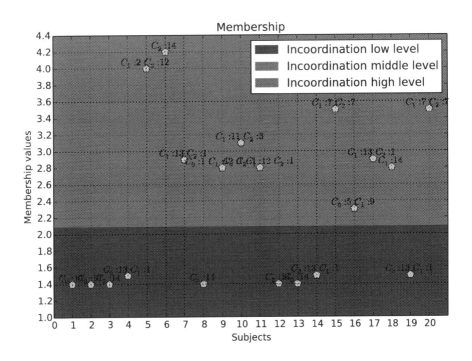

Fig. 4. Belonging subjects to the categories

4   Conclusions

A cluster analysis of responses to FNT protocols has been achieved, by applying a sequence of machine learning techniques to data obtained from the NeuroScreening Coordination software. We found in the first place that this computational methodology is able to discriminate not only between responses of healthy subjects and the ones of SCA2 patients, but also between responses of SCA2 patients at the first stage of the disease with different degrees of incoordination. Moreover we defined a membership function to establish the subjects' category based on the classification of their responses. These results reveal that this methodology and the implemented clustering procedure can be used to accurately evaluate the incoordination stages of healthy subjects and SCA2 patients, and offer a method to evaluate the impact of therapies and the progression of ataxia symptoms. This study can be extended by enlarging the dataset, eventually including SCA2 patients at the second stage of the disease. Such an experiment would provide an additional endorsement for the proposed strategy, and support the generality of our approach.

Acknowledgments. This work has been partially supported by the Universities of Holguín and Málaga through the joint project titled *"Mejora del equipamiento para la evaluación de la rehabilitación de enfermedades neurológicas de especial prevalencia en el oriente de Cuba"*. Also, we would like to thank the Agencia Española de Cooperación Internacional para el Desarrollo (AECID) and the Campus de Excelencia Internacional Andalucía Tech of University of Málaga because of the funding granted to this project. Finally, the feedback of the anonymous reviewers is gratefully acknowledged.

References

1. Arthur, D., Vassilvitskii, S.: k-means++: The advantages of careful seeding. In: Proceedings of the Eighteenth Annual ACM-SIAM Symposium on Discrete Algorithms, pp. 1027–1035. Society for Industrial and Applied Mathematics (2007)
2. Cheng, Y.: Mean shift, mode seeking, and clustering. IEEE Transactions on Pattern Analysis and Machine Intelligence **17**(8), 790–799 (1995)
3. Comaniciu, D., Meer, P.: Mean shift: A robust approach toward feature space analysis. IEEE Transactions on Pattern Analysis and Machine Intelligence **24**(5), 603–619 (2002)
4. Feys, P.G., Davies-Smith, A., Jones, R., Romberg, A., Ruutiainen, J., Helsen, W.F., Ketelaer, P.: Intention tremor rated according to different finger-to-nose test protocols: A survey. Archives of Physical Medicine and Rehabilitation **84**(1), 79–82 (2003)
5. Fodor, I.K.: A survey of dimension reduction techniques. Technical Report UCRL-ID-148494, Lawrence Livermore National Laboratory (2002)
6. Gagnon, C., Mathieu, J., Desrosiers, J.: Standardized Finger-Nose Test Validity for Coordination Assessment in an Ataxic Disorder. The Canadian Journal of Neurological Sciences **31**(04), 484–489 (2004)

7. Gazulla, J. (ed.): Spinocerebellar Ataxia. InTech (2012)
8. Grimaldi, G.: Cerebellar motor disorders. In: Manto, M., Schmahmann, J.D., Rossi, F., Gruol, D.L., Koibuchi, N. (eds.) Handbook of the Cerebellum and Cerebellar Disorders, pp. 1595–1625. Springer, Netherlands (2013)
9. Jain, A.K., Murty, M.N., Flynn, P.J.: Data clustering: a review. ACM Computing Surveys 31(3), 264–323 (1999)
10. Kalyani, P.: Approaches to partition medical data using clustering algorithms. International Journal of Computer Applications 49(23), 7–10 (2012)
11. Klinke, I., Minnerop, M., Schmitz-Hübsch, T., Hendriks, M., Klockgether, T., Wüllner, U., Helmstaedter, C.: Neuropsychological features of patients with spinocerebellar ataxia (SCA) types 1, 2, 3, and 6. The Cerebellum 9(3), 433–442 (2010)
12. Lanzino, D.J., Conner, M.N., Goodman, K.A., Kremer, K.H., Petkus, M.T., Hollman, J.H.: Values for timed limb coordination tests in a sample of healthy older adults. Age and Ageing 0, 1–4 (2012)
13. Louis, E.D., Applegate, L.M., Borden, S., Moskowitz, C., Jin, Z.: Feasibility and validity of a modified finger-nose-finger test. Movement Disorders 20(5), 636–639 (2005)
14. Montgomery, P.C., Connolly, B.H.: Clinical applications for motor control. SLACK, Thorofare (2002)
15. Patel, S., Sherrill, D., Hughes, R., Hester, T., Huggins, N., Lie-Nemeth, T., Standaert, D., Bonato, P.: Analysis of the severity of dyskinesia in patients with Parkinson's disease via wearable sensors. In: BSN 2006: International Workshop on Wearable and Implantable Body Sensor Networks, Proceedings, pp. 123–126. IEEE Computer Society (2006)
16. Pedregosa, F., Varoquaux, G., Gramfort, A., Michel, V., Thirion, B., Grisel, O., Blondel, M., Prettenhofer, P., Weiss, R., Dubourg, V., Vanderplas, J., Passos, A., Cournapeau, D., Brucher, M., Perrot, M., Duchesnay, E.: Scikit-learn: Machine learning in Python. Journal of Machine Learning Research 12, 2825–2830 (2011)
17. Rousseeuw, P.J.: Silhouettes: A graphical aid to the interpretation and validation of cluster analysis. Journal of Computational and Applied Mathematics 20, 53–65 (1987)
18. Saute, J.A.M., Donis, K.C., Serrano-Munuera, C., Genis, D., Ramirez, L.T., Mazzetti, P., Pérez, L.V., Latorre, P., Sequeiros, J., Matilla-Dueñas, A., Jardim, L.B.: Ataxia rating scales-psychometric profiles, natural history and their application in clinical trials. The Cerebellum 11(2), 488–504 (2012)
19. Swaine, B.R., Sullivan, S.J.: Reliability of the Scores for the Finger-to-Nose Test in Adults with Traumatic Brain Injury. Physical Therapy 73(2), 71–78 (1993)
20. Schmitz-Hübsch, T., du Montcel, S.T.: Scale for the assessment and rating of ataxia: development of a new clinical scale. Neurology 66(11), 1717–1720 (2006)
21. Velázquez-Mariño, M.: NeuroScreening Coordination: sistema para la evaluación de los trastornos de la coordinación de los miembros superiores. Tesis en opción al título académico de máster en matemática aplicada e informática para la administración. Universidad de Holguín, Cuba (2014)
22. Velázquez-Pérez, L., de la Hoz-Oliveras, J., Pérez-González, Z., Hechavarría, P., Herrera-Domínguez, H.: Evaluación cuantitativa de los trastornos de la coordinación en pacientes con ataxia espinocerebelosa tipo 2 cubana. Revista de Neurología 32(7), 601–606 (2001)

23. Velázquez-Pérez, L., Sánchez-Cruz, G., Santos-Falcón, N., Almaguer-Mederos, L.E., Escalona-Batallan, K., Rodríguez-Labrada, R., Paneque-Herrera, M., Laffita-Mesa, J.M., Rodríguez-Díaz, J.C., Aguilera-Rodríguez, R., González-Zaldivar, Y., Coello-Almarales, D., Almaguer-Gotay, D., Jorge-Cedeno, H.: Molecular epidemiology of spinocerebellar ataxias in Cuba: Insights into SCA2 founder effect in holguin. Neuroscience Letters **454**(2), 157–160 (2009)
24. Xu, R., WunschII, D.: Survey of clustering algorithms. IEEE Transactions on Neural Networks **16**(3), 645–678 (2005)

Exploiting Neuro-Fuzzy System for Mobility Prediction in Wireless Ad-Hoc Networks

Mohamed Elleuch[1(✉)], Heni Kaaniche[1], and Mohamed Ayadi[2]

[1] National Engineering School of Sfax (ENIS), University of Sfax, Sfax, Tunisia
mohamed_tn_sf@yahoo.fr, heni.kaaniche@enis.rnu.tn
[2] High School of Communication in Tunis, Ariana, Tunisia
mohmaed.ayadi@supcom.rnu.tn

Abstract. Ad-hoc mobile wireless network is characterized by a very dynamic environment. However, the major obstacle to be resolved is to sustain the links of continuity and improved routing performance. In this paper, we propose a predictor based Neuro-fuzzy for the prediction of mobility. It predicts the trajectory of an ad-hoc mobile node in order to improve routing performance by reducing overhead and the number of broken connections. It allows estimating the stability of paths in Ad-Hoc mobile wireless networks. Using an Adaptive Neuro-Fuzzy Inference System (ANFIS) to predict the trajectory of an ad-hoc mobile, we demonstrate the effectiveness of the proposed predictor by testing it on a time series prediction problems.

Keywords: Location · Prediction · Mobility · ANFIS · Ad-hoc · Routing

1 Introduction

Mobile ad-hoc networks (MANETs) depict complex distributed systems that comprise wireless mobile nodes. These nodes can freely and dynamically self-organize into arbitrary and temporary ad-hoc network topologies, thus allowing people and devices to seamlessly internetwork in areas with no preexisting communication infrastructure [1]. This implies that each node acts as a router as well as a host since it is responsible for routing information between its neighbors, which contributes to the routing and maintenance of the network connectivity.

MANET nodes are typically characterized by their limited memory resources, processing and power as well as high degree of mobility. In such networks, the wireless mobile nodes may dynamically enter the network as well as leave the network. Due to the limited transmission scope of wireless network nodes, multiple jumps are usually required for a node to exchange information with any other node in the network. Thus, in a MANET, the used routing protocol is of paramount importance because it determines how a data packet is transmitted over multiple jumps from a source to a destination node.

The creation of the road must be performed rapidly, with minimal overhead. The routing protocol must also adapt to frequent changes in network topologies caused by nodes mobility. Various routing schemes have been proposed for MANETs [2-6].

© Springer International Publishing Switzerland 2015
I. Rojas et al. (Eds.): IWANN 2015, Part II, LNCS 9095, pp. 536–548, 2015.
DOI: 10.1007/978-3-319-19222-2_45

The majority of these solutions are based on finding the shortest path in term of distance or delay. However, the shortest path may break up quickly after its establishment. Indeed, due to node mobility, certain links on the shortest path may fail as soon as the path is established. This failure causes connection interruption and data loss, if the phase of rediscovery of routes does not happen quickly. However, the rediscovery phase of routes involves a significant overhead. Thus, routing based on the selection of the shortest path leads to the degradation in the routing quality of service. This is why stable paths are worth being exploited for routing packets instead of shortest paths [7-8]. The paths stability estimation can be made by predicting nodes future locations, which we call mobility prediction.

H. Kaaniche and F. Kamoun [9] presented a system based on the Neural Network for mobility prediction in Ad-Hoc networks. This system consists of a multi-layer and recurrent neural network using back propagation through time algorithm for training. Another method which is used extensively is Hidden Markov Model (HMM). R. Nagwani and D. Singh Tomar [10] presented a HMM approach for predicting the mobility of nodes to reduce the overhead head of routing for route discovery. L. Ghouti et al [11] proposed a new scheme which is based on a single feed-forward layer architecture known as the Extreme Learning Machine (ELM) to predict the node mobility in a mobile ad-hoc network. J. Martyna [12] proposed a new fuzzy neural controller for a mobility and bandwidth prediction in mobile Ad-hoc and sensor networks. In this study, we propose a new method for mobility prediction in Ad-Hoc networks. This method is based on a Neuro-fuzzy system.

In this paper, we give in section 2 a definition of mobility and location prediction. In section 3, we justify the importance of location prediction in routing for Ad-Hoc Networks. Sections 4 and 5 explain the basic definition of ANFIS (Adaptive Neuro-Fuzzy Inference System). Suitable system to Neuro-fuzzy base for mobility prediction is presented in this section. Section 6 illustrates the effectiveness of the proposed mobility predictor. Finally, some concluding remarks are presented in Section 7.

2   Definition of Mobility and Location Prediction

Mobility prediction of a node is the estimation of their future locations. Many location prediction methods are proposed in literature [13]. The main advantage of location prediction is to allocate, in advance, the convenient next access point before the mobile terminal leaves its current one, in order to reduce the interruption time in communication between terminal mobiles. In order to improve routing performances in Ad-Hoc networks, the location prediction is used to estimate link expiration time [14,15]. W. Su et al. and A. Agarwal et al. proposed two different methods for mobility prediction [14,15]. However, these methods assume that nodes move according the Random Waypoint Mobility (RWM) model [16]. As a result, node mobility prediction moving according to other models can lose its accuracy and efficiency.

In the present work, we are interested in developing the same technique used with Kaaniche and Kamoun [9] for mobility prediction in Ad-Hoc networks which is independent from the mobility model used by nodes. We assume that each node in the Ad-Hoc network is aware of its location. Most commonly, the node will be able to learn its location using an on board GPS (Global Positioning System) receiver. So, it can periodically record its geographical location. All the recorded locations, define the node trajectory. At time tk, this trajectory can be noted by 3 time series: $(X_{tk-i}, Y_{tk-i}, Z_{tk-i})\{i=0..p\}$, where p is the number of locations recorded before the current time t_k. Knowing its previous locations $(X_{tk-i}, Y_{tk-i}, Z_{tk-i})\{i=0..p\}$, a mobile node can estimate their future ones $(X_{tk+i}, Y_{tk+i}, Z_{tk+i})\{i=0..N\}$, where N is the number of predicted locations. The Mobility prediction of a mobile node is presented in Fig. 1.

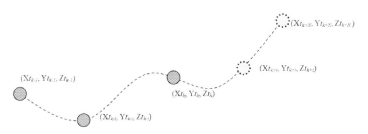

Fig. 1. Location prediction of a mobile node

Time series is a set of observations from past until present. Time series prediction is expected to estimate future observations. Location prediction is a particular case of time series prediction. In fact, the mobile node's trajectory defined by locations $(X_{tk-i}, Y_{tk-i}, Z_{tk-i})\{i=0..p\}$ represents three time series: (X_{tk-i}), (Y_{tk-i}) and $(Z_{tk-i})\{i=0..P\}$. So, location prediction problem can be resolved by the prediction of these three time series to obtain (X_{tk+i}), (Y_{tk+i}) and $(Z_{tk+i})\{i=0..N\}$. In this study, we present a time series prediction technique based on neuro-fuzzy system and provide multi-step prediction to be exploited for mobility prediction.

3   Mobility Prediction and Routing in Ad-Hoc Networks

To give the importance of mobility prediction in routing Ad-Hoc networks, we present a simple scenario in Fig. 2. As can be seen from this figure, the Ad-Hoc networks contain four nodes which are A, B, C and D. A is stable, B moves rapidly away from A and D. C moves slowly towards A. A has data packets to send to D. Therefore it broadcasts, the "Route Request" packet and finds that to reach D, it must pass either through node B or through C. If A chooses B as intermediate node, then the communication will not last a long time since the link (A,B) will be rapidly broken, due to the mobility of B. But if A takes into consideration the mobility of B and C, it will choose C as intermediate node because the expiration time of the link (A,C) is superior to that of (A,B), since C has chance to remain in A transmission range, more than B. The fact that A chooses C as next jump to reach D contributes to the selection of the path which has the greatest expiration time or the most stable path. According to this example, it is clear that selecting the most stable path in terms of average length of validity can avoid future link failure, which improves routing. This solution is based on the prediction of location of mobile nodes in an Ad-hoc network.

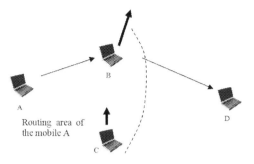

Fig. 2. Simple mobility scenario in an Ad-Hoc network

A road which is the shortest path may rupture quickly after its establishment because of nodes mobility. Breaks roads generate degradation in the quality of routing service. This is why we chose to design a route selection metric that is based on its stability. The stability of a road is measured by its period of validity which corresponds to the maximum PVP (Period of Validity of Path). This period is determined from the set of Periods of Validity PVL (PVP = min PVL) of each Link constituting the road.

To calculate the period of validity of a link, we take the general case of two mobile nodes A and B as can be seen from Fig. 3. At time t, the node A wants to calculate the period of validity of its link with B, at this time, A and B are found at the following positions A (X_{At}, Y_{At}) and B (X_{Bt}, Y_{Bt}). The link (A, B) remains valid as the distance between the two nodes remains below a threshold R, which corresponds to the coverage area of A. The more the two mobile nodes are moving, the more the distance between them changes. Eventually at a certain time tf corresponding to positions A (X_{Atf}, Y_{Atf}) and B (X_{Btf}, Y_{Btf}), the distance becomes equal to R. The period (tf - t) constitutes the period of validity of link (A, B).

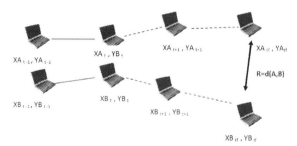

Fig. 3. Locations prediction for determining the period validity of the link

In conclusion, mobility prediction can improve considerably routing by selecting stable route and consequently, avoiding road failure. In the following section, we present a time series prediction technique based on Neuro-fuzzy system and providing multi-step prediction. This technique can be applied to a location time series.

4   Neuro-Fuzzy System Architecture for Prediction

According to literature, time series prediction techniques can be classified in two categories: statistical techniques [17] and techniques based on artificial intelligence tools such as neural networks, fuzzy systems and Neuro-fuzzy systems [18]. In particular, Neuro-fuzzy hybrid system, show their efficiency to model the dynamic behavior of the system. The efficiency of the neural-fuzzy prediction technique depends on the choice of the architecture and the training algorithm. In this work, the Adaptive Neuro-Fuzzy Inference System (ANFIS) is used.

5   ANFIS Modeling Background

5.1   Fuzzy Sets and Logic

Instead of Aristotelian logic to two values (1 or 0) in the treatment of logic states, the concept of fuzzy sets and logic is introduced by Zadeh [19]. Fuzzy approach considers the cases where the linguistic uncertainties play a role in the control mechanism of the phenomena involved. Here, the uncertainties do not mean random, probabilistic and stochastic variations, which all are based on the numerical data. Zadeh has motivated his work on fuzzy logic with the observation that the key elements in human reflection are not numbers, but the levels of fuzzy sets. The main idea of fuzzy logic is the distribution of partial membership of any object to different subsets of the universal set instead of belonging to a unique set completely. Partial membership to a set can be described numerically by a membership function which takes values between 0 and 1 inclusive. Therefore, if the shape of uncertainty happens to arise because of vagueness or ambiguity, then the variable is fuzzy and can be represented by a membership function. Even measures that are carefully performed as crisp quantities can be fuzzified. Order to simplify the calculations, usually the membership function is adopted as linear for practical applications [20].

5.2   Adaptive Neuro-Fuzzy Inference System

The adaptive Neuro-fuzzy inference system, first introduced by Jang [21], is a universal approximator and, as such, is capable of approximating any real continuous function on a compact set to any degree of accuracy [22]. ANFIS constitute a new improved tool and data-guided modeling approach to determine the behavior of complex dynamic systems defined imprecisely [23]. An ANFIS aims to generate a systematic manner unknown fuzzy rules from a data set offered as input/output. Thus, in parameter estimation, where the given data are like the system associates measurable system variables with an internal system parameter, a functional mapping may be constructed by ANFIS that nearly identifies the process of estimation of the internal system parameter [24]. Although the fuzzy inference system has a representation of structured knowledge in the form of fuzzy "if-then" rules, it does not have the adaptive capacity to deal with a changing environment. Thus, neural network learning concepts have been incorporated into the fuzzy inference systems, resulting in the modeling of adaptive Neuro-fuzzy. Adaptive inference system is a network which is composed of a number of interconnected nodes. Each node is characterized by a node function with fixed or adjustable settings. The network is "learning" the behaviour of the available data during the training phase by adjusting the parameters of the node functions to fit that data. The basic learning algorithm, the back-propagation, aims to minimize a set measure or a defined error, usually the sum of squared differences between the desired and the actual model outputs [25].

The architecture of the ANFIS modeling that is used in the current study is based on the first-order Takagi–Sugeno model. The fuzzy based modeling was first investigated by Takagi and Sugeno [26]. This modeling is presented in the next paragraph.

Basics Aspects of ANFIS
This section presents the basic architecture of ANFIS and the method that is used for hybrid learning.

For simplicity, consider a fuzzy inference system with two rules of Takagi–Sugeno type [27]:

Rule 1: *If* x_1 is A_1 and x_2 is B_1, ***then*** $y_1 = p_1 x_1 + q_1 x_2 + r_1$

Rule 2: *If* x_1 is A_2 and x_2 is B_2, ***then*** $y_2 = p_2 x_1 + q_2 x_2 + r_2$

The ANFIS architecture for implementation of these two features is shown in Fig. 4. In this figure a circle indicates a fixed node, whereas a square indicates an adaptive node. The operations of the nodes in the same layer of ANFIS are of the same family of functions, as described below. (In subsequent, $O_{k,i}$ denotes the *ith* output node of the layer k).

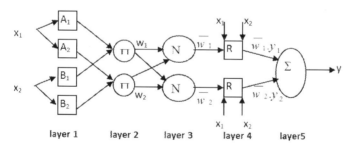

Fig. 4. ANFIS network architecture

Layer 1: each node in this layer corresponds to a linguistic label and the output node is equal to the value of membership in this linguistic label. The parameters of a node can change the shape of the membership function used to characterize the linguistic label. For example, the function of the *ith* node is given by:

$$O_{1,i} = \mu_{A_i}(x_1), \quad i=1,2 \text{ and,} \qquad O_{1,i} = \mu_{B_{i-2}}(x_2), \quad i=3,4 \qquad (1)$$

With $\mu_{A'_i}(x) = \exp\left(-\dfrac{(x_i - m_{ik})^2}{b_{ik}^2}\right)$

Where $\{m_{ik}, b_{ik}, i=1,2\}$ is the set of parameters (called premise parameters).

Layer 2: Each node in this layer calculates the firing power of each rule:

$$O_{2,i} = w_i = \mu_{A_i}(x) \times \mu_{B_i}(y), \quad i=1,2 \tag{2}$$

Layer 3: the *ith* node of this layer calculates the ratio of the firing strength of the *ith* rule to the sum of all the firing strength:

$$O_{3,k} = \overline{w}_i = \frac{w_i}{w_1 + w_2}, \quad i=1,2 \tag{3}$$

Layer 4: Node i in this layer has the following node function:

$$O_{4,i} = \overline{w}_i f_i = \overline{w}_i \left(p_i x_1 + q_i x_2 + r_i\right), \quad i=1,2 \tag{4}$$

Layer 5: The single node in this layer computes the overall output as the sum of all input signals:

$$O_{5,1} = y = \sum_{i=1}^{n} O_{4,i} = \sum_{i=1}^{n} \overline{w}_i f_i = \frac{\sum_i w_i f_i}{\sum_i w_i}, \quad i=1,2 \tag{5}$$

It can be observed that there are two adaptive layers in this ANFIS architecture, especially the first layer and the fourth layer. In the first layer, there are two modifiable parameters $\{m_{ik}, b_{ik}\}$, which are related to the input membership functions. These parameters are the known as premise parameters. In the fourth layer, there are also three modifiable parameters $\{p_i, q_i, r_i\}$, pertaining to the first order polynomial. These parameters are known as consequent parameters.

The task of the learning algorithm for this architecture is to settle all the changeable parameters mentioned above to make the ANFIS output correspond to the training data. When the premise parameters of the membership function are fixed, the output of the ANFIS model can be written as:

$$f = \overline{w}_1(p_1 x_1 + q_1 x_2 + r_1) + \overline{w}_2(p_2 x_1 + q_2 x_2 + r_2) \tag{6}$$

$$f = (\overline{w}_1 x_1).p_1 + (\overline{w}_1 x_2).q_1 + (\overline{w}_1).r_1 + (\overline{w}_2 x_1).p_2 + (\overline{w}_2 x_2).q_2 + (\overline{w}_2).r_2$$

This is a linear combination of the changeable consequent parameters p_1, q_1, r_1, p_2, q_2 and r_2. A hybrid algorithm combining the least squares method and the gradient descent method is adopted to identify the optimal values of these parameters [28]. The hybrid algorithm is composed of a forward pass and a backward pass. The least squares method (forward pass) is used to optimize the consequent parameters with the premise parameters fixed. Once the optimal consequent parameters are found, the backward pass starts immediately. The gradient descent method (backward pass) is used to adjust optimally the premise parameters corresponding to the fuzzy sets in the input domain. The output of the ANFIS is calculated by employing the consequent parameters found in the forward pass. The output error is used to adapt the premise parameters by means of a standard back propagation algorithm. It has been shown that this hybrid algorithm is highly efficient in training the ANFIS [29].

Selection of the ANFIS Structure for Prediction

The selection of the numbers of input neurons (Ne) and membership function (MF), can be done by heuristic methods (Fig. 5 and table 1). The method that we adopted is as follows: We consider thirty location time series based on RWM model, containing 350 patterns each one: the first patterns (nearly two-thirds of time series) are used for training, the rest of patterns for the test (generalization). Each series is applied to the fuzzy network while varying Ne and MF in the interval [2..10] and [2..5], respectively. For each combination of the couple (Ne, MF), we calculate the training error and the generalization error (RMSE). The following observations were then retained:

- A great value of Ne can lead to an under learning and a very small value of Ne can lead to an overfitting.
- A great value of MF can affect the prediction accuracy.
- A very good training, i.e. training with a very small error can affect the generalization ability of the Neuro-fuzzy network, this is why we choose the generalization error as criterion in the selection of the Ne and MF parameters and not the training error.
- The variation of MF affects greatly the prediction accuracy, starting from MF equal to 3. The choice of an elevated value of MF will increase the number of parameters to be optimized and training delay, without improving considerably the generalization ability. This is why, we choose MF equal to 3.

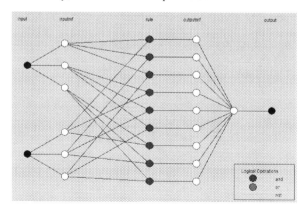

Fig. 5. The structure of ANFIS for the prediction

Table 1. Parameters of ANFIS

Fuzzy Inference System	
Input number	2
Membership functions	3 functions per input
T-norm	Product
Type inference	Linear Sugeno
Number of parameters	29 (12 premises et 17 consequents)

Error Analysis of Different Path Calculations

In order to evaluate the obtained results, the Root Mean Square Error (RMSE) values were evaluated using the following expression:

$$RMSE = \sqrt{\frac{1}{k} \times \sum_{k=1}^{k=K} (x_k - \hat{x}_k)^2}$$

6   Testing the Selected Predictor Architecture

In this study, we have considered an Ad-hoc mobile node which moves according to RWM mobility model with a varying speed in [0..20] (see Fig. 6). Its coordination is recorded each 10s, starting from 0s (initial time), until 3500s. So we obtain two location time series x(t-i) {i= 0. 350} and y(t-i) {i= 0. 350}. In order to predict the mobile node movement (trajectory), we have tested the predictor on his two location time series. The first 200 coordinates are used for training and the rest for generalization. Training as well as generalization (test) are done on a horizon of three steps.

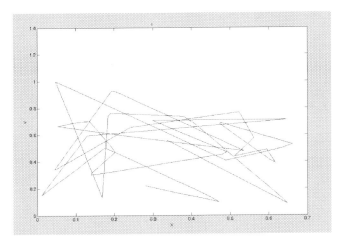

Fig. 6. Example of movement of a mobile according to the RWM model

The parameters Ne and MF are fixed respectively at 2 and 3. Fig. 7 illustrates the test of the neuronal predictor on the two series x and y. The forecasted trajectory of the mobile is deduced from predicting x and y as can be seen from Fig. 8–(a).

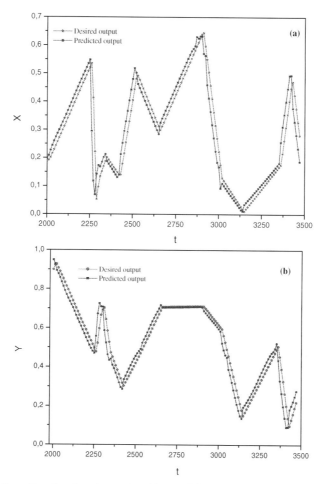

Fig. 7. Prediction of tow location time series: (a) test of time series X and (b) test of time series Y

Fig. 8 shows the prediction performance of the proposed Neuro-fuzzy system for RWM model in comparison with Recurrent Neural Network model [9] and ELM-based technique [11]. It is clear that the universal approximation capability of ANFIS enables them to track the nodes mobility once adequate training is achieved.

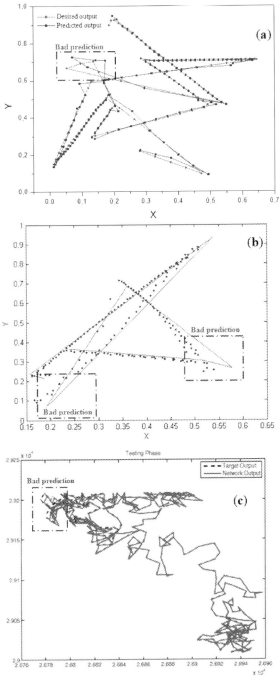

Fig. 8. Prediction of the mobile trajectory: (a) Prediction based on ANFIS, (b) Prediction based on Recurrent Neural Network [9], (c) Prediction based on ELM technique [11]

7   Conclusion

In this study, we exploited the location to predict future trajectory using a Neuro-fuzzy technique which combines advantages of contributions provided by artificial neuronal networks and fuzzy logic. In examining the effectiveness of the predictor in the prediction of mobility, we have tested Neuro-fuzzy predictor on time series describing locations of an Ad-hoc mobile node, moving following the model RWM. This prediction ensures the reduction of loss rate and reduced of ruptures communication links. And subsequently, it improves the quality of service by guaranteeing less energy consumption, less controls for transmitting messages and it can significantly improve the routing in wireless Ad-hoc networks.

In future work, the proposed prediction technique will be extended to predict routing tables (with estimated time of expiry link between tow neighbored nodes) that would reduce the exchange of data in MANET and extend the life of the node battery.

References

1. Corson, S., Macker, J.: Mobile ad hoc networking (MANET): Routing protocol performance issues and evaluation considerations. In: RFC (2002)
2. Perkins, C.E., Royer, E.M., Das, S.R.: Ad Hoc on-demand distance vector (AODV) routing. In: IETF Internet Draft, Internet Engineering Task Force (2002)
3. Park, V., Corson, S.: Temporally-ordered routing algorithm (TORA) version 1, functional specification. In: IETF, Internet Draft, draft-ietf-manet-tora-spec-02.txt (1999)
4. Jacquet, P., Minet, P., Laouiti, A., Viennot, L., Clausen, T., Adjih, C.: Multicast optimized link state routing. In: Rapport de Recherche, Institution IETF, note: draft-ietf-manet-olsr-molsr-01.txt (2001)
5. Ko, Y.-B., Vaidya, N.H.: Location-aided routing (LAR) in mobile Ad Hoc networks. In: Proceedings of ACM/IEEE MOBICOM 1998, Dallas, Texas, pp. 66–75 (1998)
6. Karp, B., Kung, H.: GPSR: Greedy Perimeter Stateless Routing for wireless networks. In: Proceedings of the Sixth Annual ACM/IEEE International Conference on Mobile Computing and Networking (MobiCom), pp. 243–254 (2000)
7. Su, W., Lee, S.J., Mario, G.: Mobility prediction and routing in Ad Hoc wireless networks. In: Proceedings of IEEE MILCOM (2000)
8. Gerharz, M., de Waal, C., Martini, P., James, P.: Strategies for finding stable paths in mobile wireless Ad Hoc networks. In: Proceedings of IEEE Conf. on Local Computer Networks, Bonn, Germany (2003)
9. Kaaniche, H., Kamoun, F.: Mobility Prediction in Wireless Ad Hoc Networks using Neural Networks. Journal of Telecommunications 2(1), April 2010
10. Nagwani, R., Singh Tomar, D.: Mobility Prediction based Routing in Mobile Adhoc Network using Hidden Markov Model. International Journal of Computer Applications 59(1), 39–44 (2012)
11. Ghouti, L., Sheltami, T.R., Alutaibi, K.S.: Mobility prediction in mobile Ad Hoc networks using extreme learning machines. In: The 4th International Conference on Ambient Systems, Networks and Technologies (2013). Published by Elsevier (2013)

12. Martyna, J.: A mobility and bandwidth prediction controller using an adaptive neuro-fuzzy inference system for mobile Ad Hoc and sensor networks. In: Chien, B.-C., Hong, T.-P., Chen, S.-M., Ali, M. (eds.) IEA/AIE 2009. LNCS, vol. 5579, pp. 429–438. Springer, Heidelberg (2009)

13. Kumar, V., Venkataram, P.: A Prediction Based Location Management Using Multi-Layer Neural Networks. J. Indian Inst. of Science **82**(1), 7–21 (2002)

14. Su, W., Lee, S.J., Mario, G.: Mobility prediction and routing in Ad Hoc wireless networks. In: Proceedings of IEEE MILCOM (2000)

15. Agarwal, A., Das, S.R.: Dead reckoning in mobile Ad Hoc networks. In: Proceedings of IEEE Wireless Communications and Networking Conference (WCNC), New Orleans (2003)

16. Camp, T., Boleng, J., Davies, V.: A Survey of Mobility Models for Ad Hoc Network Research. Wireless Communication & Mobile Computing (WCMC): Special issue on Mobile Ad Hoc Networking: Research, Trends and Applications **2**(5), 483–502 (2002)

17. Voitcu, O., Wong, Y.: On the construction of a nonlinear recursive predictor. Science B.V. Journal of Computational and Applied Mathematics (2004)

18. Ymlu, S., Gurgen, F., Okay, N.: A comparison of global, recurrent and smoothed-piecewise neural models for Istanbul stock exchange (ISE) prediction. science B.V. Proc. Pattern Recognition Letters **26**, 2093–2103 (2005)

19. Zadeh, L.A.: Fuzzy sets. Inform. Control **8**(3), 338–353 (1965)

20. Tigdemir, M., Karasahin, M., Sen, Z.: Investigation of fatigue behaviour of asphalt concrete pavements with fuzzy-logic approach. Int. J. Fatigue **24**, 903–910 (2002)

21. Jang, J.R.: ANFIS: adaptive-network-based fuzzy inference system. IEEE Trans. Syst. Man. Cybern **23**(3), 665–685 (1993)

22. Jang, J.R., Sun, C.T., Mizutani, E.: Neuro-Fuzzy and Soft Computing: A Computational Approach to Learning and Machine Intelligence. Prentice-Hall, Upper Saddle River (1997)

23. Haykin, S.: Neural Networks - A Comprehensive Foundation. Macmillan College Publishing, New York (1998)

24. Kisi, O.: Suspended sediment estimation using neuro-fuzzy and neural network approaches. Hydrol. Sci. J **50**(4), 683–696 (2005)

25. Vassilopoulos, A.P., Bedi, R.: Adaptive neuro-fuzzy inference system in modeling fatigue life of multidirectional composite laminates. Comp. Mater. Sci **43**, 1086–1093 (2008)

26. Takagi, T., Sugeno, M.: Fuzzy identification of systems and its applications to modeling and control. IEEE Trans. Syst. Man Cybern **15**, 116–132 (1985)

27. Takagi, T., Sugeno, M.: Derivation of fuzzy control rules from human operation control actions. In: Proceedings of the IFAC Symposium on Fuzzy Information, Knowledge Representation and Decision Analysis, pp. 55–60 (1983)

28. Sun, Z.-L., Au, K.-F., Choi, T.-M.: Neuro-fuzzy inference system through integration of fuzzy logic and extreme learning machines. IEEE Transactions on Systems, Man, and Cybernetics **37**(5), 1321–1331 (2007)

29. Übeyli, E.D.: Lyapunov exponents/probabilistic neural networks for analysis of EEG signals. Expert Systems with Applications **37**(2), 985–992 (2010)

A New Method for an Optimal SOM Size Determination in Neuro-Fuzzy for the Digital Forensics Applications

Andrii Shalaginov[✉] and Katrin Franke

Center for Cyber- and Information Security Norwegian Information Security Laboratory, Gjøvik University College, Teknologivn. 22, 2815 Gjøvik, Norway
{andrii.shalaginov,katrin.franke}@ccis.no
https://ccis.no

Abstract. The complexity of the fuzzy classification models in Digital Forensics is considered to be one of the most significant aspects that influence a decision making process. We focus on criteria for an optimal SOM size and amount of rules to be derived that results in accurate and interpretable model. In this paper, we proposed a new method for the SOM size determination based on the data exploratory analysis. Experiments showed that the proposed method gives an accuracy on the Android malware detection up to 92% while decreasing the number of recommended rules from 189 to 24 in comparison to Vesanto method for an optimal SOM size. This is an important step for automated training of Neuro-Fuzzy that will result in human-understandable model that will be used in Digital Forensics process.

Keywords: Neuro-fuzzy · Self-organizing map · Optimization · Computational forensics · Malware detection

1 Introduction

The crucial property of the Soft Computing (SC) methods applied in malware detection is interpretability of the model [24] that will be used further to explain and interpret found evidences. One of the most widely used methods is Neuro-Fuzzy (NF). This method consists of two stages according to Kosko [19]: allocation of fuzzy rules parameters by means of Self-Organizing Map (SOM) and tuning by means of Artificial Neural Network (ANN). NF learns automatically from data and produces human-understandable classification model based on fuzzy logic [5]. However, the main challenge is the results SOM clustering. As per today, this unsupervised procedure mostly follows the structure of the grid defined by the analyst. Alternatively, the SOM size changes iteratively over learning like it is done in Growing SOM [3]. This is an important issue in Digital Forensics application since this has to be done without manual support. Moreover, the size of SOM influence the fuzzy rules construction that needs to be

© Springer International Publishing Switzerland 2015
I. Rojas et al. (Eds.): IWANN 2015, Part II, LNCS 9095, pp. 549–563, 2015.
DOI: 10.1007/978-3-319-19222-2_46

understandable and easy interpretable. This paper studies criteria of the optimal grid size in SOM. We concentrate our attention on dataset analytic rather than pre-processing and alteration of the data.

SOM can be organized as fixed-size or growing topology according to Valova in the "SOMs for Machine Learning" [29, Chater 2] in order to find an optimal size that better fits presented data. We will concentrate our attention on the fixed-size SOM since it does not require additional knowledge about heuristic in data. Also the amount of computational resources is lower when fixed-size rectangular SOM is used. There exist several challenges to find an optimal size since this has a direct influence on the set of the fuzzy rules. Each rule is a characterisation of similar data samples derived by means of SOM. Too general clusters will cause underfitting, while too specific cluster will cause underfitting of the model [8]. From the other side, well-packed clusters will give more empty clusters when the size of SOM is big. The author in [11] stated that the maximal grid size should be equal to $5 \cdot \sqrt{N}$, where N is the size of the dataset. This can be defied as a "rule of thumb" is SOM size definition. It can be seen that with such metric the number of nodes converges to infinity when dealing with Big Data. The data modification is not favourable due to the loss of original properties by adding abstraction level or transformation [24]. Furthermore, smaller set of rules is more appropriate for data representation in a Court of Law.

Since the most commonly used empirical measure $5 \cdot \sqrt{N}$ in SOM toolbox[1] [27], there is a need though for more precise definition of the number of nodes. At this point there are several options how the proper size can be estimated. The Growing SOM [29, Chater 2] learns from data iteratively adding new nodes. However, this process might give too big topology of SOM resulting in overfitting [8]. Moreover, the SOM can be run several times to estimate the average number of clusters after learning as a simple approach. Since SOM learning process is based on the random component it will give similar results that slightly differ from the optimal size. Finally, we can say that in the research [15] the biggest eigenvalues of dataset were used to tune this parameters. The eigenvalues ratio shows how well the data flattened and elongated [11]. In other words, the SOM grid needs to be spanned respectively. This is used in Factor Analysis to determine the proper number of factors to be used with respect to the covariance fraction. However, this does incorporate the information about the structure and dependency in the data. So, estimation of optimal SOM size is done mostly empirically. The optimal heuristic boundary of empty nodes is 5-10% after the SOM learned.

In this paper, we proposed a new method for optimal SOM grid size determination based on the exploratory data analysis. The averaged Pearson Correlation Coefficient was used as a factor that influence the SOM grid size besides the eigenvalues ratio in the method by Vesanto [27], [28]. The proposed method complies with the Digital Forensics requirements such that reasonable complexity and interpretability. Furthermore, it is important to keep the size of the model reasonable for malware analysis. We also evaluated the influence on the

---

[1] http://www.cis.hut.fi/somtoolbox/

performance fuzzy patches construction methods with corresponding membership functions (MF): simple rectangular, elliptic proposed by Kosko in 1997 [19] and based on the Gaussian MF. The main advantage of the proposed method is utilization of the exploratory data analysis and not the alteration of the dataset that contradicts with Digital Forensics Process [14]. The resulting NF model shows a good performance not only on Android malware detection problem, yet on other versatile datasets as will be shown later on.

The paper is organized as following. The Section 2 gives an overview of the used techniques, including Vesanto method, for the definition of a proper size of the SOM size. The optimization of the optimal number of nodes with respect to interpretability concerns is given in the Section 3. The experimental design and results analysis are pointed out in the Section 4. Finally, discussions and conclusions with remarks about future work are given in the Section 5.

2   Estimation of the SOM Size

In this Section we concentrate on the 1st stage of the NF method, which is clustering by means of SOM. The simple rectangular, a Kosko [19] and Gaussian MF methods are the most commonly used methods for fuzzy patches and MF construction. The first method uses 1st and n^{th} order statistics in a cluster assuming it fits rectangular region and then simple triangular MF to characterise each region. The Kosko method, instead, employs elliptic modelling and then derives corresponding parameters of the triangular MF. So, if the input data sample is $X_i = \{A \in R^M\}$ with corresponding set of features $a = \{x_0, \ldots, a_M\}$ then it can be characterized as a point in M-dimensional space. The whole set of N data samples is therefore contains in an M-dimensional ellipsoid (hyperelipsoid) with a radius α and center c_i. Further, to include the correlation between the features we need to include the covariance matrix that also defines the rotation of the elliptic region in the Kosko method:

$$(x - c)^T \, \Sigma^{-1} \, (x - c) = \alpha^2 \tag{1}$$

where Σ^{-1} is a positive definite inverted symmetric covariance matrix. The α is defined according to the degree of freedom df of the model. The number of df is chosen to be $3 * M - 1$ since for each of the dimension the statistical model will include mean, angle and spread around mean. So, the number α is set to be the same for every fuzzy patch as in the Kosko method. To simplify the MF construction, Dickerson and Kosko in [10] defined a rotated rectangular region, in which the hyperelipsoid in inscripted. As a result, the following triangular-based MF is constructed:

$$\mu_j(X) = \begin{cases} 1 - \dfrac{|x_j - c_j|}{p_j}, & |x - c_j| \le \dfrac{p_j}{2} \\ 0, & otherwise \end{cases} \tag{2}$$

where the μ^j defines the MF of j feature and the projection of circumscribed hypperrectangular on i axis is

$$p_i = 2 \cdot \alpha \cdot \sum \frac{|cos\gamma_{ij}|}{\sqrt{\lambda_i}} \tag{3}$$

where angle between the i axis and j eigenvector e: $\gamma_{ij} = arccos(e_j(i))$.

As for the Gaussian MF in the elliptic fuzzy patches, the parameters from the Kosko patches were used from the Equation 1 in order to derive the MF parameters. Piegat in [21] presented the angle between the axes of hyperellipsoid and features as a possible solution to increase the precision of the MF by Σ^{-1}:

$$\mu_j(X) = e^{-\frac{1}{2}(x-c)^T \Sigma^{-1} (x-c)} \tag{4}$$

The rectangular topology of SOM is used in this work since it will give rectangular mapping of the region, which is more appropriate for the fuzzy terms extraction. The reason for this is that it provides more abrupt clusters grouping the samples within edges rather than overlapping with better connectivity as studied by Baltimore in the [6]. Below we give the insight into existing methods and schemes. To authors awareness, there have not been proposed a method yet to find an optimal size of the SOM that complies with the Daubert Standards for the models to be tested and interpreted according to the stated requirements [12]. The main problem is to present fuzzy rules from SOM clustering without original data alteration in a forensically-sound manner. Most of the existing methods generates a big number of specific rules.

2.1 Modification of Topology

One of the possible techniques to find an optimal size of the SOM (number of nodes) is iterative process of topology modification. In most of the methods this process starts with a small number of nodes, while constantly adding new nodes until some quality stopping criteria is fulfilled. One of the first modification of Kohonen SOM was described by Alahakoon et al. in 1998 in the research [3]. In this work the problem of the optimal size definition was first addressed by introducing the Growing SOM (GSOM). So, authors proposed to use additional growing phase that adds new node with specific coordinates until the growing conditions are no longer satisfied. New nodes are added to the boundary while entire structure is preserved. Then, the weights of this node are given from the bounded values. The method has a good speed on the small dataset since the initial number of nodes in the topology is only 4. Further analysis of this method for knowledge discovery was performed in the paper [2] in 2000 by introduction the Growing Hierarchical SOM (GHSOM). In the paper, it was stated that the general guideline is to generate a smaller map first and then expand analysis process in each cluster. Yet it is necessary to define the growth threshold GT and the spread factor SF. This requires additional knowledge and work done by data analyst. So, this method requires specific knowledge of the dataset that malware analyst may not have. Another challenge is that the fuzzy relationships can not be represented properly since the boundaries on the different layers are crisp. To deal with this challenge, several other methods were proposed for fuzzy rules derivation while using SOM with modification of the topology. The first one, Adaptive Self-Organzing Neural Network (GSFNN) was proposed by Qiao in 2007 [22]. However, it requires intense computation process together with a bunch of additional parameters that need to be defined empirically by additional data analytic. So, our concern is complexity of the methods mentioned above and since the growth is not bounded, the structure may be too complex. As results, this will take much time to tune the parameters on the 2nd stage of the NF method. At this point we will consider fixed-size models to be more suitable to use in NF methods than changing the topology, since then we will not be able to control the granularity of the fuzzy sets.

2.2 Based on the Dataset Size

The next definition of the grid size refers to the dataset size. It was mentioned that the number of nodes should be no more than the number of samples in dataset. This can be explained as non-empty SOM node should consist of at least one sample. In [11,23,27] the applicability of the quantity metric defined by Vesanto in 2000 for the SOM Toolbox in Mathlab [28] discussed that is empirically determined as the optimal size of the SOM:

$$S = 5 \cdot \sqrt{N} \tag{5}$$

Definition 1: Initial size of the SOM is equal to $S = 5 \cdot \sqrt{N}$. The boundaries of the size are defined as following according to SOM Toolbox:

$$S_{lower} = 0.25 \cdot S = 2.5 \cdot N^{0.54321} \tag{6}$$
$$S_{upper} = 4 \cdot S = 20 \cdot N^{0.54321} \tag{7}$$

Furthermore, the number of nodes in each of 2 dimensions of the rectangular SOM are calculated as following mapping to integer numbers:

$$S_a = \left\lceil \sqrt{S} \right\rceil, S_b = \left\lfloor \frac{S}{S_a} \right\rfloor \tag{8}$$

This is according to SOM toolbox documentation [28] and measured empirically. We can see the following problems with this measure. Initially, when the number of data samples is huge, the size of SOM will grow infinitely:

$$\lim_{N \to \infty} S = \infty \tag{9}$$

Then, this SOM likely will give most of the nodes in empty state. Finally, with huge number of derived regions we can not make simple and understandable classification model.

2.3 Using Variance

Another possible way of determination of SOM size is analysis of the spread of the data, for example the measure of variance. Principle Component Analysis extracts the components that can be characterised by the distribution of variance [25]. It means that by means of eigenvalues for each corresponding eigenvector the characteristics of the distribution can be measured. This technique is used in Factor Analysis [1] to see how well the factors fit the model and whether they influence it at all. As defined in the Equation 5, size of the SOM may be changed in case if variance on the first components with the biggest eigenvalues take the majority of the variance in the dataset [11, 27].

Definition 2: The distribution of the data along the principle components characterise the way of how the most correlated components are distributed in the data. By applying the following formula:

$$S' = 5 \cdot \sqrt{N} \cdot \frac{e_1}{e_2} \tag{10}$$

where e_1 and e_2 are two eigenvalues with the biggest values from the work by Vesanto [27]. This can be interpreted as following data spanning. If both vectors are nearly equal, then the more general cluster will be composed by the number of nodes S. Otherwise, more nodes will be used in S' if the data are stretched along one of the components, which means significant correlation.

2.4 Interpretability Concerns

As mentioned in the Section 1, the biggest challenge in building fuzzy rules model is seeking for an optimal trade-off between the accuracy and interpretability. In the research [17], Ishibuchi stated it as a multi-objective problem as following for the fuzzy-based system F:

$$\max_{S} \left(Accuracy(F), \ Interpretability(F) \right) \tag{11}$$

However, it is not easy for forensics analyst to define the appropriate parameters, especially when dealing with Big Data. So, in this paper several experiments were presented using the size of the fuzzy-based method in the range from 2x2 to 5x5. Moreover, it was mentioned an application of Genetic Algorithm for the optimization in the Equation 11. The paper concluded that there are many obstacles and no unique solutions are available for the optimal size of the fuzzy-based systems. More expensive study of the interoperability measures was done by Gacto et al [13]. The authors mentioned a need for a trade-off between the accuracy and the interpretability. The accuracy can be measured easily, while the interpretability measures are not easy to define. One of the main features used to study their interpretability were number of rules, number of MF, etc. According to this paper, the number of conditions in the rules should not exceed 7 ± 2, which is a boundary that human brain can handle. However, the number of features sometimes can be much more than this limit. So, the only way to limit the complexity of the model is to shorten the number of rules. Another work [16] on the combination of accuracy and interpretability is done by Herrera et al. The authors tried to lower the number of rules by means of Taylor series approximation. The authors showed that usage up to 3 MF in each dimension can be considered as sufficient for a better accuracy on a dataset. Here we can state that the drop of accuracy in the fuzzy systems with higher number of rules exists because of a spread and over-fitting of the training data. According to Alonso [4], the maximum number of rules acceptable by user should be greater or equal that 10^3 times number of classes. Castellano in [7] presented A Priori Prunning for the human-understandable rules selection from NF. There were extracted only 66 rules were extracted from more than 200 thousands of the rules produced by the grid partition. So, there is need to bound the grid size to improve the generalization and it can be seen that it is reasonable to use less than 5^2 rules for the human-understandable model.

2.5   Correlation

One of the main factors that influences the goodness of fit of the fuzzy model to the data is correlation between the features [18]. We can state that if there is a correlation between the features present in the data then the bigger amount of fuzzy patches is required to cover the area. However, this amount is still lower than in the case when all possible combinations of MFs are constructed. The basic linear dependencies can be easily revealed by means of widely-used Pearson Correlation Coefficient (PCC) r. It is applied when the fast and tentative information about the relations in the dataset is needed. The PCC between two variables is calculated as it is shown in the Equation 12.

$$r = \frac{Cov(XY)}{\sigma_X, \sigma_Y} \tag{12}$$

where the σ denotes the standard deviation for each attribute and $Cov(X, Y)$ is a covariance between two attributes. The complexity can be defined as following. It requires $1 \cdot N$ computations of the mean $\bar{x}$, N computations for the covariance between and $2 \cdot N$ computations for the variance. Overall computational complexity is $O(M \cdot N)$ on the M-dimensional dataset with N samples. Additionally, it does not require to store any supplementary matrices.

The Big Data that needs to be analysed in the Digital Forensics has complex dependencies [14]. Different correlation techniques can be employed to incorporate mentioned challenges, including confounding variables and non-monotonic dependencies. According to the study [9], there are several metrics of the correlation such that PCC, Spearman, Distance Correlation (DC) and Maximal Information Coefficient (MIC). The DC is a new measure and was proposed in [26] by Szekely et al in 2007 and is based on the Euclidean distances covariance between the variables. The Spearman and PCC correlations are defined as linear. Spearman does not differ much from PCC though more iterations required to find and store ranking meta-data. From the other hand, DC and MIC have been defined as non-linear metrics. PCC, DC and MIC metrics were successfully used in the associations discovery in the large astrophysical databases by Martinez-Gomez et al [20]. According to the work, DC is more effective than MIC and better reveals associations in the datasets.

$$dCorr = \frac{dCov(X,Y)}{\sqrt{dVar_X \cdot dVar_Y}} \tag{13}$$

The distance covariance $dCov(X,Y)$ and variance $dVar_X^2$ are defined as the following measure of the element-wise Euclidean distance matrix $A(a_{ij} = ||X_i - X_j||)$ and $B(b_{ij} = ||Y_i - Y_j||)$ [26]:

$$dCov(X,Y)^2 = \frac{1}{N^2} \sum_{i,j=1}^{N} A_{ij}B_{ij}, \ dVar_X^2 = \frac{1}{N^2} \sum_{i,j=1}^{N} A_{ij}^2 \tag{14}$$

Considering complexity, this method needs N^2 Euclidean distance computations between every sample's attribute value, N^2 computations for covariance and $2 \cdot N^2$ computations for variance. In overall, the complexity can be defined as $O(M \cdot N^2)$ for the input data. Also, $2 \cdot N^2$ of memory structures required to store distance matrices A and B on every iteration. Despite the fact that DC is non-linear measure, it was shown in the [20] that there is also a sharp dependency between it and the PCC. It makes DC overhead $O(M \cdot N) < O(M \cdot N^2)$ less reasonable when using for Big Data analytic in comparison to the PCC results.

So, it can be seen that PCC and DC are the most promising measures to be used in the determination of the optimal SOM size. Even considering the fact that PCC was defined as liner one, it has a strong compliance with the defined non-linear DC. At this point, we can state that DC is less efficient on the huge datasets dues to computational complexity. Moreover, the DC takes more space for meta data when $N >> M$ than for dataset itself. At this point, it is more efficient to use PCC for the estimation of the dependencies in the dataset.

3   Optimal SOM Size Based on the Data Analytic

In this Section, the insight will be given into how the size of the SOM can be determined using the data analytic. Exploratory data analysis provides a fast way of getting the high level information about the data properties. Considering this option, there will be presented an automated procedure that will give an optimal size of SOM, before it is initialized and learned. The main advantage is that no alteration will be performed in the data. This is an important fact since according to the Digital Forensics Process requirements [14] the data must be preserved without changing a bit during the investigation process. As it was mentioned before, the biggest eigenvalues ratio does not give sufficient information about the number of required fuzzy rules.

3.1   Proposed Algorithm for Optimal Number of Nodes

Considering mentioned obstacles and limitations for determination of the SOM size mentioned in the Section 2, we will use measures of non-linear correlation to find the best grid size with respect to reasonable interpretability and accuracy.

Proposition 1: Measure of association between the variables can be used to determine the number of regions that will be clustered in SOM according to the Equation 11 using absolute averaged value of PCC 12:

$$S|_{max\ Accuracy\ (F)} \propto |\bar{r}| \tag{15}$$

Proof. In real life applications, the data do not require usage of all rules with all combinations of the MF. Instead, two contrary cases can be considered. The first one is when the data are random and attributes are uncorrelated, which means that the less specific fuzzy patches will fit the data better. In this case, mentioned correlation metrics in the Section 2, including $r \to 0$. The second case is when the data are more deterministic and follow some patterns. It means that more fuzzy patches is required to describe the dependency specifically. Ideally, when the $|r|$ tends to reach 1, more specific MF are required to fit the data, rather than rectangular patches or even Kosko patches proposed in 1997 [19].

Proposition 2: The interpretability *Interpretability* (F) in the the Equation 11 can be formulated as a range of number of rules that can be perceived by analyst without additional remembering efforts. The constraints are placed to limit the SOM size in order to make it understandable:

$$\begin{cases} S_{min} \geq nC \land S_{min} \geq 2^2 \\ S_{max} \leq N^M \land S_{max} \leq 5^2 \end{cases} \tag{16}$$

where nC is a number of classes in the classification problem, since at least such amount of rules required to distinguish the data samples. $S_{min} = 2^2$ corresponds to the minimum SOM size used in GSOM [3]. $S_{max} = 5^2$ corresponds to the maximal amount of the rules that can be used according to the previously studied researches. Greater amount of rules will bring difficulties in understanding. $S_{max} \leq N^M$ is the number of combinations of all the MF in the rules.

Proposition 3: The optimal size of the SOM grid can be defined using the limitations mentioned before and degree of randomness α:

$$S = S_{min} + \alpha \cdot (S_{max} - S_{min}) \tag{17}$$

Proposition 4: The degree of randomness in the Equation 17 will be calculated using the mean absolute PCC and metrics defined earlier:

$$\alpha = \frac{e_0}{e_1} \cdot |\bar{r}| \cdot nC = \frac{e_0}{e_1} \cdot \frac{\sum_{i \neq j}^{M} |r_{ij}|}{M^2 - M} \cdot nC \tag{18}$$

subject to the following constraints:

$$0 \leq \alpha \leq 1.0 \tag{19}$$

where e_0 and e_1 are the 1st and the 2nd biggest eigenvalues that determines the components with the highest variance value. Then, the parameters of each 2D dimensions of the SOM are calculated as following using the Equations 17, 18 and 8:

$$S_a = \left\lceil \sqrt{S} \right\rceil = \left\lceil \sqrt{S_{min} + \alpha \cdot (S_{max} - S_{min})} \right\rceil, \; S_b = \left\lfloor \frac{S}{S_a} \right\rfloor \tag{20}$$

Remark. The modification of the Vesanto method was proposed in order to meet the interpretability requirements in the Computational Forensics method. The proposed scheme does not depend on the number of instances since it is targeted on dealing with Big Data problems. Moreover, the properties of the features such that normalization, number of features, constant values and highly-correlated pairs will not influence the proposed method, which will be shown further on. The $|r|$ is required as a mean measure a dataset, since only average value of PCC is important and not the direction (negative, positive).

The new analytic scheme for the determination of the optimal SOM grid size for the 1st stage of NF that complies with the requirements of Digital Forensics Process:

1. Calculate all values of the PCC r_{ij} pair-wise between the attributes. Estimate the absolute mean of the PCC $|\bar{r}| = \frac{\sum_{i \neq j}^{M} |r_{ij}|}{M^2 - M}$.
2. Perform eigendecomposition of Correlation Matrix based on the calculated earlier PCC. Calculate the eigenvalues ratio $\frac{e_0}{e_1}$.
3. Calculate the optimal SOM size using the Equation 20 with the constraints defined in the Equation 19.

4   Experimental Results

This Section described the experimental design, used datasets and analysis of the results. We will provide a comparison between the method by Vesanto and our proposed method in terms of classification and regression accuracy.

4.1   Experimental Design

To evaluate the performance of the proposed scheme we implemented two stages of the NF method as described by Kosko in the [19]. Also, we used two types of the fuzzy patches and corresponding MF. The first one is simple rectangular patches with triangular MF. The second one is Kosko patches with corresponding triangular MF based on the projection of the ellipses. Since SOM is a method influenced by the random initialization of the nodes, the bootstrap aggregation was performed. To ensure the consistency of the results, 10 samples were randomly generated during SOM clustering and the best sequence was chosen to be used afterwards in the experiments. At the 1_{st} stage of NF, the fuzzy patches parameters are derived according to the derived number of clusters from the proposed method. During the 2nd stage, the classification rules are tuned. Moreover, we estimate the accuracy of the NF using two types of measures. (1) regression-based real-value accuracy comparison of the defuzzified output of NF using the following metric: Mean Absolute Percent Error $MAPE = \frac{1}{N} \sum_{i=1}^{N} |\frac{y_i - d_i}{d_i}| \cdot 100\%$, where y_i - the output of of the gravity deffuzifier [19], d_i - the actual class of the sample, N - number of given data samples. (2) discrete classification accuracy comparison using the selected rules by min-max principle in cross-validation: $Acc = \frac{nP}{N}$, where nP - number of properly classified samples.

4.2   Datasets

The proposed method was tested on the malware analysis dataset. Moreover, there have been chosen several datasets to demonstrate the proof of concept from the UCI Machine Learning Repository[2]. Besides the properties of the datasets are presented in the Table 1, the ratio of the biggest eigenvalues $\frac{e_0}{e_1}$ from non-zero mean set (Correlation Matrix) and absolute averaged Person Correlation Coefficient $|r|$ are given.

Table 1. The properties of the datasets used in the experiments based on the data obtained from the statistical programs PSPP and Weka

Dataset	Feat.	N	Const.	Norm.	$\bar{r}$	e_0	e_1	$\frac{e_0}{e_1}$
Climate Sim.Crash.	18	540	No	Yes	0.0091	1.0797	1.0759	1.0035
Fertility	9	100	No	Yes	0.1204	1.8654	1.4530	1.2838
Banknote Auth.	4	1372	No	No	0.4255	2.1799	1.2931	1.6857
Mob. malw.	36	596	Yes	No	0.1224	6.3839	4.3715	1.4603

The first three datasets are publically available and represent different domains. The last one is manually composed dataset of features from static and dynamic tests of the the mobile 'malware' and 'benign' applications. The Android platform was chosen because of its popularity and since user can install of 3rd party applications that might containt malicious payload. It consists of 36 versatile numerical features from 604 applications that characterize particular a property such that CPU load, memory usage, API calls, etc. The datasets are chosen to be with two classes as in the malware detection problem.

Below in the Figure 1, the visualization of the dependency between the features in the given earlier datasets are given. It can be seen that in case of independent features, the number of fuzzy patches can be decreased by covering more instances. From the other side, some non-monotonic dependencies require an increase in number of fuzzy patches to reduce the error.

---

[2] https://archive.ics.uci.edu/ml/datasets/

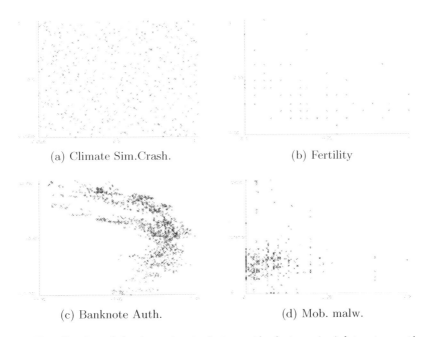

(a) Climate Sim.Crash.   (b) Fertility

(c) Banknote Auth.   (d) Mob. malw.

Fig. 1. Visualization of the dependencies between the features in 4 datasets mentioned earlier in Weka. The colors are blue and red denotes both classes.

At this point, we can expect that the the Banknote Authentication dataset might require the biggest number of more specific rules to build the classification model considering the non-linear and non-monotonic dependencies in the Figure 1.c.

4.3   Experimental Results

The hypothesis was made that the bigger size of SOM gives more unstable results due to spread of data. The trade-off has to be found to reduce the complexity and improve the specificity. To evaluate the defined hypothesis that the bigger bigger size of SOM gives more unstable results due to spread of data, we performed comparison of cross-validated results with bootstrapped training data. The SOM size over 5x5 was not considered, since the amount of rules above this amount will makes the model complicated to perceive. The performance was evaluated using the regression performance metric "MAPE, %" and classification one "Acc(rules), %". The results are presented in the Table 2 for the all given dataset. The 'Extr. rules' row denotes the number of actually extracted rules (clusters) from the SOM, while 'proposed' denotes the recommended number of rules according to the used method.

Additionally, we measured the execution time for the malware dataset as it is shown in the Table 3. The SOM using size from the proposed method trains faster due to the lower amount of rules. The C++ OpenMP implementation with

Table 2. Performance comparison (regression, classification) of the proposed method on the four datasets, including the number of proposed and extracted rules

Method	"Rule of thumb"		Vesanto		Proposed	
	MAPE,%	Acc,%	MAPE,%	Acc ,%	MAPE,%	Acc ,%
Dataset: Climate						
Simple MF	13.3848	23.5185	´13.3825	23.5185	38.3623	22.7778
Kosko MF	11.1754	91.4815	45.4244	91.6667	45.6569	92.0370
Gaussian MF	5.9791	69.6296	5.8009	76.6667	3.8818	96.4815
Extr.rules (proposed)	**61 (116)**		**61 (117)**		**6 (6)**	
Dataset: Fertility						
Simple MF	14.7705	88.0000	18.5313	88.0000	68.8893	84.0000
Kosko MF	6.0117	88.0000	10.2212	85.0000	8.9175	89.0000
Gaussian MF	6.2184	92.0000	6.5119	92.0000	4.8512	92.0000
Extr.rules (proposed)	**9 (50)**		**9 (64)**		**7 (11)**	
Dataset: Banknote Authentication						
Simple MF	22.2296	55.5394	21.8893	55.5394	5.9234	95.5539
Kosko MF	27.1998	98.9796	22.4358	98.9067	22.7381	96.2828
Gaussian MF	1.7491	100.0000	15.7715	99.6356	18.0398	99.9271
Extr.rules (proposed)	**94 (185)**		**97 (312)**		**17 (25)**	
Dataset: Malware						
Simple MF	38.4787	35.1261	39.0162	40.0000	40.1554	41.1765
Kosko MF	41.5120	58.4874	34.5776	56.8067	41.3487	58.4874
Gaussian MF	11.7330	84.5378	11.5823	80.8403	5.1865	91.9328
Extr.rules (proposed)	**48 (122)**		**42 (189)**		**16 (24)**	

Table 3. Time in seconds, required to learn the NF mode using the optimal SOM size according to Vesanto and proposed method on the mobile malware dataset

Method	Simple	Kosko	Gaussian
Vesanto	0.5955	0.9097	0.2968
Proposed	0.4581	0.7202	0.1116

Eigen, STL and Boost libraries was used on 8-threaded Intel Core i7-3632QM CPU - 2.20GHz with 8GB DDR3.

Thus, It can be seen that the size of the SOM depends, first of all on the statistical properties of the data set. It is obvious that in most cases the bigger SOM size causes the degradation of the accuracy with a decrease in interpetability.

5   Conclusions

This paper addresses the problem of optimal SOM size determination. It was studied how the trade-off between the accuracy of the resulting NF model and the interpretability can be achieved with respect to the malware analysis in Digital Forensics. We have proposed a new, an efficient and a fast approach to derive an optimal number of nodes in SOM to be clustered using data analytic rather than alteration of the dataset. Experiments on different datasets proved the applicability of the method using simple triangular, Kosko and Gaussian MF with corresponding rectangular and elliptic fuzzy patches configuration. The proposed method decreases the number of effective rules from 42 in Vesanto method, to 16 in case of Android malware detection based on versatile features collected from static and dynamic tests. Additionally, the classification accuracy growth due to a smaller spread of the clusters and achieves 92%.

Acknowledgments. The authors are grateful to the colleges, Ambika Shrestha Chitrakar and Gaute Wangen, for helpful comments and valuable suggestions on earlier drafts that substantially helped to improve and clarify the manuscript.

References

1. Ahn, S.C., Horenstein, A.R.: Eigenvalue ratio test for the number of factors. Econometrica **81**(3), 1203–1227 (2013)
2. Alahakoon, D., Halgamuge, S., Srinivasan, B.: Dynamic self-organizing maps with controlled growth for knowledge discovery. IEEE Transactions on Neural Networks **11**(3), 601–614 (2000)
3. Alahakoon, D., Halgamuge, S., Srinivasan, B.: A self-growing cluster development approach to data mining. In: 1998 IEEE International Conference on Systems, Man, and Cybernetics, vol. 3, pp. 2901–2906, October 1998
4. Alonso, J.M., Cordn, O., Quirin, A., Magdalena, L.: Analyzing interpretability of fuzzy rule-based systems by means of fuzzy inference-grams. In: World Congress on Soft Computing (2011)
5. Altyeb Altaher, A.A., Ramadass, S.: Application of adaptive neuro-fuzzy inference system for information secuirty. Journal of Computer Science **8**(6), 983–986 (2012)
6. Baltimore, R.: An Analytic investigation into self organizing maps and their network topologies. Ph.D. thesis, Rochester Institute of Technology (2010)
7. Castellano, G., Fanelli, A.M., Mencar, C.: Discovering interpretable classification rules from neural processed data (2002)
8. Chattopadhyay, M., Dan, P.K., Mazumdar, S.: Application of visual clustering properties of self organizing map in machine-part cell formation. Appl. Soft Comput. **12**(2), 600–610 (2012). http://dx.doi.org/10.1016/j.asoc.2011.11.004
9. Clark, M.: A comparison of correlation measures. University of Notre Dame, Tech. rep. (2013)
10. Dickerson, J.A., Kosko, B.: Fuzzy function approximation with ellipsoidal rules. Trans. Sys. Man Cyber. Part B **26**(4), 542–560 (1996). http://dx.doi.org/10.1109/3477.517030
11. Estévez, P., Príncipe, J., Zegers, P.: Advances in Self-Organizing Maps: 9th International Workshop, WSOM 2012. AISC, vol. 198. Springer, Heidelberg (2012). https://books.google.no/books?id=vHgnfKFpIFUC
12. Feldman, E.R.: Criteria for admissibility of expert opinion testimony under daubert and its progeny. Tech. rep, Cozen OConnor (2001)
13. Gacto, M., Alcal, R., Herrera, F.: Interpretability of linguistic fuzzy rule-based systems: An overview of interpretability measures. Information Sciences **181**(20), 4340–4360 (2011). http://www.sciencedirect.com/science/article/pii/S0020025511001034, special Issue on Interpretable Fuzzy Systems

14. Guarino, A.: Digital forensics as a big data challenge. In: ISSE 2013 Securing Electronic Business Processes, pp. 197–203. Springer (2013)
15. Hasan, S., Shamsuddin, S.M.: Multistrategy self-organizing map learning for classification problems. Computational Intelligence and Neuroscience **2011**, 11 (2011)
16. Herrera, L., Pomares, H., Rojas, I., Valenzuela, O., Prieto, A.: Tase, a taylor series-based fuzzy system model that combines interpretability and accuracy. Fuzzy Sets and Systems **153**(3), 403–427 (2005). http://www.sciencedirect.com/science/article/pii/S0165011405000333
17. Ishibuchi, H., Nojima, Y.: Discussions on interpretability of fuzzy systems using simple examples (2009)
18. Jin, Y.: Fuzzy modeling of high-dimensional systems: complexity reduction and interpretability improvement. IEEE Transactions on Fuzzy Systems **8**(2), 212–221 (2000)
19. Kosko, B.: Fuzzy Engineering. No. v. 1 in Fuzzy Engineering. Prentice Hall (1997). http://books.google.no/books?id=8QwoAQAAMAAJ
20. Martínez-Gómez, E., Richards, M.T., Richards, D.S.P.: Distance correlation methods for discovering associations in large astrophysical databases. The Astrophysical Journal **781**, 39 (2014)
21. Piegat, A.: Fuzzy Modeling and Control. STUDFUZZ, vol. 69. Physica-Verlag, Heidelberg (2001). http://books.google.no/books?id=329oSfh-vxsC
22. Fei Qiao, J., Gui Han, H.: An Adaptive Fuzzy Neural Network Based on Self-Organizing Map (SOM). INFTECH, April 2010, iSBN 978-953-307-074-2
23. Schmidt, C.R.: Effect of irregular topology in shperical Self-Organizing Maps. Ph.D. thesis, San Diego State University (December 2008)
24. Singh, R., Kumar, H., Singla, R.: Review of soft computing in malware detection. Special Issues on IP Multimedia Communications (1), 55–60 (2011), full text available
25. Smith, L.I.: A tutorial on principal components analysis. Tech. rep., Cornell University, USA (February 26, 2002). http://www.cs.otago.ac.nz/cosc453/student_tutorials/principal_components.pdf
26. Szkely, J., Rizzo, M.L., Bakirov, N.K.: Measuring and testing dependence by correlation of distances
27. Vesanto, J., Alhoniemi, E.: Clustering of the self-organizing map (2000)
28. Vesanto, J., Himberg, J., Alhoniemi, E., Parhankangas, J.: Self-organizing map in matlab: the som toolbox. In: Proceedings of the Matlab DSP Conference, pp. 35–40 (2000)
29. Zhang, Y. (ed.) Machine Learning. INFTECH (February 2010), isbn 978-953-307-033-9

SVRs and Uncertainty Estimates in Wind Energy Prediction

Jesús Prada[✉] and José Ramón Dorronsoro

Universidad Autónoma de Madrid, Cantoblanco, Madrid, Spain
jesus.prada@estudiante.uam.es, jose.dorronsoro@uam.es

Abstract. While Support Vector Regression, SVR, is one of the algorithms of choice in modeling problems, construction of its error intervals seems to have received less attention. On the other hand, general noise cost functions for SVR have been recently proposed. Taking this into account, this paper describes a direct approach to build error intervals for different choices of residual distributions. We also discuss how to fit these noise models and estimate their parameters, proceeding then to give a comparison between intervals obtained using this method. under different ways to estimate SVR parameters as well as the intervals obtained by employing a full SVR Bayesian framework. The proposed approach is shown on a synthetic problem to provide better accuracy when models fitted coincide with the noise injected into the problem. Finally, we apply it to wind energy forecasting, exploiting predicted energy magnitudes to define intervals with different widths.

1 Introduction

Support vector regression, SVR, [1] has been widely used in regression problems such as stock market [2], wind energy [3] or solar radiation [4] forecasting. Classical SVR, however, does not give probability intervals to address the uncertainty in the predictions and, in fact, error interval estimation for SVR has received a somewhat limited attention in the literature. Notice that here approaches such as the well known ones for linear regression under Gaussian models completely break down, not only by the difficulty of ensuring normal random variables but, above all, by the fact that the familiar analytic estimates of the linear coefficients are simply impossible in SVR and, of course, less so, any asymptotic analysis.

In [5], a Bayesian interpretation of SVR is described and then used to propose methods to determine, first, SVR parameters by maximizing an evidence function and, second, to derive probability intervals for predictions. A drawback of these methods is that they modify the classical SVR formulation of the problem to solve, and hence existing SVR software, such as the popular LIBSVM [6] cannot be used, at least without modifying it first.

A more direct approach is proposed in [7], which assumes prediction errors to follow a specific probability distribution that, in turn, is used to define the probability intervals. Zero-mean Gaussian and Laplace families are proposed in [7] as noise models and fitted by maximum likelihood estimation, MLE, using out-of-sample residuals of SVR models; optimal SVR parameters are obtained simply

© Springer International Publishing Switzerland 2015
I. Rojas et al. (Eds.): IWANN 2015, Part II, LNCS 9095, pp. 564–577, 2015.
DOI: 10.1007/978-3-319-19222-2_47

by cross validation. In this paper, we follow this methodology to give probability intervals under the assumption of both zero-mean Laplace and Gaussian distributions, as well as for their non–zero mean counterparts plus the Beta and Weibull distributions. A difficulty with this approach is that it assumes that the residual distribution is independent of the predicted value and, therefore, probability intervals have exactly the same width for all input instances. General error models for SVR other than the well known ϵ–insensitive loss have been proposed in [8]. This suggest that noise distribution might be different across particular problems and it should be reflected in the particular SVR model to be used. If the assumption is true and the underlying noise distribution is accurately estimated, one should expect a reduction in interval prediction errors. We study if the proposed method can estimate this noise distribution.

Our main contributions can be summarized as follows:

– We enlarge, as mentioned, the noise models considered in [7].
– We discuss Newton–Rapshon maximum likelihood estimates for the Beta and, particularly, Weibull distributions, as well as the definition of uncertainty intervals for them.
– We show on a synthetic problem how the proposed approach is able to pair the models fitted to the residuals with the specific noise injected on the problem targets. We also compare these results to the ones obtained by a statistic test for distribution hypotheses.
– We apply the methods proposed to the estimation of uncertainty intervals for the wind energy prediction of peninsular Spain, where we also consider the use of different intervals according to predicted energy magnitudes, showing how a two group data split results in more accurate intervals.

The rest of this paper is organized as follows. Section 2 briefly reviews both classical and Bayesian SVR formulations. In Section 3 there is an in-depth description of the proposed approach for error interval estimations and experiments are carried in Section 4, where we also consider four publicly available regression datasets besides the already mentioned synthetic and wind energy problems. The paper ends with a short section on conclusions and pointers for further work.

2   Support Vector Regression

2.1   Classical SVR Review

Given a sample $D = \{(x_i, y_i) : 1 \leq i \leq N\}$ of inputs $x_i \in R^n$ and targets $y_i \in R$, the SVR problem is that of minimizing the loss function $L(w, b, \xi)$ defined as

$$L(w, b, \xi) = \frac{1}{2}\|w\|^2 + C \sum_i (\xi_i + \xi_i^*), \tag{1}$$

over w, b and ξ subject to the constraints $-\xi_i - \epsilon \leq w \cdot x_i + b - y_i \leq \xi_i^* + \epsilon$, $\xi_i, \xi_i^* \geq 0$ .This is known as the SVR primal problem. It can also be seen as a variant of the standard L_2 regularized regression where instead of the familiar $z_i^2 = (y_i - w \cdot x_i - b)^2$ square error, we use the ϵ–insensitive loss function [9]

$l_\epsilon(\delta) = [\delta]_\epsilon - \max(0, |\delta| - \epsilon)$. i.e., we allow an ϵ-wide, penalty-free "error tube" around the model function $w \cdot x + b$. To solve the primal problem, it is transformed [1] using standard Lagrangian analysis into the so–called dual problem; moreover, instead of working with a simple linear model $w \cdot x + b$, the well known kernel trick is used to arrive to a non–linear model $w \cdot \Phi(x) + b$ where $\Phi(x)$ is a high (and possibly infinite) dimensional projection of the original x. We thus obtain the final model as

$$f(x) = b^* + w^* \cdot \Phi(x) = b^* + \sum \gamma_i^* \Phi(x_i) \cdot \Phi(x) = b^* + \sum_i \gamma_i^* k(x_i, x) . \quad (2)$$

We take the usual choice of a Gaussian kernel $k(x, x') = \exp\left(-\gamma \|x - x'\|^2\right)$. It is clear that model performance will be highly dependent on the choices of the C, ϵ, γ parameters which, in turn, will affect the residuals $\delta_i = y_i - f(x_i)$ whose distribution we want to estimate. One way to derive optimal C, ϵ, γ values is by standard cross–validation, CV, or validation over a fixed set. A perhaps more principled alternative could be to follow a Bayesian framework to derive their optimal values. We briefly review this approach next.

2.2   A Bayesian Framework for SVR

We assume the x_i, y_i in the sample D to be related through a model $y_i = f(x_i) + \delta_i$ where the δ_i follow random i.i.d. values. In the Bayesian approach of [5], $f = [f(x_1), f(x_2), ..., f(x_N)]$ is taken as the realization of a random field with a known prior probability. The posterior probability of f given D can then be derived by Bayes' theorem as

$$P(f|D) = \frac{P(D|f)P(f)}{P(D)} \propto P(D|f)P(f), \quad (3)$$

where the conditional probability of D given f is $P(D|f) = \prod_{i=1}^{N} P(y_i - f(x_i)) = \prod_{i=1}^{N} P(\delta_i)$ and $P(\delta_i)$ is often assumed to be of the exponential form, i.e., $P(\delta_i) \propto \exp(-C \cdot l(\delta_i))$ with $C > 0$ a normalizing constant and l a certain loss function. Putting this together, we arrive at

$$P(D|f) = \prod_{i=1}^{N} P(\delta_i) \propto \exp\left(-C \cdot \sum_{i=1}^{N} l(\delta_i)\right) = \exp\left(-C \cdot \sum_{i=1}^{N} l(y_i - f(x_i))\right). \quad (4)$$

going back to the prior $P(f)$, in [5] it is assumed to be of the form

$$P(f) = (2\pi)^{\frac{N}{2}} |K|^{-\frac{1}{2}} \exp\left(-\frac{1}{2} f^T K^{-1} f\right), \quad (5)$$

and also the following structure for $K_{i,j}$

$$K_{i,j} = K_{x_i, x_j} = cov[f(x_i), f(x_j)] = \kappa_0 \exp\left(-\frac{\kappa}{2} \|x_i - x_j\|^2\right) + \kappa_b, \quad (6)$$

with κ_0, κ and κ_b appropriate positive constants. The parameters in the prior and the likelihood are $\theta = (C, \epsilon, \kappa, \kappa_0, \kappa_b)$. κ_0 is fixed in [5] as the target variance. The optimal values of the other hyperparameters can be inferred by maximizing

$$P(\theta|D) \propto P(D|\theta)P(\theta) \ . \tag{7}$$

A common assumption now is that $P(\theta)$ is rather insensitive to the values of θ and can thus be ignored.

An optimal θ^* can then derived by maximizing the evidence function $P(D|\theta)$ for which, by Jensen's inequality, we have

$$-\log P(D|\theta) = -\log \int P(D|f,\theta)P(f|\theta)df \leq \int \left[-\log P(D|f,\theta) - \log P(f|\theta)\right]df,$$

and approximating the integral by the sample average we arrive to the problem

$$\min_\theta \left\{ C \sum_{i=1}^{N} l_{\epsilon,\beta}(y_i - f(x_i)) + \frac{1}{2}f^T K^{-1}f \right\} \ . \tag{8}$$

[5] derives in this way analytic approximations to $-\log P(D|\theta)$. Moreover, it proposes to replace the standard non–differentiable ϵ–insensitive SVR loss by a smooth version, the so–called soft insensitive loss function, SILF.

Turning our attention to probability intervals, as stated in [7] we have

$$1 - s = \int_{-\infty}^{p_s} p_\zeta(z)dz = \int_{-\infty}^{+\infty} \int_{-\infty}^{p_s-f} p(\delta)d\delta p_{f|D}(f)df, \tag{9}$$

where

$$p(\delta) = \frac{\exp\left(-C \cdot l_{\beta,\epsilon}(\delta)\right)}{\int \exp\left(-C \cdot l_{\beta,\epsilon}(\delta)\right)d\delta}, \ p(f(x)|D) = \frac{1}{\sqrt{2\pi}\sigma_t} \exp\left(-\frac{(f(x) - \hat{f})^2}{2\sigma_t^2}\right), \tag{10}$$

where $\sigma_t^2 = K(x,x) - K_{F,x}^T K_{F,F}^{-1} K_{F,x}$ and $K_{F,x}$ the vector containing all $K(x_i,x)$, with $i \in F = \{i|0 < \alpha_i < C \text{ or } 0 < \alpha_i^* < C\}$ and $K_{F,F}$ the corresponding submatrix.

3   Uncertainty Estimates

A natural way to address uncertainty estimates is by assuming that the residual distribution follows some parametric model $p(\delta;\Theta)$ and to use the sample $\Delta = \{\delta_i\}$ to derive a maximum–likelihood (ML) estimate for Θ from the likelihood function $L(\Theta;\Delta) = \prod_1^N p(\delta_i;\Theta)$. This approach, in which the uncertainty estimates will be independent of the patterns x, is followed in [7] using zero–mean Laplace and Gaussian models. In this work we extend this method to other possible residual models, namely the non–zero mean Gaussian and Laplace distributions (to address possible error biases), and the Beta and Weibull distributions. A drawback of the approach in [7] is the independence between error intervals and patterns; to alleviate this we also propose to split the uncertainty analysis according to the $\hat{f}(x)$ values.While other choices would be possible, we select these distributions because previous works, such as [10], have shown their usefulness to model wind speed and wind power production. We review next their main properties.

3.1   Density Functions

Recall that the general Laplace density with mean μ and standard deviation σ is $p(z) = \frac{1}{2\sigma} e^{-\frac{|z-\mu|}{\sigma}}$ with $z \in R$ and the μ, σ Gaussian is $p(z) = \frac{1}{\sqrt{2\pi}\sigma} e^{-\frac{(z-\mu)^2}{2\sigma^2}}$. The Beta$(\alpha, \beta)$ distribution is given for $x \in [0,1]$ by

$$p(z) = \frac{1}{B(\alpha, \beta)} x^{\alpha-1}(1-x)^{\beta-1}, \alpha, \beta > 0, \tag{11}$$

where $B(x, y) = \frac{\Gamma(x)\Gamma(y)}{\Gamma(x+y)}$ and $\Gamma(x)$ are the beta and gamma functions respectively. Finally, the Weibull(λ, k) distribution[1] is given by

$$p(z) = \frac{\kappa}{\lambda} \left(\frac{z}{\lambda}\right)^{\kappa-1} e^{-\left(\frac{z}{\lambda}\right)^{\kappa}}, \; z \in [0, \infty), \lambda > 0, \kappa > 0 \;. \tag{12}$$

3.2   Log–likelihood Parameter Estimation

Given an independent sample δ_i to which we want to fit one of the above distributions, we can estimate its parameters Θ by maximizing the log-likelihood, $l(\Theta; \delta_1 ... \delta_n) = \sum_{i=1}^{n} \log p(\delta_i|\Theta)$, where p represents the density function, which we do by solving $\nabla_\Theta l = 0$. For the general Gaussian and Laplace cases we obtain the well known values

$$\hat{\mu}_G = \sum_{i=1}^{n} \frac{\delta_i}{n}, \; \hat{\sigma}_G = \frac{\sum_{i=1}^{n}(\delta_i - \hat{\mu}_G)^2}{n}, \; \hat{\mu}_L = m_{\delta_i}, \; \hat{\sigma}_L = \frac{\sum_{i=1}^{n}|\delta_i - \hat{\mu}_L|}{n} \; ; \tag{13}$$

here m_{δ_i} denotes the median of the δ_i residuals [17]. For the Beta distribution, the gradient equations are

$$0 = \frac{\partial l}{\partial \hat{\alpha}} = n \left(\frac{\Gamma'(\alpha+\beta)}{\Gamma(\alpha+\beta)} - \frac{\Gamma'(\alpha)}{\Gamma(\alpha)}\right) + \sum_{i=1}^{n} \log \delta_i, \tag{14}$$

$$0 = \frac{\partial l}{\partial \hat{\beta}} = n \left(\frac{\Gamma'(\alpha+\beta)}{\Gamma(\alpha+\beta)} - \frac{\Gamma'(\beta)}{\Gamma(\beta)}\right) + \sum_{i=1}^{n} \log(1 - \delta_i), \tag{15}$$

where the δ values are scaled into $[0,1]$. Denoting $\phi(x) = \frac{\Gamma'(x)}{\Gamma(x)}$, we have

$$\frac{\sum_{i=1}^{n} \log \delta_i}{n} = \phi(\alpha) - \phi(\alpha+\beta) = F_1(\alpha, \beta),$$

$$\frac{\sum_{i=1}^{n} \log(1-\delta_i)}{n} = \phi(\beta) - \phi(\alpha+\beta) = F_2(\alpha, \beta) \;. \tag{16}$$

We use the well known Newton–Raphson method to solve (16) which leads to the following iterative scheme [11] to derive a sequence α_j, β_j:

$$\frac{\sum_{i=1}^{n} \log \delta_i}{n} = F_1(\alpha_j, \beta_j) + (\alpha_{j+1} - \alpha_j)\left(\frac{\partial F_1}{\partial \alpha}\right)_{(\alpha_j, \beta_j)} + (\beta_{j+1} - \beta_j)\left(\frac{\partial F_1}{\partial \beta}\right)_{(\alpha_j, \beta_j)}$$

$$\frac{\sum_{i=1}^{n} \log(1-\delta_i)}{n} = F_2(\alpha_j, \beta_j) + (\alpha_{j+1} - \alpha_j)\left(\frac{\partial F_2}{\partial \alpha}\right)_{(\alpha_j, \beta_j)} + (\beta_{j+1} - \beta_j)\left(\frac{\partial F_2}{\partial \beta}\right)_{(\alpha_j, \beta_j)}. \tag{17}$$

---

[1] We only consider the case $z \geq 0$ in the Weibull distribution as $p(z) = 0$ for $z < 0$.

As initial values (α_0, β_0), pivotal for the efficient convergence of Newton-Raphson's method, we use the ones proposed in [11], namely

$$\alpha_0 = \frac{m_1(m_1 - m_2)}{m_2 - m_1^2}, \quad \beta_0 = \frac{\alpha_0(1 - m_1)}{m_1}, \tag{18}$$

where m_1, m_2 are the first and second order momenta of the residuals.

Finally, the gradient equations for the Weibull distribution are

$$0 = \frac{\partial l}{\partial \hat{\lambda}} = -\frac{n}{\lambda} - \frac{\kappa n - n}{\lambda} + \frac{\kappa}{\lambda^{\kappa+1}} \sum_{i=1}^{n} \delta_i^{\kappa} = \frac{\kappa}{\lambda}\left(n - \frac{1}{\lambda^{\kappa}} \sum_{i=1}^{n} \delta_i^{\kappa} \right),$$

$$0 = \frac{\partial l}{\partial \hat{\kappa}} = \frac{n}{\kappa} + \sum_{i=1}^{n} \log \delta_i - n \log \lambda - \sum_{i=1}^{n} \left(\frac{\delta_i}{\lambda} \right)^{\kappa} \log \frac{\delta_i}{\lambda}. \tag{19}$$

The solution of the first equation of (19) is

$$\lambda = \left(\frac{1}{n} \sum_{i=1}^{n} \delta_i^{\kappa} \right)^{\frac{1}{\kappa}}. \tag{20}$$

Plugging this λ value into the second equation of (19) we get

$$\frac{\sum_{i=1}^{n} \log \delta_i}{n} = \frac{\sum_{i=1}^{n} \delta_i^{\kappa} \log \delta_i}{\sum_{i=1}^{n} \delta_i^{\kappa}} - \frac{1}{\kappa} = G(\kappa), \tag{21}$$

which we can solve again through Newton–Raphson's, obtaining the iterates

$$\kappa_{j+1} = \kappa_j + \frac{1}{G'(\kappa_j)} \left(\frac{\sum_{i=1}^{n} \log \delta_i}{n} - G(\kappa_j) \right). \tag{22}$$

This time the initial value κ_0 is chosen empirically through experimentation; in our case $\kappa_0 = 1$ seems to ensure a fast convergence.

3.3   Uncertainty Intervals

Given a pre-specified probability $1 - 2s$, we want to find in the Gaussian and Laplace cases an error interval for which s is the percentage of residuals above and below the upper and lower interval limits. As stated before, we assume the conditional distribution of y given x to depend on x only through the predicted value $\hat{f}(x)$; as a consequence, intervals have the same width for each (x_i, y_i). More precisely, in the zero–mean Gaussian and Laplace cases, if p_s is the upper s–th percentile of the density p, the interval would be $(-p_s, p_s)$ where we can derive p_s solving $1 - s = \int_{-\infty}^{p_s} p_\zeta(z)dz = \Phi(p_s)$, i.e., $p_s = \Phi^{-1}(1 - s)$, with Φ the distribution associated to p. For a Laplace $p(z) = \frac{1}{2\sigma}e^{\frac{|z|}{\sigma}}$ we simply have $p_s = -\sigma \ln(2s)$. For the non zero μ mean Gaussian and Laplace the interval will be $(\mu - (p_s - \mu), \mu + (p_s - \mu))$. On the other hand, since we restrict the Beta and Weibull densities to positive values, we take for them intervals of the form $(0, p_{2s})$ where $1 - 2s = \int_0^{p_{2s}} p(z)dz$.

Table 1. Interval errors for the artificial data set with s=0.1

Noise	Test	LAP	LAP*	LAPm	GAU	GAUm	BET	WEI
Noise free	GAUm/WEI	3.1	3.1	3	1.5	1.4	2.5	**1.1**
Laplace	LAP/LAPm	**0.4**	**0.4**	**0.4**	1.4	1.4	1.0	1.5
Gaussian	GAUm/WEI	2.8	2.8	2.9	0.2	**0.1**	1.2	0.5
Beta	BET	3.0	3.0	3.0	1.1	1.2	**0.2**	0.8
Weibull	WEI	3.5	3.5	3.7	1.6	1.4	1.3	**0.2**

3.4   Hypothesis Testing

Among the advantages of the log–likelihood approach above is the availability of a test for rejecting a model p_0 against another p_1 [12]. In fact, if $\{Z_i\}_{i=1}^t$ are independent random samples following a density $\frac{1}{\sigma}p\left(\frac{z}{\sigma}\right)$, we can define the statistic

$$T(p_0, p_1, \{z_i\}_{i=1}^t) = \frac{\int_0^\infty \tau^{t-1} p_1(\tau z_1) p_1(\tau z_2)...p_1(\tau z_t) d\tau}{\int_0^\infty \tau^{t-1} p_0(\tau z_1) p_0(\tau z_2)...p_0(\tau z_t) d\tau} . \tag{23}$$

Then we reject the hypothesis $H_0 : p = p_0$ against $H_1 : p = p_1$ when it is bigger than a threshold c_α associated to a given significance level α. This defines [12] a most powerful test which is invariant under scale transformation. At a given significance level, α, when H_0 is true the probability of rejecting this hypothesis is α, i.e., it holds that $P_0(T(p_0, p_1, \{z_i\}_{i=1}^t) > c_\alpha) = \alpha$. We solve this by numerical integration. Specifically, we use the SAGE function sage.gsl.integration.numerical_integral with its default values, i.e., adaptive integration and absolute and relative error tolerances equal to 10^{-6}.

4   Experiments

We consider three different data scenarios. The first one is a synthetic problem to which we add different types of noise. The second scenario corresponds to four public datasets: **abalone** from Statlog [13], **space_ga** from StatLib [14], and **add10** as well as **cpusmall** from Delve [15]. In the final scenario we deal with a real wind energy prediction problem. In all of them we fit residual models according to Gaussian and Laplace densities with both zero (GAU, LAP) and non–zero (GAUm, LAPm) mean as well as Weibull (WEI) and Beta (BET) distributions. We also consider a version of the Laplace model fitting proposed in [7], LAP*, where we discard those δ_i residuals that exceed $\pm M$ times the residual standard deviation σ_δ; M ranges from 3 to 5 depending on the problem.

4.1   SVR Parameter Selection and Interval Accuracy Metrics

As mentioned, we work with Gaussian SVR models, for which parameter selection is done using two different methods. The first one is either 5–fold cross

Table 2. Interval errors for SVR models with CV parameters for public datasets

Dataset	s	LAP	LAP*	LAPm	GAU	GAUm	BET	WEI
abalone	s=0.1	**2.8**	3.8	4.0	9.2	12	3.2	2.9
	s=0.05	0.6	**0.0**	0.8	1.9	0.8	1.6	0.5
add10	s=0.1	0.4	0.4	**0.2**	0.6	0.8	0.4	3.0
	s=0.05	**0.9**	**0.9**	**0.9**	1.9	1.9	1.7	1.8
space_ga	s=0.1	9.4	9.0	9.0	3.4	**2.2**	8.6	6.8
	s=0.05	3.0	3.9	2.0	0.7	**0.5**	3.9	4.3
cpusmall	s=0.1	7.0	6.4	**0.2**	14.2	14.8	0.8	4.3
	s=0.05	4.1	3.9	**0.2**	6.9	6.5	**0.2**	2.7

validation or validation over a fixed dataset and the second one the Bayesian approach of [5], for which there is code available[2]. In the first case we have simply performed a zoomed grid search range to arrive to a (C, ϵ, γ) parameter set giving the smallest mean absolute error, MAE, over the validation set or through CV. As initial point, $(C_0, \epsilon_0, \gamma_0)$, we use the parameters obtained after applying the approach in [16]. We recall that the MAE over $\{(x_i', y_i')\}_{i=1}^N$ is given by

$$MAE = \frac{1}{N} \sum_{i=1}^{N} |\hat{f}(x_i') - y_i'| \ . \tag{24}$$

We have used MAE instead of the more common mean square error, MSE. Strictly speaking, the ϵ–insensitive loss function of SVR is the best choice but notice that it depends on the ϵ parameter whose optimal value we want to determine. Our MAE choice avoids this circularity while being close to the ϵ–loss, agreeing with it for large deviations. In all SVR experiments we use the well known LIBSVM software [6].

To test the error err_s^M of the error intervals estimated according to a certain noise model M that corresponds to a pre–specified probability $1-2s$, we compare as in [7] the percentage of the residuals δ_i^{test} lying in the estimated error interval I_s^M derived using M with the expected number, $(1-2s) \times N'$, with N' the test sample size, i.e.,

$$err_s^M = \frac{100}{N'} |\# \ of \ \delta_i^{test} \in I_s^M - (1-2s) \times N'| \ . \tag{25}$$

4.2   Artificial Data and Results

Here we create an artificial dataset $\{(x_i, y_i)\}$ with $y_i = 2\cos x_i + 3\sin(2x_i) + n_i$ where x_i are uniformly distributed over $[0, 2\pi]$ and noise n_i is generated according

---

[2] http://www.gatsby.ucl.ac.uk/~chuwei/code/bisvm.tar

Table 3. Summary of interval errors and best noise models for public datasets

Dataset		CV	BAYESIAN	BSVR
abalone	Best	LAP/WEI	LAP*	-
	Mean	**1.70**	3.55	4.95
add10	Best	LAPm	WEI	-
	Mean	**0.55**	1.40	4.00
space_ga	Best	GAUm	GAUm	-
	Mean	**1.35**	2.00	1.80
cpusmall	Best	LAPm	LAPm	-
	Mean	**0.20**	1.75	2.5

to four distributions: 1. Laplace noise with $\mu = 0$ and scale $\sigma = 1$; 2. Gaussian noise with $\mu = 0$ and $\sigma = 1$; 3. Beta noise with $\alpha = 1$ and $\beta = 2$; 4. Weibull noise with scale $\lambda = 1$, and shape $\kappa = 5$. We randomly generate 5,000 x_i values of which 4000 are used for parameter selection and training, and 1000 as test set. The goal here is to test whether intervals built under a concrete noise assumption for the residuals perform better than the others when precisely that kind of noise is present in the data. For this scenario, SVR parameter selection is done by 5–fold CV.

The five rows in Table 1 correspond to data generated according to the four noise models considered as well as noise free targets. Columns from the third on contain the err_s^M for $s = 0.1$ using as noise models M the $LAP, LAP^*, LAPm,$ $GAU, GAUm, BET$ and WEI distributions respectively; boldface values highlight the models with smallest err_s^M. As it can be seen, in all noisy cases the underlying noise model yields the smallest error; on the other hand, the Weibull model yields the smallest error for the noise free case, closely followed by $GAUm$. The second column summarizes the results of applying the statistic $T(p_0, p_1)$ for pairwise testing of a model against the others. Again, for each target noise type modeled by p_0 the alternate hypothesis of a different p_1 can be rejected in all cases except that of Gaussian noise, where a Weibull model cannot be rejected.

4.3 Public Datasets Results

Here we consider the three approaches mentioned above to build error intervals, namely *1)* SVR with parameters obtained by CV and uncertainty intervals as in Section 3.3; *2)* SVR with Bayesian parameters as described in Section 2.2 and uncertainty intervals as in Section 3.3; *3)* direct Bayesian intervals obtained through the procedure in [5]; we refer to it as BSVR in what follows. For each one of the public datasets we employ 1000 instances as test and the rest as training.

Columns in Table 2 from the third on contain for each dataset the errors err_s^M for $s = 0.1$ and $s = 0.05$ associated to noise models $LAP, LAP^*, LAPm,$

$GAU, GAUm, BET$ and WEI, respectively, when parameters of the SVR are set through CV. Errors (not shown) obtained when Bayesian SVR parameters are used or using BSVR approach are higher; this can be observed in Table 3 that summarizes the best residual model and the mean of the $err_{0.1}^M$ and $err_{0.05}^M$ errors. As it can be seen, CV error means are about one percentage point below those of the other methods; moreover, either the GAU or LAP models, with or without means, yield the most accurate intervals.

4.4   Wind Energy Results

Finally, in this section we consider SVR uncertainty estimates in the prediction of wind energy. We use as input features the numerical weather predictions, NWP, of surface wind, temperature, pressure and 100 meter wind provided by the European Centre for Medium-Range Weather Forecasts (ECMWF) with a $0.25°$ resolution for a rectangular area that contains the Iberian peninsula. Since the ECMWF forecasts are given at three hour intervals we interpolate them linearly to hourly values. The prediction target here is the hourly total wind energy production of peninsular Spain. We work with data from January 1st 2011 to December 31th 2013. Notice that the problem has a natural time structure; because of this we use data from 2011 as training data, that from 2012 as a validation set and that from 2013 as test. In our experience this data structuring yields better model parameters than those obtained applying random cross-validation. We consider five different approaches, namely:

1. M_1^C: compute residuals on the validation set of a SVR model with parameters found by a grid search using the fixed validation set previously described, and define a unique uncertainty interval for the entire test as in Section 3.3.
2. M_2^C: compute the residuals as in M_1 but split the validation set V into two groups V_1 and V_2 of equal size as follows: find the median y_m^V of the estimates $\hat{f}(x_i)$ in V and let $V_1 = \{(x_i, y_i) : \hat{f}(x_i) \leq y_m^V\}$ and $V_2 = V - V_1$. We divide then the residuals' set R into R_1 and R_2, where $R_1 = \{\delta_i : (x_i, y_i) \in V_1\}$ and $R_2 = R - R_1$ and finally fit two different error models to the residuals in R_1 and R_2 and build two different uncertainty intervals I_1, I_2 as in Section 3.3.
3. M_4^C: Same method as in M_2^C but this time with 4 subsets and intervals.
4. M_1^{B1}: Same method as in M_1^C but computing the residuals of the validation set using as SVR parameters the ones obtained by the Bayesian approach described in Section 2.2.
5. M_1^{B2}: Compute uncertainty intervals using the Bayesian procedure in [5] in its entirety.

   Notice that procedure M_1^C yield a constant uncertainty interval for all input patterns x, while methods M_2^C and M_4^C yield either two or four intervals depending on the relationship between $\hat{f}(x)$ and y_m^V. In principle, these intervals adjusted to the magnitude of energy forecasts are a rather sensible choice in wind energy predictions, as the prediction errors are very dependent on forecast values. A comparison between M_1^C and M_2^C intervals with $s = 0.1$ for January 2013 is shown in Figure 1.

Table 4. Interval errors intervals for wind energy prediction

model	s	LAP	LAP*	LAPm	GAU	GAUm	BET	WEI	BSVR
M_1^C	0.1	6.0	5.9	5.0	5.6	5.5	5.0	**4.5**	-
	0.05	5.4	5.3	3.9	3.3	3.4	3.8	**3.2**	-
M_2^C	0.1	5.1	4.8	2.4	3.2	**2.2**	4.7	2.6	-
	0.05	5.4	5.3	3.8	2.9	**1.0**	2.3	2.7	-
M_4^C	0.1	8.7	8.3	**7.4**	**7.4**	**7.4**	9.0	7.7	-
	0.05	7.6	7.5	**5.3**	6.7	**5.3**	6.2	6.3	-
M_1^{B1}	0.1	8.2	8.1	7.8	6.1	5.8	6.9	**5.5**	-
	0.05	7.8	7.7	5.5	4.5	**4.3**	5.2	4.4	-
M_1^{B2}	0.1								**6.9**
	0.05								**5.7**

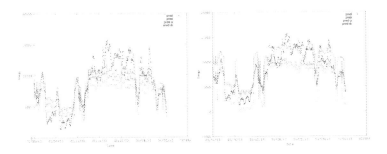

Fig. 1. M_1^C and M_2^C uncertainty intervals for January 2013 data of wind energy problem with $s = 0.1$. Figure shows real production (prod), prediction given by the model (pred), lower bound of prediction interval (pred-a) and upper bound (pred+b)

Table 5. Summary of interval errors for wind energy prediction

	M_1^C	M_2^C	M_4^C	M_1^{B1}	M_1^{B2}
Best	WEI	GAUm	GAUm	WEI	-
s=0.1	4.5	**2.2**	7.4	5.5	6.9
s=0.05	3.2	**1.0**	5.3	4.4	5.7

Columns in Table 4 contain errors err_s for $s = 0.1$ and $s = 0.05$ associated to each of the models above. For the M_i^C and M_1^{B1} models we give err_s^M values for all uncertainty models considered; these are omitted for the M_1^{B2} model, as it follows a different approach; boldface values highlight the smallest errors. Table 5 summarizes that information showing for each approach the best residual model and the corresponding $err_{0.1}$ and $err_{0.05}$ errors. As it can be seen, best results are clearly achieved with the M_2^C approach using $GAUm$ models after dividing into two subsets the residuals of an SVR whose parameters are obtained by validation over 2012 data. This best performance is followed by that of the M_1^C and M_1^{B1} approaches; on the other hand, dividing the residuals into 4 groups results in a lower accuracy. This is probably due to working with too many groups which makes it difficult to find a core of common patterns inside them. Finally, the purely Bayesian approach M_1^{B2} yields again the lowest accuracy.

5   Conclusions and Further Work

Support Vector Regression, SVR, is one of the most used tools in non–linear regression and modeling problems and, as such, uncertainty estimates of SVR predicted values are of great importance. A particularly clear example is SVR–based wind energy prediction, whose intermittency and wide fluctuations make necessary to define appropriate levels of rolling reserve; good uncertainty estimates are an obvious tool for this. In this work we have broadly followed the approach in [7], considering noise model distributions that are fitted to the residuals of SVR models whose C, ϵ and γ parameters have been found directly by CV or validation over a fixed set, or under a Bayesian perspective; as in [7], we have also considered the entirely Bayesian approach to define uncertainty intervals proposed in [5]. We have enlarged this set up by adding to the noise models considered in [7] non–zero mean versions of Gaussian and Laplace noise, as well as Beta and Weibull variants.

A first general conclusion is that, in agreement with [7], purely Bayesian interval estimates are poorer than those obtained by fitting noise distributions to residual values; moreover, interval error estimates are more accurate when SVR parameters are chosen by CV or validation over a fixed set. Since this is SVR specific only to the extent that SVRs are the underlying model, it suggests that direct residual–based fitting of error models should also be a useful approach when non–linear regressors are built under other alternative paradigms.

Moreover, we have shown over a synthetic example how this approach is able to resolve the true underlying noise model; this is the case for the four noise distributions considered, even when taking into account that the ϵ–insensitive SVR loss does not entirely corresponds to any of them. Of course, the true noise model depends entirely on the sample data and not, in principle, on the loss function used. On the other hand, the loss function somehow addresses a particular noise structure, which suggests that perhaps shifted versions of the Gaussian and Laplace densities would yield more accurate uncertainty intervals. For instance, the ϵ–insensitive loss $[\delta]_\epsilon$, that corresponds to a uniform noise density $p_\epsilon(\delta) = 1/2(1+\epsilon)$ when $|\delta| \leq \epsilon$, and $p_\epsilon(\delta) = e^{|\delta|-\epsilon}/2(1+\epsilon)$ when $|\delta| > \epsilon$, could possibly yield tighter uncertainty intervals. Similarly, this also suggests that loss functions other than the ϵ–insensitive one should be considered to build an SVR model when noise from a particular distribution is suspected. Of course, not every conceivably noise density could be so addressed, but SVR models for Beta and Weibull–inspired losses are considered in [8]. In this vein, other candidates could be the Cauchy, Logistic or Voigt distributions.

Furthermore, the proposed technique for building error intervals is not exclusive for SVR models, with the approach being independent of the model chosen to solve the regression problem and the noise assumptions presumed. Thus, testing accuracy of intervals result of applying other regression models, such as Random Forests or Neural Networks, is a clear line of further work.

Finally, a drawback of the residual fitting approach is that error intervals are built independently of the $\hat{f}(x)$ regressor values. This is particularly so in problems such as wind energy prediction, where model errors are usually much higher for large energy values. As mentioned in the Introduction, the well known analytic estimates for linear regression under Gaussian models that yield error estimates dependent on x are simply impossible for SVR.

Despite this fact, the splitting techniques we use that yield two energy level–based uncertainty regimes give much better results than a single interval procedure. However, this is still rather coarse and a finer grain split of the sample residuals into four regimes but using a single SVR model gives worse results. Because of this, it might be worth trying to consider distinct SVR models for each residual regime and fit noise distributions to the residuals of each model, instead of working with an unique SVR model for the entire dataset. We are currently working on these and other related issues.

Acknowledgments. This paper was developed with partial support from Spain's grants TIN2013-42351-P and S2013/ICE-2845 CASI-CAM-CM and also of the Cátedra UAM–ADIC in Data Science and Machine Learning. Authors also gratefully acknowledge the use of the facilities of Centro de Computación Científica (CCC) at UAM. They also thank Red Eléctrica de España for kindly supplying wind energy data.

References

1. Shawe-Taylor J., Cristianini N.: An introduction to Support Vector Machines and other kernel-based learning methods. Cambridge University Press (2000)
2. Yang, H., Chan, L., King, I.: Support vector machine regression for volatile stock market prediction. In: Yin, H., Allinson, N., Freeman, R., Keane, J., Hubbard, S. (eds.) IDEAL 2002. LNCS, vol. 2412, pp. 391–396. Springer, Heidelberg (2002)
3. Kramer, O., Gieseke, F.: Short-term wind energy forecasting using support vector regression. In: Corchado, E., Snášel, V., Sedano, J., Hassanien, A.E., Calvo, J.L., Ślęzak, D. (eds.) SOCO 2011. AISC, vol. 87, pp. 271–280. Springer, Heidelberg (2011)
4. Gala Y., Fernandez, A., Díaz J., Dorronsoro J.R.: Support vector forecasting of solar radiation values. In: Pan, J.-S., Polycarpou, M.M., Woźniak, M., de Carvalho, A.C.P.L.F., Quintián, H., Corchado, E. (eds.) HAIS 2013. LNCS (LNAI), vol. 8073, pp. 51–60. Springer, Heidelberg (2013)

5. Chu, W., Keerthi, S.S., Ong, C.J.: Bayesian support vector regression using a unified loss function. IEEE Transactions on Neural Networks **15**, 29–44 (2004)
6. Chang, C., Lin, C.: LIBSVM: a library for support vector machines. ACM Transactions on Intelligent Systems and Technology (TIST) **2**, 27 (2011)
7. Lin, C., Weng, R.: Simple Probabilistic Predictions for Support Vector Regression. National Taiwan University, Taipei (2004)
8. Hu, Q., Zhang, S., Xie, Z., Mi, J., Wan, J.: Noise model based ν-support vector regression with its application to short-term wind speed forecasting. In: Neural Networks, vol. 57, pp. 1–11. Elsevier (2014)
9. Pontil, M., Mukherjee, S., Girosi, F.: On the noise model of support vector machines regression. In: Arimura, H., Jain, S., Sharma, A. (eds.) ALT 2000. LNCS, vol. 1968, pp. 316–324. Springer, Heidelberg (2000)
10. Celik, A. N.: A statistical analysis of wind power density based on the Weibull and Rayleigh models at the southern region of Turkey. In: Renewable Energy, vol. 29, pp. 593–604. Elsevier (2004)
11. Gnanadesikan, R., Pinkham, R.S., Hughes, L.P.: Maximum likelihood estimation of the parameters of the beta distribution from smallest order statistics. In: Technometrics, vol. 9, pp. 607–620. Taylor & Francis Group (1967)
12. Lehmann, E.L., Romano J.P.: Testing statistical hypotheses. Springer Science & Business Media (2006)
13. UCI Machine Learning Repository. https://archive.ics.uci.edu/ml/datasets.html
14. StatLib Datasets Archive. http://lib.stat.cmu.edu/datasets/
15. Delve Datasets. http://www.cs.utoronto.ca/delve/data/datasets.html
16. Cherkassky, V., Ma, Y.: Practical selection of SVM parameters and noise estimation for SVM regression. In: Neural Networks, vol. 17, pp. 113–126. Elsevier (2004)
17. Johnson, N.L., Kotz, S., Balakrishnan, N.: Continuous Univariate Distributions, vol. 2. John Wiley & Sons (1995)

Search for Meaning Through the Study of Co-occurrences in Texts

Nicolas Bourgeois[1], Marie Cottrell[1], Stéphane Lamassé[2], and Madalina Olteanu[1(✉)]

[1] SAMM, Université Paris 1 Panthéon-Sorbonne,
90, rue de Tolbiac, 75013 Paris, France
{nicolas.bourgeois,marie.cottrell,madalina.olteanu}@univ-paris1.fr
[2] PIREH-LAMOP, Université Paris 1 Panthéon-Sorbonne,
1, rue Victor Cousin, Paris, France
stephane.lamasse@univ-paris1.fr

Abstract. In this paper, we combine several tools used in text-mining in order to study both the lexicon and the semantic structure of a set of medieval texts. On the one hand, the study of occurrences (Principal Component Analysis, Topic Models, Self-Organizing Maps, Hierarchical Cluster Analysis) allows a wide scope of tools to extract and display information from big data. On the other hand, the study of co-occurrences (words belonging to a sentence, a paragraph) allows to keep track of the structure of each text, but is more tedious to handle and often leads to messy visualizations. Here we use the SOM algorithm to reduce the size of the data (clustering, removal of fickle information) while preserving the semantic structure ; then we can rely on classical but slower algorithms (HCA, graph representation) to purpose data visualization.

Keywords: Text mining · SOM · Graphs

1 Introduction, State of the Art

With the development of the WEB sites, social networks, big data depositories and real time information flows, text mining has become an important focus in several fields such as statistics and computer science. Many different techniques can be used to extract lexical and semantic information from texts of various sources (literary, technical, political, scientific, and so on). We will briefly recall two of them in this introduction: counting frequencies and topic modeling. Those two, actually most of the more popular ones also, consider each text as a simple bag of words and pay little attention to its structure.

On the contrary, the study of co-occurrences focuses on the articulation between words inside of a text: sentences, paragraphs, distances. We will recall the general principle of that technique, and its main drawback also which is the difficulty to display and analyse results.

The main goal of this paper is to adapt some tools that had been developed to improve display, robustness or efficiency of frequency analysis, in order to

© Springer International Publishing Switzerland 2015
I. Rojas et al. (Eds.): IWANN 2015, Part II, LNCS 9095, pp. 578–591, 2015.
DOI: 10.1007/978-3-319-19222-2_48

cope with this drawback. In particular we hope to produce sound results on the analysis of co-occurrences for a corpus of scientific medieval texts.

1.1 Words Frequencies

The most used and the most popular method consists in the computation of the absolute frequencies (or *number of occurrences*) for each word, eventually lemmatized[1]. Then the words can be sorted by decreasing value of their frequencies, so that each text is linked to the set of words which have the higher importance for it. In this case, each text is viewed as a bag of words, and the structuring into sentences or paragraphs is neglected. The data consist of a contingency table where entry $a_{i,j}$ is the number of occurrences of word i in text j. It is in this frame that The Factorial Component Analysis (see Benzecri [1992] or Lebart et al. [1984]) is used in order to provide simultaneous representations of both the words and the texts, allowing the study of the proximity between words, between texts, between texts and words.

This method is useful to put in evidence some clusters of words, some associations between texts and clusters of words, but it is impossible to get a global representation of these associations. That is due to the limitations of the Factorial Component Analysis projections, (see for example Bourgeois et al. [2015]) where we propose an improvement to overcome this drawback. But in any case, the semantic aspect is definitely lost.

1.2 Topic Models

Topic Model is a statistical model which assumes that there exists hidden groups of words, called *topics*, which are meaningful but that the observer cannot address directly. The actual texts we can observe are built through a process of picking words from one or several of these topics. This approach recalls somehow of the so-called Mixture Models (Titterington et al. [1985]).

The aim of the observer is to do some reverse engineering and to deduce the topics from the texts, which can be very different depending on the exact statistical model chosen. Popular methods in that field include PLSI or LDA (see Blei [2012]). In this frame also, the semantic disappears as each text is considered as a simple bag of words. Actually, since the reversion operation is usually tough compared to the generation process, a topic model is probably not the easiest way to study language structures.

1.3 Co-occurrences Analysis

The analysis of co-occurrences (Martinez and Salem [2003], Martinez [2012]), helps to cluster the words without breaking the links with semantic analysis. It

---

[1] Lemmatization consists in gathering the different inflected forms of a given word as a single item: for example *choose, chose, chosen* will be associated to the item *choose*. Note that the lemmatization is a much easier task for English language than for other languages which use the declension and the conjugation in their grammar.

deals with the concordances between words and not only with their number of occurrences.

The texts are split according to a given level of segmentation: sentences, paragraphs, sections, for example. In the following, we decide to use the *paragraphs*. All the words are considered, with or without lemmatization, including numbers, whatever the texts they belong to. The only excluded forms are the punctuation signs. All the texts are gathered into a unique text.

For each given word, called *pole p*, and for each other word *w*, we count how many times *w* is present in a paragraph which contains *p*. This number is the number of co-occurrences of *w* with the pole *p*.

For example, let us take the sentence:

One *approach* to understand the *evolution* of science is the study of the *evolution* of the language used in a given field.

In this sentence, if *approach* is chosen as *pole*, the word *evolution* is its co-occurrent twice, while if we invert the role of *approach* and *evolution*, the word *approach* is co-occurrent of *evolution* once.

In the sequel, the main object of our study is the co-occurrence matrix C. It is a square matrix $T \times T$, where T is the total number of considered words (or lemmas if lemmatization is applied). Each word is seen as a pole as well as a co-occurrent. We denote by $n_{i,j}$ **the number of times where word j is inside a paragraph which contains word i**. In this notation, i is the pole, j is one of its co-occurrents. Note that this matrix is not symmetric, since according to the definition, $n_{i,j} \neq n_{j,i}$.

This co-occurrence matrix allows to define a weighted and directed graph where the words are the vertices and the values $n_{i,j}$ are the weights of the edges. For an elementary example, the reader may refer to Figure 1.3.

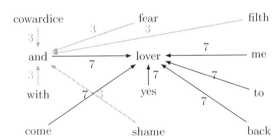

Fig. 1. Graph of co-occurrences from the 2-paragraphs quote "With fear and filth and cowardice and shame./ Yes and lover, lover, lover, lover, lover, lover, lover come back to me". For the sake of readability, edges of weight 1 have been omitted.

This graph shows all the links between all the lexical forms (seen as pole or co-occurrent) and takes into account the context and the semantic structure of the texts. A great difficulty appears immediately: this graph is usually unreadable and it is impossible to extract a meaningful information. Notice for example that every paragraph is a clique (a complete cluster) over all the words it contains. Hence the need to simplify the complete graph in order to decrease its complexity.

1.4   Towards a More Compact and More Readable Graph

The aim of the manuscript is to look for a smaller graph, sparsely connected, where most of the useful information from the initial graph would still be preserved. This goal can be achieved by defining groups of strongly connected words that are subsequently taken as vertices and by studying the connections between these *new* vertices. On this reduced graph, the groups of words themselves and the relationships between the groups should be easier to visualize and to interpret. In summary, the idea is to use clustering statistical methods to simplify the original graph.

A first idea would be to use hierarchical cluster analysis or clustering methods based on graph theory as those we used in the last part of Bourgeois et al. [2015]. The problem is that the data we want to deal with is actually too big for those algorithms to be efficient - those ones having running time asymptotically equivalent to n^3 or worse. For this reason, we decided to achieve clustering through the use of Kohonen algorithms whose running time are asymptotically bounded within $O(n^2)$, which remains tractable for a graph of $\sim 10^3$ vertices.

2   Material and Methods

2.1   The Material

The corpus (Lamassé [2012], Bourgeois et al. [2015]) includes eight treaties on arithmetical education written in the 15th century, in vernacular language (French). Their purpose was to teach merchants about arithmetics, but also to develop theory and knowledge. One can find a detailed description of their properties and their historical importance in the article mentioned above. Table 1 below describes some elements of the lexicometric characteristics of the corpus and shows its main quantitative imbalance.

Some pre-processing was necessary for these texts. First, the punctuation signs and the *hapax*[2] were removed. After this first step, the data contained 1545 remaining words, including some numbers that are used in the demonstrations or as page numbers.

Second, the co-occurrence matrix was computed. A threshold was used in order to keep the most significant co-occurrences: if $n_{i,j}$ or $n_{j,i} < 5$, both values were set to 0. The value 5 for the threshold was chosen as a compromise between the need for complete information and the need for denoised data. We considered that all co-occurrences smaller than the threshold were random and could be removed. A significantly larger threshold would have lead to a sparser graph,

---

[2] A *hapax* is a word which appears only once in the corpus.

Table 1. Corpus of texts and main lexicometric features

Manuscripts and Title	Date	Author	Number of occurrences	Words	Hapax
Bibl. nat. Fr. 1339	ca. 1460	anonyme	32077	2335	1229
Bibl. nat. Fr. 2050	ca. 1460	anonyme	39204	1391	544
Cesena Bibl. Malest. S - XXVI - 6. Traicté de la praticque	1471?	Mathieu Préhoude?	70023	1540	635
Bibl. nat. Fr. 1346. Commercial appendix of Triparty en la science des nombres	1484	Nicolas Chuquet	60814	2256	948
Méd. Nantes 456	ca. 1480-90	anonyme	50649	2252	998
Bibl. nat. Arsenal 2904,Kadran aux marchans	1485	Jean Certain	33238	1680	714
Bib. St. Genv. 3143	1471	Jean Adam	16986	1686	895
Bibl. nat. Fr. Nv. Acq. 10259	ca. 1500	anonyme	25407	1597	730

easier to analyze and cluster but suffering from loss of information. Table 2 contains a small sample of the co-occurrence matrix. One should note that it is not a symmetric matrix.

Table 2. Co-occurrences of eight words extracted from the co-occurrence matrix

	centre center	cercle circle	chacune each	diviser divide	endroit on right	part piece	partie part	raisons calculations, problems
centre	0	23	0	0	0	11	0	0
cercle	20	0	0	0	0	11	0	0
chacune	0	0	0	5	0	0	0	0
endroit	0	0	0	12	0	0	0	0
partie	0	0	0	0	0	0	0	5
raisons	0	0	0	0	0	0	7	0

Third, the co-occurrence matrix was used to build a weighted and directed graph, where the words represent the vertices and the non-negative co-occurrences represent the weights of the directed edges. At a closer look, the resulting graph contained a large connected component with 1517 words, twelve separate couples of words and one set with four words (all disconnected from each other). The rest of the paper focuses on the largest connected component.

2.2   Robust KORRESP on the Co-occurrence Matrix

In this first approach, the links between the words in the data set are considered as directed and each word appears both as pole and co-occurrent. Since each word has a double meaning, we shall interpret the co-occurrence matrix C as a contingency table crossing the poles (rows) with their co-occurrents (columns). Factorial Correspondence Analysis (FCA) is in this case the most used and apparently adapted technique. However, due to the limitations of FCA (in terms of projection quality, for example), we propose to use a variant of the SOM algorithm which deals with contingency tables (see Oja and Kaski [1999] for other applications of SOM to text mining). See Cottrell et al. [1998], Bourgeois et al. [2015] for a definition of this variant of SOM, called KORRESP.

After the convergence of the training step, rows and columns are simultaneously classified and projected on the map. In our example, one shall be able to see proximities between poles, between co-occurrents, between poles and co-occurrents. The goal of Korresp is similar to FCA but its main advantage is that it is not necessary to examine several projection planes: the whole information can be read on the map.

However, Korresp has the drawback of being a stochastic algorithm, and apparent contradictions between several runs can be troublesome. Nevertheless, this drawback can be used to improve the interpretation and the analysis of relations between the studied words. Our hypothesis is that repetitive use of Korresp may allow to identify pairs of words that are strongly attracted/repulsed and fickle pairs.

More precisely, if we consider several runs (at least 50) of the SOM algorithm, for a given size of the map and for a given data set, we observe that most of the pairs are almost always neighbor or not neighbor. But there are also pairs whose associations look random. These pairs are called *fickle* pairs. This question was addressed by Bodt et al. [2002] in a bootstrap frame and used for text mining in Bourgeois et al. [2015].

After having identified the fickle pairs, we define the fickle words as being those which belong to an important number of fickle pairs. The most fickle words are then removed and the remaining words are then projected on one of the trained maps, hereafter "reference map". In this way, we obtain a map which displays only the robust neighbor relationships.

As each cluster of Korresp is represented by its code-vector, the map provides a new graph where the vertices are the code-vectors. This representation considerably decreases the number of vertices and the complexity of the original graph. Moreover, each code-vector represents a set of words which are very close. Furthermore, hierarchical classification (HAC) may be applied to the code-vectors of the map reducing thus the number of final clusters.

2.3 Robust Relational SOM on the Co-occurrence Matrix

The second algorithm used for exploring the co-occurrence matrix is based on the online version of relational SOM, RSOM (Olteanu and Villa-Vialaneix [2015]). Relational SOM is a generalization of the original SOM algorithm for numerical data. It only requires as input a distance or a dissimilarity matrix between the data. Hence, this method may be applied to any complex data (time series, graphs, texts, etc...) as long as a dissimilarity can be computed.

One of the underlying hypothesis of RSOM is the symmetry of the dissimilarity matrix. The co-occurrence matrix C resuming the texts is not symmetric by construction. Moreover, it translates the strength of the link between the words and hence measures the similarity between them. Matrix C was thus transformed into a symmetric dissimilarity. By doing this, the direct dependency between poles and co-occurrents was lost. However, we shall see in the next sections that these operations allowed to highlight other aspects and properties of the data set and that the two algorithms are complementary.

Let us briefly explain how the dissimilarity matrix D was computed. First, a new symmetric matrix R was computed from C, where each entry $r_{i,j}$ is

$$r_{i,j} = \frac{1}{2}(n_{i,j} + n_{j,i}).$$

Second, the matrix R was used for building an undirected weighted graph. The vertices of the graph are the words in the texts, regardless of whether they are poles or co-occurrents. The weights, which will later on be used as transition costs, were computed as the inverses of the elements of R. For example, the weight of the edge linking vertex i to vertex j is

$$\omega_{i,j} = 1/r_{i,j}.$$

Third, the dissimilarity matrix D between the vertices is calculated using the shortest-path distance on the weighted graph. The dissimilarity between vertices i and j is then:

$$d_{i,j} = \min_{u_1,u_2\dots u_p | u_1=i, u_p=j} \sum_{k=1}^{p-1} \omega_{u_k, u_{k+1}}$$

In order to compare the results of the two approaches (directed links between words versus symmetric links between words), a robust version of RSOM was performed. Similarly to Korresp, RSOM was run with fifty different initializations. The fickle pairs of words and the most fickle words were computed and removed. Then, the remaining words were projected on one of the fifty maps, considered as "reference map". Furthermore, hierarchical classification (HAC) was applied to the code-vectors of the map reducing thus the number of final clusters.

3   Results

Both methods were performed on the selected data, corresponding to the largest connected component of the original graph (1517 words). For each of the robust versions, fifty different initializations of squared grids with 15×15 units were trained. Since both SOM algorithms, Korresp and RSOM, are stochastic or online implementations, the number of iterations was fixed at five times the number of input samples (15 000 iterations for each Korresp map and 7 500 iterations for each RSOM map). The main results and some insights on the specific analysis of the texts are listed below.

3.1   Robust Korresp

The Korresp algorithm performed in its robust version provided the list of the fickle words, ordered in descending order. According to the histogram in Figure 2, the most fickle words were removed from the analysis (around 950 words). The least fickle words with regard to the Korresp algorithm are listed in Table 3 below.

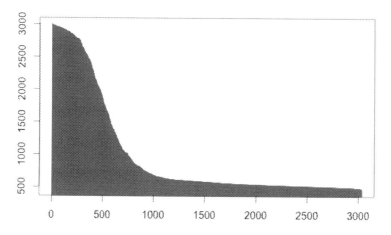

Fig. 2. Robust Korresp: fickle words with the number of their fickle pairs

Table 3. Robust Korresp: the twenty-five least fickle words, or the most stable ones. A star means it is the 'pole' form that is fickle.

$partement^*(484)$	$quant^*(484)$	$simple^*(484)$	$demander(484)$	$per(484)$
$excepte^*(483)$	$toutes^*(483)$	$on(483)$	$r3^*(482)$	$mon(482)$
$sommes^*(481)$	$tes^*(481)$	$value^*(480)$	$dont(480)$	$dire^*(479)$
$loing(478)$	$121(477)$	$hommes^*(476)$	$moien(474)$	$105^*(472)$
$partiteur^*(472)$	$appartenir(472)$	$148385(471)$	$mais^*(468)$	$tenir^*(464)$

The reference Korresp map, containing the remaining least fickle words, and combined with a hierarchical clustering in eleven superclasses is represented in Figure 3. The smoothed distances between the code-vectors are available in Figure 4. According to these two figures, clusters are generally close from one another, except for one region in the large green class, which may suggest the existence of two sub-classes within it.

The clusters of the reference Korresp map and the related super-classes are not homogeneous in terms of size and content. For example, let us study the eleventh super-class, colored in light green and situated in the upper-right corner

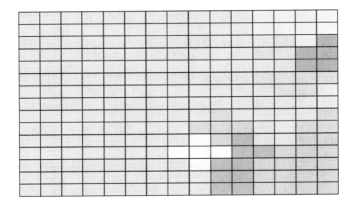

Fig. 3. The robust Korresp. The colors are the super-classes which are determined through a HCA.

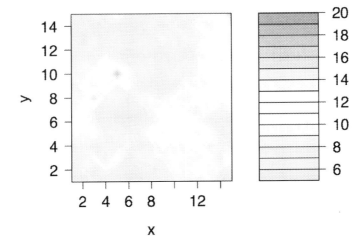

Fig. 4. Robust Korresp: smoothed distances between prototypes

of the map. It is composed of two clusters only, 224 and 225 (the grid is indexed starting from the lower-left corner ; the cluster in the lower-left corner is 1, the cluster in the upper-left corner is 15, etc...). The contents of this super-class are displayed in Table 4.

The clustering and its projection on the map can be further improved. In order to do so, we built a reduced weighted and undirected graph whose vertices are the clusters of SOM and whose edge-weights are computed as follows

$$A_{I,J} = A_{J,I} = \sum_{i \in I, j \in J} (n_{i,j} + n_{j,i}),$$

Table 4. Robust Korresp: the contents of super-class 11, composed of clusters 224 and 225

Cluster 224	Cluster 225
*prix**	147
	surpris
	*juste**

where I and J design two clusters, $i \in I$ indexes the words in cluster I and $j \in J$ indexes the words in cluster J. The thickness of the line which joins I and J is proportional to the value of this weight. Hence, the thicker is the edge, the stronger is the link between the clusters. The reduced graph is represented in Figure 5. Since the data used for training the maps was drawn from the largest connected component of the original graph, the reduced projected graph is very connected. Since our goal was to improve the visualisation of the original data by reducing it, two different thresholds were applied to the edges and only the most representative ones were plotted.

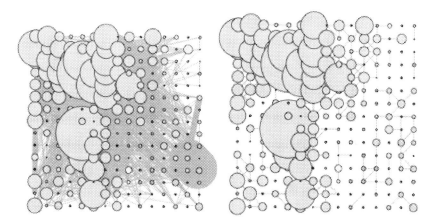

Fig. 5. Robust Korresp: projected graph. Left: only the edges with weights larger than the mean weight value are drawn. Right: only the edges of the four nearest neighbors - left, right, below, above - are drawn. The radius of the clusters is proportional to their size; the width of the edges is proportional to their weight.

According to Figure 5, the reference map is composed of a large super-class which contains most of the words while some very specific super-classes emerge very clearly. Since the Korresp algorithm performs on the co-occurrence matrix, it allows to understand the status of words better, poles or co-occurrents.

The first super-class, dark green on the map, is the largest and contains 1871 terms of the 2077 which were clustered. For this large class, it is interesting to zoom further and look within each cluster. For example, cluster 1 deals with the designation of numbers, cluster 11 concerns the computation of interest rates, while cluster 30 concerns ratios of numbers.

The rest of the super-classes contain much fewer terms and are very specific. For example, super-class 2 (light yellow) which is composed of ten poles and three co-occurrents. The three co-occurents (*vous, gectons, mer*) are specific to abacus computation. Super-class 4 (dark red) is very small, three poles and three co-occurrents. This class is specific to the proof and the verification of results (*vraye, certaine, preuve, prouver, bien, venue*).

We also notice here the effect of the absence of lemmatization, since some close words appear in the same cluster (*compagnon, compaignon*).

3.2   Robust Relational SOM

The histogram of the fickle words provided by the robust version of RSOM and ordered in descending order is plotted in Figure 6. We notice that in this case, most of the words have a high index of fickleness. This seems to be mainly due to the fact that the syntax of the texts was suppressed when symmetrizing the co-occurrence matrix.

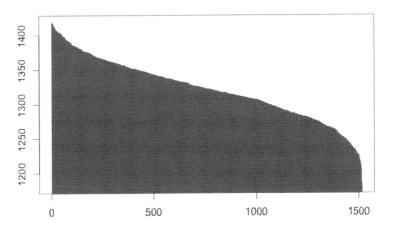

Fig. 6. Robust RSOM: fickle words with the number of their fickle pairs

The least fickle words with regard to the RSOM algorithm are listed in Table 5 below. The first remark one can make is that the most fickle words in this case are generally numbers (Roman or Arabic) and pronouns. The least fickle words are generally nouns and verbs. Hence, it appears that the use of RSOM and the symmetrization of the co-occurrence matrix is more specifically adapted for highlighting lexical aspects, while the directed co-occurrence matrix used together with Korresp seemed to be more specific for the semantics of the texts. Before building the reference RSOM map, around fickle 200 words were removed.

The reference RSOM map, combined with a hierarchical clustering in eight superclasses is represented in Figure 7. The smoothed distances between the code-vectors are available in Figure 8. According to these two figures, several

Table 5. Robust RSOM: the twenty-five least fickle words, or the most stable ones

condition(1227)	*cubbe*(1227)	*gnomon*(1227)	*marcz*(1227)	*mon*(1227)
precedente(1227)	*racine*(1227)	*sur*(1225)	*appellee*(1224)	*douzaine*(1224)
nous(1224)	*gaigneroye*(1223)	*superparticuliere*(1220)	*droit*(1219)	*mars*(1218)
layne(1217)	*precedentes*(1216)	*secondement*(1215)	*draps*(1213)	*noter*(1210)
car(1206)	*disain*(1205)	*gangne*(1186)	*nommer*(1183)	*notable*(1181)

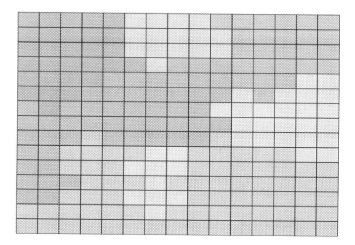

Fig. 7. The robust relational SOM. The colors are the super-classes which are determined through a HCA.

large distances colored in dark pink suggest the existence of several super-classes. The composition of the super-classes in terms of SOM clusters appears more homogeneous than previously and one may expect well-balanced super-classes. When studying the contents of the super-classes, the number of elements in each of them is between 100 and 280.

Similarly to robust Korresp, a reduced graph with the SOM clusters represented as vertices is build and projected on the map, as illustrated in Figure 9. We shall remark here that, since the lexical aspect is more powerful here than the semantics, clusters are almost all linked together, even after thresholding the edges to plot.

Let us now take a closer look at some of the super-classes. The first super-class (dark green, lower-left part of the map) appears to contain the lexical ensembles which structure the text in a mathematical problem. For instance, clusters 1 16 are clearly containing the words used in the statement of a mathematical problem. The second super-class (orange, left) groups mathematical terms such as sums, numbers and quantities. The third super-class (blue, upper-left) contains almost exclusively Arabic numbers. In the fourth super-class, we find

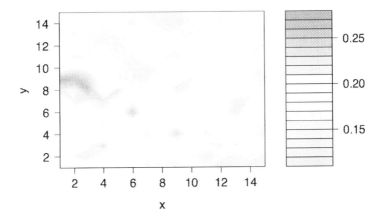

Fig. 8. Robust relational SOM: smooth distances between prototypes

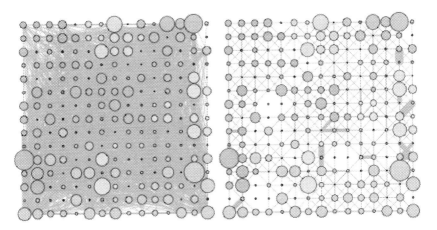

Fig. 9. Robust RSOM: projected graph. Left: only the edges with weights larger than the 3rd quartile are drawn. Right: only the edges of the four nearest neighbors - left, right, below, above - are drawn. The radius of the clusters is proportional to their size; the width of the edges is proportional to their weight

words which design numbers (*digit, articles, composes, mixte*) and the specific technique of abacus calculation (*gectons, gect*).

4   Conclusions and Perspectives

The detailed description of the projection results on the two reference maps was not addressed here since, on the one hand, the number of pages is limited and, on the other hand, the specificities of the corpus of documents would only passionate the specialists of the discipline. However, let us remark that Korresp seems more likely to be adapted to semantic issues and that it appears to isolate and highlight very specific groups of words. At the same time, RSOM seems more likely to provide lexical insights on the studied texts. These characteristics are due to the fact that Korresp is trained on the original co-occurrence matrix C, which is directed, while RSOM is trained on a symmetric dissimilarity matrix computed from C.

Although each of them has its specificities, both methods, Korresp and RSOM, may be viewed as useful tools for simplifying, reducing and visualizing complex graphs, of high dimensionality. However, one of the current drawbacks of the two algorithms is the computational time, especially for RSOM where the code-vectors are supposed to be linear combinations of the original input data. The latter hypothesis could be weakened, especially as the co-occurrence matrices are generally very sparse. Hence, the next step of our research is to adapt the two algorithms to very large and sparse matrices and test them with big text-mining data.

References

Benzecri, J.-P.: Correspondence Analysis Handbook. Marcel Dekker, New York (1992)

Blei, D.: Probabilistic topic models. Communications of the ACM **55**(4), 77–84 (2012)

Bourgeois, N., Cottrell, M., Deruelle, B., Lamassé, S., Letrémy, P.: How to improve robustness in kohonen maps and display additional information in factorial analysis: Application to text mining. Neurocomputing **147**, 120–135 (2015)

Cottrell, M., Fort, J.-C., Pagès, G.: Theoretical aspects of the SOM algorithm. Neurocomputing **21**, 119–138 (1998)

de Bodt, E., Cottrell, M., Verleysen, M.: Statistical tools to assess the reliability of self-organizing maps. Neural Networks **15**(8–9), 967–978 (2002)

Lamassé, S.: Les traités d'arithmétique médiévale et la constitution d'une langue de spécialité. In: Ducos, J. (ed.) Sciences et langues au Moyen Âge. Actes de l'Atelier franco-allemand, Paris, janvier 27–30, 2009, pp. 66–104. Universitätesverlag, Heidelberg (2012)

Lebart, L., Morineau, A., Warwicki, K.: Multivariate Descriptive Statistical Analysis Correspondence Analysis and Related Techniques for Large Matrices. Wiley, New York (1984)

Martinez, W.: Au-delá de la cooccurrence binaire...poly-cooccurrence et trames de cooccurrence. Corpus **11**, 191–216 (2012)

Martinez, W., Salem, A.: Contribution à une méthodologie de l'analyse des cooccurrences lexicales multiples dans les corpus textuels. Thèse doctorat (2003)

Oja, E., Kaski, S.: Kohonen Maps. Elsevier (1999)

Olteanu, M., Villa-Vialaneix, N.: On-line relational and multiple relational som. Neurocomputing **147**, 15–30 (2015)

Titterington, M., Smith, A., Makov, U.: Statistical Analysis of Finite Mixture Distributions. Wiley (1985)

Evaluation of Fitting Functions for the Saccade Velocity Profile in Electrooculographic Records

Rodolfo García-Bermúdez[1,2], Camilo Velázquez-Rodríguez[2],
Fernando Rojas[3,4]([✉]), Manuel Rodríguez[3], Roberto Becerra-García[2],
Michel Velázquez-Mariño[2], José Arteaga-Vera[1], and Luis Velázquez[5]

[1] Facultad de Ciencias Informáticas, Universidad Laica Eloy Alfaro de Manabí,
Manta, Ecuador
{rodolfo.garcia,jose.arteaga}@live.uleam.edu.ec
[2] Grupo de Procesamiento de Datos Biomédicos (GPDB), Universidad de Holguín,
80100 Holguín, Cuba
{cvelazquezr,idertator,mvelazquez}@facinf.uho.edu.cu
[3] Departamento de Arquitectura y Tecnología de Computadores,
Universidad de Granada, Granada, Spain
{frojas,manolo}@ugr.es
[4] Centro de Investigación en Tecnologías de la Información y de las Comunicaciones
(CITIC-UGR), Universidad de Granada, Granada, Spain
[5] Centro para la Investigación y Rehabilitación de Ataxias Hereditarias,
Holguín, Cuba
cirahsca2@cristal.hlg.sld.cu

Abstract. A saccade is an ocular movement that is characterized by speed and precision. The velocity profile of this movement is used to extract the maximum speed value, that is one of the most important features of the saccade. A gamma function was used by other authors to describe the waveform shape of the velocity profile. However, this function does not present an optimal profile description in records of patients suffering from Spinocerebellar Ataxia type 2. In order to find a function that better describes the velocity profile, this contribution compares the fittings of several functions through visual and numerical analysis. Results showed a better performance of the partial sums of gaussian series and a gaussian fit function.

Keywords: Velocity profile · Saccade · Eye movements · Ataxia · SCA2 · Curve fitting · Gaussian series · Nonlinear regression

1 Introduction

Ocular movements allow to define with better quality objects in the visual field of the person, as they stabilize the image of interest in the maximum keenness visual zone [13]. Eye movements are useful to identify dysfunctions in a wide range of neurological conditions, including a group of diseases with a severe impact in quality and duration of life, like Parkinson and hereditary ataxias

© Springer International Publishing Switzerland 2015
I. Rojas et al. (Eds.): IWANN 2015, Part II, LNCS 9095, pp. 592–600, 2015.
DOI: 10.1007/978-3-319-19222-2_49

[2, 4, 10], providing to neurologists with an useful tool for the exploration of neural functions [8].

A classification of ocular movements used in medical literature [9] divide them into two categories: stabilization movements, that are meant to maintain and stabilize an image in the retina; and the saccadic, referred to the movement of the eyes inside of the visual field to bring objects of interest to the vision area. Nystagmus, smooth pursuit and fixation are in the first category, while the second category includes saccades and vergence movements [5]. The saccade is a rapid movement, steplike rotation of the two eyes which can sometimes be accompanied by a head movement allowing a wider visual analysis of the object. Saccades are not exclusively aimed at objects that suddenly appear, but scanning saccades are also generated when the eyes are exploring the environment [7].

Duration, amplitude, maximum speed and start and end points of the movement are features that distinguish saccades from other eye movements. In order to obtain the value of maximum speed is necessary to differentiate the eye movement signal resulting in a saccade velocity profile. The highest value of this obtained signal is the maximum speed value.

We are currently working in the development of a saccadic simulation engine which, by taking into account the specific characteristics of the ocular movements in patients of ataxia SCA2, would be able to make the generation of artificial saccadic records, so that it could develop and test digital signal processing techniques and machine learning algorithms for clustering and classification purposes, amongst many other possible applications to aid medical specialists in the diagnosis and assessment of patients condition. This aim requires a robust model with a minimal set of descriptors which, if possible, should be directly associated to the parameters of biological features of interest to the clinical diagnosis.

Van Opstal and Van Gisbergen [12] and more recently Coughlin [3] amongst others fit the velocity profiles with a certain density function to characterize their shapes. Modelling these velocity profiles for healthy subjects is not an specially difficult task, since an uniform shape is present for most of the individuals and usually the presence of distortions like blinks or extra ocular movements are treated by discarding these saccades.

However, velocity profiles of patients suffering from Spinocerebellar Ataxia type 2 (SCA2) and other degenerative diseases are severely distorted by the presence of tremor, involuntary movements, and noise; additionally there exists a high variability in the shape of the profiles, beyond the noise and the distortions mentioned before, caused by the effects of the disease on the generation of saccades by the oculomotor plant. This provokes distorted relationships amongst maximum velocity, amplitude, duration and skewness of the saccades which are reflected on its waveform.

This paper shows an study and assessment about this density function and other fitting functions in order to find which one fits better the saccade velocity profile wave shape from saccades of SCA2 patients electrooculographic records.

2   Materials and Methods

2.1   Electrooculographic Records

The electrooculographic records were obtained by the medical staff of the Centre for the Research and Rehabilitation of the Hereditary Ataxias (CIRAH) at Holguín, Cuba. A two-channel electronystagmograph (Otoscreen, Jaeger-Toennies, Hochberg, Germany) was used to record saccadic ocular movements.

Nineteen (19) records of SCA2 patients confirmed by clinical and molecular diagnostic were selected, along with 29 records of asymptomatic SCA2 patients with confirmation by molecular diagnostic and 23 records of healthy subjects. Saccades were identified using an automated algorithm based in a velocity threshold of 10 degrees per second. A manual process of visual inspection was carried out in order to eliminate saccades considered anticipatory (latencies lower than 100 ms) or presenting artifacts like blinkings, excessive noise, muscle or extra ocular movements in the close time before onset, or in the next fixation. In the next step, the mean of amplitude, duration and latency were calculated and saccades with deviations higher than 20% were excluded, as it is described in [11]. The stimulus and patient response data were automatically stored in ASCII files as comma separated values (CSV) by the Otoscreen electronystagmograph, according to its user manual specifications. All the data processing was implemented in MATLAB (ver.8.1.0.604), the data from records of subjects were imported and saved as .MAT files (one for each experiment) in order to be used as the input data for the experiments.

A numerical differentiation algorithm was used to obtain the velocity profile for every saccade in the registers, due to the numerical nature of the electrooculographic stored data. There are several well-known mathematical methods that present good results in numerical differentiation. In this paper we focus in the methods that use curve fitting instead of interpolation in their procedure, known as Lanczos [1], specifically Lanczos 7 or Lanczos of 7 points, described by (1):

$$f' = \frac{f_1 - f_{-1} + 2(f_2 - f_{-2}) + 3(f_3 - f_{-3})}{28h} \tag{1}$$

where h is the step of sampling, which in the obtained records has a value of $h=4.88$ ms. Previous visual inspection of the saccades established valid saccades and some minor corrections were made to the start and end points of that ocular movement. These points were used to extract the velocity profile of the saccade in the differentiate electrooculogram.

2.2   Fitting Functions

In order to estimate the velocity profiles with a given fitting function, we start from the point of view described in [3,12] which is that the density function can be described by the gamma distribution function. Equation (2) shows the gamma function, which is the first function that we incorporate to our study:

$$f(x) = a \left(\frac{x}{b}\right)^{c-1} e^{-\frac{x}{b}} \tag{2}$$

where a and b are scaling constants for velocity and duration, respectively, and c is the waveshape parameter which determines the degree of asymmetry.

The two other functions analyzed in this work are part of the gaussian models. This model can describe curves in many areas of science and engineering named gaussian peaks. As equation (3) shows, the model is expressed by a gaussian series:

$$f(x) = \sum_{i=1}^{n} a_i e^{\left[-\left(\frac{x-b_i}{c_i}\right)^2\right]} \tag{3}$$

where a is the amplitude, b is the location of the centroid and c is related to the width of the gaussian peak. Partial sums of this series have shown excellent performance in other curves similar to the velocity profile in the ocular saccade. Moreover in [6] authors used this series to model the pulse component, obtained by the application of independent component analysis to horizontal saccades [11].

Finally, an adaptation of a gaussian fit as described in Equation (4) was evaluated.

$$f(x) = a + b e^{-\left(\frac{x-c}{d}\right)^2} \tag{4}$$

where a, b, c and d are the fitting parameters.

Addition of a represented a remarkable improvement in the gaussian function performance to fit the sacadic velocity profiles, which it was caused by an observed displacement in most of the saccades. Nevertheless, this parameter can be removed after fitting because it does not represent any change in the profile waveform. Therefore, this function can be considered to need 3 parameters to be calculated, thus being identical to a gaussian series of one term (gauss1 from now), but showing better performance.

In order to obtain the optimal parameters of the model for every profile of saccadic velocity, the Matlab builtin function $fit()$ was used, and the Trust-Region-Reflective Least Squares Algorithm was selected accordingly to the suggestions Matlab help for gaussian models fitting. Figure 1 shows how the three proposed fitting functions (gamma, three terms gaussian serie and modified gauss) model a given velocity profile for an specific saccade. As it can be visually figured out, all of them achieve a good approximation, showing a small deviation from the original data points and no overfitting.

A further goal of this work is to apply these models for implementing a saccadic simulation engine, integrating the velocities profiles in that simulation prototype. As gamma and gauss functions can be integrated by numeric or analytical integral solutions, the following equations show the results of the function $int()$ of Symbolic Math Toolbox (Matlab) for gamma and modified gaussian functions:

$$f(x) = -a\, b\, igamma(c, x/b) \tag{5}$$

$$f(x) = \frac{\sqrt{\pi}\, b\, erfi((c-x)\sqrt{-1/d^2})}{2\sqrt{-1/d^2}} \tag{6}$$

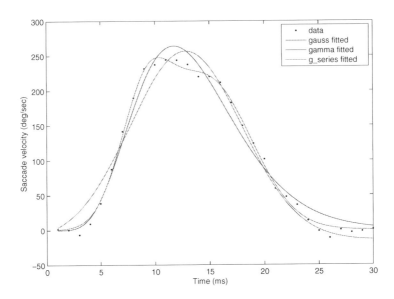

Fig. 1. Fitness functions modeling a velocity profile record

where *igamma* and *erfi* represent the incomplete gamma function and the imaginary error function respectively defined by the Symbolic Math Toolbox.

In Figure 2 an original saccade profile data and the noise free evaluation of the integral of the fitted modified gaussian function are shown, with the term a removed.

2.3   Statistics Metrics

There are several numerical metrics that can evaluate the goodness of the fitting made by a determined function in a certain data. In this work we made use of two well-known metrics in curve fitting applications, as the Root Mean Squared Error (RMSE) and the squared Pearson correlation coefficient (R-square). RMSE, which is also known as the standard error of the regression is described by Equation (7):

$$RMSE = \sqrt{\frac{\sum_{i=1}^{n}(y_i - \hat{y}_i)^2}{n}} \tag{7}$$

where n is the amount of points in the data, y_i represent the ith point of the data and $\hat{y}_i$ is the ith value estimated by the model in evaluation.

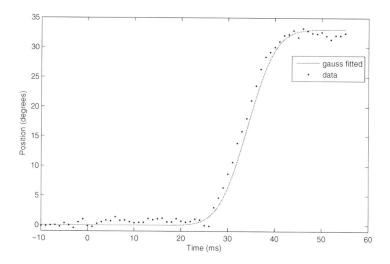

Fig. 2. Integral of fitness modified gaussian function modeling a position profile record

The squared Pearson correlation coefficient, also called R-square in linear regression, can be expressed by Equation (8):

$$Rsquare = \frac{\sum_{i=1}^{n}(\hat{y}_i - \bar{y})^2}{\sum_{i=1}^{n}(y_i - \bar{y})^2} \quad (8)$$

where $\bar{y}$ is the mean of the data, the others elements are the same as in (7).

As part of a further study in progress, this metrics can be modified in order to strengthen the importance of minimizing the number of parameters in the curve fitting model, as these models better adapt to biological processes and avoid overfitting. A simple weighted penalty factor depending on the number of parameters may be enough for this purpose.

3   Results

As the input data set, a total of 4500 velocity profiles were obtained. A subset of 100 records were chosen randomly, following this random process a normal distribution. Afterwards, a curve fitting using functions described in the previous section was applied to this subset. In the case of the gaussian series, partial sums 1, 2 and 3, which we denote as gauss1, gauss2 and gauss3 were calculated. The partial sums of the gaussian series with a degree higher than 3 have many parameters to be adjusted in a small data set as it is the velocity profile which can result in overfitting.

In order to rank the fitting functions a simple visual analysis of the obtained results and graphics was enough. This ranking was made manually by the authors, by examination of every figure to assign them a value from 1 to 5, where the function who get 1 made a better fit and the one getting 5 made the worst fit to the data. When no substantial differences were appreciated among several functions, same rank was assigned to them for this record. This is summarized in Table 1.

Table 1. Ranking of fitting functions

Fitting functions	1	2	3	4	5
gamma	0.0%	4.45%	11.11%	31.11%	53.34%
gauss	4.45%	24.44%	62.22%	6.67%	2.22%
gauss1	2.22%	8.89%	31.11%	55.56%	2.22%
gauss2	28.89%	66.67%	4.45%	0.0%	0.0%
gauss3	82.22%	17.78%	0.0%	0.0%	0.0%

Table 1 reveals that the best fitting results are achieved by gauss2 and gauss3, while the worst function describing the velocity profiles is the gamma function.Supporting this visual analysis, a numerical analysis based on the goodness-of-fit statistics previously described in Section 2.3 was carried out. Figure 3 shows the results for Root Mean Squared Error (RMSE) and the correlation coefficients of Pearson for the different fitting functions.

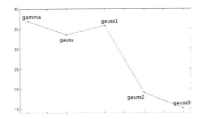 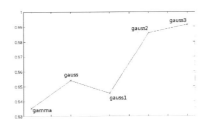

Fig. 3. RMSE (left) and Pearson correlation coefficient (right) results of the fitting functions

Regarding the gaussian series, as the numerical analysis shows, better results are achieved as the number of terms in the series is increased. The gamma function registered the worst results, while the modified gauss function stands between gauss1 and gauss2. Metrics from gauss3 are slightly better than gauss2. However, the number of fitting parameters to be estimated is an important factor to be minimized which may result in the selection of gauss2 fitting function in practice.

4   Conclusions

This contribution analyzes the performance of several fitting function in order to model the velocity profile of a human saccade. A comparison was made between a gamma function [12], a modified gauss function and the first three partial sums of the gaussian series.Results suggest that the best trade off between the goodness of fitting and the number of parameters of the functions under analysis to model the velocity profile is gauss2, since the values of the visual and numerical analysis are very similar to gauss3 and better than the rest, and when compared to gauss3, gauss2 offers the advantage of having to compute a lower number of parameters. Additionally the capability of gauss2 and gauss3 to model multi-peaks profiles could make it suitable to represent saccades with glissades, which are very usual in some of the neurological disorders affecting oculomotor system.

On the other hand, the modified gauss function shows an excellent behavior in fitting single saccades composed of only one peak, these results show the suitability of this function to model the saccadic ocular movements with a reduced set of parameters. Beyond the advantage of its simplicity, as it only needs three 3 parameters to be calculated, these parameters can be directly associated to biological features like maximum velocity, peak timing and duration of the saccade. This is specially important by considering the possible further use of this model for simulation of saccades or the application of methods of machine learning for clustering and classification of subjects.

Further research is necessary to evaluate the ratio between the skewness of the second partial sum of the gaussian series and other features of the saccade ocular movement and establish a relationship among them. Also, the use of this model with classification purposes looks promising in order to establish relationships amongst saccadic characteristics and the condition of the subject.

Acknowledgments. This work has been supported by the Project of Excellence "Proyecto Motriz de la Junta de Andalucía: Sistemas de cómputo avanzados en aplicaciones del ámbito de biotecnología y bioinformática" (P12-TIC-2082).

References

1. Becerra García, R.A.: Plataforma de procesamiento de electrooculogramas. Caso de estudio: pacientes con Ataxia Espinocerebelosa tipo 2. Ms. thesis, Universidad de Holguín (2013)
2. Caspi, A., Zivotofsky, A.Z., Gordon, C.R.: Multiple Saccadic Abnormalities in Spinocerebellar Ataxia Type 3 Can Be Linked to a Single Deficiency in Velocity Feedback. Investigative ophthalmology & visual science **54**(1), 731–738 (2013). http://www.iovs.org/content/54/1/731.short
3. Coughlin, M.: Calibration of two dimensional saccadic electro-oculograms using artificial neural networks. PhD, Griffith University, Queensland (2002)

4. Fahey, M.C., Cremer, P.D., Aw, S.T., Millist, L., Todd, M.J., White, O.B., Halmagyi, M., Corben, L.A., Collins, V., Churchyard, A.J., Tan, K., Kowal, L., Delatycki, M.B.: Vestibular, saccadic and fixation abnormalities in genetically confirmed Friedreich ataxia. Brain **131**(4), 1035–1045 (2008). http://brain.oxfordjournals.org/cgi/content/abstract/131/4/1035

5. García, R.: Procesamiento de registros oculares sacádicos en pacientes de ataxia SCA2. Aplicación del Análisis de Componentes Independientes. Phd. thesis, Universidad de Granada (2010)

6. García, R., Velázquez, C., Rojas, F., Becerra, R., Velázquez, M., López, L., Velázquez, L.: A comparison of two fitting functions for sacadic pulse component mathematical modelling. In: Proceedings of ANNIIP 2014, Vienna, Austria, pp. 88–94, September 2014

7. Goffart, L.: Saccadic eye movements. Encyclopedia of Neuroscience, pp. 437–444. Academic Press, Oxford (2009). http://www.sciencedirect.com/science/article/B98GH-4TVBCX5-YV/2/aea673d7d084e8c143439f1e2cd07ecd

8. Langaas, T., Mon-Williams, M., Wann, J.P., Pascal, E., Thompson, C.: Eye movements, prematurity and developmental co-ordination disorder. Vision Research **38**(12), 1817–1826 (1998). http://www.sciencedirect.com/science/article/B6T0W-3T3JBK5-X/2/0e0918534c89a50d4b525a94d26c8dda

9. Lukander, K.: Measuring gaze point on handheld mobile devices. Ph.D. thesis, Helsinki University of Technology, Helsinki, Finland (2004)

10. Pötter-Nerger, M., Govender, S., Deuschl, G., Volkmann, J., Colebatch, J.G.: Selective changes of ocular vestibular myogenic potentials in Parkinson's disease. Movement Disorders (2014). http://onlinelibrary.wiley.com/doi/10.1002/mds.26114/full

11. Rojas, F., García, R., González, J., Velázquez, L., Becerra, R., Valenzuela, O., San Román, B.: Identification of saccadic components in spinocerebellar ataxia applying an independent component analysis algorithm. Neurocomputing (2012). http://www.sciencedirect.com/science/article/B6V10-45M6G4R-5/2/be1c7b9ddf4b59d44f5f07e02f93e80b

12. Van Opstal, A., Van Gisbergen, J.: Skewness of saccadic velocity profiles: A unifying parameter for normal and slow saccades. Vision Research **27**(5), 731–745 (1987). http://www.sciencedirect.com/science/article/B6T0W-484M5BN-6B/2/38501f6bfeb1e12e0cde519aef29c08c

13. Velázquez, Luis: Ataxia espino cerebelosa tipo 2. Principales aspectos neurofisiológicos en el diagnóstico, pronóstico y evaluación de la enfermedad. Ediciones, Holguín (2006)

esCam: A Mobile Application to Capture and Enhance Text Images

J. Pastor-Pellicer, Maria Jose Castro-Bleda[✉], and J.L. Adelantado-Torres

Departamento de Sistemas Informáticos y Computación,
Universitat Politècnica de València, Valencia, Spain
mcastro@dsic.upv.es

Abstract. Taking high resolution photos with mobile devices anytime anywhere is becoming increasingly common. Therefore, images of all kinds of text documents are recorded. This work presents esCam, an application for Android platform, whose goal is to preprocess the images of those text documents, in particular, perspective correction and image cleaning and enhancing. What truly differentiates our application is that esCam focuses on treatment of text that may appear in the image, using neural networks. These preprocessing steps are needed to make easier the digitalization and also to benefit subsequent steps such as document analysis and text recognition.

1 Introduction

Since the recent emergence of mobile platforms (like iOS, Android, Windows Phone, etc.), mobile terminals have experienced a huge increase in computational power and other features like screen and camera resolution. Furthermore, these devices include many tools that can be used anytime, anywhere. Therefore, we can take pictures of any type of documents: printed or manuscripts, ancient documents, or the most recent ones. All this without the need for specific devices such as scanners or high resolution cameras. However, taking snapshots with mobile devices, smartphones or even tablets, has disadvantages in comparison with other methods:

- Resolution: even though resolution of camera mobile phones is constantly increasing, these cameras still have lower resolutions than other specific image capture devices.
- Quality: brightness, shadows, noise and others, are examples of issues of snapshots taken by mobile devices, because they are not taken under ideal conditions.
- Perspective: when mobile devices are used to take snapshots, often we use our hands to hold the device, so the image is commonly taken from a bad perspective or unwanted areas appear within the snapshot.

In particular, when dealing with handwritten or printed documents, it is very important to be able to detect and extract the area of text with relevant information for the user, and to store it in a digital format accessible from any

© Springer International Publishing Switzerland 2015
I. Rojas et al. (Eds.): IWANN 2015, Part II, LNCS 9095, pp. 601–604, 2015.
DOI: 10.1007/978-3-319-19222-2_50

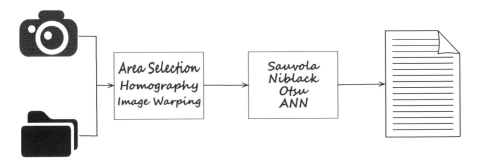

Fig. 1. esCam application workflow

Fig. 2. esCam application menu

device [7]. It is also likely, specially when dealing with ancient documents, that artifacts such as noise, gloss, or perspective distortion appear [3]. These effects must be corrected before applying a character recognition system [2,6].

Despite the existence of similar applications in the market, none of these offers text enhancement filters like ours. The goal of this work is to design an application to use classical binarization algorithms like Otsu, Niblack, or Sauvola [8,10], as well as other more specific techniques like convolutional neural networks [4,9]. This demostration presents "esCam", the developed Android application to capture and enhance images of text documents in mobile devices, freely available at [1].

2   Application Description

Figure 1 shows a workflow diagram of the application and Figure 2 shows its initial menu. Figure 3 illustrates actual screenshots of the main phases of the application. It can be observed that, regardless of the image loading and saving steps, the image processing is divided in two main stages: perspective correction, and image cleaning and binarization.

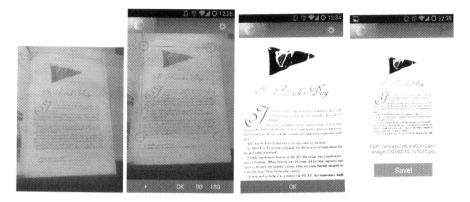

Fig. 3. Screenshots of esCam app (from left to right): Original image; Area selection; Normalized perspective image; Cleaned and enhanced image

Load and Save Images. Like other similar applications, the image can be taken by a camera or have been previously stored in memory. This is useful for the user, who can take a snapshot using the device camera for later image processing.

Perspective Correction. The perspective correction is a key step, especially when the images are taken with mobile cameras. Four coordinates suffice to correct perspective, that is, the four corners of the region that the user wants to focus on. In our application, the user can adjust these four points. Once the points are fixed, image warping correction is applied by means of homography. In addition, the user can correct the orientation of the image before image warping using proper rotations.

Filtering. The following preprocessing steps are image cleaning and binarization. When dealing with documents, particularly with text, it is desirable to have a high ratio between foreground text and background. Applying binarization filters have several advantages: first, it helps human reading and benefits the recognition for most OCR engines, and, secondly, it reduces the storage size of the image. Our application includes the following binarization filters:

- Niblack/Sauvola filtering [10] calculates a threshold using a sliding window in a greyscaled version of the image. Threshold calculation is based on the local mean and standard deviation of that window.
- Otsu method [8] is based in the idea of separate background and object. It uses statistical methods as the variance among grey tones in the image.
- A Neural Networks technique presented in [2] is also included: a multilayer perceptron returns a clean and enhanced pixel, given an input of this pixel plus a context. The neural network has been trained with the corpus Noisy-Office [11], freely available at UCI Machine Learning Repository [5], http://archive.ics.uci.edu/ml/datasets/NoisyOffice.

3   Conclusions

The esCam application is presented in this demostration.[1] In addition to the Android software developing kit intended for Android apps, we used OpenCV as a library for efficient image processing. The convolutional neuronal networks are trained using the APRIL-ANN toolkit [12]. The application is devoted to the preprocessing of text images, especially images taken with mobile devices, adding tools for perspective correction and image filtering. The preprocessed images achieve a better information representation for further steps such as line extraction or text recognition.

Acknowledgments. This work has been partially supported by the Spanish Government project TIN2010-18958.

References

1. http://www.dsic.upv.es/mcastro/esCam/index_en.htm
2. España Boquera, S., Castro-Bleda, M.J., Gorbe-Moya, J., Zamora-Martinez, F.: Improving offline handwritten text recognition with hybrid HMM/ANN models. IEEE Trans. PAMI 33(4), 767–779 (2011)
3. Gatos, B., Pratikakis, I., Perantonis, S.J.: Adaptive degraded document image binarization. Pattern Recognition **39**(3), 317–327 (2006)
4. Hidalgo, J.L., España, S., Castro, M.J., Pérez, J.A.: Enhancement and Cleaning of Handwritten Data by Using Neural Networks. In: Marques, J.S., Pérez de la Blanca, N., Pina, P. (eds.) IbPRIA 2005. LNCS, vol. 3522, pp. 376–383. Springer, Heidelberg (2005)
5. Lichman, M.: UCI ML repository (2013). http://archive.ics.uci.edu/ml
6. Mori, S., Suen, C.Y., Yamamoto, K.: Historical review of OCR research and development. Proceedings of the IEEE **80**(7), 1029–1058 (1992)
7. Nagy, G.: Twenty Years of Document Image Analysis in PAMI. IEEE Trans. PAMI **22**(1), 38–62 (2000)
8. Otsu, N.: A threshold selection method from gray-level histograms. Automatica **11**(285–296), 23–27 (1975)
9. Pastor-Pellicer, J., Zamora-Martínez, F., España-Boquera, S., Castro-Bleda, M.J.: F-Measure as the Error Function to Train Neural Networks. In: Rojas, I., Joya, G., Gabestany, J. (eds.) IWANN 2013, Part I. LNCS, vol. 7902, pp. 376–384. Springer, Heidelberg (2013)
10. Sauvola, J., Pietikäinen, M.: Adaptive document image binarization. Pattern Recognition **33**(2), 225–236 (2000)
11. Zamora-Martínez, F., España-Boquera, S., Castro-Bleda, M.J.: Behaviour-Based Clustering of Neural Networks Applied to Document Enhancement. In: Sandoval, F., Prieto, A.G., Cabestany, J., Graña, M. (eds.) IWANN 2007. LNCS, vol. 4507, pp. 144–151. Springer, Heidelberg (2007)
12. Zamora-Martínez, F., España-Boquera, S., Gorbe-Moya, J., Pastor-Pellicer, J., Palacios, A.: APRIL-ANN toolkit, A Pattern Recognizer In Lua with Artificial Neural Networks (2013). https://github.com/pakozm/april-ann

[1] Freely available at http://www.dsic.upv.es/~mcastro/esCam/index_en.html, along with a tutorial and a demo video.

Computer Access and Alternative and Augmentative Communication (AAC) for People with Disabilities: A Multi-modal Hardware and Software Solution

Salvador Sancha-Ros[✉] and Esther García-Garaluz

Eneso Tecnología de Adaptación S.L., C/Gargantúa 8, 29006, Málaga, Spain
{ssancha,esther}@eneso.es

Abstract. Personal computers and smartphones in their standard form are, in general, inaccessible to people with reduced mobility. It has been necessary to design alternative interfaces and peripherals that allow them to use the technology comfortably and effectively, a difficult task given the heterogeneity of the physical and cognitive profiles. We will demostrate a solution for computer access and alternative communication for people with disabilities using three of the most effective access methods: switch-based input, head tracking and eye tracking, and custom designed software.

1 Introduction

Access to new technologies, specifically computers and smartphones, is no longer a luxury, but rather a fundamental part of belonging to an increasingly connected society. This is particularly true for people with disabilities, since these technologies allow them to communicate with others freely (regardless of the distance or their own handicaps), study, telecommute or enjoy some leisure time.

But the standard computer interfaces and peripherals are in general inaccessible for people with reduced mobility. A computer mouse, for instance, cannot be used comfortably by someone without fine manual dexterity or a condition that makes such tasks tiresome, let alone someone without upper limb mobility. The industry has created a number of adaptations for the most common peripherals, but they do not cover many of the existing physical and cognitive profiles. The problem is a tough one, because the characteristics of users with disabilities are extremely heterogeneous. The problem of universal access remains unsolved [1][2].

We must aim for multi-modal solutions that can "read" the needs of the user, both physical and cognitive, and adapt themselves automatically in order to offer the most accessible and intuitive environment. It is a challenging task, but the reward is well worth it.

In this demonstration we will show a global solution for computer access and alternative and augmentative communication (AAC) targeted to severely affected people. The solution is both hardware (as explained in Section 2) and software (Section 3). In Section 4 we will describe the system as whole and the exact behavior we intend to demonstrate.

© Springer International Publishing Switzerland 2015
I. Rojas et al. (Eds.): IWANN 2015, Part II, LNCS 9095, pp. 605–610, 2015.
DOI: 10.1007/978-3-319-19222-2_51

2   Adapted Hardware for Computer Access

The variety of affectations among people with reduced mobility makes it hard to develop a universal solution for computer access that is both comfortable and productive. As a general rule, it is more effective (in the sense of "more generally usable") to design devices that are controlled by those body parts whose mobility is typically preserved: the head and the eyes; but in some cases it can be very efficient to control a computer with one or two switches if the user is able to activate them.

2.1   Switch-Based Access

Even in the most severe cases, it is not uncommon that a person with reduced mobility can consciously control some body part, such as a finger or foot. If that is the case, an adapted interface based on a simple switch can be an efficient solution.

There is a great variety of commercial switches specifically designed for this task. Based on the type and degree of mobility of the user, the most appropriate switch can be a pushbutton, a string, a lever, or a sponge activated by pressure. Whichever the choice, from a designer point of view the problem consists of adapting the interface so that a single bit of information can control such a complex system as a computer. This problem has two facets: software (which we will talk about in Section 3) and hardware.

From a hardware point of view, we need to translate the switch activation to commands that can be understood by the computer. Most commercial switches use a mini-jack plug. We have designed encore [3], a configurable multi-input USB switch interface. This device, which is commercially available and commonly used by many handicapped users, transforms switch activations into simulated mouse clicks, keyboard presses, joystick motion or button presses, or even generic events via a custom HID interface.

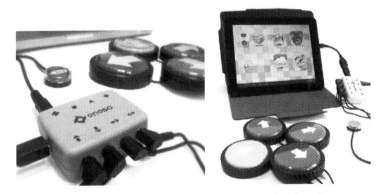

Fig. 1. enCore. Configurable multi-input USB switch interface.

2.2   Access with Head Movements

People who have suffered spinal cord injuries may still preserve some degree of mobility of their head. This is also the case among people with brain palsy, multiple sclerosis and other types of disabilities. This is potentially a much more sophisticated input method than a simple switch activation, but it also presents additional challenges, including the need for calibration and involuntary-movement handling.

Most of the alternatives that have been created fall in one of these groups: artificial vision, infrared tracking and inertial sensors. We have developed enPathia, a patented system that allows people with reduced mobility in their upper extremities to control a computer with gentle head movements. enPathia[3][4] is based on inertial sensors, and can be optionally used with switches to offer a multi-modal solution for computer access.

Fig. 2. enPathia. Head tracking system.

Our system enPathia is multi-platform. The calibration process only takes a few seconds, and it adapts itself automatically to the type and range of mobility of the user.

2.3   Access with Eye Movements

Voluntary eye movement is almost always preserved, even with degenerative diseases such as Amyotrophic Lateral Sclerosis (ALS)[5]. As such, it is the only available input for communication and interaction in the most severe cases of reduced mobility.

Fig. 3. The Eye tribe eye tracking system

Solutions based on this type of mobility have traditionally been extremely expensive, the reason being, among others, that the technology required to track the eye movements with accuracy is sophisticated. Fortunately, a number of devices have been released recently that bring eye-tracking technology to a much more affordable level. For example, The Eye Tribe Tracker[6] is below the 200€ range.

3   Adapted Software for Computer Access

All the previously presented hardware solutions require (or benefit hugely from) an adaptation of the underlying software. In general, the information rate or the accuracy provided by the peripheral will be too low to use the standard WIMP (window, icon, menus and pointer) interfaces [7].

3.1   Software Adaptations

If the peripheral provides only one or two bits of information (as when using switches) at a low rate, a common solution is using scanning: the elements of the interface are sequentially highlighted, either manually or automatically, and the user generates an input at the right time in order to activate one. This system can benefit from scanning strategies, e.g. rows-then-columns, to increase the productivity.

If the user gets to control the mouse pointer (e.g. with head or eye tracking), there is a number of features that can be implemented to make the interface more usable. Auto-clicking, element selection by hovering or simplifying the desktop are some of them. Our system enPathia, for instance, implements some of these strategies.

Eye-tracking systems have, in all cases, low resolution. This makes it very difficult to use them to control the mouse pointer directly. Filtering the output of the tracker helps, but it is usually necessary to use other mechanisms such as zooming interfaces or simplified layouts.

In general, relatively complex tasks such as double or right clicks, drag-and-drop are complicated to perform with adapted peripherals, so the designer should try to avoid them.

3.2   Grid-Based Systems

Grid layouts are a common solution for adapted computer access. On the one hand, they are inherently suitable for scanning, and on the other, they can feature bigger and regularly-sized selectable elements which are easier to activate with low-accuracy devices as head and eye trackers.

Grid systems are based on the traditional communication boards, which are simply tables of standard or custom-made symbols used as an aid for communication. By bringing them to the world of computers, these systems have widely increased their range of application.

We have developed Verbo, a grid-based application for alternative and augmentative communication, computer access, social networking, home automation and other common tasks. Verbo runs on MS. Windows and Android and is compatible with a large number of hardware adaptations.

Fig. 4. Accessing Verbo with row-column scanning

Within Verbo, the user can create multi-page boards containing interactive cells. Cells can be configured to read a text using a text-to-speech engine, "jump" to a different page, use the device hardware (e.g. take a photograph), publish a message on a social network, etc., and they can have a custom appearance comprising text and/or images. By presenting the information in an organized grid, it is much simpler to access it using head or eye trackers or switches, and the look and feel of the interface can be adapted to the cognitive level of the user.

4   Demonstration

We propose to demonstrate how users with physical or cognitive disabilities can use a computer and a tablet PC with a combination of adapted hardware and software. Using a suite comprising our software Verbo, peripherals as enPathia, enCore and The Eye Tribe tracker we will provide a multi-modal system that is widely accessible.

5   Current Status

All the presented hardware has already been developed, and some of the system have even been commercially available for some time now. The AAC side of Verbo is completed, and we are currently working to implement complete computer access, social networks, home automation and other features.

References

1. Keates, S., Hwang, F., Langdon, P., Clarkson, P.J., Robinson, P.: The user of cursor measures for motion-impaired computer users. Universal Access in the Information Society **2**, 18–29 (2002)
2. Trewin, S.: Extending keyboard adaptability: An investigation. Universal Access in the Information Society **2**, 44–55 (2002)
3. Eneso Tecnología de Adaptación S.L. http://www.eneso.es. March 2015
4. Universidad de Málaga, Eneso Tecnología de Adaptación S.L. Dispositivo de control accesible de sistemas electrónicos y mecánicos mediante la monitorización del movimiento de una parte del cuerpo humano. Sancha-Ros, S. et al. Spanish Patent ES 237861 A61F4/00 (2006.01), 29 October 2009
5. Yorkston, K.M., Miller, R.M., Strand, E.A.: Management of speech and swallowing in degenerative diseases, 2nd edn. PRO-ED, Austin (2003)
6. The eye tribe. http://theeyetribe.com. March 2015
7. Hinckley, K.: «Input Technologies and Techniques» (PDF). Microsoft. Consultado el 14 de diciembre de 2011. «Researchers are looking to move beyond the current "WIMP" (Windows, Icons, Menus, and Pointer) interface [...]». http://research.microsoft.com/en-us/um/people/kenh/papers/InputChapter.pdf

Invited Talks to IWANN 2015

The Shared Control Paradigm for Assistive and Rehabilitation Robots

Cristina Urdiales García[(✉)]

ISIS group, ETSI Telecommunications, University of Malaga, 29071 Malaga, Spain
cristina@dte.uma.es

1 Extended Abstract

One of the major risks of disability is a loss of autonomy that, in extreme, may lead to institutionalization. Lack of human resources for caregiving has led to designing robots to assist people in need. Assistive robotics are meant to help people cope with Activities of Daily Living (ADL). Most ADL are heavily affected by issues related to ambulation [1], so much effort in assistive robots has focused on robotic wheelchairs, rollators, walkers and even canes. These devices typically provide monitorization, physical support and help to cope with hazardous and/or complex situations. However, it is of key importance to provide just the right amount of help to people with disabilities. According to clinicians, an excess of assistance may lead to frustration and/or loss of residual skills. Lack of assistance, however, may lead to unacceptable risks and/or failure to accomplish the desired task. Hence, help must be adapted to each specific user.

The main problem with adapting assistance is that we need to establish how much help a given person needs at each specific situation. This need is a function of his/her disability profile. Unfortunately, it's hard to measure disability in a quantitative fashion. Disability is defined as difficulty or dependency in carrying out activities necessary for independent living, including roles, tasks needed for self-care and household chores and other activities important for quality of life [2]. There are several clinical indexes and scales related to different aspects of disability, like MMSE, Barthel, CIRS, IADL, etc. However, they cover only partial aspects of disability. Furthermore, skills are evaluated in a can do/can not do fashion, because it is hard to quantify to what degree a person can achieve a given task. Besides, from a practical point of view, these scales are typically estimated via questionnaires that take more than half an hour to be compiled and require intervention of professionals to be analyzed. For all these reasons, users fill them a few times at most, so they only reflect the condition of the user at the time they were completed.

Actually, to adapt help in mobility to a specific person, we would need to know : i) how the user copes with the problem at hand, which is related to the local environment; ii) how far he/she is from what we could consider an average performance; and iii) how to provide the minimum required amount of help in a safe, ergonomic way. A secondary, yet important concern is to minimize the cognitive load of the whole process, so the user is up to dual tasking.

© Springer International Publishing Switzerland 2015
I. Rojas et al. (Eds.): IWANN 2015, Part II, LNCS 9095, pp. 613–616, 2015.
DOI: 10.1007/978-3-319-19222-2_52

1.1   Metrics

In order to determine how well a person is coping with a navigation situation, it is necessary to establish a set of meaningful metrics. Task metrics are a popular choice [3]: they are task dependent, quantitative and objective. Typical metrics include Degree of Success, Task completion time, Number of collisions, Distance travelled, Deviation with respect to canon trajectory; etc. However, these are global metrics, so they might not be helpful to decide how a person is performing locally. Other metrics are focused on specific skills like Wheelchair propulsion, Negotiation of kerbs, Ascending slopes or Performing a wheelie[4]. However, they are typically obtained in a established obstacle course (e.g. VFM, TAMP, VST). Hence, they present similar problems that questionnaires. Alternatively, in order to evaluate local performance, we can focus only in local metrics. In this sense, safety, smoothness and directness are the best local estimators of performance according to Navigation Functions. They correspond to the overall skills of keeping away from obstacles, keeping a smooth, comfortable trajectory and moving towards a goal, respectively.

1.2   Standards

The main problem to establish how far from a standard user a person with disabilities performs at a given task is that no standard user exists. Typical approaches to check how far a person with disabilities is from standard performance include comparing his/her metrics to: i) a person with no physical nor cognitive disabilities [5] -usually researchers or students-; ii) to a previous try by the same user in the same test environment [6] -disregarding the learning effect-; and iii) to a robot navigation algorithm [7]. However, statistics can help with this problem at local level, where the number of possible navigation situations is reduced [8]. Given a large enough number of tests with a varied enough number of people, it is possible to estimate an average performance via clustering. In [9], we extracted such a profile using data from 3 years of tests performed by inpatients at a rehabilitation hospital. This study provided a set of clusters corresponding to each local situation found by volunteers during tests. In each clusters, we had information about what every person did to cope with the situation every time they faced them and how efficient their solution was. The prototype of each class was calculated as the average of these solutions weighted by their respective efficiencies. In brief, it represented the most frequent and efficient solution. The set of prototypes provides information about how an average user would respond to every potential indoor navigation situation.

1.3   Collaborative Control

In order to provide help, robots operate under the Shared Control paradigm [10]. Shared control ranges from *safeguarded navigation* [11] to autonomous navigation, where people simply choose the destination[12]. Traditionally, it is either the robot or the person in charge of control at any given time. However, a special branch of shared control known as collaborative control allows both robot

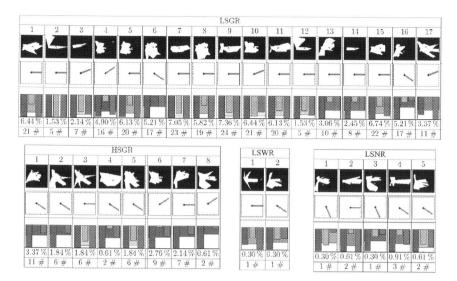

Fig. 1. Classes resulting for each of the 4 non-empty bins

and human to be in charge at the same time [13][14][15][16]. This approach prevents users from giving up on difficult situations and, hence, losing residual skills. A typical approach is to weight user's and robot's commands according to local performance metrics and then to combine them into an emergent vector. Adaptation is improved if we continuously modulate this combination using the difference between the user's performance and a standard. In these case, we can provide less assistance in average and, yet, achieve better results.

References

1. Erickson, W.: 2012 Disability status report: United States. Cornell University Employment and Disability Institute (EDI), Ithaca, NY (2014)
2. Fried, L.P., Ferrucci, L., Darer, J., Williamson, J.D., Anderson, G.: Untangling the concepts of disability, frailty, and comorbidity: implications for improved targeting and care. Journals of Gerontology Series A-Biological Sciences & Medical Sciences **59**(3), 255–263 (2004)
3. Cooperstock, J., Pineau, J., Precup, D., Atrash, A., Jaulmes, R., Kaplow, R., Lin, N., Prahacs, C., Villemure, J., Yamani, H.: Smartwheeler: A robotic wheelchair test-bed for investigating new models of human-robot interaction. In: Proc. of the IEEE Conference on Intell. Robots and Systems (IROS), San Diego, USA (2007)
4. Kilkens, O., Post, M., Dallmeijer, A., Seelen, H., van der Woude, L.: Wheelchair skills tests: a systematic review. Clinical Rehabilitation **17**(4), 418–430 (2003)
5. Carlson, T., Demiris, Y.: Collaborative control for a robotic wheelchair: Evaluation of performance, attention, and workload. IEEE Transactions on Systems, Man, and Cybernetics, Part B: Cybernetics **42**(3), 876–888 (2012)

6. Nguyen, A., Nguyen, L., Su, S., Nguyen, H.: Shared control strategies for human - machine interface in an intelligent wheelchair. In: 2013 35th Annual International Conference of the IEEE Engineering in Medicine and Biology Society (EMBC), pp. 3638–3641, July 2013

7. Li, Q., Chen, W., Wang, J.: Dynamic shared control for human-wheelchair cooperation. In: 2011 IEEE International Conference on Robotics and Automation (ICRA), pp. 4278–4283, May 2011

8. Minguez, J., Osuna, J., Montanor, L.: A divide and conquer strategy based on situations to achieve reactive collision avoidance in troublesome scenarios. IEEE Trans. on Robotics (2009)

9. Urdiales, C., Fernández-Espejo, B., Annicchiaricco, R., Sandoval, F., Caltagirone, C.: Biometrically modulated collaborative control for an assistive wheelchair. IEEE Trans. on Neural Systems and Rehabilitation Eng. 18(4), 398–408 (2010)

10. Chipalkatty, R., Droge, G., Egerstedt, M.: Less is more: Mixed-initiative model-predictive control with human inputs. IEEE Transactions on Robotics 29(3), 695–703 (2013)

11. Tomari, M.R.M., Kobayashi, Y., Kuno, Y.: Development of smart wheelchair system for a user with severe motor impairment. Procedia Engineering 41, 538–546 (2012); International Symposium on Robotics and Intelligent Sensors 2012 (IRIS 2012)

12. Frese, U., Larsson, P., Duckett, T.: A multigrid algorithm for simultaneous localization and mapping. IEEE Transactions on Robotics 21(2), 1–12 (2005)

13. Urdiales, C., Fernández-Carmona, M., Peula, J., Annicchiaricco, R., Sandoval, F., Caltagirone, C.: Efficiency based modulation for wheelchair driving collaborative control. In: 2010 IEEE International Conference on (nominated best paper) Robotics and Automation (ICRA), Anchorage, USA, pp. 199–204, May 2010

14. Carlson, T., Demiris, Y.: Human-wheelchair collaboration through prediction of intention and adaptive assistance. In: IEEE International Conference on Robotics and Automation, ICRA 2008, pp. 3926–3931, May 2008

15. Li, Q., Chen, W., Wang, J.: Dynamic shared control for human-wheelchair cooperation. In: 2011 IEEE International Conference on Robotics and Automation (ICRA), pp. 4278–4283, May 2011

16. Vanhooydonck, D., Demeester, E., Huntemann, A., Philips, J., Vanacker, G., Brussel, H.V., Nuttin, M.: Adaptable navigational assistance for intelligent wheelchairs by means of an implicit personalized user model. Robotics and Autonomous Systems 58(8), 963–977 (2010)

Author Index

Printed in the United States
By Bookmasters